Lecture Notes
in Computational Science
and Engineering

20

Springer
Berlin
Heidelberg
New York
Barcelona
Hong Kong
London
Milan
Paris
Tokyo

Timothy J. Barth
Tony Chan
Robert Haimes
Editors

Multiscale and Multiresolution Methods

Theory and Applications

With 118 Figures, 11 in Color

Springer

Editors

Timothy J. Barth

NASA Ames Research Center
NAS Division
Moffett Field, CA 94035, USA
e-mail: barth@nas.nasa.gov

Tony Chan

Department of Mathematics
University of California
Los Angeles, CA 90095-1555, USA
e-mail: chan@math.ucla.edu

Robert Haimes

Department of Aeronautics and Astronautics
MIT - 37-467
Cambridge, MA 02139
e-mail: haimes@orville.mit.edu

Cataloging-in-Publication Data applied for

Die Deutsche Bibliothek - CIP-Einheitsaufnahme

Multiscale and multiresolution methods : theory and applications /
Timothy J. Barth ... ed.. - Berlin ; Heidelberg ; New York ; Barcelona ;
Hong Kong ; London ; Milan ; Paris ; Tokyo : Springer, 2002
(Lecture notes in computational science and engineering ; Vol. 20)
ISBN 3-540-42420-2

Mathematics Subject Classification (2000):
35B27, 42C40, 65D15, 65T60, 68U10, 76M50, 80M40, 94A08

ISSN 1439-7358
ISBN 3-540-42420-2 Springer-Verlag Berlin Heidelberg New York

Springer-Verlag Berlin Heidelberg New York
a member of BertelsmannSpringer Science + Business Media GmbH
http://www.springer.de

Printed in Germany

Cover Design: Friedhelm Steinen-Broo, Estudio Calamar, Spain
Cover production: *design & production*
Typeset by the authors using a Springer TEX macro package

Printed on acid-free paper SPIN: 10653067 46/3142/LK - 5 4 3 2 1 0

Preface

Many computationally challenging problems ubiquitous throughout science and engineering exhibit multiscale phenomena so that the prospect of numerically computing or even representing all scales of action is either very expensive or completely intractable. Some examples of practical interest include fluid turbulence at large Reynolds number, weather forecasting, flow through porous media, spray combustion and detonation, structural analysis of composite and foam materials, many-body galaxy formation, large scale molecular dynamic simulations, *ab-initio* physics and chemistry, terabyte data mining, large scale data visualization and analysis, and a multitude of others.

The computational challenge has several origins. For many of the cited multiscale problems, one seeks to compute as many scales as possible but quickly finds that the algorithmic complexity of conventional algorithms rises too steeply with the number of degrees of freedom. For another class of multiscale problems, one does not actually desire the fine scale information, but owing to nonlinearity in the modeled physics it is found that the effect of fine scale information on coarse scales must be included to achieve quantitative predictive capability. Compounding the computational problem is the fundamental question of optimal data representation for multiscale problems where it is known that even modern wavelet basis representations can yield overall suboptimal algorithmic complexity, e.g. problems containing embedded manifolds of discontinuity or discontinuous derivatives.

An arguable conclusion is that these multiscale problems will remain computationally expensive or completely intractable for the foreseeable future unless new algorithmic paradigms of computation are developed which fundamentally embrance the multiscale nature of these problems.

The Yosemite Educational Symposium (YES) was conceived by the organizers with the goal of examining these problems and discussing ongoing developments. YES was held in the fall of 2000 midst the surroundings of the scenic Yosemite valley. The symposium was successful in bringing together leading international specialists with diverse backgrounds and differing perspectives on multiscale methods. The afternoon break-out times permitted cross-fertilization of ideas and inspired discussion about new avenues for future research.

The organizers are indebted to the generous financial support of the following organizations: the National Science Foundation CISE/ACIR, NASA Ames Research Center IT/HPCCP, IBM T.J. Watson Research Center and Silicon Graphics. The organizers also acknowledge Springer-Verlag for agreeing to publish this volume in a quality book form. For those who attended, it is clear that much of the logistical success of YES was due to the efforts of the symposium administrator Ms. Ping M. Lee.

The volume consists of both invited and contributed articles covering a range of topics related to multiscale and multiresolution methods. The volume begins with opening remarks by Professor Gilbert Strang on multiscale and multiresolution methods setting the stage for the remainder of the volume. The first invited article by Professor Achi Brandt gives his comprehensive perspective on multiscale techniques in scientific computing. The remaining articles then delve more deeply into many important multiscale topics ranging from wavelet data analysis to numerical homogenization of PDEs.

The YES Organizers

Timothy Barth
Tony Chan
Robert Haimes
July 2001

Multiscale and Multiresolution Methods

The presence of two or more widely different length scales in a single problem is a sure sign that innovative analysis and algorithms are needed. The discrepancy in the length scales will defeat conventional computations. The fine scale and coarse scale must be treated differently, one approach cannot resolve them both. This volume is an early step in the computational analysis of equations with multiple scales.

Multiscale problems are familiar in fluid mechanics (boundary layers and eddies). They are unavoidable in nuclear physics and chemistry. This conference also brought digital signal processing into the center of the discussion, because wavelet techniques for compression are inherently multiscale. The reader will see two communities, CFD and DSP, trying to explore common ground.

Surely wavelets and multigrid techniques and hierarchical finite elements share key inspirations. For wavelets, the basis functions are local (but not as local as finite elements). The classical wavelet basis comes from translating and rescaling as in $w(2^j x - k)$. An expansion in this basis separates the details at different scales ($j = 0, 1, 2, 3, 4$ would be typical, and then the averages at the coarsest scale contain most information about the image). Multigrid moves back and forth between scales as it solves a problem on the fine scale.

One significant difference between these methods: Finite elements provide an effective and very flexible technique to model an *unknown function* (the solution to a PDE). Wavelets are more often used to capture a *known function* (a still image or a video sequence). Where multigrid solves equations in an extremely efficient way, Mallat's algorithm changes from a fine basis to two coarse bases (and inversely). The purposes are different and therefore finite elements look quite different from wavelets. But they share this key idea of *multiscale.*

The importance of that idea to science and engineering is creating a new generation of algorithms. In this part of science, computations don't necessarily lag behind analysis. We compute so as to understand. The problems are multiscale, and it will be multiscale algorithms that solve them.

Gilbert Strang
Department of Mathematics
Massachusetts Institute of Technology
Cambridge, Massachusetts

Contents

Invited Papers

Contributed Papers

Appendix

Part I

Invited Papers

Multiscale Scientific Computation: Review 2001

Achi Brandt**

The Weizmann Institute of Science, Rehovot 76100, Israel

Abstract. Most of the fundamental problems in physics, chemistry and engineering involve computation too hard even for future supercomputers, if conventional mathematical approaches are used. The reason is always a product of several complexity factors associated with the wide range of space and time scales characteristic to such problems. Each of these complexity factors can in principle be removed by various multiscale algorithms, i.e., employing separate processing at each scale of the problem, combined with interscale iterative interactions. A wide range of multiscale computational methods is described, emphasizing main ideas and inter-relations between various fields. The reported areas include: top-efficiency multigrid methods in fluid dynamics; inverse PDE problems and data assimilation; feedback optimal control; PDE solvers on unbounded domains and on adaptable grids; wave/ray methods for highly indefinite equations; rigorous quantitative analysis of multigrid; many-eigenfunction problems and ab-initio quantum chemistry; fast evaluation of integral transforms on adaptive grids; multigrid Dirac solvers; fast inverse-matrix and determinant calculations and updates; multiscale Monte-Carlo methods in statistical physics, including the renormalization multigrid (RMG) methods; molecular mechanics (including fast force summation, fast macromolecular energy minimization, and Monte-Carlo methods at equilibrium, both for macromolecules and for large ensembles of small molecules); combination of small-scale equilibrium with large-scale dynamics; image processing (edge detection and picture segmentation); tomography (medical imaging and radar reconstruction); efficient, general and highly accurate algebraic multigrid (AMG) and homogenization schemes; fast practical graph algorithms; data clustering; and multiscale approaches to global optimization.

1 Introduction

1.1 Multiscale computation

Despite their dizzying speed, modern supercomputers are still incapable of handling many most vital scientific problems. This is primarily due to *the*

** Research supported by AFOSR and the Materials and Manufacturing Directorate, AFRL, Wright-Patterson Base, contract No. F33615-97-D-5405, by the European Office of Aerospace Research and Development (EOARD) of the US Air Force, Contract F61775-00-WE067, by Israel Ministry of Science, Culture and Sport grant 9680, by Israel Absorption Ministry, Project No. 6682, by Israel Science Foundation grant No. 696/97, and by the Carl F. Gauss Minerva Center for Scientific Computation at the Weizmann Institute of Science.

scale gap, which exists between the microscopic scale at which physical laws are given and the much larger scale of phenomena we wish to understand.

This gap implies, first of all, a huge number of *variables* (e.g., atoms or gridpoints), and even a much larger number of *interactions* (e.g., one force between every pair of atoms). Moreover, computers simulate physical systems by moving *one variable at a time*; as a result, each such move must be extremely small, since a larger move would have to take into account all the motions that should in parallel be performed by all other variables. Such a computer simulation is particularly incapable of moving the system across large-scale *energy barriers*, which can each be crossed only by a large, and unknown, simultaneous motion of very many variables.

This type of computational obstacle makes it impossible, for example, to calculate the properties of nature's building blocks (elementary particles, atomic nuclei, etc.) from a certain known underlying theory — and thereby to confirm the theory itself. Likewise, such obstacles are the main bottleneck in the drive to *computerize chemistry*: to replace expensive experiments with computer simulations, yielding detailed understanding of molecular structures and interactions, creating the ability to design materials and processes, with enormous potential benefits for medicine, biotechnology, agriculture, material sciences, industrial processing, etc. Similar scale-born slowness factors and barriers, multiplying each other, plague many other engineering and scientific endeavors. All would be greatly facilitated if unlimited computing power were available — or if much better algorithms could be devised.

Just building ever faster machines will not do, in fact. With current computational methods the needed amount of computer processing often increases too steeply with the rise in problem size, so that no conceivable computer will be adequate. Completely new mathematical approaches are needed.

Past studies have demonstrated that all scale-born complexities can be effectively overcome, or drastically reduced, by *multiscale* ("multi-resolution", "multilevel", "multigrid", etc.) algorithms.

Indeed, any many-variable problem defined in the physical space can have an approximate description at any given length scale of that space: a continuum problem can be discretized at any given resolution; average motions of a many-particle system can be represented at any given characteristic length; etc. The multiscale algorithm recursively constructs a *sequence* of such descriptions at increasingly larger (coarser) scales, and combines local processing (relaxation of equations, simulation of statistical relations, etc.) at each scale with various inter-scale interactions. Typically, the evolving solution (or the simulated equilibrium) on each scale recursively dictates the *equations* (or the Hamiltonian) on coarser scales while supplying large-scale corrections to the *solutions* (or configurations) on finer scales. In this way large-scale changes are effectively calculated on coarse grids, based on information previously gathered from finer grids.

As a result of such multilevel interactions, the fine scales of the problem can be employed very sparingly, and sometimes only at special and/or representative small regions. Moreover, the inter-scale interactions can eliminate all kinds of scale-associated difficulties, such as: slow convergence (in minimization processes, PDE solvers, etc.); critical slowing down (in statistical physics); ill-posedness (e.g., of inverse problems); large-scale attraction basin traps (in global optimization and statistical simulations); conflicts between small-scale and large-scale representations (e.g., in wave problems); numerousness of long-range interactions (in many body problems or integral equations); numerousness of long-range (non-local) eigenfunctions (e.g., in quantum chemistry); the need to produce many fine-level solutions (e.g., in optimal control) or very many fine-level independent samples (in statistical physics); etc. Also, the evolving large-scale equations bring out the large-scale dynamics, or the macroscopic equations, of the physical system, which is often the very objective of the entire calculation.

Since the local processing (relaxation, etc.) in each scale can be done in parallel at all parts of the domain (e.g., at all cells of a given lattice), the multiscale algorithms, based on such processing, are ideal for implementation on massively parallel computers. Indeed, many problems cannot be efficiently solved by such computers without employing a multiscale procedure. For example, to fully parallelize a time-dependent calculation (i.e., to compute for earlier and later times simultaneously), a multiscale (multigrid) algorithm must be used (see Sec. 3.1).

Often, a combination of several multiscale approaches can benefit one particular problem in *many* different ways (see examples in Secs. 4.3, 9 and 14.2 below. Also the different multiscale algorithms discussed in Secs. 11, 12 and 13 are all parts of solving the same "grand challenge" problem of elementary particles).

Multilevel computation has evolved into a discipline by itself, having its own internal development, gradually increasing our understanding of the many types of multiscale interaction, their modes of operation and domains of application. Various underlying relations and algorithmic ideas are carried back and forth between widely varying types of problems.

1.2 Do you need multiscale algorithms?

A multiscale computation is usually considerably more complicated than the more common algorithms. Also, for small problems it is often more expensive. So when do you need to go into this extra trouble? The most obvious and important sign for such a need is that you have a *computational bottleneck* associated with the increasing "size" of the problem. If the computational cost rises more than linearly with the number of variables, or if the number of variables is so large that even linear-scaling algorithms would be too expensive, then a multiscale approach, or a combination of several multilevel procedures, may well be the answer.

To be sure, not *every* difficult computational task can be usefully multiscaled. Intractable, undecidable and other impossible problems typical to theoretical computer science (see for example [101]) cannot be efficiently solved by *any* algorithm. Interestingly, however, these are mostly *man-made* problems. The computational tasks in natural sciences and engineering are not of this type. Their complexity usually results from a multitude of variables (particles, picture elements, a discretized function, etc.), most of which are usually positioned in some low dimensional spaces. The experience is that all problems of this type can benefit from multiscaling, which yields either low-complexity (normally linear-scaling) solvers or "macroscopic equations", i.e., the means for large-scale coarse simulations, derived from computations in just small fine-scale windows. The various sections of this review gives many examples of both these possibilities.

Note also that some problems can benefit from multiscaling because this is the best way to *formulate* the problem, or some parts of it (see Sec. 18.1).

1.3 The present survey

The present report has been written as a thorough updating and modification of [38], many parts of which had previously appeared in [35]. The first chapters summarize important recent techniques, and some less known older ones, in the field of multigrid PDE solvers, assuming a general familiarity with its basic elements. (For introductory books, see [67] and [153]; or at least read the "Elementary acquaintance with multigrid" in [24]; see also the basic insight described in Sec. 17 below.) Some algorithms and concepts are explained in more detail than others, mainly because they are more recent. Later chapters introduce a variety of other fields of multiscale computation, including fast matrix multiplication, integral and integrodifferential equations, statistical physics, chemistry, image processing and tomography.

In particular, the report surveys the main ideas, current developments and future perspectives in the following directions.

1. New top-efficiency multigrid methods for steady-state fluid dynamics at all Mach and Reynolds numbers, and other non-elliptic stationary PDE systems (see Sec. 2 below).
2. Multilevel approaches to time-dependent partial-differential equations, emphasizing a fast method for solving an implicit-time-step system of equations (sometimes faster than an explicit time step), and parallel processing and grid adaptation across both space and time (see Sec. 3).
3. *Grid adaptation* techniques exploiting multigrid structures and creating a one-shot solver-adaptor (Sec. 6.1). Similar techniques for treating problems in *unbounded domains*, costing essentially the same as in bounded domains.
4. Direct multigrid solvers for inverse problems, including system identification (e.g., impedance tomography; see in Sec. 16.2) and data assimilation

(in atmospheric simulations — Sec. 4), showing multiple benefits of several kinds of multiscaling employed in one problem. The solution of an ill-posed problem can often cost far less than its well-posed counterpart.

5. Optimal control: Feedback control via very fast updating of open-loop solutions, based on their multiscale representations (Sec. 5).
6. Optimal location of singularities of PDE systems (e.g., finding the minimal-total-energy location of the nucleons in electronic structure calculations), integrated into a one-shot multigrid solver (Sec. 9.1).
7. Top-efficiency multigrid algorithms for highly indefinite (e.g., standing wave) problems, featuring ray equations (geometrical optics) at the limit of large scales with wave equations at small-scale regions where ray formulations break down (Sec. 7).
8. Multigrid solvers for the Dirac equations arising in quantum field theory (Sec. 11).
9. Compact multiresolution representation of the inverse matrix of a discretized differential operator; fast updating of the inverse matrix and of the value of the determinant upon changing an arbitrary term in the matrix itself; with application to the QCD fermionic interaction (Sec. 12).
10. Collective multiscale organization of eigenbases and $O(N \log N)$ calculation of N eigenfunctions of a differential operator, e.g., the Schrödinger operator in condensed-matter electronic-structure calculations (Sec. 9.2).
11. Calculation of the N roots of the secular equation in $O(N)$ operations.
12. Multiscale Monte-Carlo algorithms for eliminating both the critical slowing down and the volume factor in increasingly advanced models of statistical physics (Sec. 13).
13. Multigrid Monte-Carlo approaches for solving the high-dimensional (several-particle) Schrödinger equation by real-time path integrals (Sec. 18).
14. Introducing multiscale computations to many-particle (macromolecule or many-small-molecule) calculations, including fast evaluation of forces, fast convergence to ground states, fast Monte Carlo simulations and large time steps, with application to molecular mechanics (Sec. 14); a new approach to molecular dynamics, based on stochastic implicit time steps (Sec. 14.8).
15. Multigrid methods for fast dense-matrix multiplications, integral transforms and for integro-differential equations, on adaptable grids, with applications to tribology (Sec. 10).
16. Multiscale methods for the fast evaluation and inversion of the Radon transform and other line-integral transforms (Sec. 16.1); applications to medical tomography and radar reconstruction.
17. Multiscale algorithms for early vision tasks such as surface reconstruction, edge and fiber detection (Sec. 15.1) and segmentation (Sec. 15.2).
18. Multilevel clustering and other graph algorithms (Sec. 15.3).
19. Rigorous quantitative theory for predicting the performance of multigrid solvers (Sec. 8).
20. New efficient, general and accurate approaches for coarsening or homogenizing algebraic systems of equations (Sec. 17).

21. Multilevel strategies for solving global optimization problems that harbor many local minima and nested multiscale attraction basins, including multilevel approaches for *formulating* fuzzy optimization problems (Sec. 18).
22. Some thoughts about wavelets (Sec. 19).

2 Steady-State Fluid Dynamics

2.1 Objective: textbook multigrid efficiency

An efficient multigrid algorithm for steady-state incompressible viscous flows in two dimensions appeared already in 1972 [18], a relatively efficient multigrid solver for a compressible inviscid transonic flow was demonstrated in 1975 [145], and a fully efficient solver for a *system* of *several* coupled differential equations, characteristic of computational fluid dynamics (CFD), was presented already in 1978 [40]. However, in the decades that followed, the development in this area has not been fully satisfactory. In particular, the efficiency of solvers for non-elliptic steady-state systems (such as Euler and high-Reynolds Navier-Stokes equations) has lagged several orders of magnitude behind the ideal efficiency that had been attained for general elliptic systems. Although the main reasons for this inefficiency have also been understood for a long time (see for example [22]), the recommended cures seemed complicated, and code developers opted for partial efficiency. The leading multigrid method has been based on multi-stage pseudo-time-stepping relaxation schemes [105], [104]. Although such schemes can be optimized to damp high-frequency errors [157], the resulting algorithms are still relatively slow, because some *intermediate* (neither high-frequency nor very smooth) "characteristic components" cannot adequately be reduced by coarse grids (cf. [22], [64]). Other multigrid solvers were based on incomplete LU decomposition (ILU) and related relaxation schemes [159], [155], [144]. While such schemes give excellent results in some cases, they cannot cure the aforementioned trouble of characteristic components in general transonic flows, especially in three dimensions. (Also, much of the efficiency of ILU schemes depends on their sequential marching, hence the performance on massively parallel machines will drastically diminish.) The same is true for other methods (e.g., based on defect corrections) which seem not even to identify that basic trouble.

More generally, all these attempted solution methods have failed to decompose the solution process into separate treatments of each factor of the PDE principal determinant, and therefore did not identify, let alone treat, the separate difficulties associated with each such factor. The fact is that, in a typical CFD problem, each of these factors may have different ellipticity measures (some are uniformly elliptic, others are non-elliptic at some or all of the relevant scales) and a different set of characteristic surfaces, requiring for top efficiency different relaxation/coarsening procedures.

Thus, the objective of the recent work has been to develop and demonstrate methods that solve non-elliptic *steady-state* problems in general, and high-Reynolds stationary flow problems in particular, at the same "*textbook multigrid efficiency*" attained for uniformly elliptic systems. This means, typically, to obtain an $O(h^2)$ approximation to the differential solution on a grid with meshsize h at a cost of just few (less than 10) "*minimal work units*", this unit being the amount of operations involved in the simplest discretization of the differential problem on the meshsize-h grid. The methods, again as in the elliptic case, will allow local refinements (cf. Sec. 6.1) and high degree of parallel processing. (For general remarks about *time-dependent* problems, see Sec. 3).

2.2 Problem decomposition

As shown in the past (see [24], [28] and [64]), to obtain that "textbook" multigrid efficiency for any discretized partial differential system of equations (PDE), it is necessary and usually (with proper boundary treatment) also sufficient to attain that efficiency for each factor of the PDE principal determinant. Each such factor is a scalar differential operator of first or second order, so its efficient solution is a vastly simplified task. The way for separating the factors is by a *distributed* (and possibly also *weighted*) relaxation scheme in which to each factor there corresponds a "*ghost*" discrete function. The latter can be directly relaxed for its corresponding factor, dictating a resulting pattern of changes to be distributed to the *actual* discrete functions (see details in [24, §3.7] and also in [163], and examples in Secs. 17–20 of [24]). To obtain the top efficiency, the relaxation of each ghost function should incorporate an essential part of an efficient multigrid solver for its corresponding operator: sometimes this is just the relaxation part of that solver, sometimes this may even be the entire solver (applied at some proper subdomain).

For the *incompressible* Euler and Navier-Stokes equations, the relevant factors are the Laplace and the convection (or convection-diffusion) operators. The former's multigrid solver is classical; the latter's can be based on downstream relaxation [64], with additional special procedures for recirculation flows [65], [166]. Indeed, incorporating such procedures into the relaxation schemes for the appropriate ghost functions yields very efficient solvers for incompressible flows even at high Reynolds numbers and at second-order accuracy [64]. The same procedures will also yield efficient solvers for compressible flows at low Mach numbers, where the relevant factors are similar.

The most important remaining factor of flow systems for which no general adequate multigrid solver has been developed until recently is the "*full potential*" operator

$$(u\partial_x + v\partial_y + w\partial_z)^2 - a^2\Delta \ , \tag{2.1}$$

where (u, v, w) is the flow velocity vector and a is the speed of sound. This operator appears as a factor in the principal determinant of the 3-D *compressible* Euler equations. Its *Mach number* is the ratio $M = (u^2 + v^2 + w^2)^{1/2}/a$.

In the deep *subsonic* case ($M \leq .7$, say) the operator (2.1) is uniformly elliptic, hence a usual multigrid V-cycle, employing red/black Gauss-Seidel relaxation at all levels, yields top-efficiency solvers. When M approaches 1, however, the operator becomes increasingly anisotropic, and classical multigrid algorithms severely degrade, due to the above-mentioned difficulty with characteristic components. (An exception is the case where the anisotropy directions are aligned with grid directions. For example, if $u^2 + v^2 \ll w^2$, full efficiency can still be obtained by employing z-plane block relaxation).

In the deep *supersonic* case (e.g., $M \geq 1.3$) the full potential operator is uniformly hyperbolic (with the stream direction serving as the time-like direction), and an efficient solver can be obtained using downstream relaxation, marching in the time-like direction. If the equations are of higher-order and/or not strictly upstream, a *predictor-corrector* marching can provide the same approximation order, hence fast convergence of smooth components; this has been shown by detailed experiments and mode analyses [78]. This procedure no longer works as M drops toward 1, since the Courant number associated with this time-like marching approaches infinity.

Thus, the most difficult situation for solving the full potential operator is the *near sonic* regime ($.7 \leq M \leq 1.3$, say), especially in the (usual) case of *non-alignment* (e.g., when the grid is Cartesian and no velocity component is consistently much larger than the others). No "classical" multigrid approach would attain good efficiency in this case. A new approach has recently been developed, based on a piecewise semi-coarsening and some rules for adding artificial dissipation at the coarser levels. To understand this, note first that in the general scheme for solving, e.g., the Euler equations, the solution of (2.1) is only a *relaxation* step, and it is enough to confine this step to one subdomain at a time (whose size, however, is not $O(h)$ but $O(1)$). Without loss of generality we can therefore limit the discussion to the case that throughout this subdomain the velocity is, e.g., *vertically-inclined* (i.e., $w^2 \geq .3(u^2 + v^2)$, say). In this case, the multigrid solver of (2.1) will use *horizontal semi-coarsening* (coarsening only in the x and y direction), possibly together with *vertical* line relaxation. (This z-line relaxation is actually not needed on the finest levels, but may be required after several levels of semi-coarsening.) With this semi coarsening, the inherent cross-characteristic numerical dissipation at the coarse level is *smaller* than at the fine one (opposite to their relation upon *full* coarsening); we can therefore stably add artificial dissipation terms at the coarse level so that its total cross-characteristic dissipation matches the local fine-level average.

The resulting algorithm can fully exploit massively parallel processing. It can be extended to other non-elliptic operators, including the convection operator. (The aforementioned approach for the convection operator, based on downstream relaxation, is not fully efficient on massively parallel machines.)

Extensive numerical tests have been performed with the linear full-potential equation: first in 2D, then in 3D, starting with constant-coefficients, then variable. In 2D we have also carried out comprehensive half-space FMG mode analyses (cf. [24, §7.5]), achieving full agreement with the numerical tests. The results reported in [41], [42], [77] and [78] show that at any Mach number the algorithm can always attains the "textbook" efficiency.

2.2.1 Comment on semi-coarsening schemes

Instead of the *piecewise semi-coarsening* described above, another possibility is to use just one *global* semi-coarsening, but of one of the following two types (preferably the second).

A. *Total semi-coarsening*. By this we mean (e.g., in 2D) that each coarser grid is formed by omitting every other line from the next finer grid (every other vertical line *as well as* every other horizontal line), but on the remaining lines (the coarse-grid lines) leave *all* the fine-grid points (not just the intersections of the coarse-grid lines).

B. *Variable-direction semi-coarsening*. Here the coarser grid for each level is a *subset* of the total-semi-coarsening grid for that level. Simply omit from the latter all unnecessary points in regions where semi-coarsening at only one particular direction is needed (as in various anisotropic and non-elliptic cases, like those discussed above).

2.3 A road map

A group at NASA/Langley has launched a multi-year program aimed at achieving "textbook" multigrid efficiency for flows at all Mach and Reynolds numbers, using the general approach described above [151], [152].

A road map for further development has been assembled in the form of a detailed table called "Barriers to Achieving Textbook Multigrid Efficiency in CFD". It lists every foreseen kind of computational difficulty for achieving that goal, together with the possible ways for resolving the difficulty, their current state of development, and references [36].

Included in the table are staggered and nonstaggered, conservative and nonconservative discretizations of viscous and inviscid, incompressible and compressible flows at various Mach numbers, as well as a simple (algebraic) turbulence model and comments on chemically reacting flows. The listing of associated computational barriers involves: non-alignment of streamlines or sonic characteristics with the grids; recirculating flows; stagnation points; discretization and relaxation on and near shocks and boundaries; far-field artificial boundary conditions; small-scale singularities (meaning important features, such as the complete airplane, which are not visible on some of the coarse grids); large grid aspect ratios; boundary layer resolution; and grid adaptation.

3 Time-Dependent Differential Problems

In the numerical solution of time-dependent problems, to allow large time steps and/or fully adaptable discretization (cf. Sec. 3.2), *implicit* time steps must be used, hence a system of equations must be solved at each time step. Multigrid solvers for such systems are usually similar to but *simpler* than their steady-state counterparts, because these systems are easier than the steady-state equations, in various ways: they have better ellipticity measures (due to the time term); they do not involve the difficulties associated with recirculation (in flow problems); and they each come with a good first approximation (from the previous time step). A simple "F cycle" at each time step (effectively an FAS-FMG algorithm for the solution *increment*, i.e., its departure from the previous-time solution) should solve the equations much below the *incremental* discretization errors (the errors added in the current time step). Hence, the errors *accumulated* over time due to the solver are generally much below the *accumulated* discretization errors [49].

It is generally true that fully efficient multigrid methods for the steady-state equations directly yield also at-least-as-efficient methods for time-accurate integrations, where the work per implicit time step is just comparable to the work of an *explicit* time step. Moreover, in various cases (e.g., parabolic equations with steady or smoothly-varying-in-time forcing terms), the work can be substantially *smaller* than that of an explicit time step. This is due to the smoothness of solution *increments* (solution changes from a previous time or solution departures from a simple convection). Such smoothness is typically established away from the immediate neighborhood of oscillatory initial or boundary conditions. It implies that the high-frequency part of the solution changes slowly. Hence the multigrid solver applied at each time step needs to actually visit the finest levels only once per many time steps, provided that the fine-to-coarse correction τ_h^{2h} is carried from each such visit to subsequent time steps [95], [100].

3.1 Parallel processing in space-time

A unique feature of multigrid solvers is the possibility to apply parallel processing across space *and time*, i.e., to process simultaneously earlier and later time steps, whereas single-level solvers must proceed sequentially in time. (This unique feature is discussed in [21, §3.10], [27, §11] and elaborated in [156]).

To achieve that, time is treated just as another space coordinate, and the whole problem is solved by an FMG (full multigrid) algorithm, starting with a coarse grid *in both space and time*, proceeding to finer levels, with one (sometimes two) cycles at each level. At fine levels, where most of the computational work is spent, all the processes (relaxation and inter-grid transfers) can employ many processors in parallel, each one working in its own space-time subdomain.

This of course makes it possible to use *efficiently* (i.e., at a given arithmetic to communication ratio) a larger number of parallel processors than can be used when parallelization is done only across space (marching sequentially in time).

Depending on the number of processors, available storage, etc., the above algorithm will often be applied not to the entire time evolution, but to one (large) time interval at a time. (A more sophisticated multilevel time windowing is described in Sec. 4.2 below).

3.2 Grid adaptation in space-time

Simultaneous space-time multigridding also yields a very efficient way for general *space-time grid adaptation*, where both the spatial meshsize and the time step can be adapted — locally in both space and time. Just as in the case of pure spatial (i.e., steady-state) problems, the multigrid environment can provide convenient flexible structures, where discrete equations need be derived only for uniform grids (facilitating economic high-order discretizations and parallelizable and vectorizable processing), while only negligible geometric information need be stored. The multigrid algorithm also provides local refinement criteria and one-shot self-adaptive solvers; see Sec. 6.1 below.

4 Inverse Problems. Atmospheric Data Assimilation

A *direct* partial differential problem involves an interior differential equation and a set of initial/boundary conditions which stably determines a unique solution. An *inverse* problem is one in which the differential equation and/or the initial/boundary conditioned are not fully given and instead the results of a set of solution observations (measurements) are known. The latter may contain errors, and even without errors the problem is usually ill-posed: the known data may be approximated by widely different solutions.

In this chapter we use the problem of atmospheric data assimilation to illustrate the many ways in which multiscale computation can benefit the solution of inverse PDE problems. Other problems of this type are discussed in Secs. 5 and 16.2 below.

4.1 Background and objectives

A major difficulty in weather prediction is the need to assimilate into the solution of the atmospheric flow equations a continuously incoming stream of data from measurements carried out around the globe by a variety of devices, with highly varying accuracy, frequency, and resolution. Current assimilation methods require much more computer resources than the direct solution of the atmospheric equations. The reason is the full 4-D coupling: Any measurement, at any place and time, should in principle affect the solution at any other place and time, thus creating a dense $N_sN_t \times N_sN_t$ matrix of influence, where N_s is the huge number of gridpoints representing the 3-D atmosphere and N_t is the large number of time steps spanning the full period over which large-scale atmospheric patterns are correlated. As a result, not only are current assimilation methods very slow, but they are also based on highly questionable compromises, such as: ignoring the all-important spatially or temporally *remote* correlations of *large-scale* averages; limiting control to only the initial value of the flow at some arbitrarily chosen initial time, instead of controlling the numerical equations at all times; and assimilating only the data from one time interval at a time, without fully correlating with other intervals.

The objective is to develop multiscale methods that can avoid all these compromises, and can assimilate the data into the multigrid solver of the direct flow equations at small extra cost, i.e., using extra computer time smaller than that required by the direct solver by itself.

This is considered possible because: (1) Large scale averages can inexpensively be assimilated on the correspondingly coarse levels of the multigrid solver (coarse in both space and time). (2) Deviations from any large-scale average must be assimilated on some finer scale, but their correlation on that scale is local. (3) The measurements (with their representativeness errors) are generally less accurate and in most regions less resolved than the numerical flow itself, hence their assimilation should not be done at the finest numerical level. (The overall solver of an ill-posed inverse problem can sometimes cost even far less than the solver of a corresponding well-posed problem, because ill-defined high-frequencies need not be calculated: see Sec. 16.2.)

4.2 Multiscale 4D assimilation

Since the atmospheric data assimilation problem involves full 4D couplings, both forward and backward in time, it is proposed to use one full-multigrid (FMG) algorithm for the entire 4D problem (but possibly with the storage-saving windowing described below). This algorithm would be like a usual FMG solver for the *direct* 4D atmospheric equations, except that at each stage, on each level excluding the finest ones, the relaxation of the solution variable will be accompanied by relaxation of the *control variables* $\sigma(x)$ at that level (see the nature of $\sigma(x)$ in Sec. 4.3 below). Thus, in essence, large-scale averages of the solution will be assimilated on correspondingly coarse grids (coarse in both space and time).

The levels at which $\sigma(x)$ will be adjusted will depend on the local density of the measurements, their accuracy and their distance from regions where details of the solution are of interest.

Windowing. Should the 4D solution require too much storage, it is possible to reorganize it in multiscale windows, marching in time, without much loss of efficiency. That is, only a certain window (time slice) of the finest grid need be kept in memory at a time. Having relaxed over it, residuals are then transferred from this window to the coarser grids. On returning from the coarser grids more relaxation is made on the finest grid, now in a somewhat advanced window (shifted forward in time, but partly overlapping its predecessor) and so on. At the coarser levels, on increasingly coarser grids, increasingly wider (in real time, but poorer in gridpoints) windows are kept and advanced in a similar manner. The domain covered by each coarse-grid window always strictly contains all the finer ones. The coarsest windows extend very far in time, especially into the past; as far indeed as there exist data whose large-scale averages are still correlated to the solution at the time of the current finest window. At times where a coarse window exists while the next finer one has already been removed, the coarse-level equations can still retain the FAS-multigrid fine-to-coarse (τ) corrections (static or modified), thus still maintaining the fine-level accuracy of coarse-level features (cf. the "frozen τ" technique in [23, §15] and in [95]).

Some of the finest windows may be local not only in time but also in space, effecting *local refinements* at regions of greater human interest and/or regions requiring higher resolution for mathematical and physical reasons (sea straits, islands, mountains, etc.).

4.3 Multiple benefits of multiple multiscale techniques

Our studies have uncovered *many* different ways in which multiscale computational methods can contribute to data assimilation problems (and similarly to other inverse problems). The following is the full list — brought as an example of what a "*full multiscaling*" of a problem may involve.

1. *Implicit nonlinear time steps.* At the level of the underlying *direct* CFD equations, fast multigrid solvers make it possible to use implicit-time-step discretizations at full efficiency (see the general approach to time dependent problems in Sec. 3 above, and multigrid methods for shallow water and three-dimensional atmospheric models in [11], [10], [9], [164], [165], [117], [118] and [133]). This entails not only unconditional linear stability, but also avoidance of bad effects associated with *linearized* time steps (in which one would use fully implicit equations, but based on linearization around the previous-time-step solution) [9]. The unconditional stability is important for the multiscale data assimilation processes, enabling work on various temporal and spatial scales, unconstrained by various Courant numbers.

2. *Local refinements* are well known to be greatly facilitated by the multigrid algorithm, as also hinted in the algorithm description above. The multiscale environment simultaneously provides convenient flexible structures, refinement criteria and one-shot self-adaptive solvers; cf. Secs. 3.2 and 6.1.

3. *Space + time parallel processing.* Still at the level of the direct CFD equations (but similarly also at the level of the inverse (data assimilation) problem), multiscaling is a necessary vehicle to obtain parallel processing not only across space at each time step, but also across time (see Sec. 3.1 above).

4. *One-shot solution of inverse problems.* Normally, inverse problems are solved by a *sequence* of direct solutions (e.g., direct multigrid solutions), through which an iterative adjustment is made to the control parameters (the inverse-problem unknowns). For example, in the *adjoint method* for atmospheric data assimilation, a direct solver of the flow equations (marching forward in time) is followed by an adjoint solution (backward in time) that gauges the first derivatives of the data-fitness functional with respect to the initial values (the flow variables at the initial time). These derivatives then drive some adjustments of the initial values, from which another direct flow solution is next calculated, and so on. *Many* iterations are needed for this process to converge. In multigrid solvers, by contrast, one can integrate the adjustment of the inverse parameters into the appropriate stages of only *one* direct-problem solver (see Sec. 4.2 above and Secs. 5, 9.1 and 16.2 below. This general approach has been described in [23, §13] and [24, §13], with more details in [27, §8.2] and full development in [150], [2], [3]).

5. *One-shot continuation.* The assimilation problem is highly nonlinear, hence a good starting guess for the solution is important. A general way to obtain such an initial guess is by continuation (embedding), in which the problem is embedded in a sequence of problems, each requiring another application of the solver (using the previous-problem solution as the initial guess). In multigrid solvers, however, the continuation can often be integrated into just one FMG solver (see [23, S 8.3.2] or in [24]). For example, at the coarser stages of the FMG algorithm more artificial viscosity (and/or more regularization, and/or a smaller coefficient of D_t in the continuity equation) can be used, then gradually be taken out as the algorithm proceeds to finer levels. This makes the solution much easier in the first stages, from which it is then continuously dragged into the desired neighborhood. Such FMG continuation devices are often natural. For example, larger artificial viscosity would quite naturally be introduced on coarse grids, even without aiming at continuation. A natural continuation is also supplied by the inverse covariance matrix S (see below), which would be smaller on coarser FMG levels due to larger discretization-error estimates.

6. *Full flow control.* In most data assimilation approaches (such as the adjoint method described above), the control parameters (the parameters that can be changed to obtain fitness of solution to observations) are only the initial values of the solution. This makes it impossible to benefit from the details (the oscillating components) of the observations at time far removed from the initial time, because those details at those times are ill-determined by the initial values. Instead of controlling just initial values, one should really control the entire numerical solution. Namely, the control parameters

$\sigma(x)$ is a vector-valued grid function that at each point x gives the deviations in satisfying the set of flow equations. The objective function (the error functional that should be minimized) has the general form

$$E = \boldsymbol{\sigma}^T S \boldsymbol{\sigma} + d^T W d, \tag{4.1}$$

where $\boldsymbol{\sigma} = \sigma(x)$ is the vector of all control parameters, $d = \big(d(y)\big)$ is the vector of deviations of the solution u from the observation u^0 (i.e., $d(y) = (P^0 u)(y) - u^0(y)$, where P^0 is a projection from the solution space (x) to the observation space (y)), and S and W are (positive-definite) weight matrices. In a crude approximation, one can take these matrices to be diagonal, where the diagonal inverse $S(x,x)^{-1}$ is (a very rough estimate of) the expected square error in the equation at x, which is the sum of the local discretization error (conveniently estimated by the "τ correction" of the FAS multigrid solver; see [24, §8.4]) and the local modeling errors (errors in the physical assumptions embodied in the equations). The diagonal inverse $W(y,y)^{-1}$ is (a very rough estimate of) the expected square error in the *measurement* $u^0(y)$, including in particular the "representativeness error" (accidental deviation at the point of measurement from the relevant local average). More precisely, S and W should be corresponding *general* (not necessarily diagonal) inverse covariance matrices (in which case the discussion at Item 8 below is relevant).

A detailed Fourier analysis by Rima Gandlin, comparing full-flow control with initial-value control in a model case of 1D + time wave equations, has demonstrated the great advantage of the former [90].

So extensive control parameters can only be handled by a multiscale treatment. Moreover, using the methods described above the solution is expected not to be expensive, especially since the control parameters $\sigma(x)$ need not be controlled at the finest computational levels; on such levels $\sigma(x)$ can simply be interpolated from the coarser levels and kept unchanged during the relaxation (cf. Item 9 below).

7. *Unlimited correlation range.* In conventional assimilation methods, each control value interacts with a limited range of measurements: measurements at a restricted (e.g., 6 hours) time interval and sometimes only at confined distances. However, it is clear that large-scale averages of the dynamic variables interact at much larger ranges. Multiscale data assimilation makes it possible to correlate solution and measurements at any desired distance in space and time, since correlations at increasingly larger distances are calculated on increasingly coarser grids.

8. *Efficient representation of direct and inverse covariance.* There are a number of ways to derive or estimate covariance matrices and various simplification assumptions are made. However, the real covariance matrices (especially the model error covariance) are actually dense (not sparse), and thus involve huge ($N_s^2 N_t^2$, in principle) amount of information. Even when the matrix is sparse, its inverse, used in (4.1), is certainly dense. The only efficient way of *representing*, let alone computing, such huge dense matrices

and their inverses is a multiscale representation, based on their asymptotic smoothness. This would be similar to the methods described in Secs. 10 and 14.3 below and in [66] for calculating integral transforms, many-body interactions, solutions to integro-differential equations, and Kalmen filtering, all involving $n \times n$ dense matrices whose complexity (the amount of computer operations required to perform a multiplication by either the matrix *or its inverse*) is reduced to $O(n)$ by multiscale techniques.

To achieve such a low complexity it is of course necessary to assume the covariance matrices to be *reasonably smooth*. Namely, if the errors at two points, x and y, remote from each other, are correlated at all, their correlation is assumed to vary like $g_1(x)g_2(y)G(x,y)$, where $G(x,y)$ is *asymptotically smooth* (meaning that up to a certain order, p-order derivatives of $G(x,y)$ are not larger than $O(|x-y|^{-p+q})$, q being a fixed small integer). Such assumptions seem very reasonable in practice, and are certainly more accurate than neglecting distant error correlation altogether. They can also be weakened in various ways and still benefit from multiscale processing. (For example, it may be enough to assume at each point x smoothness for variations in only some directions, although the complexity may then rise to $O(n \log n)$. The processing in such cases would be akin to those in [43] and [55].)

9. *Improved regularization.* First, the multiscale solver described above is likely to require much less regularization than conventional solvers since the main ill-posedness in the problem is the long term and long range influence of fine-scale oscillations, while the multiscale large-scale interactions are mediated by coarse grids, omitting these oscillations. Secondly, attractive regularization devices are offered by the multiscale processing. For example, statistical theories of the atmospheric equations yield the relative expected energy at different scales. In a multiscale processing this can be used to properly penalize any excessive local energy at every scale, yielding an excellent regularization scheme (which could not even be *formulated* in uniscale processing). Generally, the multiscale data assimilation need not be governed by one all-embracing objective function, but can employ a collection of different directives at different scales. (Cf. Item 12 below).

10. *Fast assimilation of new data.* Normally, new observation data keep arriving and need to be assimilated into an already partly existing approximate solution; i.e., the new data should usually both modify the previous solution and extend it into a new time interval. The multiscale solver is particularly suitable for the task: The new data normally does not affect the *h–f details* of the solution in much older times; also, these details are normally no longer of interest. Hence, increasingly older times can participate in the new processing on increasingly coarser levels (still maintaining the fine-to-coarse τ corrections previously computed for them). This exactly fits into the windowing algorithm above (Sec. 4.2). The resulting ease of assimilating new pieces of data may well facilitate a *continuous assimilation policy*, with new data being assimilated much more often than today.

11. *Multiscale organization of observation data.* Either for the purposes of the multiscale assimilation procedure, or for a variety of other procedures, it is very useful to organize the observation data in a multiscale structure. This may simply mean pointers from a multiscale hierarchy of uniform grids into the set of data, with finer uniform levels introduced only where there are still more than a couple of observations per grid cell. Such data structures are commonly used to facilitate regional computations of all kinds. Beyond this, it is possible to replace many observations by their average at some larger scale, serving as a kind of *macro-observation*, its associated error estimate being of course reduced by standard rules of statistics. This can be repeated, to obtain still-larger-scale representations. Such structures may save much storage, and provide directly the needs of the multiscale assimilation algorithms.

12. *Scale-dependent data types.* Instead of simple averaging, the macro-observations just mentioned can be formed from the fine-scale data in a variety of other, often more meaningful, ways. In particular, fine-scale waves should be represented on coarse scales by their slowly-varying *amplitude.* Indeed, at large distances the wave phase is ill-posed, while its amplitude is still meaningful. (Cf. the techniques in Sec. 7 below).

5 Feedback Optimal Control

We consider a dynamical system that involves a vector x of *state functions* and a vector u of *control functions*, both being functions of time (and possibly also of space), governed by the initial-value ODE (or PDE)

$$\frac{dx}{dt} = F(x,u) \ , \qquad B_0 x(t=0) = b_0 \ , \tag{5.1}$$

where F and B_0 are vectors of known functions (or spatial operators). The *optimal control* problem is to find the control u for which this dynamical system minimizes a given *objective functional* $J(x,u)$ under various constraints, such as target-time (T) conditions of the type $B_1 x(t=T) = b_1$. In the *feedback* optimal control problem it is assumed that new initial conditions $B_o x(t)$ are continuously fed from the controlled device at all times $0 \le t < T$, requiring continuous updating of the control u. Fast numerical updates are required for *real-time* control.

In a usual approach to the feedback problem, the dynamical system is approximated by a linear-quadratic regulator (LQR), in which $F(x,u) = \mathcal{A}x + u$ and $J(x,u) = \int |Cx(t)|^2(|\mathcal{C}x(t)|^2 + |u(t)|^2)dt$, where $\mathcal{A}$ and $\mathcal{C}$ are linear operators on a suitable Hilbert space $\mathcal{H}$, $x \in \mathcal{H}$ and $u \in \mathcal{U} \subset \mathcal{H}$. Provided the system is stabilizable and $\mathcal{C}$ renders it detectable, there exists a unique nonnegative solution $\mathcal{K}$ to the *Ricatti equation*

$$(\mathcal{A}^*\mathcal{K} + \mathcal{K}\mathcal{A} - \mathcal{K}^2 + \mathcal{C}^*\mathcal{C})x = 0 \ , \qquad \forall x \in \mathcal{H} \ , \tag{5.2}$$

and $u(t) = -\mathcal{K}x(t)$ yields the desired feedback. In most cases this approach is very inefficient, either because the LQR approximation should be iterated many times, and/or because of the non-sparseness of the (discretized) operator $\mathcal{K}$, and the resulting high dimensionality of the Ricatti equation.

Our approach, by contrast, is based on a fast multigrid solver for the *open-loop* (i.e., not feedback) optimal control problem, installed in a multiscale way that allows super-fast (essentially local) updates upon feedbacks.

The multigrid open-loop solver is very efficient by itself, a one-shot solver for the nonlinear (non-LQR) problem. In fact, for various actual problems, it can share many of the potential benefits listed in Sec. 4.3 above. The super-fast *updates* are based on the observation that, upon changing the initial conditions, the change in the solution is increasingly smoother at times increasingly far from the initial. (In various actual problems, the sense of this smoothness has to be carefully understood.) This makes it possible for the multigrid *re*-solving algorithm to re-process its *fine* grids only at the very early times, while at later times only coarse levels are re-processed, with FAS fine-to-coarse defect corrections being frozen there (cf. [23, §15].) More precisely, at increasingly later times, any given multigrid level (a given timestep and, when relevant, spatial meshsize) need be re-processed increasingly more rarely. As a result, the computational cost of re-resolving is equivalent to only *local* re-processing (essentially just few steps near the initial time) of the full solver. This will usually be far less expensive than applying $\mathcal{K}$ (even just once, and even assuming the Riccati equations has already been solved).

We have tested this approach on several toy F-16 maneuvering problems (given to us by Dr. Meir Pachter of the Air Force Institute of Technology at Wright Patterson Air Force Base.) The linear dynamics includes three state and one control functions. Both quadratic and non-quadratic objectives were tested, including the L_∞ norm (the maximum absolute value) of one of the state functions.

The open-loop optimal control problem, which in this case is a two-point boundary value ODE system with 7 unknown functions, has been discretized by second-order finite differences on a staggered grid and solved by an FMG algorithm. Just two $V(1,1)$ multigrid cycles per grid proved enough to produce a solution with algebraic errors much smaller than (only few percent of) the discretization errors. In the case of the L_∞ objective, a continuation process has been integrated into this FMG solver, approaching L_∞ by L_p, with $p = 2^m$ at the m-th FMG level. Each of the relaxation sweeps included one red/black pass for each of the 7 ODEs, some of the passes being of the Gauss-Seidel type, the others — Kacmarz type. For some of the toy problems the principal part of the ODE system was scale dependent, producing boundary layers and requiring two different discretization schemes, one at fine levels the other at coarse levels, each with its own corresponding relaxation scheme.

The feasibility of the super-fast solution updates upon feedbacks has been established in our tests by monitoring the fine-to-coarse (τ) corrections. When

τ is appropriately scaled (divided by proper solution values available to the coarse-level re-solver) its values (excluding a couple of them near the initial time) turn out to change very little upon changing the initial conditions. This shows that τ can be frozen, so that re-solving can be restricted to coarse levels, as expected.

6 Adaptable Grids and PDE Solvers on Unbounded Domains

6.1 The multigrid solver + adaptor

A very substantial saving in the number of degrees of freedom needed by a discretization of a PDE to attain a given accuracy can be obtained by employing various forms of local grid adaptation. The multigrid solver, in its nonlinear FAS form, yields a particularly flexible and efficient framework for that purpose, with some additional benefits. The general description of this framework has been given elsewhere [19, §§7–9], [23, §9], [24, §9] or [6], so we only summarize the main points.

Local refinements are created by adding local *patches* of finer *uniform* grids over desired parts of a domain covered by a "parent" coarse grid. (The system is recursive: each of the "child" patches may itself contain smaller subdomains over which "grandchildren" patches of a further refinement are set). Each fine-patch solution supplies a certain *defect correction* to the equations of its parent (coarse) grid, thereby enforcing there the fine-grid accuracy. This is a natural part of the FAS multigrid solver (which introduces anyway coarser grids over finer ones to accelerate convergence, using in the process the same fine-to-coarse defect corrections). Hence, the multigrid solver solves the resulting composite discretization with the same efficiency (per degree of freedom) as solving uniform-grid equations. This composite structure is very flexible and can be highly non-uniform, while all its discrete equations are still written in terms of *uniform* grids. This makes it simple and inexpensive to use high-order approximations, while storing only a negligible amount of geometrical information. Also in this way no *unintended grid anisotropy* is introduced (in contrast to other grid generation or grid transformation methods, in which such anisotropies do enter, causing considerable complications for the multigrid solver).

The fine-to-coarse defect correction also yields, as a byproduct, precise *adaptation criteria*: a defect correction larger than a natural threshold indicates that a further local refinement is needed. Moreover, an automatic self-adaptation process can be integrated into the so-called "full multigrid" (FMG) solver: as the latter proceeds to increasingly finer levels, it can also decides (using these adaptation criteria) *where* those finer levels should be, thereby yielding a *one-shot solver/adaptor*.

It is also possible for each of the local grid patches to have its own *local coordinate system*. For example, in flow problems, the coordinate system may fit wall boundaries (or more generally: stream lines), facilitating the introduction of highly anisotropic grids in boundary layers (or particularly fine cross-stream meshsizes). In electronic structure calculations (cf. Sec. 9), a local patch at the atomic core can use spherical symmetry, as appropriate for core electrons.

An important feature of this adaptation is that often the calculation within the local-refinement patch can be done *once for all*: Although the solution in the patch changes when the parent-grid solution changes, the fine-to-coarse defect corrections usually change very little. At most one more short "visit" to the patch (e.g., one more relaxation sweep at the finer level) toward the end of the calculation will normally be needed to update the defect corrections. Alternatively, one can calculate apriori the approximately linear dependence of the defect corrections on the local parent-grid values.

6.2 PDE solvers on unbounded domains

As pointed out in [19, §7.1], problems in unbounded domains can be solved by a multigrid structure employing increasingly coarser grids on increasingly larger domains, using an FAS multigrid solver. The structure is essentially the same as described above (Sec. 6.1). We have embarked on a detailed study of how this should be done: At what rate should the domains increase with increased meshsize? What is the largest needed domain? What interpolation is needed at interior boundaries (boundaries of a grid h embedded in a larger domain covered by grid $2h$)? What multigrid algorithm should be applied?

For the Poisson equation $\Delta u = F$ we have developed, in collaboration with Jeffrey S. Danowitz, theoretical answers to these questions, then tested them numerically. Using general grid optimization equations (see [19, §8.1] or [24, §9.5] or [24, §9.3]) and the known smoothness properties of the solution, one can calculate how far out one must use a certain meshsize to maintain a certain accuracy. For example, one can find that if the domain of interest (outside which $F = 0$) has diameter d_0 and if the desired accuracy inside that domain would be obtained (had its boundary values been given) by a second-order discretization and a grid with meshsize h_0, then the diameter of each coarser grid h ($h = 2h_0, 4h_0, \ldots$) should only satisfy $d(h) \geq d_0(h/h_0)^{2/3}$ and $d(h) \geq d(h/2) + Ch\log h_0$. Without significantly departing from the desired accuracy one can cover a domain (the coarsest-grid domain) with diameter R, spending only $O(\log R)$ gridpoints, so R can easily be taken so large as to admit small enough boundary-condition errors. Employing a suitable version of the λ-FMG algorithm [24, §9.6], it has been shown that the accuracy-to-work relation typical to multigrid solvers of the *bounded*-domain problem can in this way be obtained for the *unbounded* domain, where accuracy is in terms of approaching the *differential* solution. The same can be obtained for higher-order discretizations (with another exponent in the first $d(h)$ inequality).

7 Wave/Ray Multigrid Methods

The aim is to develop advanced and general numerical tools for computing wave propagation on scales much larger than the wavelength, when there may also exist interactions with special smaller-scale inhomogeneities where ray representations (geometrical optics) would break down. Such tools can revolutionize important computations, such as: radar cross sections; wave propagation through dispersive media; seismic wave characteristics resulting from various types of explosion zones; generation and control of acoustic noise; electronic waves in condensed matter; etc.

We have developed two basic approaches relevant to the problem. One is a general multiscale solver for *integral* equations with oscillatory kernels [29], which is a very efficient way to solve wave propagation in *homogeneous* (and some *piecewise* homogeneous) media by replacing the differential equations with *boundary* integral equations.

The other approach is a fast multigrid solver for the highly indefinite *differential* equations of stationary waves in a domain containing many wavelengths, with radiation boundary conditions. The basic idea of this work had been stated long ago (see, e.g., [20, §3.2], and more details in [30]), but important algorithmic aspects had still to be worked out.

The model equation we use is the Helmholtz equation

$$\Delta u(x) + k^2 u(x) = f(x) \ . \tag{7.1}$$

Traditional multigrid solvers are not effective for this problem, because some "characteristic" components (i.e., those with wavelength close to $2\pi/k$) are non-local (their size is determined by conditions many meshsizes away) exactly on all those grids which are fine enough to approximate such components.

On each of its levels, the new solver represents the solution as

$$u(x) = \sum_j A_j(x) \exp\left(i\varphi_j(x)\right) \ . \tag{7.2}$$

At the highest (finest) level this sum includes just one term and $\varphi_j(x) \equiv 0$, so the representation includes just one function — the desired solution — and the equation for it is the usual five-point finite-difference discretization of (7.1). Increasingly lower levels of the solver employ on the one hand increasingly *coarser* grids of x to discretize each amplitude $A_j(x)$ and each eikonal $\varphi_j(x)$, and, on the other hand, correspondingly *finer* sets of "momenta" (i.e., more terms j in the above sum). The interaction between these levels has been shown to yield a solver (for the discrete equations given at the highest level) which is as efficient as the best traditional multigrid solvers for definite elliptic systems. The radiation boundary conditions are naturally enforced at the lowest level, where the representation essentially coincides with geometrical optics (ray representation, appropriate for scales much larger than the wavelength).

Details of the one-dimensional solver and a preliminary version of the two-dimensional solver were given in [122]. The current version of the two-dimensional solver, together with numerical results, is described in detail in [51]; its accuracy is analyzed in [52].

An important feature of the solver is the alignment of the grid on which $A_j(x)$ is discretized with the propagation direction of the corresponding eikonal (the direction of $\nabla\varphi_j(x)$), its meshsize growing (upon coarsening) faster in that direction than in the perpendicular directions. Specifically, if J is the number of terms in the summation (7.2) at a given multigrid level, then the propagation-direction meshsize for that level is $O(J^2k^{-1})$, while the perpendicular-direction one is $O(Jk^{-1})$. Incidentally, such oriented grids should have also been employed in [25], reducing the order of complexity stated there to the same one as in the non-oscillatory case (with an additional $O(\log n)$ factor in the case of integral transforms or integral equations defined on a curved manifold of codimension 1, e.g., a boundary).

A finite-element representation akin to (7.2) appears in [5] and [124], but only on one level, and without the above-mentioned grid alignment. Unlike that representation, the present one can be used to bridge the entire gap between the wave discretization needed at small subdomains and the ray discretization needed at the large outer regions, thus producing fully efficient fast solver, as well as the basis for the development described next.

7.1 Future plans: variable coefficients, local refinements and diffraction

The plan is to develop the solver for the variable-coefficient case $k = k(x)$, and to advance a new setting where only geometrical optics is used in most of the domain, while the wave equations, as well as intermediate levels with representations of the type (7.2), are just introduced at special restricted subdomains where geometrical optics breaks down.

Geometrical optics can certainly be used throughout large regions where $k(x)$ is either a constant or has a small relative change per wavelength. Although in the latter case the rays are curved, they can still be followed by Snell's law, or more generally by marching solutions of the eikonal equation (see, e.g., [160]). Discontinuities in $k(x)$ can also be accommodated by geometrical optics, employing the usual rules of reflection and refraction, as long as the surfaces of discontinuity have curvature radii large compared with the wavelength (assuming the number of repeated reflections is not too large).

The pure geometrical optics approach will typically break down in smaller regions (e.g., neighborhood of fast changes in $k(x)$ or large-curvature surfaces of discontinuity). It is proposed to introduce, in such regions, nested local refinements structured in the usual FAS-multigrid manner (see Sec. 6.1). The finer levels will generally use representations of the type (7.2), the finer the level the smaller the number of terms in the summation, eventually yielding a direct discretization of (7.1) on sufficiently fine grids in small subdomains; see more details in [51, §10].

Effectively this will produce ray dynamics in the large, with relations between rays modified by the finer grids in the small special regions (around an aperture, corners, edges, a radar target, etc.), yielding a general numerical tool for computing diffraction (the rays produced by small-scale disturbances; cf. [111]).

8 Rigorous Quantitative Analysis of Multigrid

Since the early days of multigrid development, the "local mode analysis" (LMA), based on heuristic local Fourier decomposition of the error function, has been the chief tool for the practical design, precise quantitative understanding, and even debugging, of the various multigrid processes. Although rigorously justified in very special cases only, the easily computable predictions of that analysis have turned out to be precise for quite general PDE boundary value problems discretized on uniform grids with quite general domains and boundary conditions. In several important cases, however, the predicted LMA convergence factors were not obtained, presumably due to the influence of boundaries, which are usually not accounted for by the local analysis; domains with reentrant corners are a notorious example.

A general rigorous framework for the local mode analysis on one hand, and for the treatment of boundaries on the other hand, has appeared [28], [32].

For general linear elliptic PDE systems with piecewise smooth coefficients in general domains discretized by uniform grids, it has been proved that, in the limit of small meshsizes, the convergence factors predicted by LMA are indeed obtained (provided the multigrid cycle is supplemented with a proper processing at and near the boundaries). That processing, it is proved, costs negligible extra computer work. Apart from mode analysis, a Coarse Grid Approximation (CGA) condition has been introduced which is both necessary and sufficient for the multigrid algorithm to work properly.

Unlike most other multigrid theories, convergence factors thus proven are not just qualitative; they are *quantitatively sharp*: they are exactly obtained (or arbitrarily closely approached) by the worst local mode.

The assumptions made by the theory are shown to be, in a sense, the weakest possible. Except for its stability, no assumption is made about the relaxation scheme: whether it is good or bad is not *assumed*, but *calculated*.

The first part of this work [32] studies the L_2 convergence in one cycle, for equations with constant coefficients. In the second part, extensions are discussed (in the form of comments) to variable coefficients, to block relaxation schemes, to many cycles (asymptotic convergence), to more levels with arbitrary cycle types (V, W, etc.), and to FMG algorithms. Various error norms and their relations to the orders of the inter-grid transfer operators are analyzed. Global mode analysis, required to supplement the local analysis in various border cases, is developed, and practical implications of the analysis, including practical ways for constructing and debugging multigrid solvers, are generally reviewed. A major emphasis is on the importance and practicality of adding partial (local) relaxation passes to the multigrid algorithm (cf. [19, App. A.9]): Theory and practice show that multigrid efficiency is greatly enhanced by adding special relaxation steps at any local neighborhood exhibiting unusually large residuals.

9 Many Eigenfunction Problems: Electronic Structures

Some important scientific problems involve the computation of a large number of eigenfunctions of a partial differential operator. As an important example of such problems we consider here the *Kohn-Sham* equation, of the Density Functional Theory (DFT) [116], central to ab-initio condensed-matter and quantum-chemistry calculations of electronic structures. (For surveys of the DFT equations and their current multigrid solvers see [13], [17].) The Kohn-Sham N-eigenfunction equation is:

$$-\Delta\psi_n(r) + V(r)\psi_n(r) = \lambda_n\psi_n(r), \qquad r \in \mathbb{R}^3, (n = 1, \ldots, N), \tag{9.1}$$

where $2N$ is the number of electrons in the system (or their number per period, in the case that V is a periodic function) which can be very large. Actually, the "effective potential" V depends on the eigenfunctions ψ_n and the nuclear positions through the relation

$$V(r) = V_{nc} + V_{el} + V_{xc}, \quad V_{nc} = -\sum_{i=1}^{N_a} \frac{2Z_i}{|r - R_i|}, \quad V_{el} = \int \frac{2\rho(r')dr'}{|r - r'|}. \tag{9.2}$$

Here N_a is the number of atoms in the system, R_i and Z_i are the position and charge of the i-th nucleon, $\rho(r)$ is the electronic density defined by

$$\rho(r) = \sum_{n=1}^{N} |\psi_n(r)|^2, \tag{9.3}$$

and $V_{xc}(r)$ is the so-called exchange and correlation potential, describing the average effects of exchange and correlation derived from the theory of the electron gas. In the *local* density approximation, $V_{xc}(r)$ depends only on $\rho(r)$ and possibly on the gradient $(\nabla\rho)(r)$.

Fast multigrid eigenproblem solvers have been developed before [57], but the ab-initio problem includes new traits and difficulties that call for new multiscale techniques, such as in the following list:

(1) *Singularities.* The nuclear potential energy harbors a singularity at each atomic nucleus (if pseudo-potential is not used). The multigrid solver

(unlike Fourier methods) allows local refinements that would remove the global inaccuracies associated with such singularities: see Sec. 6.1 above. Note that the local patches of finer grid levels can supply once-for-all (or seldom-updated) "defect corrections" that can act like, and indeed replace the use of, pseudopotentials.

Even with pseudopotentials, local refinements around nuclei, as well as *high-order discretization* everywhere, are necessary for efficiency, since high accuracies are required for predicting the energy *differences* that govern chemical processes. Because of the neighborhood of the singularity, *conservative discretization* is needed [6], which is especially tricky for high-order discretization at grid *interfaces* (the boundaries of any level of local refinement); see [12], where the FAS conservative discretization of [6] is extended to high-order schemes in three dimensions, and applications to quantum chemistry are discussed.

(2) *Unbounded or very-large-scale domains* can efficiently be treated by multigrid solvers which employ increasingly coarser grids at increasingly larger distances from the region(s) of interest (cf. Sec. 6.2 above). In particular, for electronic problems involving isolated molecules, the meshsize in the vacuum away from the molecules can grow; it can grow so fast that the total computational work is dominated by the work near (up to several inter-atomic distances away) from the nuclei.

Similarly, in problems with extended, essentially periodic structures, deviations from the periodic-structure solution due to defects or surfaces can be described on increasingly coarser grids at increasing distances from the defect or surface, while for extended *amorphous* material, increasingly coarser grids can be used at increasingly greater distances from the subdomain of interest. (This *cannot* be done without the new multiscale structure proposed in Sec. 9.2 below).

(3) *Self-consistency.* The dependence of the potential function V on the total electronic charge distribution ρ introduces a nonlinearity into the problems, which usually requires many iterative applications of a linear solver. FAS multigrid procedures can directly solve nonlinear problems, as efficiently as solving their linear counterparts [19], [23], [24]. The development of such one-shot solvers for the Schrödinger operator depends on the ability to update the self-consistent potential as the solution changes on the coarse grids. This is also related to the following issue.

(4) *Multi-integrations* are required in calculating the potential (e.g., the Hartree potential). This can be performed fast by solving auxiliary Poisson equations. Solving them by multigrid would facilitate the needed interaction between the coarse-level moves of this Poisson solver and the coarse-grid updates to the self-consistent potential in the eigenproblem solver (see #3 above).

(5) *External optimization.* In solving the electronic problem the nuclei are assumed fixed (the Born-Oppenheimer approximation), but one actually

needs to find the nuclei positions for which the electronic-solution energy together with the inter-nucleus potential yield the minimal total energy. This external optimization would normally be done iteratively, requiring solving the electronic eigenproblem many times. Again, a one-shot multigrid solver + optimizer can and should be developed, incorporating suitable nucleus moves into each of the levels of the multigrid electronic solver. A model study reported below (Sec. 9.1) has shown the feasibility of this approach and the exact multigrid techniques required for its full efficiency.

(6) *Multitude of eigenfunctions.* Even with a multigrid solver, the cost of calculating a large number N of eigenfunctions (N being the number of electrons in the system) may grow proportionally to N^3 (employing discretizations with $O(N)$ degrees of freedom), since each eigenfunction is represented separately and may need to be orthogonalized with respect to all others to ensure their distinction. A theoretical study and tests with 1D model problems, reported in Sec. 9.2 below, indicate that for periodic problems (the usual type in condensed matter calculations), it may be possible to reduce the complexity to $O(N_g \log N \log \frac{1}{\epsilon})$, by employing a multiscale collective representation of the eigenmodes. Here ϵ is the desired accuracy and N_g is just the number of grid points per periodicity cell required for adequately resolving the various features of the potential function $V(x)$.

(7) *Highly oscillatory functions*, such as the orbitals of high-energy electrons, would normally require dense grids, and would not allow effective multigrid solvers, because of the corresponding high indefiniteness of their equation. The multiscale approach described in Sec. 9.2 below effectively deals also with this difficulty.

(8) Multiscale structures may offer improved non-local representations for the *exchange correlation* potential. This aspect is largely unexplored. Certain non-local exchange-correlation operators can be translated into a system of Poisson equations, which (like the Hartree potential in #4 above) augment the Kohn-Sham system, yielding a larger PDE system that is simultaneously solvable by a multigrid algorithm.

Of all the scaling difficulties listed above, several (those numbered 1, 2, 3, 4, and partly also #5) have been dealt with in other contexts (similar difficulties in other fields). So, once multigrid solvers are introduced, the technique for treating these difficulties will already be at hand. We have therefore focused our attention mainly on #5 (see Sec. 9.1) and #6 and #7 (see Sec. 9.2).

9.1 Model problem for external optimization

A simplified model problem for the external optimization is the minimization of the two-dimensional two-atom total energy

$$\min_{z=(z_1,z_2)\in D} \left[E(z) + \lambda(z)\right] , \tag{9.4}$$

where $E(z)$ models the ("external") repulsive energy between ions located at $(0,0)$ and at (z_1, z_2), and $\lambda(z)$ is the corresponding electronic energy, modeled by the eigenvalue of the equation

$$\big(-\Delta + V(x,z)\big)\psi(x) = \lambda\psi(x) \ , \qquad x = (x_1, x_2) \in D \ . \tag{9.5}$$

We chose $V(x,z)$ that models the Coloumbic potential at x of the two-ion system, $D = [0,1] \times [0,1]$, and ψ was required to satisfy periodic boundary conditions on D (having chosen V and E also with this periodicity).

The Euler equations for minimizing (9.4) under the constraint (9.5) can be simplified (since the Lagrange multiplier coincides with ψ) to the system of equations (9.5)–(9.7), where

$$\langle\psi, \psi\rangle = 1 \ , \tag{9.6}$$

$$\frac{\partial E}{\partial z_i} + \left\langle \psi, \frac{\partial V}{\partial z_i}\psi \right\rangle = 0 \ , \qquad (i = 1, 2) \ . \tag{9.7}$$

The eigenproblem (9.5)–(9.7) was solved by a classical FAS multigrid eigen-solver [57]. The main point of the research was to find out how to include Eq. (9.7) and where to adjust z in the course of this solver. Since (9.7) is a global equation and z is a "global" unknown (unlike $\psi(x)$ it cannot be *smoothed*), it is enough to treat both of them at the coarsest level, where all the discrete equations can simply be solved simultaneously for all the unknowns, since their number is small. This would be fully efficient, provided a suitable "fine-to-coarse correction" for Eq. (9.7) is recursively calculated at each coarsening step, see [23, §5.6], except that in the FAS scheme the residual transfer is replaced by the τ_h^{2h} fine-to-coarse correction; see [23, §8.2].

The main finding of this research, done in collaboration with Ron Kaminsky, was that in the above situation (and for similarly "localized" global unknowns, whose movements may not be resolved on some of the coarse grids), a linear dependence on the global unknowns should be introduced into τ_h^{2h}; see details in [35, §6.1]. The linear terms are important in the cases where the functions $\partial V/\partial z_i$ are not resolved well enough on the coarse level to yield there the correct dependence of $\langle\psi, (\partial V/\partial z_i)\psi\rangle$ on variations in z. This generally happens when V has a singularity (or a particularly large local variation on the scale of the grid h) which moves with z. Fortunately, exactly in these cases, τ_h^{2h} can be calculated in a small neighborhood of the singularity to a good approximation.

With this simple change, the one-shot solver for the external optimization problem (9.2)–(9.4) has attained essentially the same convergence factors as in solving Poisson equation, costing only a fraction more.

This can straightforwardly be generalized to any number of unknown point locations (e.g., nuclei) in terms of which a PDE is formulated along with a corresponding number of additional "global" conditions, such as (9.7). At coarse levels where the separation between two such points becomes smaller than the meshsize, new equations can be formulated for their *collective* motion (keeping constant the position of one relative to the other).

9.2 $O(N \log N)$ calculation of N eigenfunctions

What is the amount of computation needed to calculate the N lowest eigenfunctions of a differential operator discretized on N_g gridpoints? A usual multigrid eigensolver [57] would need $O(N^2 N_g)$ operations, since each eigenfunction needs to be orthogonalized with respect to each other. Under favorable conditions, these orthogonalizations can be performed on the coarse grids [73], [74], possibly reducing the cost to $O(NN_g)$. However, all of these methods are adequate only for N sufficiently small, $N \ll N_g$. For large N, with eigenfunctions featuring variations on the scale of the grid, coarser grids cannot be used in such "naive" ways. For large eigenvalues the eigenproblem is highly indefinite, hence methods akin to those in Sec. 7 above should be used for coarsening.

A new approach (pointed out in [30]) is being developed for calculating N eigenfunctions of a differential operator discretized on N_g gridpoints in $O(N_g \log N \log \frac{1}{\varepsilon})$ computer operations, ε being the accuracy. This approach is based on the observation that "neighboring" eigenfunctions are distinguishable from each other only at large enough scales, and hence, in suitable representations, one can use a common description of their details at finer scales, and progressively separate them out only on increasingly coarser grids. The core procedure is indeed similar in structure to multigrid algorithms developed for *wave* equations (see Sec. 7 above). (Recent "linear scaling" methods in electronic structure calculation, reviewed in [85] and [91], are based on a localization assumption, typically solving a localized problem in $O(NN_\ell^2)$ operations, where N_ℓ is the number of atoms in the localization radius. The new approach would solve such a problem in $O(N \log N_\ell)$ operations.)

The feasibility of obtaining the $O(N_g \log N)$ efficiency has first been demonstrated by Oren Livne for *one-dimensional* problems [119], [120]. Moreover, that work has also shown that the developed *multiscale eigenbasis* (MEB) structure can be used to expand a given function in terms of the N eigenfunctions, again at the cost of just $O(N_g \log N)$ operations. This has been extended to *general 1D linear differential operators*. It constitutes a vast generalization of the Fast Fourier Transform (FFT), whose basis functions are the eigenfunction of discretized differential operators with constant coefficients, periodic boundary conditions with 2^ℓ uniformly spaced gridpoints. The new $O(N \log N)$ expansion is in terms of the eigenfunctions of a general operator with general boundary conditions and a general number of gridpoints. Also, summations like (9.3), with N terms summed at each of N_g gridpoints, are performed at the same $O(N_g \log N)$ cost.

To be sure, these algorithms have been based on formulations unique to the one-dimensional case. In particular, in 1D it has been possible to avoid solving highly indefinite boundary-value problems, thus skipping the more involved mechanism required for coarsening such problems (see Sec. 7). The extension to higher dimensions (discussed in [120, §6]) is far from trivial, and intimately related to the extension of the wave/ray multigrid methods to variable coefficients (see Sec. 7.1) and to general matrices (see Sec. 17.2.2).

A work related to the computation of the eigenvalue of a matrix and the fast updating of the singular-value decomposition of matrices is reported in Sec. 10.2 below.

10 Matrix Multiplication, Integral Transforms and Integrodifferential Equations

Multilevel algorithms have been developed for the fast evaluation of integral transforms, such as

$$Gu(x) = \int_{\Omega} G(x,y)u(y)dy \ , \qquad \Omega \subseteq \mathbb{R}^n \ , \ x \in \Omega' \subseteq \mathbb{R}^n \ , \tag{10.1}$$

and for the solution of corresponding integral and integro-differential equations. They exploit in various ways the smoothness properties of the kernel $G(x,y)$. For an $M \times N$ evaluation (i.e., x being discretized with M gridpoints, y with N) of either the transform *or its inverse* to a certain accuracy ϵ, these algorithms require $O\big((M+N)\log\frac{1}{\epsilon}\big)$ operations in the case of the Gaussian $\big(G(x,y) = \exp(-(x-y)^2/r^2)\big)$ or the potential-type $\big(G(x,y) = |x-y|^{-\alpha}$ or $G(x,y) = \log|x-y|\big)$ transforms, and $O\big((M+N)\log(\min(M,N))\log\frac{1}{\epsilon}\big)$ operations in the case of the Fourier $\big(G(x,y) = \exp(ixy)\big)$ or the Laplace $\big(G(x,y) = \exp(-xy)\big)$ transforms, for example. More generally, the algorithms yield *fast matrix multiplication* rules for dense matrices that include large low-rank submatrices. (See [53], [29], [158] and references in [158]). The same algorithms can also be used for the fast $\big(O(N\log\frac{1}{\epsilon})\big)$ summation of all the forces that N particles exerts on each other (see Sec. 14.3 below).

10.1 Adaptive grids. Integrodifferential equations

In their original form, the fast algorithms for evaluating (10.1) for potential-type kernels relied for their efficiency on the (asymptotic) smoothness of the *discrete* kernel (the matrix) and thereby on *grid uniformity*. However, in actual applications, e.g., in contact mechanics [158], in many cases large solution gradients as well as singularities occur only locally, and consequently a substantial increase of efficiency can be obtained by using *nonuniform* grids.

A new discretization and evaluation algorithm has been developed more recently in collaboration with Kees Venner. It relies on the (asymptotic) smoothness of the *continuum* kernel only, independent of the grid configuration. (Asymptotic smoothness roughly means that $G(x,y)$ is smooth except possibly near $x = y$; cf. [29].) This will facilitate the introduction of local refinements, wherever needed. Also, the new algorithm is faster: for a d-dimensional problem only $O(s^{d+1})$ operations per gridpoint are needed, where s is the order of discretization and d is the dimension. See [62], [63]: Numerical results were obtained for a model problem in which u has a singularity where its derivative is unbounded. It has been demonstrated that with the new fast evaluation algorithm on a *non-uniform* grid one can restore the regular work to accuracy relation (where accuracy is measured in terms of approximating the *continuum* transform), i.e., obtain the same efficiency as for the case without a singularity.

In combination with a multigrid solver, the fast evaluation algorithm also yields a fast solver for *integral and integrodifferential equations* [53], [158]. The main special feature of this multigrid solver is the distributive relaxation: a combination of several local changes to the solution is introduced at a time, such that the effect of the changes on the integral equation at any far point is negligible (due to the asymptotic smoothness of the kernel). A full multigrid (FMG) solver can be organized so that all its evaluations of integrals of the form (10.1), except for one, use reduced accuracy, hence costing much less. The cost of such a solver is only a fraction above the cost of just *one* (fast) evaluation of the involved integral transform.

The plan is to develop a multigrid solver for integro-differential equations discretized on adaptive grids, based on the new discretization and evaluation algorithm. As previously developed for PDE systems (see Sec. 6.1), self-adaptation criteria based on the local fine-to-coarse defect corrections (τ) are planned, as well as full integration of the grid adaptation process into the solver.

10.2 Secular equations. Discontinuous softening

A special case that involves dense-matrix multiplications is the computational task of finding all the roots $\{\lambda_j\}_{j=1}^N$ of the *secular equation*

$$1 + \sigma v(\lambda) = 0 \, , \qquad v(\lambda) = \sum_{k=1}^{N} \frac{u_k}{d_k - \lambda} \tag{10.1}$$

where $d_1 < d_2 < \cdots < d_N$ are real, $\sigma > 0$ and $u_k > 0$ for all k. This problem has various applications in numerical linear algebra, such as subspace computations [80], [84], solving constrained least-squares type problems [83], [92], updating the singular value decomposition of matrices [68], and modifying the symmetric eigenvalue problem [75]; see survey of literature and solution methods in [125], [126]. The equations should often be solved many times as a subproblem of a larger one. All existing solution methods cost $O(N^2)$ operation, since just the direct evaluation of $\{v(\lambda_j)\}_{j=1}^N$ costs that much. In joint work with Oren Livne, we used our multilevel dense-matrix-multiplication algorithm to solve the problem in $O(CN)$ operations, where C depends logarithmically on the desired accuracy [121]. The kernel here is $G(d,\lambda) = 1/(d-\lambda)$.

The fast algorithms for matrix multiplication and integral transforms described above are based on the *smooth* softening of singular (but asymptotically smooth) kernels (such as $G(d,\lambda) = (d-\lambda)^{-1}$). For high-order approximations, the softener (i.e., the modified kernel in the "softened neighborhood" of the singularity) is a high-order polynomial. This polynomial has to be calculated for each pair of variables in the softened neighborhood (e.g., each pair (d_k, λ_j) such that $|d_k - \lambda_j|$ is less than the "softening radius"). In a d-dimensional problem, and with a target accuracy ε, the softened neighborhood of each variable (e.g., each d_k) should include at least $O\Big(\big(\log \frac{1}{\varepsilon}\big)^{d+\eta}\Big)$ neighbors (neighboring values of λ_j), where $\eta \geq 0$ depends on smoothness properties of G and the order of the polynomial is $O\big(\log \frac{1}{\varepsilon}\big)$; hence the total work turns out to be $O\Big(N\,\big(\log \frac{1}{\varepsilon}\big)^{d+\eta+1}\Big)$. As shown in [121], this work can be reduced to $O\Big(N\,\big(\log \frac{1}{\varepsilon}\big)^{d}\Big)$ by choosing a *simple* (e.g., zero) but discontinuous softener. This requires a substantially more complicated algorithm, and can be advantageous only for a low dimensional (in particular: one dimensional, such as (10.1)) problem and high prescribed accuracy. The discontinuous softening is particularly advantageous to 1D kernels such as $(d-\lambda)^{-1}$ that are harder to soften (compared with $|d-\lambda|^{-1}$, the more common type).

11 Dirac Solvers

A major part of lattice quantum field calculations is invested in the inversion of the discretized Dirac operator M^h appearing in the fermionic action. Solutions of systems of the form

$$M^h \phi^h = f^h \tag{11.1}$$

are many times called for, either for calculating propagators or for the fast update of $\det M^h$ (see Sec. 12).

In the Euclidean staggered lattice formulation [138], the discrete equation at the gridpoint z is defined by

$$(M^h\phi)(z) = \frac{1}{h}\sum_{\mu=1}^{d} \eta_\mu(z)\big[U(z+\tfrac{1}{2}e_\mu)\phi(z+e_\mu) - U^\dagger(z-\tfrac{1}{2}e_\mu)\phi(z-e_\mu)\big] + m_q\phi(z)\,,$$

where h is the meshsize of the grid, $\phi = \phi^h$, d is the number of dimensions, m_q is the (given) quark mass, and e_μ is a vector of length h pointing in the μ-th coordinate direction. η_μ are complex numbers of modulus 1, and may be chosen as $\eta_1(z) = 1$, $\eta_2(z) = (-1)^{n_1}$, $\eta_3(z) = (-1)^{n_1+n_2}$ and $\eta_4(z) = (-1)^{n_1+n_2+n_3}$ for the gridpoint $z = h(n_1,\ldots,n_d)$, n_ν being integers. $U(z+\frac{1}{2}e_\mu)$ is the gauge field value defined on the directed link $(z, z+e_\mu)$. The inversely directed link $(z, z-e_\mu)$ carries the gauge field $U^\dagger(z-\frac{1}{2}e_\mu)$, where $\dagger$

denotes the Hermitian conjugate of the matrix. Each $U(z+\frac{1}{2}e_\mu)$ is an element of the model's unitary gauge group.

In collaboration with Michael Rozantsev, we have investigated two such models: $U(1)$ and $SU(2)$. In the $U(1)$ model, the gauge group elements are complex numbers of modulus 1, and $\phi^h(z)$ and $f^h(z)$ are complex numbers. (In the case of a trivial gauge field ($U \equiv 1$) in 2D, the $U(1)$ operator M^h reduces to the well known Cauchy-Riemann system.) In the $SU(N_c)$ model the gauge group elements are unitary complex $N_c \times N_c$ matrices whose determinant is 1, and $\phi^h(z)$ and $f^h(z)$ are complex N_c-vectors. See more about these models in [162], [112], [113], [114], [147], and about a multigrid approach to related, simplified models in [108] and [109].

These systems, despite their linearity and good ellipticity measures, are very challenging, due to their topology-induced singular (or nearly singular) eigenmodes and their disordered and non-commutative coefficients (the gauge field). The disorder results from the probabilistic physical rules by which the gauge field is determined, and from the "gauge freedom", i.e., the fact that those rules determine the field only up to arbitrary "gauge transformations". The latter are arbitrary multiplication of each $\phi^h(z)$ by an element of the gauge group and corresponding changes of the gauge field U so that (11.1) is still satisfied. Such changes do not change the physical content of the field.

11.1 Geometric multigrid solvers

Our first approach, based on red/black Kacmarz relaxation (since all equations in the Dirac system are first order), on pre-coarsening gauge smoothing and on multiscale iterate recombination, had previously been applied to the two-dimensional ($d = 2$) $U(1)$ model (see general description in [31], and full account in [130]). More recently we have been working on the $U(1)$ and $SU(2)$ gauge models in 4D [131], [132].

For the 4D-$U(1)$ gauge model, general conditions have been formulated under which the gauge field can be smoothed *globally* by gauge transformations, hence a fully efficient multigrid solver can, and has been, constructed. An important concept in this multigrid solver (as in any other geometric multigrid Dirac solver, for any model in any dimension) is to distinguish between different *species of unknowns* and between different *species of equations*. They can best be distinguished at the limit of low-temperature (well-ordered) gauge fields, for which each *species of unknowns* forms a grid function that must be constant for the homogeneous equations to be satisfied, and each *species of equations* forms a grid function which changes smoothly upon a smooth change of any one species of unknowns. The multigrid fine-to-coarse transition must transfer each fine residual to a coarse equation of the same species. Similarly, the coarse-to-fine transition must interpolate a correction to a fine unknown from course unknowns of the same species. It is also important, in the fine-to-coarse gauge averaging, to distinguish different *species of gauge links*: Two links are in the same species if they join the same species of unknowns.

The conditions for global gauge smoothing are *not* satisfied, however, in two kinds of topological situations. In the first kind, the total topological charge over the domain does not vanish. In this case the field can still be smooth *semi-globally*, i.e., it can be smoothed everywhere except for a certain local neighborhood which can easily be shifted away to any other place by gauge transformations, so that good intergrid transfers can be formulated *locally*. This is enough for obtaining nearly top multigrid efficiency.

The second topological case is more severe, featuring a *local* topological object, i.e. gauge-field discontinuities which cannot be shifted away. In such cases, and in many other cases (e.g., non-abelian models), it has been found that global or semi global gauge smoothing is not feasible. A general procedure has then been developed by which just *local* gauge smoothing at a time (over just 5 gridpoints in each direction) allows local constructions on each grid of the fine-to-coarse residual transfers and the coarse-to-fine correction interpolations required for the multigrid cycle. The local smoothing, which can be applied in any model, is done in a unique way, so that the resulting inter-grid transfers come out gauge invariant. Also, a general gauge-invariant procedure for the fine-to-coarse averaging of the gauge field itself has been constructed, based on transport averaging similar to that of [16], [14], [15]. This averaging has the advantage of reproducing on the coarse level various local topological objects, facilitating good coarse-grid approximations.

The local topological objects in the 4D-$U(1)$ model are "*monopol loops*", and they can also appear in a $U(1)$ component of a 4D-$SU(2)$ configuration. In the $SU(2)$ case these loops would persist for only few Monte-Carlo passes in gauge simulations at the critical temperature, and it is believed that in both $U(1)$ and $SU(2)$ only short loops are physically probable at critical or lower temperatures. If not treated, these loops lead to critical slowing down (CSD) of the multigrid solver (i.e., the larger the grid the more computational work per unknown is required). The number of slowly converging components introduced by the loops is small, however, so they can be eliminated by recombining iterants (taking linear combinations of results of the latest multigrid cycles so as to minimize the residual L_2 norm; which can also be done on coarser levels of the multigrid hierarchy; see [130], [59]) together with local relaxation passes added around the local discontinuities. With these devices, and with the local-gauge intergrid transfers and transport gauge averaging mentioned above, the multigrid convergence is still slower than in the absence of loops, but it seems free of CSD (at least when applied to gauge fields which are physically probable at critical or lower temperatures) [131]. We suspect that with wider regions of local relaxation the better efficiency may be restored; unfortunately, our domains were not wide enough for testing this.

Indeed, a severe problem in the work on these 4D models was the huge amount of computer time needed to produce reasonably sized, well equilibrated gauge fields on which to test our solvers: the Monte Carlo processes for producing these fields were far too slow. A general method to overcome this problem has only recently been devised (see Sec. 13.2 below). We hope to obtain by such a method larger 4D gauge fields for testing our Dirac solvers.

11.2 Algebraic multigrid solvers

Increasing complexities have been accumulating in the geometric-multigrid approaches described above: the different species, the treatment of various topological structures, the need for local smoothing of very disordered fields, etc. In particular, it has been found that for full efficiency the *geometric* coarsening (in which a coarse-level gridlines (or grid hyperplanes) consist of every other gridline (hyperplane) of the next-finer level) must be supplemented with simultaneous relaxation of various gauge-dependent strongly-coupled local sets of fine-level variables. Such sets can be identified by *compatible* relaxation sweeps (see Sec. 17). However, it became increasingly clear that *algebraic* multigrid (AMG) methods (see again Sec. 17) may be more convenient for treating at once all these difficulties. So, as already suggested in [31], we have returned to the development of AMG Dirac solvers.

Our first AMG Dirac solver has been based on the highly-accurate coarsening techniques of [37] (see Sec. 17.1 below). This solver, briefly described in [37, App. C] and detailed in [132], has been tested for the Schwinger model (two-dimension, with $U(1)$ gauge). We use Kacmarz or least square relaxation and distributive coarsening (i.e., in the notation of Sec. 17.2 below: either $(P = I,\ M = A^T)$ or $(P = A^T,\ M = I)$, both in relaxation and in defining coarse variables), which is usually a very good distribution for first-order PDE systems. The coarse-level set of variables is first selected *geometrically* (taking every fourth fine-level ghost variable, in a certain fixed 2D pattern). Then this set is enhanced using the compatible relaxation tool (see Sec. 17), thereby adding another 10%–20% of the ghost variables to the coarse level. The coarse-level equations have been derived using either a 3×3 or 5×5 coarse-grid stencil, each including also all those extra coarse variables added (following the compatible relaxation test) at the corresponding region. The coarse-to-fine interpolation of corrections has been done by several passes of compatible relaxation. Recombination of iterants has also been employed.

The different tests we ran, on a 32×32 grid, proved that *all and each one* of the above devices is necessary for fast convergence in the more difficult cases. Very good asymptotic convergence rates have been obtained (e.g., a convergence factor of 0.2 to 0.3 per two-level cycle) with the 5×5 coarse stencil even for *hot* (practically random) gauge fields, provided some 15% extra points were added to the coarse level and upto 8 iterants were recombined. For *critical* gauge fields only a couple of iterants needed to be recombined.

Satisfactory as these results are, they can only serve to demonstrate the potential of the AMG approach. However, the coarsening method employed here, even though local in principle, is far too expensive, especially for the 5×5 stencil. Even with this size of stencil some iterant recombinations have proved necessary, showing that the lowest lying eigenmodes are not yet sufficiently well approximated. There is no hope to efficiently use this approach for large 4D models. A far less expensive coarsening, that can well approximate even the near-zero modes, is the improved "bootstrap" AMG approach described in Sec. 17.2 below.

12 Fast Inverse-Matrix and Determinant Updates

In parallel to the development of the multigrid fast Dirac solvers (Sec. 11), collaborating again with Michael Rozantsev, methods have been developed for using multigrid solvers for constructing an inexpensive structure of the *inverse* Dirac matrix, allowing fast self-updating upon each change in the matrix itself (each gauge update). This will allow fast updating of the fermion propagators and the associated determinant (needed at each step of the Monte Carlo simulations of the unquenched gauge field). The approach was first described in [26, §12]; the substantially improved version presented here exploits the development described in Sec. 17 below.

For a large d-dimensional lattice with $N = L^d$ sites and meshsize $h = O(L^{-1})$, the storage of the Dirac inverse matrix $(M^h)^{-1}$ would require $O(N^2)$ memory and $O(N^2)$ calculations, even for fully efficient multigrid solvers. Using the following special multigrid structure, they can be respectively reduced to $O\big(N(\log\varepsilon^{-1})^{q_1}\big)$ and $O\big(N(\log\varepsilon^{-1})^{q_2}\big)$ where ε is the relative error allowed in the calculations and q_1 and q_2 depend only on d. More important, the structure will allow very fast *self-updating*.

The implemented multigrid structure first calculates and stores $O(\log L)$ *accurate algebraic coarsening* levels of the operator M^H, from the given finest ($H = h$) to the coarsest possible ($H = O(1)$). Each M^H is constructed from the next finer one in the manner of [37] (see Secs. 11.2 above and 17.1 below): for accuracy ε, the stencil of M^H at each point involves some $n_c = O\big((\log\varepsilon^{-1})^{q_2}\big)$ neighbors (so each row of M^H has n_c non-zero terms). (With the much more efficient methods of Sec. 17.2, substantially smaller n_c will be needed.)

Secondly, the proposed structure calculates and stores at each level enough "central terms" of $(M^H)^{-1}$. Such central terms in each column of $(M^H)^{-1}$ are terms corresponding to variables neighboring to the equation associated with that column (e.g., the n_c neighbors participating in the equation will suffice). The central terms of $(M^H)^{-1}$ can easily be constructed from those of the next coarser level, since the latter are all one needs in a two-level multigrid cycle for solving the relevant systems (at the finest level, for example, each of these systems has the form (11.1), with f^h being the delta function corresponding to the calculated column of $(M^H)^{-1}$). See details in [132].

This structure can be immediately updated upon changes in the gauge field. Indeed, each local change in the gauge field, if done in a properly distributive manner (i.e., so that some moments of the fields are kept unchanged) has only local effect on the propagators. Since the calculation of the latter can be regarded as solving by multigrid the system (11.1) with $f^h = \delta_{x,y}$, the effect of each local change can be calculated just by local relaxation sweeps around the change on some of the finest levels. More global changes will similarly be introduced (in a distributive manner) at coarser levels of the gauge field Monte Carlo simulations. The cost per update is $O(1)$, i.e., independent of lattice size.

With $(M^h)^{-1}$ thus monitored, one can inexpensively calculate changes in $\log detM^h$. For a *small* change δM^h in the gauge field

$$\delta \log detM^h = Tr\big((M^h)^{-1}\delta M^h\big) \ , \tag{12.1}$$

which can be computed locally, based on the central terms of $(M^h)^{-1}$. For *larger* changes one can locally integrate (12.1), since the local processing also gives the dependence of $(M^h)^{-1}$ on δM^h. Again, the amount of calculation per update does not depend on the lattice size.

Simplified model. The approach described above has first been developed for model matrices with a simplified structure: matrices M^h arising from discretizing on a lattice with meshsize h the random diffusion equations $Lu = f$, where

$$Lu(x,y) = \frac{\partial}{\partial x}\left[a(x,y)\frac{\partial}{\partial x}u(x,y)\right] + \frac{\partial}{\partial y}\left[b(x,y)\frac{\partial}{\partial y}u(x,y)\right] \ ,$$

and the discrete analogs of the diffusion coefficients $a(x,y)$ and $b(x,y)$ have random values, uniformly distributed in $(0,1)$. Excellent accuracies were obtained, but they required very expensive (though *local*) coarsening: 5×5 and even 7×7 stencils [132]. Far less expensive algorithms, based on the BAMG methods of Sec. 17.2, are possible.

13 Monte Carlo Methods in Statistical Physics

The general goal is the systematic development of advanced multigrid Monte-Carlo (MC) methods in statistical mechanics, molecular dynamics, quantum mechanics and quantum field theory.

A Monte Carlo simulation aimed at calculating an average of a certain observable is called "***statistically optimal***" if it achieves accuracy ϵ in $O(\sigma^2\epsilon^{-2})$ random number generations, where σ is the standard deviation of the observable. This is just the same order of complexity as needed to calculate, by statistical sampling, any simple "pointwise" average, such as the frequency of "heads" in coin tossing. The goal is to attain such an optimal performance in calculating much more complicated averages in statistical physics, including in particular thermodynamic limits, i.e., limits approached by the averages of system observables when the system size tends to infinity.

Two basic factors usually prevent naive Monte Carlo calculations of a thermodynamic limit from being optimal, even when $O(\sigma^2\epsilon^{-2})$ independent samples are indeed enough to average out their deviations down to $O(\epsilon)$ accuracy. First, to achieve an $O(\epsilon)$ approximation to the thermodynamic limit, each sample should be calculated on a system of sufficiently large *volume*, that is, a system whose linear size L grows with ϵ^{-1}; typically $L \sim \epsilon^{-\rho}$ for some $\rho > 0$. So in d physical dimensions, the required simulation volume for each sample is $L^d = O(\epsilon^{-\rho d})$. This factor is called the *volume factor*. The second factor is the *critical slowing down* (CSD), i.e., the increasing number n of MC passes needed when L grows in order to produce each new (essentially independent) sample; usually $n \sim L^z$, where z is typically (at least at the critical temperature) close to 2. As a result of these two factors, the cost of calculating the thermodynamic limit to accuracy ϵ rises as $O(\sigma^2\epsilon^{-2-\rho d-\rho z})$. (Additional complexity factors, that multiply these two, arise in quantum field theory from propagator calculations and fermionic interaction; separate research for eliminating them is described in Secs. 11–12 above.)

Two different multiscale approaches have been developed for treating these two complexity factors. They are respectively described in Secs. 13.1 and 13.2 below, the latter being more generally applicable (e.g., for molecular-dynamics calculations: see Sec. 14). Each of these approaches generates a sequence of increasingly coarser descriptions ("levels") of the simulated system. The coarser levels basically perform three different tasks:

(i) *Acceleration* of the Monte Carlo simulations on the finer levels (to eliminate the CSD). This is in general similar to the multigrid *convergence* acceleration in PDE solvers.

(ii) *Gathering statistics*: large-scale fluctuations can be cheaply averaged out through coarse-level MC, by cycling enough between these levels (much more than usual in multigrid PDE solvers) before returning to finer levels. Indeed, averaging out *fine-scale* fluctuations does not require *many* returns to the fine levels, since such fluctuations are largely averaged out in any *one* fine-level configuration.

(iii) Increasingly larger computational domains can be simulated cheaply by using increasingly coarser levels: The finest level covers only a relatively small domain, or small "windows"; a coarse level created from it can then switch to a larger domain.

The elimination of both the volume factor and the CSD factor implies *ideal performance* (statistical optimality). It also implies that on sufficiently coarse level the derived description allows *true macroscopic simulation* of the system, i.e., such that does not require operations at finer levels.

Statistical optimality was first demonstrated in calculating various thermodynamic limits in Gaussian models with constant coefficients, and also in calculating the critical temperature of the Ising model [48], [86]. This led to several years of efforts to develop interpolation-based multigrid algorithms, with cycles similar to those of multigrid PDE solvers. These are reported in Sec. 13.1 below. The increasing complicated nature of the interpolation rules and the coarse-level Hamiltonians required to treat advanced nonlinear models has later brought us to favor renormalization-type methods, which are described in Sec. 13.2.

13.1 Interpolation-based methods

These methods imitate multigrid solvers of discretized differential minimization problems, where the *Hamiltonian* (or the *energy* functional which should be minimized) is automatically defined on increasingly coarser grids by recursively specifying, level after level, coarse-to-fine interpolation rules. They also use the same type of cycles, except that the relaxation sweeps (each composed of a sequence of local minimization steps) are replaced by Monte Carlo sweeps (local steps of simulating the probability distributions induced by the Hamiltonian). The *cycle index* (specifying how many times one switches from a given multigrid level to the next coarser level per each switch to the next finer level) in statistical multigrid algorithms for computing large-scale observables will be larger than usual in multigrid solvers, to allow cheap averaging of large-scale fluctuations.

To obtain statistically optimal algorithms, as defined above, it is necessary to choose the coarse-to-fine interpolation so as to obtain full physical mobility at the coarse level: poor interpolation would not allow accessing mutually independent samples at the coarse level without in-between visits to the fine level for Hamiltonian updating. In addition, for statistical optimality the resulting coarse-level Hamiltonians need to be computable in a bounded complexity per coarse-level degree of freedom; they cannot be left just expressed in terms of finer-level variables.

Statistical optimality was first demonstrated for *Gaussian models with constant coefficients* [86], [48]. It was shown there, for the one-dimensional Gaussian model, that the susceptibility thermodynamic limit can be calculated to accuracy ϵ in about $4\sigma^2\epsilon^{-2}$ random number generations, while the average energy per degree of freedom requires $3\sigma^2\epsilon^{-2}$ generations for a similar accuracy. It was also found that the algorithmic flow (as determined by the multigrid *cycle index*) should generally depend on the observable being calculated. In the *two-dimensional* Gaussian model, the susceptibility limit can be measured to accuracy ϵ in about $20\sigma^2\epsilon^{-2}$ random number generations. In the one-dimensional *massive* Gaussian model, the susceptibility limit can be calculated in less than $8\sigma^2\epsilon^{-2}$ random generations, essentially independently of the mass size, although the algorithm flow may change with that size [45].

For the *variable-coupling Gaussian models*, it was shown that in order to reach ideal performance, the algorithm should employ during the multigrid cycle *weighted* interpolation and *variable* sampling (the Monte Carlo process should sample more frequently regions with smaller coupling values because such regions have larger contributions to observable fluctuations). Such algorithms have been implemented for *strongly* discontinuous cases in one and two dimensions. ("Strongly" means that the couplings may change by orders of magnitude between adjacent regions.) For the one dimensional variable-coupling Gaussian model, the susceptibility limit is calculated to accuracy ϵ in less than $8\sigma^2\epsilon^{-2}$ random number generations. In the two-dimensional variable-coupling Gaussian model, that limit can be measured in less than $20\sigma^2\epsilon^{-2}$ random generations [46]. These results are independent of the maximal ratio between strong and weak couplings, unlike the severe extra slowness that large such ratios can inflict on pointwise Monte Carlo.

The development of an optimal algorithm for the variable-coupling Gaussian model provides an important tool for general nonlinear models, where non-constant couplings stochastically emerge at coarser levels of the multigrid Monte Carlo processing.

Doubts have however been raised whether ideal MC performance can really be obtained for nonlinear models, where large-scale fluctuations are highly correlated with small-scale fluctuations. By applying the new analysis methods to the nonlinear *anharmonic crystal model* we have shown, and confirmed by actual simulations, that, down to a certain (small) ϵ, performance similar to that of the Gaussian models can still be obtained (although it requires careful choice of the multigrid cycling parameters [87], [47]). Such a performance is realizable because the large-scale fluctuations depend only on some *averages* of the small-scale fluctuations, and these averages are approximated well enough at any *single* fine-level configuration used at coarsening.

For a sufficiently small ϵ, however, and for models sufficiently dominated by the anharmonic term, both the analysis and the numerical tests show that ideal performance can no longer be obtained by a multigrid process which employs weighted linear interpolation. In fact, the analysis shows that no interpolation in the form of a linear operator can obtain ideal performance for all ϵ. We have therefore introduced another type of interpolation, the *minimization interpolation*.

This interpolation is best defined in terms of the Full Approximation Scheme (FAS; cf. [61, §7]), where the coarse-grid variables represent the *full* current configuration (i.e., the sum of a coarsened representation of the current fine-grid configuration and the current coarse-grid correction) instead of just the current coarse-grid correction. To define a value u_0 at a fine-grid point based on coarse-grid values $(u_1, u_2, \ldots)$, the minimization interpolation method is first to calculate $U_0(u_1, u_2, \ldots)$, defined as the value of u_0 that would be obtained by some, exact or approximate, local Hamiltonian minimization with the values of $(u_1, u_2, \ldots)$ being held fixed. Then, to retain statistical detailed balance, the FAS minimization-interpolation value is defined by

$$u_0 = U_0(u_1, u_2, \ldots) + \widetilde{u}_0 - U_0(\widetilde{u}_1, \widetilde{u}_2, \ldots) \ . \tag{13.1}$$

where the $\widetilde{u}_i$ are the values of the variables *at coarsening*, i.e., at the last transition from the fine level to the current coarse one.

Two-level unigrid experiments with the anharmonic crystal model have shown that the volume factor, along with the CSD, can be completely eliminated with an *exact* minimization interpolation. However, this interpolation creates a complicated coarse-level Hamiltonian, so we next designed simple *approximate* minimization interpolations, employing polynomial best fit. This yields a simple (fourth-order polynomial) coarse level, allowing the recursive construction of still coarser levels and application of complete *multi*-level cycles, which do indeed demonstrate the desired ideal MC performance [87], [47].

The situation is less convenient in more advanced physical models, where topological structures are present, because large-scale topologies may be correlated to specific fine-scale features, such as vortex centers. Also, linear-like interpolation of spinors is problematic.

A partial elimination of the volume factors in measuring susceptibility for Ising models was previously obtained by the three-spin coarsening technique [31], [48], as well as full elimination of that factor (namely, ideal MC performance) in determining that model's critical temperature [48].

Various attempts to attain ideal performance for two-dimensional nonlinear *σ models* (several of which are described in [142]) have failed. Nevertheless, we have developed a variety of new *stochastic* coarsening procedures by which at least *partial* elimination of the volume factor can be achieved. These procedures include: a detailed-balance way to associate the introduction of linear (or linear-like) interpolation with a certain probability for reducing adjacent coupling strength; smart choice of the interpolation in a neighborhood depending on local features at coarsening; stochastic simplification of the derived coarse-grid Hamiltonian in ways which do not destroy the statistical detailed balance; and introduction of less restrictive stochastic interpolations [142]. Most of the developed schemes are applicable to specific cases of XY and Manton's models, while some are universal for any $O(N)$ model.

Specially devised two-grid numerical experiments have demonstrated that the designed techniques are capable of eliminating the volume factor almost *completely* at low temperatures of the XY and Manton's model, and *partially* in the $O(4)$ model as well as in the critical region of the XY model. The non-optimality of the latter results have been attributed to the insufficient accuracy in representing and sampling some of the statistically important features by means of currently employed interpolation and stochastic coarsening procedures.

This led us to an attempt to introduce the FAS minimization interpolation (13.1) also to the XY model. It yielded an improved, but not yet statistically optimal, performance. The reason for non-optimality has been shown to be the bias introduced by the FAS correction $\widetilde{u}_0 - U_0(\widetilde{u}_1, \widetilde{u}_2, \ldots)$. For example, if the coarse configuration at coarsening $(\widetilde{u}_1, \widetilde{u}_2, \ldots)$ happens to be locally *non-smooth*, then the corresponding FAS correction is likely to be large, preventing the coarse-level system from efficiently sampling *smooth* configurations. A way around this difficulty is to replace (13.1) by

$$u_0 = U_0(u_1, u_2, \ldots) + \frac{Q_0(u_1, u_2, \ldots)}{Q_0(\widetilde{u}_1, \widetilde{u}_2, \ldots)} \left[\widetilde{u}_0 - U_0(\widetilde{u}_1, \widetilde{u}_2, \ldots)\right] , \qquad (13.2)$$

where $Q_o(u_1, u_2, \ldots)$ is a characteristic size of the likely fluctuations in u_0 given $(u_1, u_2, \ldots)$. More precisely, the interpolation (13.2), like (13.1), is suitable in case u_i are real variables; it has modified forms to suit other types of variables, such as XY.

13.2 Renormalization multigrid (RMG)

The increasing complexity of the coarse Hamiltonians produced by the interpolation-based techniques has led more recently to a new type of algorithms. They combine renormalization-like derivation of increasingly coarser "descriptions" of the system, with multigrid-like coarse-to-fine Monte-Carlo accelerations.

This RMG approach has already yielded optimal performances (eliminating both the CSD and the volume factors) for certain observables in the Ising model, and it can in principle be extended to arbitrary models, since it is not based on cluster-type (such as Swendsen-Wang and Wolff) algorithms. In fact, we have already preliminarily used this approach in several models of molecular dynamics: a model of polymers (see Sec. 14.6) and models of fluids (see Sec. 14.7). Moreover, the approach has inspired a similar, equally-general procedure for the coarsening of *deterministic* equations (see Sec. 17).

For simplicity, the RMG techniques are surveyed here mainly in terms of the 2D Ising model, where they were first developed by Dorit Ron [60]. In this model, each fine level configuration U consists of an unbounded two-dimensional lattice of Ising spins u_i (variables taking either the value $+1$ or -1), with a probability distribution $P(U) \sim \exp\left(-\mathcal{H}(U)/k_\beta T\right)$, and Hamiltonian $\mathcal{H}(U) = -J\sum_{\langle i,j\rangle} u_i u_j$, where $\langle i,j\rangle$ runs over all distinct pairs of nearest-neighbor spins.

Coarsening. Generally, each coarse-level "description" in RMG consists of two items: The coarse-level *variables*, and the *statistical rules* that govern their probability distributions. The rules will generally be expressed in terms of *conditional probability* (*CP*) *tables* (whose far-reaching generality will be discussed in Sec. 13.2.2 below).

Each coarse-level *variable* will be defined in terms of a small local set of next-finer-level variables. For example, in 2D-Ising with *majority rule* blocking, the coarse level consists again of a 2D array of Ising spins (± 1 variables), each of which being a "block spin", i.e., its sign representing the sign of the sum of the four spins in a corresponding $b \times b$ block of the next finer level. (The sign of zero is taken to be + or −, each with probability 1/2. Usually $b = 2$.)

There is no unique way to choose the set of coarse-level variables. Indeed, given the full description (variables + CP table) of the next finer level, many different choices can be equally good. There exists however a general *criterion to gauge the adequacy of any candidate course set.* This criterion is the speed of equilibration of *compatible Monte Carlo* (CMC) runs. A CMC is a Monte Carlo process on the fine level which is restricted to the subset of fine-level configurations whose coarsening (e.g., $b \times b$ blocking) coincides with a given *fixed* coarse-level configuration. A *consistently fast* CMC equilibration (i.e., CMC with very short average decorrelation time, averaging being over an ensemble of the fixed coarse configuration) implies that the fine-level equilibrium can be produced from the coarse-level equilibrium just by local processing, which is indeed the main desired property of coarsening.

The fast CMC equilibration implies that the set of coarse variables enjoys the *near locality property*. This is the property that the *conditional* probability distribution of a coarse variable at a point A, *given fixed states of all other coarse variables*, depends mainly on the closest neighbors: the average dependence decays exponentially with the distance from A. (For example, in 2D-Ising if the neighborhood of A is changed only at points at distances larger than r from A, the conditional probability to have +1 at A given its entire neighborhood can change at most by $O\big(\exp(-c_\ell r)\big)$, with some constant c_ℓ.)

The strength of near locality (the rough size of c_ℓ) can be directly measured. Strong near locality has been measured, e.g., for the 2D-Ising with the majority coarsening. This yields the possibility to construct CP tables for coarse levels by the following quite general branching procedure.

The CP table for each coarse level is derived by running an efficient MC simulation at the next finer level, during which appropriate statistics is gathered. In 2D-Ising, statistics are gathered for estimating $P_+(u_1, u_2, \ldots, u_m)$, the probability for any block spin to be +1 given that its (ordered set of) neighboring block spins are having the values $(u_1, u_2, \ldots, u_m)$.

The size m of the considered neighborhood is *variable*: If only a small amount of statistics is gathered, only the four nearest neighbors (u_1, u_2, u_3, u_4) are considered. With more statistics, all *eight* closest (nearest and next-nearest) neighbors $(u_1, \ldots, u_8)$ are considered. Further on, when the amount of statistics *for a particular neighborhood* $(u_1, \ldots, u_8)$ is sufficiently large, that neighborhood is *split*, i.e., statistics is gathered for its "*child neighborhoods*": These are neighborhoods $(u_1, \ldots, u_8, u_9, \ldots, u_{12})$ with the same $(u_1, \ldots, u_8)$ as in the "parent" neighborhood, and with $(u_9, \ldots, u_{12})$ representing values of the four *subsequent* neighbors (those which are exactly two meshsizes away). Children with enough statistics may further be split into grand-children, and so on. The general rule is to split a neighborhood when (and only when) some of its children have enough statistics to make the difference between their P_+ values significant (i.e., larger than their standard deviations). Between several candidate splits of a neighborhood, the one with the largest *spread* (average child deviation) should be adopted.

The CP (e.g., P_+) table represents the coarse-level transition probabilities. Indeed, it is all one needs, and exactly what one needs, to run an MC simulation on that level (the level of blocks). Also, due to the adaptability in the size of the neighborhoods, this method for calculating the coarse transition probabilities is *statistically optimal*, in the sense that it will yield $O(\varepsilon)$ relative accuracy in calculating large-scale averages when the amount of statistics (the total number of random-number generations in producing the CP table) has been $O(\varepsilon^{-2})$. This claim has been confirmed in a sequence of numerical tests [60]. It may depend on having a fully efficient Monte-Carlo simulation at the fine level. Which is the next topic.

Monte-Carlo Acceleration. For a given finite lattice with a given CP table, suppose now that the CP tables for all its coarser levels (the level of blocks, the level of blocks of blocks, etc.) are also given. Then a new equilibrium configuration of the given action on the given lattice can efficiently be produced using the following "*coarse to fine equilibration*" (CFE) cycle.

First an equilibrium is easily obtained at the coarsest level of the finite lattice, by few MC passes with the corresponding CP table. From this, an equilibrium in the next level will be derived, and so on, until the target level (the given lattice) will be reached. To obtain an equilibrium in any level of spins given an equilibrium of its blocks, we use "*stochastic interpolation*", which is simply a small number of CMC passes. If the coarse-level (the block) CP table has not been fully accurate, the CMC passes should be followed by a small number of regular MC passes, a process we call "*post relaxation*".

A particular advantage of this equilibration process is the ability to cheaply produce very far regions of the same equilibrium configuration, without having to produce (at the fine levels) all the regions in between. This yields a very efficient way to calculate *far correlations*. (See also below about the role of "windows".)

Fast iterations. Since the derivation of a coarse CP table depends on efficient simulations at the finer levels, which in turn depends on employing the coarse CP table, iterating back and forth between these processes is in principle needed. However, these iterations converge very fast, since these processes only *weakly* depend on each other. Indeed, a very good first approximation to the P_+ block-spin tables is already obtained by just *local* equilibration at the spin level, produced by just a limited number (independent of the lattice size) of regular Monte Carlo passes (even starting from a completely random configuration). And, on the other hand, a very good approximation to critical-temperature equilibrium at the fine level can be obtained from a CFE cycle even with crude approximations to the coarse level CP tables (as long as they are kept critical: see below), provided a small number of post-relaxation sweeps is added.

In fact, an extremely simple way to obtain an approximate equilibrium on a given lattice with a critical action is by a CFE cycle employing *this same action* at all levels, plus post relaxation at each level. It can be shown that the required number of post relaxation sweeps is small whenever the convergence to a fixed point of the renormalization flow is fast. For the 2D-Ising model we have confirmed, in a sequence of numerical tests, the good approximation to equilibria obtained in this simple way [60].

Windows. Although just local equilibration is enough to produce good approximations to the CP table in the Ising model, in many other models it may suffer from *too low statistics* (or no statistics at all) for certain neighborhoods that are not *generally* rare, but that happen to be rare in the simulated regions (see example in Sec. 14.7.2). Hence, simulations at some coarse levels may run into regions whose simulation is inaccurate due to poor statistics in the CP table. In such a situation, and exactly at those regions, temporary *local* returns to finer levels should be made, in local *windows*, to accumulate more CP statistics relevant to the local conditions there. This can be done by first interpolating the window to the finer level: using CMC fine-level passes over the window, a correct equilibrium is generated in its *deep* interior (far enough from its borders); regular MC passes can then be done in that deep interior, to accumulate the desired statistics.

Generally then, the CFE technique would mostly be applied in windows rather than on the entire domain, its main role really being to supply rich samples of neighborhoods.

Errors. The principal sources of errors in the RMG processes are the finite statistics, the truncated size of the neighborhoods for which each CP table is calculated, and the finite size of the lattice employed at each level.

The latter type of error is easily removed: arbitrarily large lattices can be used due to the fast equilibration, while the P_+ calculations have nearly-local nature at all levels; and the simulation at any coarse level can be extended to its own desired domain size, since it is done directly, not through simulations at a finer level.

The finite-statistics errors are well controlled so as to keep all of them, at all levels, at the same optimal order ε, where the total amount of statistics is $O(\varepsilon^{-2})$. The truncation errors are also kept at $O(\varepsilon)$, by adjusting the neighborhood sizes; it is estimated that the size of the considered neighborhoods should grow proportionately to $\log(\varepsilon^{-1})$. The only trouble is the error enhancement from level to level, which is discussed next.

Back to criticality. In critical-temperature calculations of the CP tables, errors introduced at any level are magnified in the level derived from it (the next coarser level), and so on, due to the strong divergence of the renormalization flow away from the critical surface. To hold back this magnification, a mechanism should be added at each level to project the CP tables back onto the critical surface. Such a "criticalization" mechanism also facilitates calculating renormalization flows toward a fixed point when the critical temperature of the initial (finest-level) Hamiltonian is not known in advance.

The criticalization of a given CP table can be done by multiplying the temperature by a suitable factor $1/\theta$, i.e., by raising each probability to the power θ, then normalizing. (Normalization is not really needed for MC simulations with this CP tables.) The criticalization factor θ can be estimated in a number of ways. See details in [60]. Another, very accurate type of criticalization can be done near the fixed point (see next).

Fixed point algorithm. The fixed point of the renormalization group is approached by a sequence of coarsening steps, as described above, with a criticalization factor applied to each new CP table in the sequence. At each step the accuracy is raised in every respect: The total amount of statistics is much increased, accompanied by a (slower) increase in the lattice linear size and a (logarithmically slow, as dictated by the amount of statistics) increase in the size of the neighborhoods.

Critical exponents. The calculation is done in terms of the vector p of the entries of the CP table. Each stage of the fixed point algorithm can be regarded as a transformation $\mathcal{T}$, transforming p into $\mathcal{T}p$. The algorithm converges to the fixed point $p_* = \mathcal{T}p_*$. The correlation length critical exponent can immediately be derived from the largest eigenvalue λ_* of the equation

$$\mathcal{T}(p_* + q) = p_* + \lambda q \ , \qquad \| q \| \ll 1 \ .$$

Denoting by q_* the normalized eigenvector associated with λ_*, at each stage of the fixed point algorithm, increasingly more accurate values for λ_* and q_* are calculated (for the exact procedures, see [60]).

If p is near the fixed point p_*, further criticalizations and better approximations to p_* can be obtained by iterations, in each of which p is replaced with $(\lambda_1 \mathcal{T}_k p - \mathcal{T}_{k+1} p)/(\lambda_1 - 1)$, where λ_1 is the current approximation to λ_* and $\mathcal{T}_k p$ is the CP table calculated for $b^k \times b^k$ spin-blocks while running spin simulations with the CP table p; in particular $\mathcal{T}_1 = \mathcal{T}$. For convergence one should use $k > (\lambda - 1)^{-1}$; we used $k = 2$.

Once the CP table at the fixed point has been accurately calculated, the CFE cycle can be operated with the same p_* table being used at all levels, cheaply producing large equilibrated configurations (or very far regions of the same equilibrium configuration, as noted above). This can be used in a variety of ways for highly accurate calculations of various other critical exponents (a work in progress, by Ron and Swendsen).

13.2.1 Extension to continuous-state models

Initial steps of applying the above coarsening and acceleration techniques to the XY model are reported in [142]. Each 2×2 block spin is here defined to be the *average* of its four constituent spins, *without normalization* (whereby the original XY group of length-1 vectors is *not* preserved at the coarse levels). Compared with the ± 1 majority spins discussed above, each coarse spin here contains much more information; as a result, much smaller neighborhoods are needed in the probability tables to attain a given truncation accuracy. Still, these tables are more complicated than the Ising P_+ tables, since they should describe a *continuous* distribution, conditioned on *continuous* neighboring values.

To accumulate continuous-variable statistics, one partitions the range of this variable into *bins*: Counting the number of MC hits in each bin gives an estimate for the *integral* of the continuous variable over that bin. From those integrals, the value of the variable at any particular point can be *interpolated* (by a polynomial whose integrals over several adjacent bins fit the estimates). The same is true for a *vectorial* variable, such as the one representing the entire (truncated) neighborhood, whose bins may each be a tensor product of *elementary* bins, one elementary bin per each real variable participating in describing the neighborhood. More generally, the bins of the neighborhood are constructed *adaptively*, similar to the adaptively branching neighborhoods in the Ising case above, except that here a bin can be split into several bins in *two* ways: either by adding another variable to the description of that particular neighborhood, or by refining the current elementary binning of one or several of the existing variables.

The set of tests with the XY model reported in [142] clearly indicates that ideal MC performance free of the volume and CSD factors can be obtained in calculating various thermodynamic limits, such as the two-point correlation and the scaled susceptibility.

Future plans. The intention is to extend the RMG techniques to more advanced physical problems, possibly including gauge field models such as $U(1)$, $SU(2)$ and $SU(3)$. Together with the methods of Secs. 11–12 above, it is hoped ultimately to obtain ideal MC performance also with unquenched fermionic interactions.

13.2.2 Generalizations. Driven systems

As explained above, by a proper choice of coarse variables (checked by the CMC equilibration speed) the property of *near locality* is obtained for the next coarser level (the level of blocks), which allows the construction of that level's *CP table*.

Notice that the near locality property indirectly holds even in the case of long-range interactions, such as electrostatic or gravimetric interactions. Indeed, each such interaction can be decomposed into the sum of a *smooth* part and a *local* part (where "smooth" and "local" are meant relative to the particular scale of the next coarser level). All the smooth parts can be transferred (anterpolated) directly to the coarse level (cf. Secs. 14.3 and 14.6.1), hence it is only the local parts that remain to be expressed on the coarse level. For that expression the near-locality property still holds.

The CP table is a much more general representation of the "dynamics" (the transition probabilities) of models than the Hamiltonian representation, in the same way that, for deterministic models, systems of (differential or discrete) *equations* are much more general than variational (energy minimization) problems.

It has been found by Ron and Swendsen [129] that the CP representation of the renormalization-group transformation $\mathcal{T}$ described above provides an excellent test for the validity of the more common Hamiltonian representation. They have also developed a method based on the CP representation that allows them to produce a stable calculation of larger sets of renormalized coupling constants than either the Swendsen [148] or the Gupta-Cordery [97] methods, thus reducing the effects of truncation in renormalization-group calculations.

The CP table is particularly useful when even the given fine-level system is non-Hamiltonian. Such systems abound. An important case is that of *time-dependent* systems, such as driven diffusive systems [136]. The CP representation of such systems can be renormalized in both space and time, at various space/time coarsening ratios, yielding long-time and large-scale dynamics of the system.

A particular type of such renormalization leads to a fine-level CP table for the steady state of the driven system. The idea is to construct CP tables for the dependence of fine-scale fluxes on neighboring fluxes at the same time level t and on average densities at a previous time $t-\delta t$, where δt is doubled at each further renormalization transformation (together with a corresponding increase of the scale at which the densities are averaged). At the limit of such transformations, the steady-state CP tables should emerge.

13.2.3 Low temperature algorithms for frustrated system

Frustrated systems are those where conflicting influences arise from different terms of the Hamiltonian; e.g., some terms tend to align neighboring spins with respect to each other, while others tend to anti-align them. For such systems, especially at low temperatures, simple blocking (such as $b \times b$ blocks with the majority rule) are inefficient, yielding slow CMC equilibration. In this situation, correct coarsening can gradually be identified, for increasingly larger scales, by a gradual decrease of the temperature. Cf. Sec. 14.7.3 below. See also Sec. 18 for the limit case of zero temperature (strict minimization).

14 Molecular Mechanics

14.1 Background and objectives

Molecular mechanics (or dynamics) is a major tool of theoretical chemistry, with immense practical potential in medicine, material design and biotechnology. The Born-Oppenheimer approximation to the potential energy $E(r)$

as function of the n atomic positions $r = (r_1, r_2, \ldots, r_n)$ can be imagined as the objective functional of these calculations, the electrons being implicit. Explicit approximations to $E(r)$ as a sum of various few-atom interactions are derived by accumulated computational experience, compared with finer-scale calculations (such as those discussed in Sec. 9 above) and with molecular measurement data (crystal structure geometries, vibrational spectroscopy, heats of formation, etc.). The most common few-atom interactions are of the following two kinds (see a typical example in Sec. 14.6): (1) *The bond interactions* between chemically-bonded atoms, including three types: length (distance) interaction between 2 atoms, angle interaction between 3 atoms and torsion interaction between 4 atoms. The first is much stronger than the second, which in turn is much stronger than the third. (2) *Non-bond interactions*, including the short-range Lennard-Jones and hydrogen-bond terms and the long-range Coloumbic potential.

The aim of the calculations is usually either *statics* (finding the configuration r which minimizes E), *dynamics* (calculating trajectories $r(t)$ which satisfy Newton's law $-\nabla E(r) = M\ddot{r}$, where M is the diagonal matrix of masses), or *equilibrium statistics* (average properties under the probability distribution $P(r) \sim \exp\big(-E(r)/k_B T\big)$), where k_B is the Boltzmann constant and T is the absolute temperature).

The computing cost of current molecular dynamics algorithms rises very steeply with problem size, restricting the modeling efforts to relatively small molecular ensembles and to time intervals many orders of magnitude smaller than needed. Preliminary model studies conducted by [33], [34] have indicated that this steep rise in cost can be radically reduced by combining several types of multiscale approaches.

The research objective is to develop these approaches and demonstrate their ability to perform the above computational tasks in computing times that rise only *linearly* with the number n of atoms in the system. Also, the aim is to show the possibility to blend statistical approaches in the small (for the high-frequency molecular oscillations) with deterministic dynamics or statics in the large (see Sec. 14.8 below). The long term goal is to establish the computational tools for the development, scale by scale, of material "descriptions" at increasingly larger scales, each description being either in the form of a Hamiltonian, or more generally in the form of conditional probability tables for properly chosen coarse-level variables (cf. Secs. 13.2 above and 14.7 below). Such tools would facilitate and encourage an interactive, scale-by-scale development, by chemists and computational scientists, of computer libraries of ab-initio material descriptions.

14.2 Complexity factors and research strategy

The enormous complexity of molecular calculations is the product of several factors that multiply each other, including:

(1) A very large number of *long-range (electrostatic) interactions* that need to be summed up to calculate the energy difference associated with each move of one atom.

(2) *Tiny time steps* (and similarly tiny steps in Monte Carlo simulations and in energy minimization) enforced by the strong chemical bonds.

(3) *Multiscale attraction basins*: The energy functional of the many particle problem includes a multitude of local minima. Moreover, each small-scale basin resides, together with similar basins, inside a larger-scale basin, which itself resides within a still-larger-scale basin, and so on. Conventional algorithms (in search for the global minimum, or in Monte-Carlo simulations at some finite temperature), even when capable of escaping some small-scale basins (e.g., by simulated annealing), remain practically trapped in larger-scale ones.

(4) *Thermal fluctuations*. In equilibrium statistics, to obtain accuracy ϵ in calculating a thermodynamics quantity, one has to average over $O(\epsilon^{-2})$ independent configurations. The computational cost of producing each such independent configuration by a Monte-Carlo process is very large due to the large number of degrees of freedom, *multiplied* by the *product* of the three aforementioned complexity factors.

To investigate multiscale techniques to deal with these obstacles, a systematic study has been undertaken of model problems which include only one or two obstacles at a time. Moreover, unlike the common methodology of starting a research on macromolecular algorithms with small molecules and advancing to increasingly larger ones, the development of multiscale techniques necessarily employs at each stage molecules of an *arbitrary* (large) size n, starting with very simple potential functionals and advancing to increasingly more complicated ones, progressing also from simple geometries (e.g., stretched homogeneous chains, then simple helices) to increasingly more realistic ones. At each stage just one new type of force is added, and the study objective is to still obtain the linear ($O(n)$) complexity. This research strategy is necessary since linear complexity and large-scale processing are indeed our ultimate aims, and since at small molecular systems the advantages of multiscaling cannot be observed.

14.3 Fast summation of electrostatic interactions

Direct summation of all the electrostatic interactions between n particles costs Cn^2 computer operations, where C is around 10. Instead, several methods exist to sum the forces in just $C_1 n$ operations (see, e.g., survey [94]), although note that in three dimensions $C_1 > 10^4$, so these methods become advantageous only for $n > 10^3$. A multiscale method for fast summation, suggested in [29] (based on an idea described earlier in [23, §8.6], [26, App. A] and [53], and related to the methods discussed in Sec. 10 above), is being used. It is based on a decomposition of the two-particle potential into a local part and a smooth part, the latter being evaluated at larger scales (interpolated from coarser grids), where a similar decomposition is being recursively used. An important advantage of this approach is that it gives the kind of multiscale description of the force fields which is needed for the efficient multiscaling of atomic *motions* — in statics, dynamics and equilibrium calculations (see for example the description of the electrostatic calculations in Secs. 14.6.1 and 14.7.4 below), or for solving *equations* (e.g., polarization equations).

Several important new developments by Bilha Sandak, reported in [135], include: (i) Generalization of the method to fields generated by *dipoles*, in addition to those created by charges. (ii) Substantially higher accuracy for negligible additional CPU time. This has been obtained by introducing enhanced interpolation orders and longer softening distances *at the coarser levels*, and by correcting for some false *self-interaction*, i.e., the residual interaction of a charge with itself, caused by the multiscale calculations. (iii) Efficient *software* for general use has been implemented.

14.4 Fast macromolecular energy minimization

Energy minimization may serve here two somewhat different objectives: one in statics, the other in dynamics. In statics, the objective is to calculate the lowest energy $E(r)$, yielding the most stable conformations of the molecular structure. In dynamics, the objective is the solution of the system of equations arising at each time step of *implicit* dynamics simulations. "Implicit" refers to the method which evaluates the forces $-\nabla E(r)$, at each time step, partly or wholly in terms of the particle *arrival* positions, i.e., positions r at the *end* of the step. This method ensures stability of very large time steps, but it does not yield the arrival positions explicitly. Instead, they should be calculated by solving a large system of equations. (Also, this method damps molecular vibrations at scales not resolved by the large time step; we return to this point in Sec. 14.8 below.) Solving the implicit system of equations is equivalent to minimizing an *augmented* energy functional, identical to $E(r)$ except for an additional quadratic *kinetic term* (cf., e.g., [128] and also the functional H in Sec. 14.8 below). For large time steps this additional term is locally very small, but its large-scale effect is still profound.

The macromolecular energy minimization problem is somewhat similar to the minimization problem encountered in structural mechanics, for which very efficient multigrid solvers have been developed. Of these, the closest to the ones needed in molecular mechanics are the *algebraic multigrid* (AMG) solvers (cf. Sec. 17 below), which do not assume that the problem arises from PDE or that the unknowns are really placed on a grid. The methods we have developed for molecular energy minimization follow the general AMG outline: coarser levels are constructed each by taking a suitable subset of the next-finer-level degrees of freedoms; a coarse-to-fine interpolation of displacements is defined based on the fine-level couplings and current configuration; the coarse-level set of equations (or rather, the coarse-level Hamiltonian) is derived based on this interpolation and on the current residual forces at the fine level; and the algorithm consists of relaxation (local minimization) sweeps at all levels with fine-to-coarse transfers of residual forces and coarse-to-fine interpolation of displacements. The molecular forces, however, are much more involved than those of structural mechanics (exhibiting severe nonlinearities and large variations in strength of the different types of interactions), so very systematic development of all these algorithmic components was required.

Our first stage of developing multiscale molecular energy minimizers, in collaboration with Dov Bai, was described in [33]. More advanced techniques for more advanced models are described in [35, §11.4] and [38, §14.4]. However, these studies inevitably led to the conclusion that macromolecular energy minimization is unnecessarily complicated. The energy barriers are much more easily traversed by multiscale methods equipped with the stochasticity introduced by the natural (e.g., room) temperature. Indeed, nearly all practical problems are actually given at finite temperatures (including dynamic problems; cf. Sec. 14.8). Moreover, it is unlikely that at finite temperatures the material stays exactly at the attraction basins of the minimal energy. For these reasons, our interest has shifted to finite temperature calculations, discussed in the following sections. Fortunately, some of the tools acquired in the study of minimization techniques, such as coarsening in terms of a combination of internal and cartesian coordinates, have proved very useful also for the finite-temperature algorithms.

Notice also that finite-temperature algorithms lead themselves to powerful minimization techniques: see Sec. 14.7.3.

14.5 Monte-Carlo methods at equilibrium: General

To calculate equilibrium statistics, an atom-by-atom Monte-Carlo process is usually performed. In this process, each atom in its turn changes position stochastically, according to the probability density distribution $P(r)$. Making repeated sweeps of this process, one can calculate the desired statistics on the sequence of produced configurations.

To calculate accurate averages of some observable, however, an extremely long sequence of configurations is needed. There are two basic causes for this complexity: (1) Due to the local nature of the Monte-Carlo process, only very slowly it affects large-scale conformational features,

hence extremely many Monte-Carlo sweeps are needed to produce each new, statistically independent configuration. (2) Many such independent samples are needed to average out the deviation observed at each of them.

For some very simple lattice problems, multigrid Monte-Carlo algorithms were developed which overcome *both* these complexity causes (see Sec. 13 above, where these two causes, which multiply each other, are called the CSD factor and the volume factor, respectively). Two complementary types of multiscale Monte-Carlo methods for the fast simulation of atomistic systems have developed based on the RMG approach (Sec. 13.2): one type for macromolecules, the other for very large, flowing collections of small molecules (gases, liquids). The methods are described in the following two subsections.

In the future the intention is to combine those two types of methods to treat macromolecules in solutions.

14.6 Multiscale Monte Carlo for macromolecules

Monte Carlo simulation of long polymers (and generally all macromolecules) is one of the most computationally intensive tasks. This is due mostly to the large variation in time scales (10^{-15} seconds to several hours) and length scales (1Å–1000Å) involved in each problem and the many energy barriers and attraction basins found at all scales. While much of the interesting behavior occurs at longer time (or length) scales, the shorter scales, where the basic equations are given, constrain the size of steps in simulations. However, by applying multiscale methods these constraints can hopefully be removed as different physical scales are resolved on corresponding appropriate computational levels.

Together with Dov Bai, multiscale MC algorithms for the united-atom polymer model of [127] were studied. The details are reported in [7] and briefly reviewed below. For alternative coarsening schemes, see [8].

The simple polymer is a non-branching long chain of n repeat units (monomers; see the comprehensive survey [82]). In the united-atom model each monomer is considered as a single unit and details of its internal structure and interactions are ignored, so mathematically the polymer is represented as a chain of points in $\mathbb{R}^3$, located at positions $r_1, r_2, \ldots, r_n$. The internal coordinates employed to describe interactions are the distances $r_{i,j} = |r_i - r_j|$, the angles θ_i (angle between the lines $\overline{r_{i-1}r_i}$ and $\overline{r_i r_{i+1}}$) and the dihedral (or torsion) angles $\phi_{i+1/2}$ (angle between the planes $\overline{r_{i-1}r_i r_{i+1}}$ and $\overline{r_i r_{i+1} r_{i+2}}$). The overall Hamiltonian (energy) functional is

$$
\begin{aligned}
H(r) = &\sum_{i=1}^{n-1} K_r (r_{i,i+1} - r_0)^2 && \text{bond-length potentials} \\
&+\sum_{i=2}^{n-1} K_\theta (\cos\theta_i - \cos\theta_0)^2 && \text{bond-angle potentials} \\
&+\sum_{i=2}^{n-2} F_\phi(\phi_{i+1/2}) && \text{bond-dihedral (torsion) potentials} \\
&+\sum_{|i-j|\geq 4} \epsilon \left[\left(\frac{\sigma}{r_{ij}}\right)^{12} - \left(\frac{\sigma}{r_{ij}}\right)^{6} \right] && \text{Lennard-Jones potentials}
\end{aligned}
$$

where $r_0 = 1.52$ Å, $K_r = 250$ Kcal/mol/Å^2, $\theta_0 = 110°$, $K_\theta = 60$ Kcal/mol, $\epsilon = 0.09344$ Kcal/mol, $\sigma = 4.5$ Å and F_ϕ is a function featuring three local minima with 1.5 to 2 Kcal/mol energy barriers between them. The relative probability associated with each configuration r is $P(r) = \exp\left(-H(r)/k_B T\right)$, where T is the absolute temperature and k_B is Boltzmann's constant. One can see that the bond-length potentials are much stronger than the bond-angle potentials which in turn are much stronger than the torsion potentials. Therefore the dihedrals are the main active degrees of freedom. The trouble is that a conventional MC simulation is extremely slow in sampling the dihedrals, first because it is constrained by the stronger bond-length and angle potentials and, more important, because each MC switch of the local minimum at which one dihedral lives is only probable provided suitable (but unknown) similar switches are simultaneously performed at a substantial number of neighboring dihedrals.

In our multiscale approach, simulations are mainly performed at coarser levels, which already average over such local attraction basins. Each coarse level consists of a reduced number $N = n/m$ of points, or "atoms"; typically $2 \leq m \leq 4$. Each coarse-level "atom" stands for the average location R_I of m next-finer level atoms. The coarse-level Hamiltonian $H^c(R) = H^c(R_1, \ldots, R_N)$ is developed by extensive, but only local, Monte-Carlo simulations, iteratively fitting coarse-level distribution functions and correlations with those found in fine-level simulations.

The strategy is to calculate local terms of the coarse Hamiltonian by using simulations (to be described below) which involve only a *local set* (typically several dozens) of neighboring points (atoms). Bond interactions between points inside the local set and those outside are ignored: this does not significantly affect the accuracy of the coarse Hamiltonian terms located sufficiently deep inside (several atomic distances from the margins of) the local set. This approach is similar to the one used in Algebraic Multigrid (AMG) and in Renormalization Multigrid (RMG; cf. the *near locality property* in Sec. 13.2 and in Sec. 17). Such a derivation of coarse Hamiltonian terms needs to be done only once for all similarly-structured molecular neighborhoods. Due to the employment of internal coordinates, and with suitable coarsening ratios $1 : m$, it turns out that most inter-coordinate coarse-level correlations can be neglected, yielding quite simple coarse-level Hamiltonians.

The coarse-level Hamiltonian includes, first of all, modified Lennard-Jones-type interactions in terms of cartesian coordinates. The exact formula for this interaction has been derived once-for-all (for a given coarsening ratio), by averaging during MC simulations over all vectorial sums of fine-level Lennard-Jones forces involved in the interaction between two coarse atoms in a given distance bin.

The rest of the coarse Hamiltonian is in terms of local coordinates: distances, angles and dihedral angles (between the *coarse* atoms, of course, which also form a chain). So the general form of the coarse Hamiltonian is

$$H^c(R) = \sum_{|I-J| \geq m_0} H^c_{LJ}(|R_I - R_J|) + \sum_k A_k H_k(R)$$

where $H_{LJ}(\rho)$ is the Lennard-Jones interaction between two coarse atoms at distance ρ from each other, m_0 is a chosen small integer ($2 \le m_0 \le 4$), A_k are coefficients to be iteratively determined (as described below), and each H_k is either a *single-internal-coordinate interaction* or a *correlation* between two such coordinates. For each coarse internal coordinate U (either a coarse length $\overline{R_I R_{I+1}}$, or a coarse angle $\overline{R_I R_{I+1} R_{I+2}}$, or a coarse dihedral $\overline{R_I R_{I+1} R_{I+2} R_{I+3}}$) several single-coordinate interaction terms of the form

$$H_k = \begin{cases} 1 \text{ if } \alpha_k \le U \le \alpha_{k+1} \\ 0 \text{ otherwise} \end{cases} \tag{14.1}$$

are in principle included in H^c; except that one can include in one unified term all the interactions that are expected by symmetry to have the same coefficient A_k (e.g., all terms of the form (14.1) associated with angles, excluding perhaps angles near the ends of the coarse polymer).

As a first approximation one can start with H^c that includes (besides H_{LJ}) only such single-coordinate terms, with coefficients A_k such that the distribution of each coarse local coordinate U is the one calculated at fine-level simulations (each fine-level configuration implying of course a value for each U). This approximation to H^c will generally fail to give the correct *correlations* between the coarse coordinates. So we next iteratively correct H^c, adding needed correlations terms and readjusting the coefficients $\{A_k\}$ to still yield the correct distributions. The first correlations to be added are those that are shown to be significant in the measurements conducted during the fine-level simulations; the most significant, it turns out, are the correlations between every angle $\overline{R_I R_{I+1} R_{I+2}}$ and each of the adjacent lengths $\overline{R_I R_{I+1}}$, $\overline{R_{I+1} R_{I+2}}$.

In each iteration a large number of MC steps over the local set are made with the coarse-level current Hamiltonian, calculating various observables and comparing them to corresponding values obtained by simulations of the local set at the fine level. The difference is then used to get a better coarse Hamiltonian, in a Newtonian iterative way which converges fast. That is, a set of corrections $\{\delta A_\ell\}$ to the set of coefficients $\{A_\ell\}$ is calculated by solving the linear system

$$\frac{1}{k_B T} \sum_\ell \left(\langle H_k \rangle \langle H_\ell \rangle - \langle H_k H_\ell \rangle \right) \delta A_\ell = \langle H_k \rangle_f - \langle H_k \rangle_c \quad \text{for all } k \,. \tag{14.2}$$

$\langle H_k \rangle_c$ is of course the average of the operator H_k calculated with the current (before correction) *coarse-level* Hamiltonian, while $\langle H_k \rangle_f$ is the corresponding average computed by *fine* level simulations. The averages on the left side of (14.2) can be calculated on either the fine or the coarse levels. (More generally, they can be just approximated, since they only serve as iteration coefficients. In fact, one can ignore most of these terms, only those should be calculated that correspond to neighboring coordinates which might be significantly correlated.) For H_k and H_ℓ that are more strongly correlated, $\langle H_k H_\ell \rangle$ is calculated *both* on the fine and on the coarse levels: In case one finds $\langle H_k H_\ell \rangle_c$ and $\langle H_k H_\ell \rangle_f$ to be too different, *in the next iteration a new correlation term* $A_m H_m$ *is added to* H^c*, where* $H_m = H_k H_\ell$.

In a small number of iterations, all the coefficients $\{A_k\}$ converge, with all needed added correlation terms, yielding H^c which very faithfully reproduces all the distribution and correlations exhibited at the fine level.

Numerical experiments have been conducted with two coarsening ratios, 1:3 and 1:4. The resulting coarse Hamiltonians did turn out to fulfill our expectations: With vastly-reduced number of degrees of freedom and allowing much larger simulation steps, it very accurately reproduced large-scale statistics: see [7] for details.

A separate study has shown that the coarsening ratio 1:3 is best, for the following reason. Unlike the 1:4 ratio, it yields fast CMC equilibration (see Sec. 14.7.2), since its coarse configurations fully determine the attraction basin in which each fine-level dihedral resides. This means that the coarsening statistics can be gathered in *windows of long chains*, not only in short local sets. (Cf. the concept of windows at Secs. 13.2 and 14.7.2). This is important since the short local sets may not give rich enough statistics, while long chains cannot be simulated efficiently at the fine level.

14.6.1 Electrostatic interactions

The next task is to test polymer models that also involve electrostatic interactions. To add long-range interactions to this scheme it is proposed, similar to the approach described above (Sec. 14.3), to decompose each two-body electrostatic potential into the sum of two parts: a smooth potential and a local potential. (See details of such decompositions, for charge and for dipolar interactions, in [135].) The charges or dipoles will then be *anterpolated* to the coarse level. (Anterpolation is the adjoint of interpolation; see, e.g., [11, §3] or [135, §3.2].) This will give good coarse-level approximation to the *smooth* potentials.

Unlike the fast summation schemes based on the same principle, in which charges/dipoles are anterpolated to a *fixed* lattice (as in [29], [135]), here they will be anterpolated to points that *move* during the coarse-level simulations. As a result, the field produced by the coarse-level charges/dipoles will continue to approximate the fine-level *smooth* potentials even under large global movements of the (coarse) molecule.

The *local* parts of the electrostatic potentials remain to be described at the coarse level. Being local, these interactions can be added to the local scheme described above, similar to the LJ interactions.

With this approach no explicit electrostatic summations are necessary, especially if the local MC simulations are made in a *distributive* manner (cf. [53]). This means, e.g., to move two particles at a time in such a way that their mass center remains unchanged. Such (and *higher-order*) distributive moves hardly affect, or are affected by, the smooth interactions. (Such distributive moves make sense only in a multilevel dynamics, where mass centers that remain fixed during the fine level motions are moved at the coarse level. This exactly is the motions of the coarse "atoms" described above.) Such distributive moves can also serve to reduce the cutoff distance for calculating LJ interactions.

14.7 Multiscale Monte Carlo for fluids

The efficient equilibrium simulation of gases and liquids at the atomic level, needed for the derivation of their large-scale behavior and macroscopic equations, is a central problem in scientific computation. Direct fine-scale MC simulations tend to be extremely inefficient due to the very slow change of various kinds of *clusters* at various scales. Especially difficult are the calculation at critical conditions, where clusters at all scales interact. Of particular interest is the simulation for *water*, by itself or as a solvent.

The main two kinds of clustering difficulties associated with water and other fluids are *positional* clustering and *electrostatic* (dipole) alignment. We will start out by studying in parallel the following two simple cases, in each of which only one kind of clustering is present. The first case will deal with the molecular motions (see Sec. 14.7.1), the second — with the molecular dipole rotations (Sec. 14.7.3). Later, these studies will lead to multiscale simulations of real water models, such as TIP4P [106], [107], in which both molecular translations and rotations, with both Lennard-Jones and dipole interactions, are considered.

14.7.1 Moving particles

Two models of single-atom molecules have been investigated in collaboration with Valery Ilyin [50]: a one-dimensional hard ball model, for which full exact thermodynamic description is known, and a two-dimensional Lennard-Jones fluid. In the latter, describing noble gases, the atoms move solely under their mutual two-atom Lennard-Jones interaction. Clusters of atoms that move together, and clusters of "*holes*" (i.e., absence of atoms inside a larger-scale atomic cluster) are stochastically formed, their likely scales depending on the physical conditions, such as temperature and density. At critical conditions all scales are likely. The larger the scale of clusters, the longer they persist in MC simulations, hence slower is their sampling.

Generally speaking, the multiscale approach here is again of the RMG type (cf. Sec. 13.2), but unlike the former cases (statistical mechanics in Sec. 13.2 and macromolecules in Sec. 14.6), here the coarse levels all *differ* from the finest one in their nature: at the finest level we have atoms at arbitrary locations, while each coarse level is defined on a uniform *lattice*. In simple cases, the value defined at each lattice point stands for the average fluid density in a lattice box around that point. The probability distribution of this density depends on neighboring-point densities, as specified in detail by a suitable *conditional-probability (CP) table*. The CP table for each level is calculated by extensive *local* simulations at the next finer level (the level of a *lattice* with half the meshsize or, eventually, the finest level of *atoms*).

The development of this multiscale structure along the lines described next (Sec. 14.7.2) is now in progress. It has been shown [50] that density fluctuations at all scales can be accurately calculated with only a bounded number of particles or gridpoints employed at each level.

14.7.2 General outline of the multiscale approach

Generally, the (scalar or vectorial) variable at each lattice point at each coarse level may represent various local averages: of density, or of electrostatic charge, or dipole moment, or energy density, etc. Also, at certain physical conditions, the locations of *moving* blobs, each carrying its own set of properties, may be added to the fixed lattice as additional degrees of freedom. As before (see Sec. 13.2), a general criterion in choosing the coarse level set of variables is the speed of equilibration of compatible Monte Carlo (CMC) runs (each such run being an MC simulation at the fine level, restricted to configurations that are compatible with a given coarse-level configuration); a fast-equilibrating CMC entails the locality property of the coarse variables and thus allows the construction of the CP tables.

The CP table "resolution" (i.e., the number of coarse neighbors on which the probability distribution of a coarse variable is conditioned, and the resolution at which each of these neighbors is tabulated) should in principle increase for, and only for, frequent neighborhoods (see the branching system in Sec. 13.2). The errors in the CP tables can be fully controlled by this resolution and by the amount of statistics gathered at the fine level in setting up the tables. These, together with the interpolation orders used in employing the tables at the coarse-level simulations, determine the accuracy of those simulations.

Because of the near-locality property, no *global* equilibration is needed; local equilibration is enough to provide the correct CP values for any neighborhood for which enough cases have appeared in the simulation. Thus, the fine-level simulation can be done in a relatively small periodicity cell. *The idea is to simulate increasingly larger volumes at increasingly coarser levels.*

However, since the fine-level canonical ensemble simulations use only a *small* periodicity cell, many types of neighborhoods that would be typical at some parts of a *large* volume (e.g., typical at parts with densities different than used in the periodicity cell) will not show up or will be too rare to have sufficiently accurate statistics. Hence, simulations at some coarse level may run into a situation in which the CP table being used has flags indicating that values one wants to extract from it start to have poor accuracy. In such a situation, a temporary *local* return to finer levels should be made, to accumulate more statistics that are relevant for the new local conditions.

To return from a coarse level to the next finer level one needs first to *interpolate*, i.e., to produce the fine level configurations represented by the current coarse level configuration, with correct relative probabilities. The interpolation is performed by CMC sweeps at the fine level; few sweeps are enough, due to the fast CMC equilibration. This fast equilibration also implies that the interpolation can be done just over a restricted *subdomain*, serving as a *window*: In the window *interior* fine-level equilibrium is reached. Additional passes can then be made of *regular* (not compatible) MC, to accumulate in the interior of the window the desired additional CP statistics, while keeping the window boundary frozen (i.e., compatible with the coarse level). The window can then be coarsened and returned to the coarse level, where simulations can now resume with the improved CP table.

Iterating back and forth between increasingly coarser levels and window processing at finer levels whenever missing CP statistics is encountered, one can quickly converge the required CP tables at all levels of the system, with only relatively small computational domains employed at each level. The size of those domains needs only be several times larger than the size of the neighborhoods being used (with a truncation error that than decreases exponentially with that size). However, somewhat larger domains may be better, since they provide sampling of a richer set of neighborhoods (diminishing the need for returning later to accumulate more statistics), and since the total amount of work at each level depends anyway only on the desired amount of statistics, not on the size of the computational domain.

Simulating at all levels in terms of such periodicity cells and windows can effectively eliminate both the volume factor and the slowing down which plague usual (one-level) MC simulations. Provided of course that the *coarsening ratios* (the ratio between a coarse meshsize and the next-finer meshsize), as well as the average number of original particles per mesh volume of the finest lattice, are all suitably low. The typical meshsize ratio is 2, typical number of particles per finest-lattice mesh is between 4 and 10. More aggressive coarsening ratios would require much longer simulations to accumulate accurate CP statistics.

The particle number density at the coarsest level is equal to some input value. Each finer-level window covers only part of the coarsest-level domain, so the particle number density may differ from the initial one. As a result of the multilevel process, the configurations produced at the coarsest level correspond to the canonical ensemble; at finer levels they yield direct accurate representation of the grand canonical ensemble [103].

At sufficiently coarse levels, this entire algorithm effectively produces *macroscopic "equations"* for the simulated system, in the form of numerical CP tables. This can yield a macroscopic numerical description for the fluid even for those (most frequent) cases where the traditional derivation of closed-form differential equations is inapplicable.

14.7.3 Low temperature procedures

The multilevel algorithm can efficiently get into equilibrium even at low temperatures by an *adaptive annealing* process. In this process the temperature is reduced step by step. At each step, upon reducing the temperature from a previous value T to a new one T', a first approximation to the CP tables of T' is obtained from those of T by raising each CP to the power T/T' (and renormalizing; actual renormalization is not necessary since only probability *quotients* are needed in using the CP table for MC simulations). Then, in just few multilevel cycles, the CP tables can easily be made more accurate, provided the quality of the set of coarse variables has not been deteriorated.

However, the type of coarse-level variables appropriate at low temperatures does generally differ from that at high temperatures. At high temperatrues the average density is an adequate coarse-level variable. At low temperatures, e.g., at the appearance of liquid drops in a gas or at the onset of piecewise crystallization, other coarse-level variables should be added, such as the average crystal direction, and/or the average density of holes, and/or the location of mass centers. Thus, in the annealing process one should monitor the quality of coarsening by occasionally checking the CMC equilibration speed. When this speed starts to deteriorate at some level, additional variables should be added at that level, with a corresponding extension of the CP table. Candidate new variables can be found by physical understanding and/or by suitably blocking highly-correlated variables at the next-finer level; then the new variables should be admitted provided they pass the CMC-equilibration-speed test. Some of the old variables may be removable, as judged again by CMC equilibration tests.

In fact, unlike the classical *simulated annealing* method (whose aim is the minimization of the energy, not the simulation of equilibrium), the chief purpose of annealing here is the gradual *identification* of the degrees of freedom that should be employed at increasingly coarser levels. At the zero-temperature limit these procedures can also yield powerful multiscale minimization procedures (see Sec. 18.2).

14.7.4 Rotating dipoles

The chosen model features a large set of electrostatic dipoles, of given strengths and *fixed* locations, rotating in their mutual fields in thermal equilibrium. Clusters of aligned dipoles tend to form, their sizes depending on the given temperature. Again, these clusters are very slow to change in ordinary MC simulations, making large-scale fluctuations extremely slow to average out.

As before (cf. Secs. 14.3 and 14.6.1), at any spatial scale the electrostatic interactions can be decomposed into the sum of a *smooth* part and a *local* part. In addition to using this decomposition for the fast summation of the dipole field, here it will also be used for accelerating the MC simulations and for cheap (coarse-level) averaging over many large-scale fluctuations.

Similar to the above (Secs. 14.7.2–3), each coarse level is defined on a grid, the mesh size being doubled at each coarsening. The vector defined at each lattice point stands for the dipole anterpolated from the next-fine-level dipoles. With this type of coarsening, the RMG methodology is again applied: a CP table at each level is derived by local MC simulations at the next finer level. Then this structure can be employed both for MC acceleration (e.g., by "half-V cycles"; see Sec. 13.2), and for calculation of the large-scale electrostatic properties.

This system is currently under development in collaboration with Bilha Sandak.

14.8 Small-scale statistics with large-scale dynamics

The multiscale structure may allow a natural combination of temperature-accurate statistical simulations at small scales with time-accurate dynamics at large scales. The following approach has been preliminarily studied.

Stochastic implicit time stepping. A first-order implicit discretization to Newtonian dynamics, leading from old positions $r^0 = r(t)$ and old velocities $v^0 = v(t)$ to new positions $r^1 = r(t + \delta t)$ and new velocities $v^1 = v(t+\delta t)$, is given by $v^1 = (r^1 - r^0)/\delta t$ and $M(v^1 - v^0)/\delta t = -\nabla E(r^1)$. This set of equations in r^1 and v^1 is equivalent to the minimization of the functional

$$H(r^1, v^1) = E(r^1) + w^T M w + \frac{1}{4}(v^1 - v^0)^T M (v^1 - v^0) ,$$

where $w = (v^1 + v^0)/2 - (r^1 - r^0)/\delta t$. In our *stochastic* dynamics, instead of minimizing H at each time step, we perform a multiscale Monte Carlo simulation with the probability density distribution

$$P(r^1, v^1) \sim e^{-\beta H(r^1, v^1)} ,$$

where $\beta = (k_B T)^{-1}$ and T is the real temperature of the system. The coarse-level moves we have used in the multiscale cycle are based on interpolation (see Secs. 13.1 and 14.4, and [35, §11.4]); an RMG approach to coarsening (cf. Secs. 13.2 and 14.6) may also be considered. At increasingly coarser scales β can be increased, to enforce practically deterministic large-scale dynamics.

This approach yields two benefits in performing very large time steps: first, it allows much easier handling of local minima. Secondly it avoids the killing of highly-oscillatory modes (those vibrations that are not resolved by the time step), which would occur if the implicit equations of a large time step were *imposed* at all scales. Instead, these modes assume stochastic amplitudes, nearly according to their equilibrium probability distribution. The desired temperature is introduced very directly in this way (with the fast atomic vibrations serving as a natural heat bath), thus getting around the need for fabricating Langevin stochastic forces.

Tests with this scheme on model problems with quadratic potential have shown the expected behavior, except that the stochastic treatment at fine levels gradually introduces deviation from deterministic evolution also at large scales. This deviation seems generally to be of the order of the discretization error. We have nevertheless learned how to control this deviation by "distributive Monte Carlo" (similar to distributive relaxation [24]), forcing fine-scale moves to be as nearly orthogonal to large-scale moves as desired.

15 Image Processing. Clustering. Graphs

15.1 Edge (or fiber) detection

Straight features. Fast multiscale approaches for some early vision tasks, such as edge detection and surface reconstruction from sparse, noisy or blurred data, have been developed in collaboration with Jonathan Dym [79]. In particular, fast multiscale methods for enhancing and detecting straight features (straight edges or straight fibers) have been demonstrated [43], [44]: They detect *all* such features, of all widths, lengths, positions and orientations, in just $O(N \log N)$ operations, where N is the number of pixels (picture elements) in the given picture. This has been achieved by constructing a hierarchical collection of numerical integrals of grey levels along straight segments of the pictures, with the lengths, positions and orientations of the segments in the collection chosen in such a way that:

(1) The collection is *rich enough*, in the sense that any other straight-segment integral over the picture can be readily interpolated from the collections's integrals; each interpolation is over a short distance, so that it is equivalent to interpolating grey levels only between nearest-neighbor pixels. Specifically, the collection includes segments of length $1, 2, 4, 8, \ldots$ (in pixel units), those of length ℓ have locational resolution which is $O(\ell)$ in the direction of the segment and $O(1)$ in the perpendicular (width-wise) direction, while their orientational resolution is $O(\ell^{-1})$ (analogously to the *Heisenberg principle* in quantum mechanics).

(2) The collection is *fast to construct*, by using shorter segment integrals to calculate the next-longer ones, scale after scale, so that the construction of each integral requires only $O(1)$ operations.

Curved featuers. For detecting smooth curved features (edges or fibers), a variety of approaches have been proposed. One good example is the *completion fields*. In this approach, the picture is described in terms of "edgels" (edge elements), i.e., short pieces of a straight edge (or fiber), defined at $N_1 = O(N)$ locations in the picture, at m different orientations in each location. The *original* value of edgel i is the response u_i to an elementary edge detector at i; that is, u is the result of a local integral transform which yields a higher value if the local picture elements do indicate existence of an edge at that particular location and orientation (and at the chosen scales of length and width, typically being, respectively, 3 and 1 times the pixel size). The *completion field* value v_j of edge j can be built from the set of all elementary responses u_i in a variety of ways (see different approaches in [161] and [98]). As a representative example for our discussion here, we can take

$$v_j = \sum_{i=1}^{N_1 m} a_{ij} u_i \ , \qquad (j = 1, \ldots, N_1 m) \ , \tag{15.1}$$

where a_{ij} expresses the "affinity" of edgels i and j: it is large if edgel j is a direct continuation of edgel i, and it falls off with their distance and orientation difference. For a given i, its "induction field" a_{ij} is qualitatively similar to the field of a magnetic dipole. It is shown in [161] that such completion fields are biologically plausible, and give eye-pleasing curves. They are particularly powerful in completing curves partly occluded by large objects. The original method however has several severe shortcomings, which can be overcome by multiscaling.

Multiscale methods can contribute to the process in two fundamental ways. First, the method as described in [161] would require $O(N_1^2 m^2)$ computer operations; multiscale methods, resembling those of Sec. 10 above, would do the same job in $O(N_1 m)$ operations, while retaining the same (very high) degree of computational parallelism. Second, and more importantly, still with this low cost, the multiscale processing can produce much better completion fields. Indeed, a fundamental flaw in the uniscale completion fields is their additivity, as in (15.1). In reality, the completion field of a long edge should be very *different* from (farther reaching and more orientation-specific than) the sum of the fields of several short edgels that compose it. In the multiscale approach, this flaw can be avoided, since completion fields can be constructed *separately* at each scale of length and width, with *scale-dependent* affinity parameters. The multi-resolution *input* of straight-edge responses required for such multiscale completion fields is exactly the kind resulting from the $O(N \log N)$ straight-feature algorithm mentioned above.

The multi-resolution of both the input (straight responses) and the output (completion fields) also involves further cost reductions. For example, as mentioned above, for *short* edgels only low orientational resolution need be used, while for long edgels a low locational resolution is needed (in the lengthwise direction). Thus, the value of $N_1 m$ mentioned above can itself be radically reduced. Moreover, the multiscale *output* of the algorithm is a very desirable structure to interact with the higher vision processes of labeling and segmentation (cf. Sec. 15.2), whether or not the latter are themselves multiscaled.

A detailed study of multiscale completion fields, their parameterization and fast implementation has been conducted with Eitan Sharon and Ronen Basri. It is summarized in [139].

Intriguing possibilities of combining the developed algorithms in a variety of ways should be investigated:

1. Iterating a multiscale algorithm, with the output of the first iteration (e.g., the set of v_j) being used in forming the input (e.g., the set of u_i) for the next iteration. This can be done in various manners: linear, nonlinear, with or without thresholding.
2. Using the output from one scale in forming the input for the next coarser scale.
3. Thresholding after the previous iteration, one can use in the next iteration several different and more complex algorithms, due to the smaller set of data. In particular, one can afford at this stage specialized algorithms, such as circle and corner detection. With further iterations, increasingly higher levels of recognition algorithms may enter.
4. Combining in various ways edge detection with picture segmentation (see next).

15.2 Picture segmentation

A basic task in pattern recognition is the decomposition of a given picture into meaningful segments. The criteria for blocking two picture elements into the same segment include similarity in color levels, absence of separating edges, etc. Quantitatively, these can be expressed in terms of *coupling* coefficients between neighboring pixels. It is not uniquely defined how to derive the segments once the coupling coefficients are given. Multiscale approaches can play *several* essential roles (somewhat analogous to their variety of roles in other areas; see for example Sec. 15.1 above).

Regarding the pixels as nodes of an electric network, and each coupling constant as the conductance (reciprocal of resistance) of a connecting wire, the approach to the segmentation problem is to define a picture segment as a block of nodes that will have approximately the same electric potential under whatever input currents applied to the network. The first possible role for a multiscale approach is in terms of a fast solver for such networks. Since the network is highly disordered, *algebraic multigrid* (AMG) solvers best fit the task (see Sec. 17).

As pointed out by Sorin Solomon, there is in fact no need to solve the electric-network problem for any particular input currents: Some of the coarse-level nodes defined by the AMG coarsening process can directly be identified with the desired picture segments. More precisely, if all the couplings of a node at any coarse level are weak (compared with its own couplings to finer-level nodes), the node can be recognized as a picture segment, containing all the pixels (finest-level nodes) which are coupled to it (through the AMG recursive coarse-to-fine interpolations).

The AMG hierarchical coarsening can indeed be viewed as a process of *iterated weighted aggregation*. In an iterated aggregation process, the elements (pixels) are blocked in small-scale aggregates, which are then blocked in larger-scale aggregates, then still larger, etc. In the *weighted* aggregation process, *fractions* of the same element can be sent into different small-scale aggregates, and similarly at all larger scales. This weighting is important in order to express the *likelihood* of elements to belong together; these likelihoods will then accumulate at the higher levels of the process, automatically reinforcing each other where appropriate.

Only *after* larger-scale aggregates have been formed, the *boundaries* for smaller-scale aggregates can be delineated more sharply, taking into account the larger-scale picture. Hence, into the bottom-up process of weighted aggregation, *up-bottom* procedures are added which at appropriate stages tighten and soften the couplings between pixels or between some fine-level aggregates, based on higher aggregation levels. (More specifically, the new couplings are based on values of local fine-level solutions, each obtained by an AMG-type coarse-to-fine interpolation of a coarser-level delta function followed by intermediate-level and fine-level local relaxation sweeps. The couplings to be strengthened are the couplings between those pixels that get values close to 1.) These up-bottom procedures serve to focus the created aggregates and sharpen the boundaries of the emerging segments.

This integrated multiscale process offers much more than simple segmentation. They in fact yield a *hierarchical segmentation*, where segments within segments can be recognized. They can also yield *scaled segmentation*, where the scale of the picture at which segmentation is desired can be specified.

More important, the multiscale weighted aggregation is free to apply new types of couplings at different levels. The coupling between larger-scale blocks (blocks which have been created by the smaller-scale aggregation, or alternatively, simple *geometric* blocks of $k \times k$ pixels), instead of (or in combination with) being induced by the fine-scale couplings (as in the AMG process), they can employ new criteria. Such criteria can include for example similarity in the *average* color levels of the blocks. More generally, all kinds of other intra-block "*observables*" can be used: the block's center of mass, its diameter, principal orientation, texture measures (being, e.g., statistics on average sizes and directions of smaller-scale sub-blocks), etc., with the number of observables per block increasing at coarser levels. For example, strong couplings can be assigned between two (not necessarily neighboring) aggregates whose principal orientations align with the direction of the line connecting their centers of mass. More important, strong couplings should be established between two neighboring aggregates whose boundaries (sharpened by the up-bottom procedures) seem to *continue* with each other (using criteria akin to those used to form a_{ij} in (15.1)). These kinds of couplings may establish affinities even between quite *distant* aggregates, promoting the appearance of *disconnected* segments, presumably signifying partly occluded objects.

Another critera for blocking at each level can be the absence of separating edges on the scale of that level. This will directly benefit from the *multiscale* edge-detection algorithms, as described above. Alternatively, it may be desired to detect the large-scale edges *from* the large-scale blocks by applying a suitable edge detector at that level (a suitable integral transform on a chosen block quantity, such as its average gray level or any other observable).

The multilevel aggregation and hierarchical segmentation algorithms are very fast. On coarse levels the number of variables is drastically reduced, so most of the work is at the initial, finest levels. On those levels the geometric ordering of the pixels and of the small aggregates still dominates and can be used to create very inexpensive processing, so that *the entire algorithm costs only several dozen computer operations per pixel.*

A detailed account of our current multiscale algorithm for image segmentation is given in [140], demonstrating its properties on several line drawings and real images.

The future strategy is to enhance the algorithm in several ways indicated above: adding more coarse-level observables, introducing various interactions with our multiscale edge detection processes, etc. Multiscale approaches to other aspects of image processing are also envisioned.

15.3 Clustering and graph algorithms

The problem of picture segmentation is a special case of the following *clustering problem*: Given a set of objects $\{1,2,\ldots,n\}$ and "affinities" (or "couplings") $a_{ij} = a_{ji} \geq 0$ between the objects $(i,j = 1,\ldots,n;\ i \neq j)$, find "clusters", i.e., disjoint subsets of objects such that objects within the same subset are "strongly coupled", directly or indirectly, while objects in different subsets are generally weakly coupled. The strength of direct coupling between i and j may be defined by the size of $a_{ij}/(a_i a_j)^{1/2}$, where $a_i = \max_k a_{ik}$; indirect strong coupling between i and j is formed by a short chain $(i = i_0, i_1, i_2, \ldots, i_k = j)$ such that $i_{\ell-1}$ is strongly coupled to i_ℓ, $(\ell = 1,2,\ldots,k)$. So defined, the clustering problem is of course fuzzy; it can be defined more precisely in various ways, although a direct precise definition in terms of the minimization of some functional can be coutnerproductive (cf. Sec. 18.1). In fact, as with the above special case of picture segmentation, the best definition can often only be done through the *multiscale clustering* process, where larger-scale affinities are defined or modified *at coarser levels*, depending on properties of intermediate aggregates.

In collaboration with Eitan Sharon and Ronen Basri, the segmentation algorithm described in Sec. 15.2 has been extended to a general clustering algorithm. The main feature distinguishing picture segmentation was its two-dimensional *locality*: affinities were only defined between *neighboring* pixels. In general clustering problems, *all* the affinities a_{ij} may be positive. To account for this situation, a general efficient way has been developed to involve increasing number of affinities at increasingly coarser levels of the algorithm.

Clustering algorithms are central to many areas of applications, including *bioinformatics* and *data mining*. In all these areas the multilevel clustering, and in particular the multilevel *definition* of affinities, have enormous potential, which we plan to demonstrate.

The clustering problem is a special case of *fuzzy graph problems*. Many other problems in this class can greatly benefit from multilevel algorithms, including such well-known problems as the *traveling salesman* (aiming at producing a close-to-optimal, not *the* optimal, route), the *transportation problem* (see [110] for an early multiscale approach), vertex ordering (see, e.g., in [4]), two-dimensional embedding or the problem of *drawing graphs nicely* (successfully multiscaled in [99]), *min-cut* or *max-flow* [72], *sparse spanners* [93], *dense subgraphs* [81], and others.

The general approach in these graph problems is that of *coarsening*: recursive transition to increasingly coarser graphs, each having only a fraction of the number of nodes and edges at the next finer graph. For each coarser graph a new problem is formed such that its solution would easily lead to an approximate solution at the next finer level. With various variations, the AMG coarsening (see Sec. 15.2, or the more general approach in Sec. 17.2) is the basic vehicle. Research along these lines has been initiated, in collaboration with Eitan Sharon and Evgeniy Bart. The emphasis is on *practical* algorithms, obtaining very good *approximate* solutions in very low *average* complexity (unlike the emphasis in theoretical comptuer science on exact solutions with low *worst-case* complexity). Linear average complexity should typically be expected from such multiscale algorithms.

16 Tomography: Medical Imaging

To develop multiscale computational methods for tomography, we have started by working on the two mathematically extreme cases: X-ray tomography, requiring the inversion of the sharp radon transform, and impedance tomography, requiring inversion of a very diffusive process.

16.1 Inverting the Radon transform and related problems

Reconstruction of a function of two or three variables from its Radon transform has proven vital in X-ray computed tomography (CT), emission computed tomography, nuclear magnetic resonance (NMR) imaging, astronomy, geophysics, and a number of other fields [76]. One of the best known reconstruction algorithms is the convolution backprojection method (CB), which is widely used in commercial medical CT devices [76] (with "rebinning" for divergent-beam projection [102]). It has also been applied to spotlight-mode synthetic aperture radar (SPSAR) image reconstruction [102]. While CB provides good reconstruction relatively efficiently, it is still too slow for some purposes, requiring large computational resources and limiting the ability of CT machines to produce real-time 3-D images or video. A faster technique sometimes used, based on direct Fourier method, yields images of much poorer quality.

For other medical imaging and radar problems which are non-uniform, existing Fourier-dependent methods (e.g., CB) are less applicable, resulting in worse performance. This includes the Positron Emission Tomography (PET), the Single Photon Emission Computed Tomography (SPECT), impedance tomography, ultrasound and similar medical imaging techniques (see, e.g., [75]), as well as non-uniform problems in CT, such as the limited-angle problem and the 3D cone-beam reconstruction.

A new multi-level approach to the inverse Radon transform (X-ray tomography) was developed by us several years ago. While the backprojection of the conventional CB raises the computational complexity of the method to $O(N^3)$ for an $N \times N$ images, we have developed a novel $O(N^2 \log N)$ multilevel backprojection algorithm and an accompanying, even less expensive, post-processing procedure [54], [55]. Tests for a number of phantoms, and measurements of point-spread functions, show that the combined method produces at least as good images as those produced by classical CB, in far less time. Further improvements, including an adjustment of the post-processing part to concrete CT machines and a stochastic device to obtain translation invariance in the multilevel backprojection, were introduced by Meirav Galun.

Fast algorithms for other fields with line-integral transforms are under development by Galun. Direct and inverse computation of line integrals of the two-dimensional SPECT is being done by multiscale computation in $O(N^2 \log N)$. (The direct algorithm is an extension of the line integral computation for the direct Radon transform [43].) Also being developed is a solver for the *limited-angle* problem, the case where the X-ray tomograph scans only part of the full 180-degree view. Based on the methods described in Sec. 10.1 above, we have developed a new type of multiscale transform which is applied in the process together with the backprojection, replacing the naive convolution which is less suitable in this case. The construction of the multiscale transform is done once for all, in off-line iterations. *Three dimensional Cone-Beam* reconstruction, used by the new generation of CT machines, can be achieved by fast multiscale solver in $O(N^3 \log N)$ complexity, using a similar off-line construction of a suitable multiscale transform together with a 3D version of our backprojection algorithm.

A more complicated problem is the *three dimensional* PET reconstruction. In this case, a typical situation is that the number of emission events is much smaller than the number of possible rays. Multiscale processes can be applied here to efficiently perform *three* types of tasks: the gathering and averaging of the event data; the backprojection; and, again, a multiscale transform to replace the convolution, constructed off-line.

16.2 Impedance tomography: inverse diffusion problem

An EIT (Electrical Impedance Tomography) device for medical use consists of a set of N electrodes attached to the chest of a patient. A small *known* current is passed between two driver electrodes. In each measurement the current is passed through a different electrode pair, while the voltage drops at *all* the electrodes is recorded. The collected data are used in order to calculate the conductivity distribution in a part of the patient's chest and then to display it on a screen, in order to detect anomalies, such as tumors.

The electrical potential satisfies the equation $\nabla(\sigma\nabla u) = 0$, where σ is the electrical conductivity. The set of measurements gives ideally (in the limit of many small electrodes and as many measurements) the *Neumann to Dirichlet* mapping: the Dirichlet (u) boundary condition resulting from any Neumann ($\partial u/\partial n$) condition. The inverse EIT problem is to calculate σ from this mapping.

The first description of the inverse EIT problem was given by Calderon [69]. Kohn and Vogelius [115] showed that under certain assumptions the conductivity of a medium is uniquely determined by the Neumann-to-Dirichlet mapping. Then Sylvester and Uhlmann [149] provided a general framework for proving uniqueness of the solution of the inverse problem. Alessandrini [1] gave a mathematical explanation for the blurriness of conductivity images and proved that the conductivity depends on the EIT data in a very weak way. Therefore the inverse problem of EIT is ill-posed, and a regularization is necessary if conductivity is to be obtained stably from data.

There exist some works on numerical methods for the relevant problems, but their number is rather sparse and even those papers do not consider the question of numerical efficiency, despite its importance for applications.

The main purpose of our work on this problem, together with Rima Gandlin, has been to demonstrate two general methodological points. First, an ill-posed problem is not necessarily difficult or expensive to solve. On the contrary: once the nature of the ill-posedness has been generally understood, the solution may even be much less expensive than solving the direct problem. For example, in the inverse EIT problem, employing local Fourier decompositions one can show that all components of wavelength λ are ill-posed at distances $r \gg \lambda$ from the boundary. Hence there is no need to use at such distances fine solution grids: all we can know about the solution can be calculated with grids whose meshsizes increase proportionality to r. Moreover, one can recombine the different measurements into N new ones, such that in the k-th measurement the electric current enters the k-th electrode and leaves *uniformly* through all other electrodes. Then it can be shown that in solving the k-th (direct) problem one needs a fine grid only near the k-th electrode, with increasingly coarser grids away from it.

The second general point is that such a careful choice of grids, in a suitable multigrid algorithm, can *replace the need for explicit regularization of the problem.*

Our first pass at the problem, employed the well known Tikhonov regularization method, reformulating the inverse problem as a variational minimization problem. The resulting Euler equations form a PDE system (for u, σ and a Lagrange-multiplier function), which make the problem suitable in principle for an effective numerical solution by multigrid methods. The FMG solvers were designed with large and then with progressively smaller regularization. Special attention has been paid to properly adapting many features of classical multigrid to the case of the problem under consideration (including intergrid communications, boundary condition treatment and coarse grid solution). In the case of large regularization, numerical experiments have demonstrated a good convergence of the developed solver, but the obtained solution is too smeared and doesn't approximate the real conductivity function too well.

At *small regularization values* the final approximation is much better, especially near the boundary. In this case, however, the system is no longer elliptic, and much more sophisticated relaxation methods are necessary, featuring a DGS scheme [24, §3.7], which effectively decomposes the system into its scalar factors. With this approach, although the multigrid cycles asymptotically slow down, the final approximation *to the conductivity* is practically obtained by just *one* multigrid cycle per grid refinement, even when approaching the smallest regularization for which solution still exists.

It took some effort [141], [88] to complete this part of the program. Then, the solution method *without regularization* was developed. Preliminary results [89] show it to give better approximations to σ than the regularized method with its many artificial parameters (the regularization coefficients, which should change over the domain), for less work (no Lagrange multipliers). However, it also turns out that without regularization the solver requires a control parameter p to be gradually decreased through the FMG algorithm, allowing in each stage only those σ changes whose "profit" (in terms of improving the approximations to the Dirichlet data) per unit change is at least p.

17 Algebraic Multigrid (AMG): New Approaches

Algebraic multigrid (AMG) algorithms are solvers of linear systems of equations which are based on multigrid principles but do not explicitly use the geometry of grids; see [23, §13.1], [56], [58], [25], [134], [146]. The emphasis in AMG is on automatic procedures for *coarsening* the set of equations, relying exclusively on its algebraic relations. AMG is widely employed for solving discretized partial differential equations (PDEs) on unstructured grids, or even on structured grids when the coarse grid can no longer be structured, or when the PDE has highly disordered coefficients. AMG can also be used (as in [56]) for many types of discrete systems not arising from differential equations.

Given any system of linear equations $Ax = b$, where A is any (possibly rectangular) matrix, the starting point for all multilevel (multigrid) fast solvers is the following insight. (For convenience we assume, without actually losing generality, that the matrix A is roughly *normalized*, i.e., the ℓ_2 norm of every row in A is roughly 1.)

For any approximate solution $\widetilde{x}$, denote by $e = x - \widetilde{x}$ the error vector, by $r = Ae = b - A\widetilde{x}$ the vector of residuals, and by $\|\cdot\|$ the ℓ_2 norm. The common feature of all local relaxation schemes is that at each step some corrections to $\widetilde{x}$ are calculated based on the values of a small number of residuals. As a result, convergence must be slow when the individual residuals do not show the true magnitude of the error, i.e., when $\|r\| \ll \|e\|$. The converse is also true (and proved in [25]): If the convergence of a *suitable* (e.g., Kacmarz) relaxation scheme is slow, then $\|r\| \ll \|e\|$ must hold. Since, for a normalized matrix A, the deeper the condition $\|Ae\| \ll \|e\|$ is satisified the more special must be the type of the error $\|e\|$, *a suitable relaxation can always efficiently reduce the information content of the error, and quickly make it approximable by far fewer variables*. (This is true even for general *nonlinear* systems.)

Thus, following a small number of relaxation sweeps, the remaining error can be approximated by a "coarser" (or "diluted") system, i.e., a system with only a much smaller number of variables (at most half the original number, for example). General approaches for first defining the *set* of coarse variables and then for deriving the *equations* they should satisfy are briefly described below. The coarse equations themselves are then (approximately) solved by a similar procedure: a small number of relaxation sweeps followed by approximating the remaining error with a still coarser system. This recursively defines the multilevel cycle, which, for a work comparable to that of just few relaxation sweeps over the finest level (the given system), can reduce the error to a small fraction (far less than .5, typically) of its pre-cycle size.

The set C of coarse variables is chosen as a *subset* of the set of fine (original) variables; or, more generally, each coarse variable is chosen to be a linear combination of a small number of fine variables (or fine *ghost* variables — a generalization explained in Sec. 17.2 below, and also in [37, App. A]). In classical AMG [56], [58], [25], [134], the set C is chosen so that each fine variable is "strongly coupled" to C. More generally, a criterion for gauging, and a practical method to control, the quality of this set can be based on sweeps of *compatible relaxation*. This is a modified fine-level relaxation scheme that keeps the coarse-level variables invariant (i.e., it keeps the fine-level configuration always compatible with the same coarse-level configuration). *The set C is guaranteed to be good when* (*and only to the extent that*) *the compatible relaxation exhibits uniformly fast convergence rates*. Where these rates are too slow, they point to variables part of which should be added to C (or, alternatively, they point to variables that should be relaxed *simultaneously*; see [37]). (An analogous criterion for coarsening *statistical* fields, involving fast equilibration of *compatible Monte Carlo*, is described in Secs. 13.2 and 14.7 above.)

The derivation of the coarse-level equations is described below for systems of *local* equations, i.e., systems $Ax = b$ whose variables $(x_1, x_2, \ldots)$ each has a location in a low-dimensional space, and whose equations each involves only few variables in a local neighborhood of that space. Generalizations exist to "sparsely positive definite" matrices, including positive-type matrices (see [25]), to "asymptotically smooth" and "asymptotically smoothly oscillatory" matrices, including electrostatic or gravimetric interactions (see [37, §11] and Secs. 13.2.2 and 14.6.1 above), and to some other types of systems. Also, the same procedures often work well for cases not belonging to any of these types.

The fast convergence of the compatible relaxation implies that the values of the coarse set of variables indeed determine, up to fast local processing, the values of the fine set. Moreover, it implies that the chosen coarse set satisfies the "*near locality*" property, i.e., the fine level solution at each point can be calculated *locally*, given just its coarse *neighborhood*, with very weak remnant dependence on coarse values outside that neighborhood: the remnant dependence decays *exponentially* (or even faster) as a function of the neighborhood radius. (Cf. the "near locality" for statistical problems, in Secs. 13.2 and 14.7 above.) For 2D discrete Poisson equations, for example, the remnant dependence tends (after enough coarsening levels) exactly to $\exp(-\pi r^2/2)$, where r is the neighborhood radius measured in meshsizes of the coarse level [137], [167].

Since each coarse variable is defined locally by few fine variables, it too depends only *nearly-locally* on all other coarse variables. Hence, an *equation* for each coarse variable in terms of other coarse variables can be derived locally, using only a *local set* of fine-level equations. The error in that coarse equation will decrease exponentially as a function of the size of that local set.

We describe below two approaches for deriving the coarse equations: In Sec. 17.1 the highly accurate derivations of [37] are mentioned, and examples for the use of very accurate coarsening are listed. A new, much more efficient general approach is detailed in Sec. 17.2.

17.1 Highly accurate coarsening

Several general methods for local derivation of highly accurate coarse equations are described and demonstrated in [37], including a method developed by Irad Yavneh. One approach is based on the tranditional *Galerkin coarsening*: the coarse-grid equation approximating $Ax = b$ is $A^c x^c = (I_c)^T b$, where $A^c = (I_c)^T A I_c$ and I_c is an accurate coarse-to-fine interpolation derived by solving a local optimization problem. (For highly non-symmetric A, see the more general form in Sec. 17.1 below.) Another approach, called *direct coarsening*, directly derives coarse equations by solving another local optimization problem. (In statistical physics, the Galerkin coarsening corresponds to the interpolation-based method (Sec. 13.1 above), while the direct coarsening is analogous to the RMG method (Sec. 13.2).)

In both these approaches one can control the coarsening *accuracy*, and the corresponding amount of computational work per coarse equation, by choosing the size of certain stencils. Although the work per equation is always in principle only $O(1)$ (i.e., it depends on the desired accuracy but not on the size of the matrix A), the actual constant can be very large, rising as some power of the size of the local set.

For the purpose of multi-level (multigrid) *cycles*, a low coarsening accuracy would usually suffice. For example, a coarse grid equation with at most 10% error for all "smooth" components (i.e., those slow to converge in relaxation) can yield a multilevel cycle with a convergence factor close to 0.1. By performing successively any number of such cycles, any desired solution accuracy can rapidly be obtained. This will usually be far more cost effective than deriving higher accuracy coarsening.

Such low coarsening accuracy can often be inexpensively obtained by the *classical AMG approach*, i.e., using the Galerkin coarsening, with x^c being a subset of x and the interpolation I_c having weights proportional to the size of the corresponding terms in A (or in A^2, or in A^TA). This approach is particularly effective for simple matrices, such as *positive-type* ones (matrix $A = \{a_{ij}\}$ such that $a_{ij} \le 0$ for all $i \ne j$ and $\sum_j a_{ij} \ge 0$ for all i).

In many other cases, however, higher degrees of coarsening accuracy, obtainable by the techniques of [37], or those of Sec. 17.2 below, are really needed. Usually in such cases, the system involves a high degree of repetitiveness, so the high cost of deriving very accurate coarsening can be afforded. Examples:

(i) Once-for-all coarsening, for the purpose of deriving the *macroscopic equations* of the given system, or *homogenizing* it.

(ii) Cases in which one needs to solve many linear systems of the form $Ax = b$, where large parts of A and b do not change from one system to the next, so re-computing those parts at fine levels can be avoided by having accurately coarsened them before. One important such case is the calculation of many desired terms (e.g., the main diagonal) of A^{-1}; this requires solving many times the system $Ax = b$, each time b being another unit vector. Moreover, in important cases (e.g., see Sec. 12 above), those desired terms of A^{-1} must be recalculated upon each change in A occuring during certain Monte-Carlo simulations.

(iii) Problems with a large number of almost-zero modes (AZMs), i.e., eigenvectors with unusually close to zero eigenvalues. Such modes often reflect some ill defined global moves, such as rigid-body motions of the entire system in problems of elasticity, or a gliding motion of two rigid bodies along their contact surface. Such AZMs also plague various disordered problems, such as Dirac equations on critical gauge fields (cf. Sec. 11). For problems with *many* AZMs, a general cure is to increase the coarsening accuracy. A small number m of AZMs (such as those associated with global rigid body motions) may still persist even at higher accuracies, but they can be eliminated by recombining $m + 1$ iterants (each being, for example, the approximate solution obtained after another multi-level cycle) so as to minimize the ℓ_2 residual norm; see, e.g., [59].

(iv) A computing environment which makes it preferable to use as few multigrid cycles as possible, such as massive parallel processing with poor inter-processor communications, or a computer with a very high-speed cache memory.

17.2 Bootstrap AMG (BAMG)

The methods described above offer highly accurate coarse equations. They are however very expensive, being practical only for highly-repetitive systems. The main flaw in these and other AMG methods is the completely *local* derivation of the equations. This cannot yield *efficient* approximation to the lowest eight modes. More generally practical coarsening methods, developed in collaboration with Irad Yavneh, are described next.

The proposed coarse equations depend on the properties of the matrix A, which can be described in terms of the relaxation scheme that goes with it. A general relaxation scheme for the system $Ax = b$ is *weighted distributed Gauss-Seidel* (WDGS), which is a Gauss-Seidel (or SOR) relaxation of the system $A'x' = b'$, where $A' = PAM$, $x = Mx'$ and $b' = Pb$. The "weighting matrix" P and the "distribution matrix" M are chosen in various ways. If for example A is symmetric and semi-definite, or even non-symmetric but with enough diagonal dominance, then one can choose $A' = A$ and $P = M = I$, the identity matrix. If the system is a discretization of a PDE system, P and M are usually determined at the differential level, or at the level of the first differential approximation to the discrete operator, based on the operator matrix (cf. Sec. 2.2 above; see [36]). If nothing better is available, one usually chooses either $P = I$, $M = A^T$ (Kacmarz relaxation) or $P = A^T$, $M = I$ (least-square relaxation). This indeed guarantees convergence of the WDGS relaxation, but that may be insufficient. What is needed is to have good "smoothing" in the generalized sense, that each value in a relaxed vector essentially depends only on its neighboring values (except possibly for a deviation that decays exponentially with the size of the neighborhood), where the neighborhood is defined either geometrically or algebraically (in terms of strongest couplings). Such a smoothing condition would not generally happen for discretization of integral or integro-differential equations. To obtain good smoothing one can then multiply either P or M by a matrix that corresponds to *differencing* (i.e., taking differences of neighboring values), raised to a sufficiently high power.

The variable x' are called "*ghost variables*" because they need not be known: the relaxation calculates changes $\delta x'$ for these variables, but those can be directly expressed as changes $\delta x = M(\delta x')$ introduced to the explicit variables x.

For simplicity we will assume below that the given system $Ax = b$ can be relaxed by Gauss-Seidel ($P = M = I$); otherwise, A, x, and b in the discussion below can be replaced by A', x' and b', respectively.

A comment is passing: in rare situations the matrix A' may have few eigenvalues with magnitudes much larger than all other eigenvalues. In this situation (and *only* in this situation) the relaxation process should use iterant recombinations (e.g., conjugate gradient or GMRES) to reduce the corresponding outlying error components, so that the relaxation parameters can be fitted to treat efficiently the majority of eigenmodes.

The coarse-level equations $A^c x^c = b^c$ proposed here are of the Galerkin type: $A^c = I^c A I_c$ and $b^c = I^c b$. Here I_c is the coarse-to-fine interpolation; i.e., if the equation $Ax = b$ is already relaxed, a good approximation to its solution is expected to satisfy $x \approx I_c x^c$. The issue treated below is how to construct I_c and the fine-to-coarse transfer I^c. They will generally be constructed in several iterations.

It can easily be shown that I_c should interpolate the low eigenvectors of A well; i.e., a vector $\widetilde{x}$ which is a combination of low-eigenvalue eigenvectors should have a vector $\widetilde{x}^c$ such that $\| \widetilde{x} - I_c \widetilde{x}^c \| \ll \| \widetilde{x} \|$. It can also be shown that $(I^c)^T$ should well interpolate low eigenvectors of A^T. We describe here the derivation of the interpolation I_c; if A (or actually A') is symmetric or nearly symmetric, $I^c = (I_c)^T$ can be used; otherwise I^c will be derived by a similar procedure, applied to A^T.

A general form of the interpolation I_c is

$$(I_c x^c)_i = \sum_{j=1}^{n_i} w_{ij} x^c_{I_{i,j}} \ . \tag{17.1}$$

The sequence $\{I_{i,j}\}_{j=1}^{n_i}$ is the ordered set of the n_i indices of coarse-level variables from which interpolation to the i-th fine-level variable is made. They are chosen in the "neighborhood" of x_i, defined either geometrically or in terms of algebraic couplings. A necessary lower bound for their number n_i is often known in advance. For example, in solving discretized PDEs, the orders m_c and m^c of I_c and I^c, respectively, should satisfy well-known rules (see [23] or [24] or [32] or [153]), so for d-dimensional problems $n_i \geq m_c + d$. Generally, one should start with a small reasonable value for each n_i, since the iterative procedure described below will indicate when n_i should increase, or when the set $\{I_{i,j}\}_{j=1}^{n_i}$ needs to be modified.

As the example of PDE systems show, to keep all the n_i's small, it is beneficial, when possible, to divide the set of variables into disjoint "*species*", both on the fine and on the coarse levels, such that the coarse variables of each species are defined in terms of the fine-level variables of the same species, and the interpolation too is defined within each species. For example, in discretized PDE systems each species corresponds to the discretization of one function.

First approximation. Let $n_c = \max(n_i)$. A first approximation to I_c will be derived from $\overline{n}_c$ relaxed solutions $x^{(k)}$, $(k = 1, \ldots, \overline{n}_c)$, where typically $n_c < \overline{n}_c < 2n_c$. Namely, each $x^{(k)}$ is a result of several fine-level relaxation sweeps on the homogeneous equation $Ax = 0$, each starting from another *random* approximation. The number of sweeps for each $x^{(k)}$ should be small (typically less than 5) since it is enough to start with a crude approximation to I_c. A first approximation to the set of interpolation coefficients $\{w_{ij}\}_{j=1}^{n_i}$ for each i is determined so that it satisfies best, in the least-square sense, the over-determined set of equations

$$x_i^{(k)} = \sum_{j=1}^{n_i} w_{ij} x_{I_{i,j}}^{(k)c} , \qquad (k = 1, \ldots, \overline{n}_c) \tag{17.2}$$

where $x^{(k)c}$ is the coarse vector corresponding to $x^{(k)}$ (see above: the coarse variables are defined in terms of the fine ones, e.g., as a subset). If the least-square procedure for a particular i does not satisfy (17.2) well enough (the least-square error is larger than a threshold), then n_i is increased and the procedure for that particular i is repeated until satisfaction is obtained. (The threshold should be chosen comparable to the size of the current local normalized residuals of the homogeneous equations.) In these iterations for a particular i, one can also try to cancel any interpolation point $x_{I_{i,j}}^c$ which turns out to have a small interpolation weight w_{ij} or which exhibits near-dependence on others (a fact naturally detected by the least-square solver).

This procedure already gives a reasonable approximation to I_c, in the sense that it well interpolates most low-eigenvector eigenvalues, except that it is not likely to be good enough for many eigenvectors with *too low* eigenvalues, because the interpolation error should be small compared with the corresponding (normalized) eigenvalue. A similar first approximation is obtained for I^c. (Another way to obtain these first approximations is of course by the traditional AMG coarsening, *when applicable*.) This yields the first approximation to the coarse-level matrix $A^c = I^c A I_c$, which can then be used in a similar way to obtain a first approximation for the next, still-coarser-level matrix. There is, however, no point usually in proceeding this way too far: It is useless to access very coarse levels, whose role is to approximate very-low-eigenvalue eigenvectors, when the latter are ill approximated already in A^c.

Improved approximations. Once several coarse levels have been so defined, they can be used to obtain much better approximations to I_c and I^c. These are defined similarly to the first approximation described above, but instead of the *relaxed* vectors $x^{(k)}$, one obtains each of these vectors by a short *multilevel* procedure: Starting from a random configuration at the coarsest currently-available level, one relaxes the homogeneous equation on that level, then interpolates it to the next-finer level, where the result is again relaxed with the (finer-level) homogeneous equation, and so on to the finest level. Each interpolation is a two-stage procedure: First one uses the already-available I_c, then the result is relaxed by *compatible* relaxation (before it will next be relaxed by a usual relaxation). Each relaxation (compatible or usual) employs just a couple of sweeps.

Having obtained in this way improved approximations to A^c and similarly to coarser matrices, one can use them to similarly obtain such matrices on more levels. Then one obtains still better approximations by repeating the above procedures once more, now with more levels and with much better accuracy. This better accuracy is achieved by adding to the above short multilevel procedure a multilevel *correction cycle* to get better approximate solution to $Ax = 0$ (but keeping at the coarsest employed level still the same, relaxed but not converged, random configuration).

Accuracy and cost. The overall cost *per unknown* of this accurate coarsening procedure is $O(\overline{n}_c n_c^2 \log \frac{1}{\varepsilon})$, where $\overline{n}_c n_c^2$ is the work needed to set up each of the least-square systems, and ε is the desired accuracy in approximating *the lowest eigenvectors*. For producing a good multigrid solver, it is enough to have ε which is small compared with the lowest normalized eigenvalues (except perhaps for few of them, whose corresponding error can be expelled by recombining iterants of the multigrid cycles).

An important advantage of the above procedure is that it keeps all n_i (and hence also n_c, $\overline{n}_c$) almost as small as possible, hence producing A^c almost as sparse as possible, saving much work in its calculation, and also in the actual operation of the multigrid solver. The latter is often the most important consideration, as the solver is re-used many times (cf. Sec. 7.2.1).

For some purposes (see item (i)–(iv) in Sec. 17.1) one may want to have a certain accuracy ε_1 in approximating also the *other, perhaps even all, eigenvectors*. For that purpose one has to increase n_c (and accordingly also $\overline{n}_c$). the likely relation is $n_c = O\left(\left(\log \frac{1}{\varepsilon_1}\right)^q\right)$; for 2D-Poisson equations, for instance, $q = 1$. The algorithm to derive accurate I_c in this case is actually simpler than the above: The relaxed vectors $\{x^{(k)}\}_{k=1}^{\overline{n}_c}$ should each be obtained by a long enough sequence of *compatible* relaxation sweeps, starting from a random $x^{(k)c}$; the sequence needs not be really long, just $O\left(\log \frac{1}{\varepsilon_1}\right)$, due to the fast convergence of such a relaxation. Despite the simplicity, this approach is of course considerably more expensive: it similarly costs $O\left(\overline{n}_c n_c^2 \log \frac{1}{\varepsilon_1}\right)$, but n_c here is usually much larger. Also, the produced multigrid solver is less efficient, since it involves heavier I_c, I^c and A^c at all levels.

Various combinations of the two approaches are also conceivable, depending on the nature of the desired accuracy. In such a combination, each $x^{(k)}$ may be produced by interpolations (including compatible relaxation) from a different level, and larger weights in the least square calculation may be attached to "smoother" $x^{(k)}s$, i.e., $x^{(k)}$ produced from coarser levels. Working with very smooth $x^{(k)}s$ and high accuracy (smaller ε) would yield increased values of $\{n_i\}$, effectively producing *higher order* interpolations and coarsening.

17.2.1 Nonlinear and repeated problems

Most problems that require very fast solvers need to be solved again and again many times over with small variations. This includes nonlinear problems, in which one repeatedly solves a linearized version, and the kind of problems listed at the end of Sec. 17.1. Most of the coarsening work described above need not be repeated each time the problem is modified. The coarsening should not be repeated at all if only the right-hand side changes (as in the cases of calculating propagators and determinants, described in Sec. 12 above). When the operator A changes only in some neighborhood (as in the case of *updating* the determinant value in Sec. 12), the coarsening computations need be repeated only at that neighborhood (plus at most several meshsizes around it, at each level); only the last iteration may sometimes have to be repeated globally.

Quasilinearity. Nonlinear problems can often usefully be written in the *algebraic quasilinear* form $A(x) \cdot x = b$, where the dependence of $A(x)$ on x is *non-principal*, by which we mean that $\| A(x+\delta) \cdot (x+\delta) - A(x) \cdot (x+\delta) \| \ll \| A(x) \cdot \delta \|$ for any small δ. For example, most nonlinear PDE systems in mathematical physics are *differentially quasilinear*, meaning that each term in the system is linear in the highest derivative included in it; in the discretization, only the dependence on the highest derivative (in each such term) is principal, so the algebraic quasilinear form comes here naturally. Unlike Newton linearizations, this quasilinear discretization is *autonomous* (independent of external information, such as an approximate solution) wherever the PDE is autonomous.

In a quasilinear system, to a very good approximation the interpolation I_c depends only on $A(x)$ and, furthermore, I_c need seldom be changed when x changes. Also the form of $A(x)$ is often simple and explicit; e.g., in CFD and other areas, each term in A depends on x *linearly*. It is then possible to transfer this form of dependence also to the coarse level, enabling the employment of an FAS-like algorithm (see [19] or [23]), where the nonlinear problem is solved *directly*, without linearizations.

17.2.2 Indefinite and eigen problems

Indefinite-like systems. For some systems, at some or all regions, from a certain level of coarsening on, the numbers $\{n_i\}$ of required interpolation points (as indicated by the inaccuracies of the least square solutions) will start to snowball, calling for multiplying $\overline{n}_c$ by a certain factor for each further coarsening level, causing swelled complexity. The typical examples are highly indefinite systems, although some definite systems exhibit similar traits (e.g., definite systems with indefinite factors, such as $A = B^T B$, where B is highly indefinite). To check this complexity, algebraic devices generalizing the wave/ray algorithms (see Sec. 7) should be developed. Namely, the vectors $\{x^{(k)}\}$ should be recombined to extract from them a small, locally nearly orthonormal set of smooth "*basic vectors*". Any relaxed error approximates a linear combination of those basic vectors, similar to (7.2) in Sec. 7 above. As pointed out in [20, §3.2], the coarse-level correction should actually be the sum of several such corrections, each prolongated by another "shape function". So instead of deriving one interpolation I_c, several such shape functions should be identified from the basic vectors. A generalized procedure is under development, closely related to the work on many-eigenfunction solvers (cf. Sec. 9.2).

Eigenfunction calculation. Note that the above coarsening scheme can directly yield very inexpensive calculations of many eigenfunctions of A. All the *lowest* eigenfunctions, for example, and quite many of them, would accurately be interpolated by the same interpolation I_c (especially with the *higher order* interpolations mentioned above). Hence, for all of them, the generalized eigenproblem $(A - \lambda B)x = 0$ can simultaneously be coarsened to the eigenproblem $(A^c - \lambda B^c)x^c = 0$, with $A^c = I^c A I_c$, $B^c = I^c B I_c$. This *joint* coarsening can be continued for several levels (using increasingly higher order interpolations, if necessary). Only on some coarse level the eigenfunctions are *separately* calculated. Similarly, to calculate all the eigenfunctions with eigenvalues close to a certain λ_0, the same process can be repeated for the matrix $A - \lambda_0 B$ instead of A, except that now the procedure described above for dealing with indefiniteness may have to be invoked. One can proceed this way to increasingly coarser levels by progressively narrowing the set of approximated eigenfunctions. This will naturally lead to the construction of a multiscale eigenbasis (MEB) for the matrix (cf. Sec. 9.2).

18 Global Optimization: Multilevel Strategies

An optimization problem is the task of minimizing (or maximizing — for definiteness we discuss minimization) a certain real-valued "objective functional" (or "cost", or "energy", or "performance index", etc.) $E(u)$, possibly under a set of constraints of the form $A(u) = 0$ and/or $B(u) \leq 0$, where $u = (u_1, u_2, \ldots, u_n)$ is a *vector* (often the discretization of one or several *functions*) of unknown variables (real or complex numbers, and/or integers, and/or Ising spins, etc.). A general process for solving such problems is the *point-by-point minimization*, in which one changes only one variable u_j (or few of them) at a time, lowering E as much as possible in each such step. More generally, the process accepts any candidate change of one or few variables if it causes a drop in energy ($\delta E < 0$). This process would usually suffer from the following two types of difficulties:

(i) *Slow convergence*: due to the localness of the process, large-scale features (e.g., smooth components) in u are slow to converge. Acceleration by multiscale (e.g., multigrid) methods is the general cure to this trouble, since it supplement the local processing with increasingly larger scale processing, based on information suitably gathered from the fine scale. This in fact is the topic of many chapters above; a fairly general efficient approach is presented in Sec. 17.2.

(ii) *False convergence*: instead of converging to the true *global* minimum of E, the process converges to the minimum of E in a certain restricted "*attraction basin*", in which the process is trapped. The basin is a set of configurations from which the *employed* process cannot proceed to configurations with lower E, although such configurations do exist. The emphasis in *global* optimization methods is the treatment of this type of trouble. In this chapter we do not attempt to fully cover this very extensive topic. We only outline some basic multilevel strategies that deal with it.

18.1 Multilevel formulations

In many, perhaps most, global optimization problems, the objective functional E is not uniquely determined by direct physical laws, but is man-constructed, somewhat arbitrarily, to give a precise meaning to a practical problem, whose original form is more fuzzy.

This, for example, is the formulation of ill-posed problems, like *inverse PDE problems* (system identification, as in Sec. 16.2 above, or data assimilation, as in Sec. 4, etc.). The solution of such problems is often uniquely and stably fixed with the aid of *regularization*, which recasts the problem into a minimization task. The same is true in formulating *optimal control* problems (see Sec. 5 above). In all these cases, the objective, or the sense in which one solution is considered to be better than another, is not exactly apriori given; it is *chosen*, with somewhat arbitrary form and parameters.

Another typical example is the problem of reconstructing pictures from blurred or noised data. It is often recast as the problem of minimizing an energy functional which is the sum of penalty terms, penalizing the reconstruction for various unwanted features, such as (i) its distance from the data; (ii) non-smoothness, except across lines recognized as "edges"; (iii) proliferation of such edges; (iv) non-smoothness of edges; etc. This combination of penalty terms creates a monstrous minimization problem, with many nested attraction basins at all scales. It is extremely difficult to solve — *and unnecessarily so*: The difficulty largely arises from taking too seriously a set of arbitrary choices. Indeed, the form and the numerical coefficients of the various terms are quite arbitrarily chosen; a picture which is slightly better than another according to one choice may well be worse according to many other, equally reasonable choices.

More generally, unnecessary computational difficulties often arise from our tradition to cast fuzzy tasks into "stationary" formulations, that is, to define as a solution a configuration which satisfies (exactly or approximately) one well-defined criterion, such as minimizing a certain functional under specified constraints. A more universal, and often far easier way is to admit a solution which is just the end product of a suitable *numerical process*, not necessarily designed to satisfy, even approximately, any one governing criterion. In reconstructing pictures, for example, features like edges and segments can be captured very satisfactorily by very inexpensive (multiscale) processes (few dozen operations per picture element; see Sec. 15 above); the results may well fit our perception even *better* than the true or approximate minimizer of the objective functional mentioned above. Similarly, for many other fuzzy problems, a numerical process can yield excellent solutions, whose only "fault" is our inability to say what stationary objective functional they (at least approximately) optimize.

While this may be fairly obvious, one can argue that the objective-functional formulation is still in principle the "true" one: if fully carefully chosen, it would precisely reflect what one would *want* to obtain, complicated or impractical as it may be. However, even this is often not the case: a numerical process can incorporate a host of driving directives that are impossible to include in one stationary criterion. *Examples*: (i) The process for detecting curved edges can employ different completion-field parameters at different scales (see Sec. 15.1). (ii) The process for detecting picture segments can introduce new affinities between emerging intermediate aggregates, based on their internal statistics (see Sec. 15.2). The same is true in more general clustering problems and various other fuzzy graph problems (see Sec. 15.3 and the example of graph drawing [99]). (iii) In solving inverse PDE problems one can apply *multiscale* regularizations, which use different penalty terms at different scales (see for example Item 9 in Sec. 4.3).

It can be seen from these examples that an important tool in *formulating* various problems is to have different, sometimes even conflicting, objectives at different scales of the problem. The multiscale processing is thus not just a method to accelerate convergence and escape false attraction basins (as discussed below), but can often also be essential for an improved *definition* of the problem.

Incidentally, even for *linear* problems multi-scale *formulations* are sometimes needed. An example is the case of wave equations with radiation boundary conditions: such conditions are most appropriately formulated at the coarsest levels of the wave/ray algorithm (see Sec. 7 above), while the differential equations themselves are discretized at the finest level.

18.2 Multilevel annealing

A general method to escape false attraction basins is to modify the strict point-by-point minimization by a process that still accepts each candidate change which lowers the energy ($\delta E < 0$), but also assigns a positive probability, proportional for example to $\exp(-\beta \cdot \delta E)$, for accepting a candidate step that *increases* the energy ($\delta E > 0$). This is similar to a Monte Carlo simulation of the system at a finite temperature T, where $\beta = (k_B T)^{-1}$ and k_B is the Boltzmann constant. This is indeed the very way by which natural materials escape various attraction basins and advance toward lower energies.

To have a reasonable chance to escape *wide* attraction basins or basins within *high* energy barriers in a tolerable computational time, a low value of β, or a high temperature, must of course be applied. This however makes it improbable to hit the true minimum. A general approach therefore is the *gradual* decrease of temperature, hoping first to escape false high-energy attraction basins, than lower-energy ones, etc. This process is called *simulated annealing*, since it simulates the common industrial process of "annealing" — obtaining low-energy materials (such as less brittle glass) by carefully gradual cooling. Variations on the theme include various procedures of alternate heating and cooling.

The simulated annealing algorithms are extremely inefficient for many physical problems, requiring exponentially slow temperature decrease to approach the true minimum. This is usually due to the multiscale structure of the attraction basins: small-scale basins reside within larger-scale ones, which reside within still-larger-scale ones etc. The small-scale basins correspond to *local* structures in the physical space; larger-scale basins correspond to larger physical structures. When the temperature is high enough to enable transition between large-scale attraction basins it would completely randomize finer-scale basins, even when they have already settled into low-energy local structures (by a previous cooling).

Clearly, the transitions between basins at various scales should be better coordinated. It should employ much lower temperatures in switching between large-scale basins, which can be achieved only if well orchestrated large-scale moves are constructed. This is done by what we will generally call "*multilevel annealing*", whose main features are described below. Its first, incomplete version appeared in [61].

18.2.1 Identifying multiscale variables or moves

In multilevel optimization, the main role of annealing is to *identify* increasingly larger-scale degrees of freedom that are acceptable to simulation at progressively lower temperatures. We describe two approaches to go about it.

One approach is to work in terms of coarse-level **variables** that are coupled to each other through temperature-dependent conditional probability (CP) tables, as in the RMG method (cf. Secs. 13.2 and 14.7.2). Gradually, as the temperature is lowered, new coarse-level variables are generally introduced, checked by the CMC-equilibration test. The procedure is like that of Monte Carlo simulation at low temperatures (see Secs. 13.2.3 and 14.7.3), except that it can be executed without strict adherence to statistical fidelity ("detailed balance"). In many cases a low-temperature-like simulation is actually more realistic than strict minimization, either because the minimization task is fuzzy anyway (see Sec. 18.1), or simply because the material whose minimal energy is sought has in reality a *finite* temperature.

Note the similarity of this procedure to the BAMG approach in Sec. 17.2, in which increasingly coarser (large-scale) variables and interpolation rules associated with increasingly lower eigenvalues (corresponding to lower temperatures here) are gradually revealed, through a process that uses coarser levels already accessible by the current interpolation rules to accelerate relaxation (or the Monte Carlo simulation here) at finer levels.

In this approach each coarse level configuration corresponds to the *equilibrium* of all fine-level configurations that are compatible with it. When the temperature is lowered, the equilibrium narrows down to the vicinity of few specific fine-level configurations. Another approach then is to work explicitly with the fine level, and to identify on it increasingly larger-scale **moves** that can be done with progressively lower temperatures. If an efficient simulation has already been obtained at some temperature T, it can be employed to identify suitable moves for a lower temperature T', assuming $T - T' \ll T'$. Indeed, the moves already identified for T are at a scale close to those required for T', hence each suitable T'-move is approximately a linear combination of just a small number of T-moves. Such combinations can be identified by calculating correlations between neighboring T-moves during Monte Carlo simulations with the temperature T. Each combination can then be "reshaped" into more precise T'-move by optimizing around it (see Sec. 18.2.2).

The work in terms of large-scale *variables* is perhaps preferable whenever the system is *highly repetitive*, so that the same coarse-level variables and CP tables can be used at all (or many) subdomains, as in the case of fluids (Sec. 14.7). The tables then can be derived in just representative small *windows* of the fine-scale system (see the description of windows in Secs. 13.2 and 14.7.2). On the other hand, the identification of *explicit large-scale moves* is perhaps more practical for systems that have different specific structures at different neighborhoods, making it too expensive to derive place-dependent CP tables. However, the explicit moves are not flexible enough, requiring the device discussed next.

18.2.2 Reshaping large scale moves

Any preassigned large-scale move is likely to bring about a substantial energy increase since its fine details would not generally quite fit the fine details produced by other large scale moves. In other words, in switching to a new large-scale attraction basins one does not generally immediately hit the lowest-energy configurations of that basin; since in the *previous* basin a process of minimization *has* already taken place, the new configuration is likely to exhibit much higher energy. Thus, only rarely the large-scale move will be accepted in a low-temperature simulation, even if the new attraction basin does harbor lower energy configurations. Therefore, before applying the acceptance test to a large-scale move, one should "*reshape*" it, or "optmize around it", by employing in the neighborhood around it a Monte Carlo simulation of smaller-scale moves. Each of these smaller-scale moves may itself need "reshaping" by local simulations around it at still finer scales. And so on. Such nested reshaping processes are needed when the energy landscape has nested attraction basins. Each of these processes can itself employ a kind of annealing (see details in [61]).

Working with the difficult discrete optimization problem of spin glasses, it was shown already in [61] that such multiscale nested optimization techniques (together with the technique of Sec. 18.2.3 below) work reasonably well ever *without* any prior identification of specialized moves at all scales (cf. Sec. 18.2.1). However, the amount of work in that case turned out to increase at least quadratically as a function of the number of spins in the system, due to the excessive nested reshaping processes that were required. Much shorter reshaping procedures will suffice with more specialized moves. (Also, as mentioned in Sec. 18.2.1, the reshaping procedure can be used to optimize the specialized moves themselves, prior to their use in the T' simulations.)

Note that the reshaping procedure (unless confined only to the prior identification of moves) does not satisfy the statistical detailed balance. It is very efficient in the search for a minimum, but cannot be used for obtaining accurate finite-temperature statistics.

18.2.3 Taming local fluctuations and genetic-type algorithms

In any sufficiently large-scale problem with local couplings (i.e., its objective functional is the sum of terms each of which depends only on a *local* set of variables, in some space), there is a large accumulation of likelihood that any stochastic simulation, even with a low temperature, will create some small-scale local fluctuations, frustrating the chance to identify *the* global minimum. Since these fluctuations are indeed likely to be local, one can eliminate them by the following simple procedure.

Keep in memory one or several of the best-so-far (BSF) configurations. Once in a while (e.g., whenever the stochastically-evolving current configuration yields a particularly low energy) compare the current configuration with each of the BSF configurations. The two compared configurations will generally have spots of just *local* disagreement, i.e., *disconnected* subsets where the values of the two configurations differ, but outside which the configurations coincide. Hence, for each such subset, separately from all other subsets, one can decide whether or not to replace the BSF values by those of the current configuration, depending which option would yield at that spot the lower energy. In this way all the BSF configurations can be replaced by better ones. The current configuration should continue its evolution from its previous value, in search for new optima. At the end, the BSF configurations can be compared to choose the best among them.

This device should apply not only to the main optimization process, but also to each of the auxiliary "reshaping" processes defined above (Sec. 18.2.2), as successfully demonstrated in [61].

Analogous devices can be used even for more general problems (not just locally coupled). The general approach can be described as a combination of multilevel annealing with genetic-type algorithms. Instead of one minimization process, a population of such processes evolve in parallel. Once in a while one of the evolving configurations (a "parent") chooses another (a "partner"), from which it borrows a combination of large-scale moves, reshaping them using its own finer multiscale moves (see Sec. 18.2.2), then (and only then) deciding whether to adopt the resulting configuration (accept it as an addition to the population or as a replacement). Each of the reshaping processes can itself be done in terms of several evolving children, and so on recursively. "Fitness" parameters can be defined in terms of the low-energy levels attained by the evolving configuration and its relatives. The choice of "partner" can be based on its fitness and criteria of compatibility with the choosing "parent".

In short, one can marry the ideas of multiscale optimization with those of genetic algorithms and study the (fuzzy) fitness of their evolving offsprings. The success is likely to be especially high for problems dominated by a multitude of local couplings.

19 What About Wavelets?

Wavelets, perhaps the currently most popular form of multiscale representation, have not been mentioned in any of the above sections. An explanation is due.

First, in all areas described above, either wavelets are not at all applicable (as in Secs. 13, 14, 15, 16, 17), or they are less developed than multigrid-type methods. Part of the reason for that is historical: Multigrid methods, in either finite-difference or finite-element formulations, have appeared long before wavelets. The excitement about wavelets is probably due not only to its mathematical elegance, but also to the fact that this has been the first form of multiscale representation encountered by several communities, hence the form through which they first discovered the great computational benefits of multiscaling. There are indeed many very effective algorithms developed with wavelets. But multiscaling had existed before, and the question is how wavelets compare with other forms of multiscale representation from the point of view of computational efficiency. The answer is that wavelets are less efficient, at least for the kind of problems surveyed in this article.

To explain, note that there are mainly two (related) differences in emphasis distinguishing wavelets from more general formulations. The first, and less essential difference is the *incremental* representation used by wavelets: while multigrid methods represent at the fine level the full function, wavelets separately represent its high-resolution (e.g., high-frequency) part. All other scales of resolution are similarly separated. This separation degrades the efficiency in treating *nonlinear* problems. (In FAS multigrid the full function is represented at all levels; as a result the solution of nonlinear problems is usually as fast and easy as solving linear problems: no linearizations, with their vast extra storage and iterations, are needed [19], [23].)

The separate representation of the high-resolution part is considered by some to be advantageous from the point of view of self-adaptive discretization: wherever that part is larger than some threshold, the need for a still-finer resolution is indicated. However, there is no special advantage here, because, in any other multilevel representation, that high resolution part can straightforwardly be estimated from the difference between the solutions at the finest level and at the next level. (The true *local* criterion for grid adaptation is solving PDE systems is the local error in the *equation*, not in the solution. In FAS multigrid methods this error is automatically given by the fine-to-coarse correction τ_h^{2h} [19], [23].)

The second, and more important, feature that distinguishes wavelets is the *orthogonality* of the levels: Each level of wavelet resolution is exactly orthogonal to all other levels. This is very pleasing mathematically, but is also a source of a certain computational inefficiency: The representation is substantially more complicated and costly than a simple multigrid representation at the corresponding order.

Although the latter does not enjoy orthogonality of levels, it does always have *near orthogonality*: even though the processing (e.g., relaxation) at each level does slightly affect other levels, this slight influence for most purposes is insignificant. For example, it does not harm the multigrid convergence rates, which fit the "smoothing rates" of the separate levels [19], [23], [28]. Moreover, in special cases where this is needed, one can make inter-level influence as small as desired by employing, e.g., *distributive* moves at each level (see for example Secs. 10 and 14.6.1 above).

Thus, the natural near-orthogonality of *any* multiscale representation makes it unnecessary, and usually more expensive, to use the exact orthogonality offered by wavelets.

References

1. Alessandrini, G., Stable determination of conductivity by boundary measurements, *Appl. Anal.* **27** (1988) 153–172.
2. Arian, A. and Ta'asan, S., Multigrid one shot methods for optimal control problems: Infinite dimensional control, ICASE Report, 1994, pp. 94–52.
3. Arian, A. and Ta'asan, S., Shape optimization in one shot, in: *Optimal Design and Control* (Boggard, J., Burkardt, J., Gunzburger, M. and Peterson, J., eds.), Birkhauser Boston Inc., 1995.
4. Ausiello, G., Crescenzi, P., Gambosi, G., Kann, V., Marchetti-Spaccamela, A. and Protasi, M., *Complexity and Approsimation*, Springer Verlag, 1999.
5. Babuška, I. and Melenk, J. M., The partition of unity method, *Int. J. Num. Methods in Engineering* **40** (1997) 727–758.

6. Bai, D. and Brandt, A., Local mesh refinement multilevel techniques, *SIAM J. Sci. Stat. Comput.* **8** (1987) 109–134.
7. Bai, D. and Brandt, A., Multiscale computation of polymer models, in [39], pp. 250–266.
8. Baschnagel, J., Binder, K., Doruker, P., Gusev, A. A., Hahn, O., Kremer, K., Mattice, W. L., Müller-Plathe, F., Murat, M., Paul, W., Santos, S., Suter, U. W. and Tries, V., Bridging the gap between atomistic and coarse-grained models of polymers: Status and perspectives, *Advances in Polymer Science*, Springer Verlag **152** (2000) 41–156.
9. Bates, J. R., Li, Y., Brandt, A., McCormick, S. F. and Ruge, J., A global shallow water numerical model based on the semi-Lagrangian advection of potential vorticity, *Quart. J. Roy. Met. Soc.* **121** (1995) 1981–2005.
10. Bates, J. R., McCormick, S. F., Ruge, J., Sholl, D. S. and Yavneh, I., A semi-Lagrangian approach to the shallow water equations, 6th *Copper Mountain Conf. on Multigrid Methods*, NASA Conference Publication 3224, Part 2, 1993, pp. 593–604.
11. Bates, J. R., Semazzi, F. H. M., Higgins, R. W. and Barros, S. R. M., Integration of the shallow water equations on the sphere using a vector semi-Lagrangian scheme with a multigrid solver, *Mon. Wea. Rev.* **18** (1990) 1615–1627.
12. Beck, T. L., Multigrid high order mesh refinement technique, *J. Comput. Chem.* **20** (1999) 1731.
13. Beck, T. L., Multiscale methods for self-consistent electronic structure in real space, in [39], pp. 90–103.
14. Ben-Av, R., *Acceleration Methods in MC Simulations*, Ph.D. Thesis, Weizmann Institute of Science, Rehovot, 1991.
15. Ben-Av, R., Brandt, A., Harmatz, M., Katznelson, E., Lauwers, P. G., Solomon, S. and Wolowesky, K., Fermion simulation using parallel transported multigrid, *Phys. Lett.* **B253** (1991) 185.
16. Ben-Av, R., Brandt, A. and Solomon, S., The Fermionic matrix, instantons, zero modes and multigrid, *Nucl. Phys.* **B329** (1990) 193.
17. Bernholc, J., Briggs, E. L., Buongiorno Nardelli, M., Fattebert, J.-L., Ramamoorthy, M., Schmidt, W. G. and Sullivan, D. J., Large-scale, multilevel solutions of Kohn-Sham equations: methodology and applications, in [39], pp. 65–89.
18. Brandt, A., Multi-level adaptive technique (MLAT) for fast numerical solutions to boundary value problems, in: *Proc.* 3rd *Int. Conf. on Numerical Methods in Fluid Mechanics* (Cabannes, H. and Temam, R., eds.), Lecture Notes in Physics **18**, Springer-Verlag, 1973, pp. 82–89.
19. Brandt, A., Multi-level adaptive solutions to boundary value problems, *Math. Comp.* **31** (1977) 333–390.
20. Brandt, A., Stages in developing multigrid solutions, in: *Proc.* 2nd *Int. Congr. on Numerical Methods for Engineers* (Absi, E., Glowinski, R., Lascaux, P. and Veysseyre, H., eds.), Dunod, Paris, 1980, pp. 23–43.
21. Brandt, A., Multi-grid solvers on parallel computers, in: *Elliptic Problem Solvers* (Schultz, M., ed.), Academic Press, New York, 1981, pp. 39–84.
22. Brandt, A., *Multi-Grid Solvers for Non-Elliptic and Singular-Perturbation Steady-State Problems*, Weizmann Institute of Science, Rehovot, November 1981.
23. Brandt, A., Guide to multigrid development, in: *Multigrid Methods* (Hackbusch, W. and Trottenberg, U., eds.), Springer-Verlag, 1982, pp. 220–312.

24. Brandt, A., *Multigrid Techniques: 1984 Guide, with Applications to Fluid Dynamics*, 191 pages, 1984, ISBN-3-88457-081-1, GMD Studien Nr. 85. Available from GMD-AIW, Postfach 1316, D-53731, St. Augustin 1, Germany, 1984.
25. Brandt, A., Algebraic multigrid theory: The symmetric case, *Appl. Math. Comput.* **19** (1986) 23–56.
26. Brandt, A., Multilevel computations: Review and recent developments, in: *Multigrid Methods: Theory, Applications and Supercomputing* (McCormick, S. F., ed.), Marcel-Dekker, 1988, pp. 35–62.
27. Brandt, A., The Weizmann Institute Research in Multilevel Computation: 1988 Report, in: *Proc. 4th Copper Mountain Conf. on Multigrid Methods* (Mandel, J. *et al.*, eds.), SIAM, 1989, pp. 13–53.
28. Brandt, A., Rigorous local mode analysis of multigrid, in: *Preliminary Proc. 4th Copper Mountain Conf. on Multigrid Methods*, Copper Mountain, Colorado, April 1989. An updated version appeared as Weizmann Institute Report, 1991. The first part has appeared as [32].
29. Brandt, A., Multilevel computations of integral transforms and particle interactions with oscillatory kernels, *Comput. Phys. Comm.* **65** (1991) 24–38.
30. Brandt, A., Multiscale computational methods: research activities, in: *Proc. 1991 Hang Zhou International Conf. on Scientific Computation*, (Chan, T. and Shi, Z.-C., eds.), World Scientific Publishing Co., Singapore, 1992.
31. Brandt, A., Multigrid methods in lattice field computations, *Nuclear Phys. B Proc. Suppl.* **26** (1992) 137–180.
32. Brandt, A., Rigorous quantitative analysis of multigrid: I. Constant coefficients two level cycle with L_2 norm, *SIAM J. Numer. Anal.* **31** (1994) 1695–1730.
33. Brandt, A., Multiscale research in molecular dynamics, Gauss Center Report WI/GC-3, April 1995.
34. Brandt, A., Proceedings of the multigrid tutorial, with applications to molecular dynamics, Weizmann Institute of Science, October 10–12, 1995.
35. Brandt, A., The Gauss Center research in multiscale scientific computation, *Electr. Trans. Numer. Anal.* **6** (1997) 1–34.
36. Brandt, A., *Barriers to Achieving Textbook Multigrid Efficiency in CFD*, ICASE Interim Report No. 32, NASA/CR-198-207647. Gauss Center Report WI/GC–10, June 1998; updated December, 1998. Appears as Appendix C in [153].
37. Brandt, A., General highly accurate algebraic coarsening schemes, *Electr. Trans. Num. Anal.* **10** (2000) 1–20.
38. Brandt, A., Multiscale scientific computation: six year summary, Gauss Center Report WI/GC–12, May 1999. Also in: http://www.wisdom.weizmann.ac.il/~achi, and in MGNET.
39. Brandt, A., Bernholc, J. and Binder, K. (Eds.), *Multiscale Computational Methods in Chemistry*, NATO Science Series, Computer and System Sciences, Vol. 177, IOS Press, Amsterdam (2000).
40. Brandt, A. and Dinar, N., Multi-grid solutions to elliptic flow problems, in: *Numerical Methods for Partial Differential Equations* (Parter, S., ed.), Academic Press, New York, 1979, pp. 53–147.
41. Brandt, A. and Diskin, B., Multigrid solvers for the non-aligned sonic flow: the constant coefficient case, Gauss Center Report WI/GC–8, 1997. *Computer and Fluids* **28** (1999) 511–549.
42. Brandt, A. and Diskin, B., Multigrid solvers for non-aligned sonic flows, *SIAM J. Sci. Comput.* **21** (1999) 473–501.

43. Brandt, A. and Dym, J., Fast computation of multiple line integrals, *SIAM J. Sci. Comput.* **20** (1999) 1417–1429.
44. Brandt, A. and Dym, J., Fast robust discontinuity detection using multiple scales, preprint.
45. Brandt, A. and Galun, M., Optimal multigrid algorithms for the massive Gaussian model and path integrals, *J. Statist. Phys.* **82** (1996) 1503–1518.
46. Brandt, A. and Galun, M., Optimal multigrid algorithms for variable-coupling isotropic Gaussian models, *J. Statist. Phys.* **88** (1997) 637–664.
47. Brandt, A. and Galun, M., Statistically optimal multigrid algorithm for the anharmonic crystal model, Gauss Center Report WI/GC–9, March 1998.
48. Brandt, A., Galun, M. and Ron, D., Optimal multigrid algorithms for calculating thermodynamic limits, *J. Statist. Phys.* **74** (1994) 313–348.
49. Brandt, A. and Greenwald, J., Parabolic multigrid revisited, in: *Multigrid Methods III* (Hackbusch, W. and Trottenberg, U., eds.), Birkhäuser Verlag, Basel, 1991, pp. 143–154.
50. Brandt, A. and Iliyn, V., Multilevel approach in statistical physics of liquids, in [39], pp. 187–197.
51. Brandt, A. and Livshits, I., Wave-ray multigrid methods for standing wave equations, *Electr. Trans. Numer. Anal.* **6** (1997) 162–181.
52. Brandt, A. and Livshitz, I., Accuracy properties of the wave-ray multigrid algorithm for Helmholtz equation, in preparation.
53. Brandt, A. and Lubrecht, A. A., Multilevel matrix multiplication and the fast solution of integral equations, *J. Comput. Phys.* **90**(2) (1990) 348–370.
54. Brandt, A., Mann, J. and Brodski, M., *A Fast and Accurate Radon Transform Inversion Scheme*, Patent assigned to Yeda Research and Development Co. Ltd., August 1995. U.S. Patent and Trademark Office Application No. 08/659,595, filed 06/06/96. U.S. Patent No. 5,778,038, granted July 7, 1998 under the title "Computerized Tomography Scanner and Method of Performing Computerized Tomography". European Patent Office Application No. 97108722.6-2305, filed 05/30/97.
55. Brandt, A., Mann, J., Brodski, M. and Galun, M., A fast and accurate multilevel inversion of the Radon transform, *SIAM J. Appl. Math.* **60** (1999) 437–462.
56. Brandt, A., McCormick, S. and Ruge, J., Algebraic multigrid (AMG) for automatic multigrid solution with application to geodetic computations, Institute for Computational Studies, POB 1852, Fort Collins, Colorado, 1982.
57. Brandt, A., McCormick, S. and Ruge, J., Multi-grid methods for differential eigenproblems, *SIAM J. Sci. Statist. Comput.* **4** (1983) 244–260.
58. Brandt, A., McCormick, S. and Ruge, J., Algebraic multigrid (AMG) for sparse matrix equations, in *Sparsity and its Applications* (Evans, D. J., ed.), Cambridge University Press, Cambridge, 1984, pp. 257–284.
59. Brandt, A. and Mikulinsky, V., Recombining iterants in multigrid algorithms and problems with small islands, *SIAM J. Sci. Comput.* **16** (1995) 20–28.
60. Brandt, A. and Ron, D., Renormalization Multigrid (RMG): Statistically optimal renormalization group flow and coarse-to-fine Monte Carlo acceleration, *J. Stat. Phys.* **102** (2001) 231–257. Also appeared as Gauss Center Report WI/GC–11, 1999, and with modification in [39], pp. 163–186.
61. Brandt, A., Ron, D. and Amit, D. J., Multi-level approaches to discrete-state and stochastic problems, in: *Multigrid Methods, II* (Hackbusch, W. and Trottenberg, U., eds.), Springer-Verlag, 1986, pp. 66–99.

62. Brandt, A. and Venner, C. H., Fast evaluation of integral transforms with asymptotically smooth kernels, Gauss Center Report WI/GC–2, 1995. *SIAM J. of Sci. Comput.* **19** (1998) 468–492.
63. Brandt, A. and Venner, C. H., Multilevel evaluation of integral transforms on adaptive grids, Gauss Center Report WI/GC–5, 1996. In: *Multigrid Methods V*, Lecture Notes in Computational Science and Engineering **3** (Hackbusch, W. and Wittum, G., eds.), Sprinter Verlag, Berlin, 1999, pp. 20–44.
64. Brandt, A. and Yavneh, I., On multigrid solution of high-Reynolds incompressible entering flows, *J. Comput. Phys.* **101** (1992) 151–164.
65. Brandt, A. and Yavneh, I., Accelerated multigrid convergence and high-Reynolds recirculating flows, *SIAM J. Sci. Comput.* **14** (1993) 607–626.
66. Brandt, A. and Zaslavsky, L. Yu., Multiscale algorithm for atmospheric data assimilation, *SIAM J. Sci. Comput.* **18** (1997) 949–956.
67. Briggs, W. L., Henson, V. E. and McCormick, S. F., *A Multigrid Tutorial*, 2nd Ed., SIAM, 2000.
68. Bunch, J. R. and Nielsen, C. P., Updating the singular value decomposition, *Numer. Math.* **31** (1978) 111–129.
69. Calderon, A., On an inverse boundary value problem, in: *Seminar on Numerical Analysis and its Applications* (Meyer, W. and Raupp, M., eds.), Brazilian Mathematical Society, Rio de Janeiro, 1980, pp. 1–7.
70. Chan, T. F. and Keller, H. B., Arc-length continuation and multi-grid techniques for nonlinear elliptic eigenvalue problems, *SIAM J. Sci. Stat. Comput.* **3** (1982) 173–194.
71. Chan, T. F. and Bank, R. E., PLTMGC: A multi-grid continuation program for parametrized nonlinear elliptic systems, *SIAM J. Sci. Stat. Comput.* **7** (1986) 540–559.
72. Cormen, Leiserson and Rivest, Introduction to algorithms, *J. ACM* (2000).
73. Costiner, S., *Simultaneous Multigrid Solvers for the Nonlinear Schrödinger Eigenvalue Problem*, M.Sc. Thesis, Weizmann Institute of Science, 1990.
74. Costiner, S. and Ta'asan, S., Adaptive multigrid techniques for large-scale eigenvalue problems: solutions of the Schrödinger problem in two and three dimensions, *Phys. Rev.* **E51** (1995) 3704–3717.
75. Cuppen, J. J. M., A divide and conquer method for the tridiagonal eigenproblem, *Numer. Math.* **36** (1981) 177–195.
76. Deans, S. R., *The Radon Transform and Some of its Applications*, John Wiley and Sons, New York, 1983.
77. Diskin, B., Multigrid algorithm with conditional coarsening for the non-aligned sonic flow, *Electr. Trans. Numer. Anal.* **6** (1997) 106–119.
78. Diskin, B., *Efficient Multigrid Solvers for the Linearized Transonic Full Potential Equation*, Ph.D. Thesis, Weizmann Institute of Science, Rehovot, June 1998.
79. Dym, J., *Multilevel Methods for Early Vision*, Ph.D. Thesis, Weizmann Institute of Science, Rehovot, 1994.
80. Faddeeva, V. N., *Computational Methods of Linear Algebra*, Dover Publications, New York, 1959.
81. Feige, U., Kortsarz and Peleg, D., The dense k-subgraph problem, *Algorithmica*, March 2001.
82. Flory, P. J., *Statistical Mechanics of Chain Molecules*, January 1989, ISBN 1569900191.

83. Forsythe, G. E. and Golub, G. H., On the stationary values of a second-degree polynomial on the unit sphere, *SIAM J. Appl. Math.* **13** (1965) 1050–1068.
84. Fuhrmann, D. R., An algorithm for subspace computation with applications in signal processing, *SIAM J. Matrix Anal. Appl.* **9** (1988) 213–220.
85. Galli, G., Linear scaling method for electronic-structure calculations and quantum molecular dynamics simulations, *Current Opinion in Solid State and Material Science* **1**(6) (1996) 864–874.
86. Galun, M., *Optimal Multigrid Algorithms for Model Problems in Statistical Mechanics*, M.Sc. Thesis, Weizmann Institute of Science, Rehovot, 1992.
87. Galun, M., *Multigrid Algorithms for Optimal Computations in Statistical Physics*, Ph.D. Thesis, Weizmann Institute of Science, Rehovot, April 1998.
88. Gandlin, R., *Multigrid Algorithms for an Inverse Problem in Impedance Tomography*, M.Sc. Thesis, Weizmann Institute of Science, Rehovot, June 1998.
89. Gandlin, R., Proposal for Ph.D. Thesis, Weizmann Institute of Science, Rehovot, 1999.
90. Gandlin, R., *Multigrid Algorithms for Inverse Problems*, Progress Report, Feinberg Graduate School, Weizmann Institute of Science, Rehovot, 2000.
91. Gödecker, S., Linear scaling electronic structure methods, *Rev. Modern Phys.* **71** (1999) 1085–1123.
92. Golub, G. H. and von Matt, U., Quadratically constrained least squares and quadratic problems, *Numer. Math.* **59** (1991) 561–580.
93. Graph spanners, *J. Graph Theory* **13**(1) (1989) 99–116.
94. Greengard, L., Fast algorithms for classical physics, *Science* **265** (1994) 909–914.
95. Greenwald, J. C., *Multigrid Techniques for Parabolic Problems*, Ph.D. Thesis, Weizmann Institute of Science, June 1992.
96. Grünbaum, F. A., Diffuse tomography: the isotropic case, *Inverse Problems* **8** (1992) 409–419.
97. Gupta, R. and Cordery, R., Monte Carlo renormalized Hamiltonian, *Phys. Lett.* **105A** (1984) 415.
98. Guy, G. and Medioni, G., Inferring global perceptual contours from local features, *Proc. DARPA Image Understanding Workshop*, 1993, pp. 881–893.
99. Hadany, R. and Harel, D., A multi-scale method for drawing graphs nicely, to appear in *Discrete Appl. Math.*. Also *Proc. 25th Int. Workshop on Graph-Theoretic Concepts in Computer Science* (*WG'99V*), Lecture Notes in Computer Science, Vol. 1665, Springer-Verlag, June 1999, pp. 262–277.
100. Haras, Z. and Ta'asan, S., Finite difference schemes for long time integration, *J. Comp. Phys.* **114**(2) (1994).
101. Harel, D., *Computer Ltd.*, Oxford University Press, 2000.
102. Herman, G., *Image Reconstruction from Projections: The Fundamentals of Computerized Tomography*, Academic Press, New York, 1980.
103. Hill, T. L., *Statistical Mechanics*, McGraw-Hill Book Company Inc., New York-Toronto-London, 1956.
104. Jameson, A., Solution of the Euler equations by a multigrid method, *Appl. Math. Comput.* **13** (1983) 327–356.
105. Jameson, A., Schmidt, W. and Turkel, E., Numerical solution of the Euler equations by finite volume methods using Runge-Kutta time stepping schemes, AIAA Paper 81-1259, 1981.

106. Jorgensen, W. L., Chandrasekhar, J., Madura, J. D., Impey, R. W. and Klein, M. L., Comparison of potential functions for simulating liquid water, *J. Chem. Phys.* **79** (1983) 926–935.
107. Jorgensen, W. L. and Jenson, C., Temperature dependence of TIP3P, SPC and TIP4P water from NPT Monte Carlo simulations: seeking temperatures of maximum density, *J. Comp. Chem.* **19** (1998) 1179–1186.
108. Kalkreuter, T., *Multigrid Methods for the Computation of Propagators in Gauge Fields*, Ph.D. Thesis, Hamburg University, Report DESY92–158, 1992.
109. Kalkreuter, T., Idealized multigrid algorithm for staggered fermions, *Phys. Rev.* **D48** (1993) 1926–1930.
110. Kaminsky, R., *Multilevel Solution of the Long Transportation Problem*, M.Sc. Thesis, Weizmann Institute of Science, 1989.
111. Keller, J. B., Geometrical theory of diffraction, *J. Optical Soc. Am.* **52** (1962).
112. Kogut, J. B., An introduction to lattice gauge theory and spin systems, *Rev. Modern Phys.* **51** (1979) 659–713.
113. Kogut, J. B., The lattice gauge theory approach to quantum chromodynamics, *Rev. Modern Phys.* **55** (1983) 775–836.
114. Kogut, J. B. and Susskind, L., Hamiltonian formulation of Wilson's lattice gauge theory, *Phys. Rev.* **D11** (1975) 395–408.
115. Kohn, P. and Vogelius, M., Determining conductivity by boundary measurements, *Comm. Pure Appl. Math.* **37** (1984) 289–298.
116. Kohn, W. and Sham, L. J., *Phys. Rev.* **140** (1965) A1133.
117. Li, Y. and Bates, J. R., A study of the behavior of semi-Lagrangian models in the presence of orography, *Quart. J. R. Met. Soc.* **122** (1996) 1675–1700.
118. Li, Y., Ruge, J., Bates, J. R. and Brandt, A., A proposed adiabatic formulation of three-dimensional global atmospheric models based on potential vorticity, *Tellus* **52A** (2000) 129–139.
119. Livne, O. E., *Multiscale Eigenbasis Algorithms*, Ph.D. Thesis, Weizmann Institute of Science, Rehovot, 2000.
120. Livne, O. E. and Brandt, A., $O(N \log N)$ multilevel calculation of N eigenfunctions, in [39], pp. 112–136. Also in the present volume.
121. Livne, O. and Brandt, A., N roots of secular equation in $O(N)$ operations, submitted to *SIAM J. Matrix Anal. Appl.*
122. Livshits, I., *Multigrid Solvers for Wave Equations*, Ph.D. Thesis (Brandt, A., supervisor), Bar-Ilan University, November 1995.
123. McWilliams, J. C., Weiss, J. B. and Yavneh, I., Anisotropy and coherent vortex structures in planetary turbulence, *Science* **264**(5157) (1994) 410–413.
124. Melenk, J. M., *On generalized finite element method*, Ph.D. Thesis, University of Maryland, 1995.
125. Melman, A., Numerical solution of a secular equation, *Numer. Math.* **69** (1995) 483–493.
126. Melman, A., A numerical comparison of methods for solving secular equations, *J. Comp. Appl. Math.* **86** (1997) 237–249.
127. Paul, W., Yoon, D. Y. and Smith, G. D., An optimized united atom model for simulations of polymethylene melts, *J. Chem. Phys.* **103** (1995) 1702–1709.
128. Peskin, C. S. and Schlick, T., Molecular dynamics by backward-Euler method, *Comm. Pure Appl. Math.* **42** (1989) 1001.
129. Ron, D. and Swendsen, R. H., Calculation of effective Hamiltonians for renormalized or non-Hamiltonian systems, *Phys. Rev. E*, to appear.

130. Rozantsev, M., *Methods for Multigrid Solution of Quantum Electrodynamics Equations*, M.Sc. Thesis, Weizmann Institute of Science, Rehovot, 1993.
131. Rozantsev, M., *Multigrid Solver for 4D Quantum Chromodynamics Equations*, Final Report, Feinberg Graduate School, Weizmann Institute of Science, Rehovot, January 1999.
132. Rozantsev, M., *Multigrid Solver for 4D Quantum Chromodynamics Equations*, Ph.D. Thesis, submitted to Feinberg Graduate School, Weizmann Institute of Science, Rehovot, February 2001.
133. Ruge, J. W., Li, Y., McCormick, S., Brandt, A. and Bates, J. R., A nonlinear multigrid solver for a semi-Lagrangian potential vorticity based shallow water model on the sphere, *SIAM J. Sci. Comput.* **21** (2000) 2381–2399.
134. Ruge, J. and Stüben, K., Algebraic multigrid, in: *Multigrid Methods* (McCormick, S. F., ed.), SIAM, Philadelphia, 1987, pp. 73–130.
135. Sandak, B. and Brandt, A., Multiscale fast summation of long range charge and dipolar interactions, in [39], pp. 6–31. *J. Comp. Chem.*, in press.
136. Schmittmann, B. and Zia, R. K. P., *Statistical Mechanics of Drivan Diffusive Systems*, Academic Press, 1995.
137. Schröder, J., Trottenberg, U. and Reuterssberg, H., Reuktionsverfahren fuer differenzengleichungen bei randwertaufgaben II, *Num. Math.* **26** (1976) 429–459.
138. Sharatchandra, H., Thun, H. and Weisz, P., Susskind fermions on a Euclidean lattice, *Nuc. Phys.* **B192** (1981) 205.
139. Sharon, E., Brandt, A. and Basri, R., Completion Energies and Scale, *IEEE Conf. on Computer Vision and Pattern Recognition* (CVPR–97), Puerto Rico, 1997. Report CS97–19, Weizmann Institute of Science, Rehovot. *IEEE Trans. on Pattern Anal. and Machine Intelligence* **22** (2000) 1117–1131.
140. Sharon, E., Brandt, A. and Basri, R., Fast multiscale image segmentation, *IEEE Conf. on Computer Vision and pattern recognition*, South Carolina, 2000, pp. 70–77.
141. Shmulyian, F., *Multigrid Algorithms for an Inverse Problem in Impedance Tomography*, M.Sc. Thesis, Weizmann Institute of Science, Rehovot, August 1995.
142. Shmulyian, S., *Toward Optimal multigrid Monte Carlo Computations in Two-Dimensional $O(N)$ Non-Linear σ-models*, Ph.D. Thesis, Weizmann Institute of Science, Rehovot, March 1999.
143. Smith, G. D., Paul, W., Yoon, D. Y. and Richter, D., Microscopic Dynamics in a Long-chain Alkane Melt: Computer Simulation and Neutron Scattering, Report, Department of Chemical Engineering, University of Missouri at Columbia, July 1996.
144. Sonneveld, P., Wesseling, P. and de Zeeuw, P. M., Multigrid and conjugate gradient methods as convergence acceleration techniques, *Proc. Multigrid Conf.* (Holstein, H. and Paddon, D., eds.), Bristol, 1983.
145. South, J. C. and Brandt, A., Application of a multi-level grid method to transonic flow calculations, in: *Transonic Flow Problems in Turbo Machinery* (Adam, T. C. and Platzer, M. F., eds.), Hemisphere, Washington, 1977, pp. 180–207.
146. Stüben, K., Algebraic multigrid (AMG): An introduction with applications, Guest appendix in [153]. A review of algebraic multigrid, *J. Comp. Appl. Math.* **128**(1–2), 2001.

147. Susskind, L., Lattice fermions, *Phys. Rev.* **D16** (1977) 3031–3039.
148. Swendsen, R. H., Monte Carlo calculation of renormalized coupling parameters, *Phys. Rev. Lett.* **52** (1984) 1165.
149. Sylvester, J. and Uhlmann, G., A global uniqueness theorem for an inverse boundary value problem, *Ann. Math.* **125** (1987) 153–169.
150. Ta'asan, S., One-shot methods for optimal control of distributed parameter systems, I. Finite dimensional control, ICASE Report 91–2, 1991.
151. Thomas, J. L., Diskin, B. and Brandt, A., Textbook multigrid efficiency for the incompressible Navier-Stokes equations: High Reynolds number wakes and boundary layers, ICASE Report No. 99–51, 1999. *Computers and Fluids*, to appear.
152. Thomas, J. L., Diskin, B., Brandt, A. and South, J. C. Jr., General framework for achieving textbook multigrid efficiency: quasi-1-D Euler example, in: *Frontiers of Computational Fluid Dynamics — 2000* (Caughey, D. A. and Hafez, M. M., eds.), World Scientific Publishing Company, Singapore.
153. Trottenberg, U., Oosterlee, C. W. and Schüller, A., *Multigrid*, Academic Press, London, 2000.
154. Tziperman, E., Yavneh, I. and Ta'asan, S., Multilevel turbulence simulations, *Europhysics Lett.* **24**(4) (1993) 239.
155. Van der Wees, A. J., Van der Vooren, J. and Meelker, J. H., Robust calculation of 3D transonic potential flow based on the nonlinear FAS multigrid method and incomplete LU-decomposition, AIAA Paper 83-1950, 1983.
156. Vandewalle, S. and Horton, G., Fourier mode analysis of the waveform relaxation and time-parallel multigrid methods, *SIAM J. Sci. Comput.* (1995).
157. Van Leer, B., Tai, C. H. and Powell, K. G., Design of optimally-smoothing multi-stage schemes for the Euler equations, AIAA Paper 89-1933, June 1989.
158. Venner, C. H. and Lubrecht, A. A., *Multilevel Methods in Lubrication*, Elsevier (Tribology Series, 37), Amsterdam, 2000.
159. Wesseling, P. and Sonneveld, P., Numerical experiments with a multiple grid and a preconditioned Lanczos type method, in: *Approximation Methods for Navier-Stokes Problems* (Rautmann, R., ed.), Lecture Notes in Math. **771**, Springer-Verlag, 1980, pp. 543–562.
160. Whitham, G. B., *Linear and Nonlinear Waves*, Wiley-Interscience, New York, 1974.
161. Williams, L. R. and Jacobs, D. W., Stochastic completion fields: A neural model of illusory contour shape and salience, *ICCV-95*, 1995, pp. 408–415.
162. Wilson, K. G., Confinement of quarks, *Phys. Rev.* **D10** (1974) 2445–2459.
163. Yavneh, I., A method for devising efficient multigrid smoothers for complicated PDE systems, *SIAM J. Sci. Comput.* **14** (1993), 1437–1463.
164. Yavneh, I. and McWilliams, J. C., Breakdown of the slow manifold in the shallow-water equations, *Geo. Astr. Fluid Dyn.* **75** (1994) 131–161.
165. Yavneh, I. and McWilliams, J. C., Robust multigrid solution of the shallow-water balance equations, *J. Comput. Phys.* **119** (1995) 1–25.
166. Yavneh, I., Venner, C. H. and Brandt, A., Fast multigrid solution of the advection problem with closed characteristics, *SIAM J. Sci. Comput.* **19** (1998) 111–125.
167. Zimare, H., *Beweis der sternenkonvergenz und untersuchung der stabilitaet beim verfahren der totalen reduktion*, Ph.D. Thesis, University of Cologne, 1980.

Wavelet-Based Numerical Homogenization with Applications

Björn Engquist[1] and Olof Runborg[2]

[1] University of California, Los Angeles, Department of Mathematics, Los Angeles, CA, 90095-1555 and Royal Institute of Technology, Department of Numerical Analysis and Computer Science, 100 44, Stockholm

[2] Princeton University, PACM, Department of Mathematics, Princeton, NJ, 08544-1000

Abstract. Classical homogenization is an analytic technique for approximating multiscale differential equations. The numbers of scales are reduced and the resulting equations are easier to analyze or numerically approximate. The class of problems that classical homogenization applies to is quite restricted. We shall describe a numerical procedure for homogenization, which starts from a discretization of the multiscale differential equation. In this procedure the discrete operator is represented in a wavelet space and projected onto a coarser subspace. The wavelet homogenization applies to a wider class of problems than classical homogenization. The projection procedure is general and we give a presentation of a framework in Hilbert space, which also applies to the differential equation directly. The wavelet based homogenization technique is applied to discretizations of the Helmholtz equation. In one problem from electromagnetic compatibility a subgrid scale geometrical detail is represented on a coarser grid. In another a wave-guide filter is efficiently approximated in a lower dimension. The technique is also applied to the derivation of effective equations for a nonlinear problem and to the derivation of coarse grid operators in multigrid. These multigrid methods work very well for equations with highly oscillatory or discontinuous coefficients.

1 Introduction

In the numerical simulation of partial differential equations, the existence of subgrid scale phenomena poses considerable difficulties. With subgrid scale phenomena, we mean those processes which could influence the solution on the computational grid but which have length scales shorter than the grid size. Fine scales in the initial values may, for example, interact with fine scales in the material properties and produce coarse scale contributions to the solution.

There are traditional ways to deal with this problem. A number of methods exist, which are based on physical considerations for a specific application, such as turbulence models in computational fluid dynamics, [44], and analytically derived local subcell models in computational electromagnetics, [42]. Geometrical optics or geometrical theory of diffraction approximations of high frequency wave propagation are other classical techniques to overcome the difficulty of highly oscillatory solutions, [28].

All these techniques result in new sets of approximative equations that do not contain the small scales, but which anyway attempt to take the effect of these scales into account. A more general analytical technique for achieving this goal is classical homogenization, which will be discussed below.

If the small scales are localized, there are some numerical procedures which are applicable. Local mesh refinement is quite common but could be costly if the small scales are very small or distributed. There are also problems with artificial reflections in mesh size discontinuities and time step limitations for explicit techniques. Numerical shock tracking or shock fitting can also be seen as subgrid models, [1]. Material interfaces can be handled by grid adaptation or the immersed interface method, [32].

In this paper we present a general procedure for constructing subgrid models to be used on a coarse grid where the smallest scales are not resolved. The objective is to find models that accurately reproduce the effect of subgrid scales and that in some sense are similar to the original differential operator. The starting point is a finite-dimensional approximation of a differential equation. Let L_ε be a linear differential operator for which ε indicates small scales in the coefficients. The solution u_ε of the differential equation

$$L_\varepsilon u_\varepsilon = f_\varepsilon, \tag{1}$$

will typically inherit the small scales from the operator L_ε or the data f_ε. Consider for example the simple model problem

$$L_\varepsilon u_\varepsilon = -\frac{d}{dx}\left(g^\varepsilon(x)\frac{d}{dx}\right)u_\varepsilon(x) = f(x), \qquad 0 < x < 1, \tag{2}$$

$$u_\varepsilon(0) = u_\varepsilon(1) = 0. \tag{3}$$

The coefficient $g^\varepsilon(x)$ may be oscillatory,

$$g^\varepsilon(x) = g(x/\varepsilon), \qquad g(y) \text{ 1-periodic}, \tag{4}$$

or have a localized sharp transition,

$$g^\varepsilon(x) = \begin{cases} g_1, & x < (1-\varepsilon)/2, \\ g_2, & (1-\varepsilon)/2 \le x < (1+\varepsilon)/2, \\ g_1, & x \ge (1-\varepsilon)/2. \end{cases} \tag{5}$$

Our goal is to find an accurate and computationally efficient finite dimensional approximation of (1),

$$\bar{L}_{\varepsilon\Delta}\bar{u}_{\varepsilon\Delta} = \bar{f}_{\varepsilon\Delta}. \tag{6}$$

Difficulties originate from the small scales or high frequencies in (1). If ε denotes a typical wave length in u_ε and h a typical element size or step size in (6), then h must be substantially smaller than ε in order to resolve the ε-scale in the numerical approximation. This can be costly if ε is small compared to the overall size of the computational domain. Ideally the discretized operator in (6) should be designed such that $\bar{u}_{\varepsilon\Delta}$ is a good approximation of u_ε even if h is not small compared to ε. This goal resembles that of classical analytical homogenization.

1.1 Classical Homogenization

Homogenization is a well established analytical technique to approximate the effect of smaller scales onto larger scales in multiscale differential equations. There are explicit formulas for some classes of problems with periodic or stochastic coefficients. The problem is often formulated as follows. Consider a set of operators L_ε in (1), indexed by the small parameter ε. Find the limit solution $\bar{u}$ and the *homogenized operator* $\bar{L}$ defined by

$$L_\varepsilon u_\varepsilon = f, \qquad \lim_{\varepsilon \to 0} u_\varepsilon = \bar{u}, \qquad \bar{L}\bar{u} = f. \tag{7}$$

In certain cases the convergence above and existence of the homogenized operator can be proved, [5].

For simple model problems, with coefficients that are periodic on the fine scale, exact closed form solutions can be obtained. For instance, with $g(y)$ positive, 1-periodic and bounded away from zero, we have for the one-dimensional example (2),

$$L_\varepsilon = -\frac{d}{dx}\left(g(x/\varepsilon)\frac{d}{dx}\right), \qquad \bar{L} = -\hat{g}\frac{d^2}{dx^2}, \qquad \hat{g} = \left(\int_0^1 \frac{dy}{g(y)}\right)^{-1}. \tag{8}$$

With the same $\hat{g}$ we get for the hyperbolic operators,

$$L_\varepsilon = \frac{\partial}{\partial t} + g(x/\varepsilon)\frac{\partial}{\partial x}, \qquad \bar{L} = \frac{\partial}{\partial t} + \hat{g}\frac{\partial}{\partial x}. \tag{9}$$

These model examples are used in Sect. 4.1 and Sect. 4.2.

In higher dimensions, the solution to (7) is a little more complicated, although for some model problems the structure of the homogenized operator can still be written down, as long as the coefficients are periodic or stochastic. In the d-dimensional elliptic case, let $G(y) : \mathbb{R}^d \mapsto \mathbb{R}^{d\times d}$ be uniformly elliptic and q-periodic in each of its arguments. Let I_d denote the unit square. It can then be shown, [5], that

$$L_\varepsilon = -\nabla \cdot \left(G\left(\frac{x}{\varepsilon}\right)\nabla\right), \quad \bar{L} = -\nabla \cdot (\hat{G}\nabla), \quad \hat{G} = \int_{I_d} G(y) - G(y)\frac{d\chi(y)}{dy}dy, \tag{10}$$

where $d\chi/dy$ is the jacobian of the function $\chi(y) : \mathbb{R}^d \mapsto \mathbb{R}^d$, given by solving the so called cell problem,

$$\nabla \cdot G(y)\frac{d\chi(y)}{dy} = \nabla \cdot G(y), \qquad y \in I_d, \tag{11}$$

with periodic boundary conditions for χ. Note that $\hat{G}$ is a constant matrix.

1.2 Numerical Homogenization

Classical homogenization is very useful when it is applicable. The original problem with small scales is reduced to a homogenized problem that is much easier to approximate numerically. See the left path in Fig. 1. The subscript Δ indicates numerical approximation. If analytical homogenization is not possible, numerical homogenization should be applied in order to arrive at a method with reasonable computational complexity. With numerical homogenization, we mean a technique to produce a discrete operator $\bar{L}_{\varepsilon\Delta}$ such that $||u_\varepsilon - \bar{u}_{\varepsilon\Delta}||$ is small and where the complexity of solving $\bar{L}_{\varepsilon\Delta}\bar{u}_{\varepsilon\Delta} = \bar{f}_{\varepsilon\Delta}$ does not grow fast with ε^{-1}.

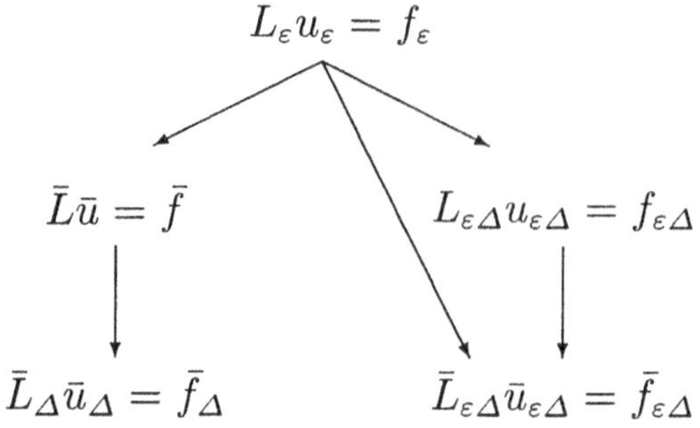

Fig. 1. Schematic steps in homogenization

The numerical homogenization can be done directly as indicated by the middle path in Fig. 1, or by first discretizing the original problem and then compressing the operator $L_{\varepsilon\Delta}$ and the data $f_{\varepsilon\Delta}$ as indicated by the right path. In this paper we shall present wavelet based methods in order to achieve numerical homogenization. We will consider the case of compressing a discretized problem but the technique could also be used for direct discretization.

There are other similar methods based on coarsening techniques from algebraic multigrid, see for example, Knapek [29] and Neuss et. al. [35,34]. In the finite element setting there are a few successful methods for numerical homogenization. In the work of Hughes and collaborators [26,25], Hou and collaborators [16,24] and also in Matache and Schwab [33], the effect of the microstructure is incorporated in a Galerkin framework.

1.3 Wavelet Projections

Given the full discrete operator on a fine grid, $L_{\varepsilon\Delta}$ in Fig. 1, we wish to find an operator of lower dimension that extracts only the coarse scales of the solution. Let V_j and W_j refer to the usual scaling and wavelet spaces of some orthogonal wavelet system. Then, for a solution in $V_{j+1} = V_j \oplus W_j$, the coarse scale is represented by V_j, and we are thus interested in the operator that yields the solution's projection onto V_j. We shall here denote the finite dimensional operator $L_{\varepsilon\Delta}$ by L_{j+1} in order to indicate scale.

Consider the equation

$$L_{j+1}U = F, \qquad U, F \in V_{j+1}, \tag{12}$$

originating from a discretization of a differential equation (1), where U approximates the solution to the continuous problem. We introduce the orthogonal transformation

$$\mathcal{W}_j : V_{j+1} \to W_j \times V_j, \quad \mathcal{W}_j U \equiv \begin{pmatrix} U_\mathrm{f} \\ U_\mathrm{c} \end{pmatrix} \qquad U_\mathrm{f} \in W_j, \qquad U_\mathrm{c} \in V_j, \tag{13}$$

and note that the linear operator $\mathcal{W}_j L_{j+1} \mathcal{W}_j^*$ can be decomposed into four operators acting between the subspaces V_j and W_j, such that (12) becomes

$$\begin{pmatrix} A_j & B_j \\ C_j & L_j \end{pmatrix} \begin{pmatrix} U_{\mathrm{f}} \\ U_{\mathrm{c}} \end{pmatrix} = \begin{pmatrix} F_{\mathrm{f}} \\ F_{\mathrm{c}} \end{pmatrix}, \qquad U_{\mathrm{f}}, F_{\mathrm{f}} \in W_j, \qquad U_{\mathrm{c}}, F_{\mathrm{c}} \in V_j. \tag{14}$$

when we apply $\mathcal{W}_j$ from the left. Block Gaussian elimination now gives an equation for U_{c}, the coarse part of the solution,

$$\bar{L}_j U_{\mathrm{c}} = \bar{F}_j, \quad \bar{L}_j = L_j - C_j A_j^{-1} B_j, \quad \bar{F}_j = F_{\mathrm{c}} - C_j A_j^{-1} F_{\mathrm{f}}. \tag{15}$$

Hence, our new "coarse grid operator" $\bar{L}_j$ is the Schur complement of $\mathcal{W}_j L_{j+1} \mathcal{W}_j^*$. We also get the homogenized right hand side, $\bar{F}_j$.

For higher dimensions, a standard tensor product extension of the multiresolution analysis allows us to use essentially the same derivation as above to obtain coarse grid operators.

Before we go on, we should note that in general $\bar{L}_j$ will not be sparse even if L_{j+1} is, because of the inverse A_j^{-1}. For the method to be efficient we must be able to approximate $\bar{L}_j$ with a *sparse* matrix. In Sect. 3.3 below we will get back to this issue and show in subsequent sections that this can indeed be done. The homogenization procedure can be applied recursively on $\bar{L}_j$ to get $\bar{L}_{j-1}$ and so on.

The great advantage of this procedure in deriving subgrid models is its generality. It can be used on any system of differential equations and does not require separation into the distinct $\mathcal{O}(\varepsilon)$ and $\mathcal{O}(1)$ scales or periodic coefficients. It can also be used to test if it is physically reasonable to represent the effect of fine scales on a coarse scale grid with a local operator.

This work is based on [2,39] and is a continuation of the work by Dorobantu and Engquist, [14]. The original ideas are from Beylkin and Brewster, [6]. See also [21] for analysis in the one-dimensional case.

2 Background Theory

In this section we shall present some background material that is helpful in understanding the ideas behind the wavelet homogenization. Most of the material is well known or has at least appeared in the literature. However, the sections on projection generated homogenization and how it relates to classical homogenization are new.

2.1 Direct Discretization

Let us first consider the simple approach of using a coarse grid even if not all scales of the original differential equation are clearly resolved. Ideally the following result should be valid,

$$\lim_{h \to 0} \big(\sup_{\varepsilon < \varepsilon_0} \| u_\varepsilon - u_{\varepsilon, \Delta} \| \big) = 0, \tag{16}$$

where h is a measure of the element size of or step size in the discretization Δ. A convergence result of the form (16) is rarely valid. One example of a problem is the typical error in phase velocity, which most finite element and finite difference methods produce. If h is not small compared to ε, oscillatory parts of a solution will propagate with $\mathcal{O}(1)$ errors. There are, however, a few problems for which a weaker version of (16) is valid.

For solutions which are highly oscillatory relative to the grid discretization, numerical techniques without phase velocity errors are needed. In [17,18] particle schemes or method of characteristics approximations of hyperbolic partial differential equations are analyzed. For a restricted class of schemes it is possible to prove convergence, or weak convergence, in L^p of the numerical approximation to the analytic solution as $h \to 0$ essentially independent of ε. *Convergence essentially independent of* ε means that a set of ratios of h/ε with an arbitrary small Lebesgue measure must be excluded to avoid resonance, see [17,18].

One simple but typical example for which a rigorous theory is possible is the method of characteristics for the Carleman equations,

$$\begin{aligned} &\tfrac{\partial u}{\partial t} + \tfrac{\partial u}{\partial x} + u^2 - v^2 = 0, \qquad && u(x,0) = a(x, x/\varepsilon), \\ &\tfrac{\partial v}{\partial t} - \tfrac{\partial v}{\partial x} + v^2 - u^2 = 0, \qquad && v(x,0) = b(x, x/\varepsilon), \end{aligned} \tag{17}$$

where $a(x,y)$ and $b(x,y)$ are 1-periodic in y and the numerical approximation is given by

$$\begin{aligned} u_j^{n+1} &= u_{j-1}^n + \Delta t\big[(v_{j-1}^n)^2 - (u_{j-1}^n)^2\big], \qquad & u_j^0 &= a(x_j, x_j/\varepsilon), \\ v_j^{n+1} &= v_{j+1}^n + \Delta t\big[(u_{j+1}^n)^2 - (v_{j+1}^n)^2\big], \qquad & v_j^0 &= b(x_j, x_j/\varepsilon), \end{aligned} \tag{18}$$

with

$$x_j = j\Delta x, \qquad t_n = n\Delta t, \qquad \Delta t = \Delta x, \qquad u(x_j, t_n) \sim u_j^n. \tag{19}$$

The homogenization theory of Tartar, [43], applies to the differential equations (17) and is also used in the convergence proof. The local truncation errors are large for $h > \varepsilon$ and a cancelation of the errors must be established. The theorem gives strong convergence in L^∞ essentially independent of ε as $h \to 0$.

The wavelet based type of homogenization was derived in order to handle wider classes of differential equations.

2.2 Comments on Classical Homogenization

Classical homogenization was briefly described in Sect. 1.1. We shall give a few more results here.

The homogenized coefficient in (10),

$$\hat{G} = \int_{I_d} G(y) - G(y)\frac{d\chi(y)}{dy}dy, \tag{20}$$

consists of two parts. One corresponds to the arithmetic average of the coefficients $G(y)$: $\int_{I_d} G(y)dy$. The other,

$$\int_{I_d} G(y)\frac{d\chi(y)}{dy}dy, \tag{21}$$

compensates for the effect the high frequency interaction has on the lower frequencies in the solution. The high frequency interaction originates from the product

$G(x/\varepsilon)\nabla u_\varepsilon$. We have the analogous decomposition of the numerically homogenized operator, see Sect. 2.4.

A rigorous convergence theory can be established for analytic homogenization, [5], and a typical result for the multidimensional problem in Sect. 1.1 with Dirichlet boundary conditions would be,

$$u_\varepsilon \rightharpoonup \bar{u}, \qquad \text{weakly in } H_0^1(\Omega), \tag{22}$$

if $G \in L^\infty(\Omega)$ is 1-periodic,

$$u^T G(y) u \geq \delta |u|^2, \qquad \delta > 0, \qquad \text{a.e. in } I_d, \tag{23}$$

and $\Omega \subset \mathbb{R}^d$ is a bounded open domain.

Analytic homogenization extends to coefficients of the form $G(x, x/\varepsilon)$. In this case the cell problem with the coefficients $G(x, y)$ has to be solved for each $x \in \Omega$. See [15] for a practical application in the field of oil reservoir modeling.

The derivation of the homogenized operator is traditionally done from an asymptotical assumption on the solution, [5],

$$u_\varepsilon(x) = \bar{u}(x) + \varepsilon u_1(x, x/\varepsilon) + \varepsilon^2 u_2(x, x/\varepsilon) + \mathcal{O}(\varepsilon^3), \tag{24}$$

where u_1 and u_2 are 1-periodic in its second set of arguments. The variables for the different scales x and $y = x/\varepsilon$, are treated independently in the derivation.

The techniques discussed here extends to other cases, including multiple scales where $G = G(x, x/\varepsilon_1, x/\varepsilon_2, \dots)$. The homogenization is then applied recursively starting from the smallest scale. However, there has to be a set of distinct scales.

2.3 Projection Generated Homogenization

In this section we describe the following approach to homogenization. Consider an equation $Lu = f$ where L is a linear operator, f a right hand side and u a solution that contains fine scales. Let P be a projection operator onto a subspace where the fine scales in the original solution do not exist. Our objective is to find the (projection generated) *homogenized* operator $\bar{L}$ such that $\bar{L}Pu = f$ for all f such that $Pf = f$. (When $Pf \neq f$ we also need to find the homogenized right hand side $\bar{f}$.) Here, we confine ourselves to the case of Hilbert spaces.

Let X be a Hilbert space of functions, typically a Sobolev space. Let $X_0 \subset X$ be a closed subspace representing the coarse part of the functions, and denote by P the orthogonal (and symmetric) projection operator in X onto X_0. Let the spaces X_0 and $X_0^\perp$ inherit the innerproduct and norm of X, so that $\|u\|_X = \|u\|_{X_0}$ when $u \in X_0$, and similar for $X_0^\perp$. In addition, set $Q = I - P$ where I is the identity operator in X, and introduce the unitary operator $\mathcal{W}$ on X defined by

$$\mathcal{W} : X \mapsto X_0 \times X_0^\perp, \qquad \mathcal{W}u = \begin{pmatrix} Qu \\ Pu \end{pmatrix}. \tag{25}$$

Let $\mathcal{L}(X, Y)$ be the set of bounded linear maps from X to Y. In the numerical finite dimensional case $L \in \mathcal{L}(X, Y)$ with $X = Y = \mathbb{R}^n$. On order to simplify the presentation we shall consider $X = Y$ also in infinite dimensions. This means that partial differential operators will be defined by a special weak formulation, see Remark 1 in Sect. 2.4. For an arbitrary operator $L \in \mathcal{L}(X, X)$, we have

$$\mathcal{W}L\mathcal{W}^* \begin{pmatrix} u \\ v \end{pmatrix} = \mathcal{W}L(P+Q)(u+v) = \begin{pmatrix} QL(P+Q)(u+v) \\ PL(P+Q)(u+v) \end{pmatrix}$$
$$= \begin{pmatrix} QL(Pv+Qu) \\ PL(Pv+Qu) \end{pmatrix} \equiv \begin{pmatrix} A & B \\ C & D \end{pmatrix} \begin{pmatrix} u \\ v \end{pmatrix} \tag{26}$$

where

$$\begin{aligned} A &= QLQ \in \mathcal{L}(X_0^\perp, X_0^\perp), & B &= QLP \in \mathcal{L}(X_0, X_0^\perp), \\ C &= PLQ \in \mathcal{L}(X_0^\perp, X_0), & D &= PLP \in \mathcal{L}(X_0, X_0). \end{aligned} \tag{27}$$

When A is invertible the following definition can be stated:

Definition 1. Suppose $L \in \mathcal{L}(X, X)$ and $f \in X$. When A in (26, 27) is invertible (one-to-one and onto), we define the *homogenized* operator $\bar{L} : X_0 \mapsto X_0$ as the Schur complement with respect to the decomposition in (26),

$$\bar{L} = D - CA^{-1}B, \tag{28}$$

and the homogenized right hand side as

$$\bar{f} = Pf - CA^{-1}Qf. \tag{29}$$

We will write $\bar{L}_{X,X_0}$ and $\bar{f}_{X,X_0}$ when there is a need to show explicitly between which spaces the homogenization step is made. From Definition 1 we immediately have

Lemma 1. *Suppose $Lu = f$, where $L \in \mathcal{L}(X, X)$, $u \in X$ and $f \in X$. If A^{-1} exists,*

$$\bar{L}Pu = \bar{f}. \tag{30}$$

Proof. Since $Lu = f$ we get

$$\mathcal{W}L\mathcal{W}^*\mathcal{W}u = \mathcal{W}f \quad \Rightarrow \quad \begin{pmatrix} A & B \\ C & D \end{pmatrix} \begin{pmatrix} Qu \\ Pu \end{pmatrix} = \begin{pmatrix} Qf \\ Pf \end{pmatrix}. \tag{31}$$

Moreover, since A is invertible, this system can be reduced with Gaussian elimination. It yields (30).

Note that $\bar{L}$ and $\bar{f}$ do not exist for all elements of $\mathcal{L}(X, X)$ since A^{-1} may not exist. For some classes of operators A is however indeed invertible.

Proposition 1. *Suppose the operator $L \in \mathcal{L}(X, X)$ is given by the weak form*

$$Lu = f \quad \Leftrightarrow \quad a[u, v] = \langle f, v \rangle, \quad \forall v \in X, \tag{32}$$

where $a[\cdot, \cdot] : X \times X \mapsto \mathbb{R}$ is a bilinear form. If for some positive constants c_1 and c_2,

$$\begin{aligned} a[u, v] &\leq c_1 \|u\|_X \|v\|_X, & \forall u, v \in X, \\ a[Qu, Qu] &\geq c_2 \|Qu\|_X^2, & \forall u \in X, \end{aligned} \tag{33}$$

then L is well-defined and $A : X_0^\perp \mapsto X_0^\perp$ is one-to-one and onto.

Proof. By the first inequality in (33), the bilinear from $a[u,v]$ is a bounded linear functional on X for each fixed u. L is then well-defined by the Riesz representation theorem. To show that A is invertible, we need to prove that for every $f \in X_0^\perp$ there is a unique $u \in X_0^\perp$ such that

$$a[u,v] = \langle f, v\rangle, \qquad \forall v \in X_0^\perp. \tag{34}$$

This follows from the Lax–Milgram lemma if (33) holds.

Some properties kept by the homogenized operator are surjectivity, symmetry and ellipticity, as seen by this proposition.

Proposition 2. *Let $L \in \mathcal{L}(X,X)$ be such that $\bar{L}$ exists. If L is onto, $\bar{L}$ is onto. If L is self-adjoint, $\bar{L}$ is self-adjoint. If L is self-adjoint and also*

$$c_1\|u\|_X^2 \le \langle Lu, u\rangle \le c_2\|u\|_X^2, \qquad \forall u \in X. \tag{35}$$

for some positive constants c_1, c_2. Then, for the same constants,

$$c_1\|v\|_{X_0}^2 \le \langle \bar{L}v, v\rangle \le \langle Dv, v\rangle \le c_2\|v\|_{X_0}^2, \qquad \forall v \in X_0. \tag{36}$$

Proof. Suppose L is onto. For any $f \in X_0 \subset X$ there exists a $u \in X$ such that $Lu = f$. By Lemma 1 then $\bar{L}Pu = \bar{f}$. From (29) we see that $\bar{f} = f$, so $\bar{L}$ is also onto. Let L be self-adjoint. Then $\mathcal{W}L\mathcal{W}^*$ is self-adjoint and therefore $A = A^*$, $D = D^*$ and $C = B^*$. When A is invertible this implies that $(A^{-1})^* = A^{-1}$. Hence, $\bar{L}^* = D^* - B^*(A^{-1})^*C^* = D - CA^{-1}B = \bar{L}$.

Next, assume (35). Take $v \in X_0$ and set $u = -A^{-1}Bv \in X_0^\perp$. Since $\mathcal{W}$ is unitary,

$$\begin{aligned} c_1\|v\|_{X_0}^2 &\le c_1\|v+u\|_X^2 \le \langle L(u+v),(u+v)\rangle = \left\langle \begin{pmatrix} A & B \\ C & D\end{pmatrix}\begin{pmatrix} u \\ v\end{pmatrix}, \begin{pmatrix} u \\ v\end{pmatrix}\right\rangle \\ &= \left\langle \begin{pmatrix} 0 \\ (-CA^{-1}B + D)v\end{pmatrix}, \begin{pmatrix} -A^{-1}Bv \\ v\end{pmatrix}\right\rangle = \langle \bar{L}v, v\rangle. \end{aligned} \tag{37}$$

This proves the left inequality in (36). Since $C = B^*$ and A^{-1} is positive by the positivity of L, we have for the right inequality

$$\begin{aligned} \langle \bar{L}v, v\rangle &= \langle (D - CA^{-1}B)v, v\rangle = \langle Dv, v\rangle - \langle A^{-1}Bv, Bv\rangle \\ &\le \langle Dv, v\rangle = \langle Lv, v\rangle \le c_2\|v\|_X^2 = c_2\|v\|_{X_0}^2. \end{aligned} \tag{38}$$

We conclude this section by showing that the homogenization is "independent of the path."

Proposition 3. *Let X_0 and X_1 be two closed subspaces of X, nested such that $X_0 \subset X_1 \subset X$. Suppose $L \in \mathcal{L}(X,X)$ is onto and that $\bar{L}_{X,X_0}$, $\bar{L}_{X,X_1}$ and $\overline{\bar{L}_{X,X_1}}_{X_1,X_0}$ all exist. If $\bar{L}_{X,X_0}$ is one-to-one, then*

$$\bar{L}_{X,X_0} = \overline{\bar{L}_{X,X_1}}_{X_1,X_0}, \tag{39}$$

$$\bar{f}_{X,X_0} = \overline{\bar{f}_{X,X_1}}_{X_1,X_0}. \tag{40}$$

Proof. Let P_j be the projection onto X_j. Take $u \in X_0$ and let $f = \bar{L}_{X,X_0} u \in X_0$. Since L is onto, there is a $w \in X$ such that $Lw = f$. By Lemma 1, we have

$$\bar{L}_{X,X_0} P_0 w = \bar{f}_{X,X_0}, \qquad \bar{L}_{X,X_1} P_1 w = \bar{f}_{X,X_1}. \tag{41}$$

Since $f \in X_0$ we must have $\bar{f}_{X,X_0} = \bar{f}_{X,X_1} = f$, by (29). Another application of Lemma 1, then gives

$$\overline{\bar{L}_{X,X_1}}_{X_1,X_0} P_0 P_1 w = \bar{f}_{X_1,X_0}. \tag{42}$$

Again, $\bar{f}_{X_1,X_0} = f$ and $P_0 P_1 w = P_0 w$ since $X_0 \subset X_1$. Hence,

$$\bar{L}_{X,X_0} P_0 w = \overline{\bar{L}_{X,X_1}}_{X_1,X_0} P_0 w = f = \bar{L}_{X,X_0} u. \tag{43}$$

But, $\bar{L}_{X,X_0}$ is one-to-one so $P_0 w = u$ and since u was arbitrary, (39) follows. Finally, letting f be an arbitrary element of X, there exists a u such that $Lu = f$ since L is onto and by Lemma 1

$$\bar{f}_{X,X_0} = \bar{L}_{X,X_0} P u = \overline{\bar{L}_{X,X_1}}_{X_1,X_0} P u = \overline{\bar{f}_{X,X_1}}_{X_1,X_0}. \tag{44}$$

This shows (40).

2.4 Relationship Between the Homogenization Approaches

In the elliptic case, there is a striking similarity between the classical homogenized operator in (10) and the the Schur complement in Definition 1, repeated here for convenience,

$$PLP - PLQ(QLQ)^{-1}QLP, \tag{45}$$

$$\nabla \left(\int_{I_d} G(y) dy \right) \nabla - \nabla \left(\int_{I_d} G(y) \frac{d\chi(y)}{dy} dy \right) \nabla. \tag{46}$$

Both are written as the average of the original operator minus a correction term, which is computed in a similar way for both operators. For the analytical case, a local elliptic cell problem is solved to get $G\partial_y \chi$, while in the projection case, a positive operator $A = QLQ$ defined on a subspace is inverted to obtain $LQA^{-1}B$. The average over the terms is obtained by integration in the analytical case, and by applying P in the projection case.

The relationship can be made more precise, and we will illustrate it by considering the one-dimensional case, with the coarse space given by the lowest Fourier modes. Similar results in one dimension have been shown for wavelet bases, see [21,41].

Consider (2) with $g^\varepsilon = g(x/\varepsilon)$ where $g \in L^\infty(\mathbb{T})$ is 1-periodic. Let $X = H^1(\mathbb{T})$ and write $\langle \cdot, \cdot \rangle$ for the H^1 innerproduct. Define the operator $L_\varepsilon \in \mathcal{L}(X, X)$ by the weak form of the equations, such that $L_\varepsilon u_\varepsilon = f$ is equivalent to

$$a[u, v] := \langle g(x/\varepsilon) \partial_x u_\varepsilon, \partial_x v \rangle_{L^2} = \langle f, v \rangle, \qquad \forall v \in H^1(\mathbb{T}). \tag{47}$$

We want to compare this with the classically homogenized operator,

$$\hat{g} \langle \partial_x \bar{u}, \partial_x v \rangle_{L^2} = \langle f, v \rangle, \qquad \forall v \in H^1(\mathbb{T}), \tag{48}$$

with $\hat{g}$ given in (8).

We start by introducing a family of ε-dependent spaces X_0^ε containing bandlimited functions,

$$X_0^\varepsilon = \left\{ u \in L^2(\mathbb{T}) \;\middle|\; u = \sum_{|k|<1/2\varepsilon} a_k e^{2\pi i k x},\ a_k \in \mathbb{C} \right\}. \tag{49}$$

Those finite-dimensional spaces are closed subspaces of both $L^2(\mathbb{T})$ and $H^1(\mathbb{T})$. The corresponding orthogonal projections are

$$P_\varepsilon u = \sum_{|k|<1/2\varepsilon} \hat{u}_k e^{2\pi i k x}, \qquad \hat{u}_k = \int_0^1 u(x) e^{-2\pi i k x} dx, \tag{50}$$

where $\hat{u}_k$ is the kth Fourier coefficient of $u(x)$.

Let us index the remaining operators by ε in the obvious way, setting $Q_\varepsilon = I - P_\varepsilon$ and

$$\bar{L}_\varepsilon = D_\varepsilon - C_\varepsilon A_\varepsilon^{-1} B_\varepsilon. \tag{51}$$

The strategy is now to explicitly construct (51) corresponding to (47) for fixed ε and to show that

$$\lim_{\varepsilon \to 0} \bar{L}_\varepsilon P_\varepsilon u = \bar{L} u \qquad \forall u \in H^1(\mathbb{T}), \tag{52}$$

where $\bar{L}$ is the homogenized operator in the classical case, given by (48). Under some additional assumptions we can then show that the solution we get from solving $\bar{L}_\varepsilon \bar{u}_\varepsilon = f$ tends to the homogenized solution $\bar{u}$ strongly in H^1.

Our notation in this section will be as follows. We let $y = x/\varepsilon$ and a y-dependent function seen as a function of x will be indicated by a superscripted epsilon, hence $g(x/\varepsilon) = g^\varepsilon(x)$. A bar above the function denotes its mean value,

$$\bar{g} = \int_0^1 g(x) dx = \hat{g}_0, \tag{53}$$

where $\hat{g}_0$ is the zeroth Fourier coefficient of g. We begin by showing some properties of the projections.

Lemma 2. *The projections P_ε and Q_ε commutes with differentiation,*

$$\partial_x P_\varepsilon u = P_\varepsilon \partial_x u, \qquad \partial_x Q_\varepsilon u = Q_\varepsilon \partial_x u, \qquad \forall u \in H^1(\mathbb{T}). \tag{54}$$

For $g \in L^2(\mathbb{T})$ and $\varepsilon = 1/n$ with $n \in \mathbb{Z}^+$,

$$P_\varepsilon g^\varepsilon P_\varepsilon u = \bar{g} P_\varepsilon u, \qquad \forall u \in L^2(\mathbb{T}). \tag{55}$$

For $u \in L^2(\mathbb{T})$,

$$\overline{P_\varepsilon u} = \bar{u}. \tag{56}$$

Finally,

$$||\partial_x Q_\varepsilon u||_{L^2} \geq \frac{\pi}{\varepsilon} ||Q_\varepsilon u||_{L^2}. \tag{57}$$

Proof. Take $u \in H^1(\mathbb{T})$ and let $\{\hat{u}_k\}$ be its Fourier coefficients. Then

$$\begin{aligned} \partial_x P_\varepsilon u &= \partial_x \sum_{|k|<1/2\varepsilon} \hat{u}_k e^{2\pi i k x} = \sum_{|k|<1/2\varepsilon} 2\pi i k \hat{u}_k e^{2\pi i k x} \\ &= P_\varepsilon \sum_k 2\pi i k \hat{u}_k e^{2\pi i k x} = P_\varepsilon \partial_x u. \end{aligned} \tag{58}$$

Since $Q_\varepsilon = I - P_\varepsilon$, we have shown (54). Moreover,

$$P_\varepsilon g^\varepsilon P_\varepsilon u = P_\varepsilon \sum_\ell \sum_{|k|<n/2} \hat{g}_\ell \hat{u}_k e^{2\pi i(k+n\ell)x} = \hat{g}_0 \sum_{|k|<n/2} \hat{u}_k e^{2\pi i k x} = \bar{g} P_\varepsilon u, \tag{59}$$

since $|k + n\ell| \geq n/2$ when $\ell \neq 0$ and $|k| < n/2$. This proves (55). Next, for $u \in L^2$,

$$\overline{P_\varepsilon u} = \langle P_\varepsilon u, 1\rangle_{L^2} = \sum_{|k|<1/2\varepsilon} \hat{u}_k \langle e^{2\pi i k x}, 1\rangle_{L^2} = \hat{u}_0 = \overline{u}, \tag{60}$$

showing (56). Finally, from (57),

$$\|\partial_x Q_\varepsilon u\|_{L^2}^2 = \sum_{|k|\geq 1/2\varepsilon} 4\pi^2 k^2 \hat{u}_k^2 \geq \frac{\pi^2}{\varepsilon^2} \sum_{|k|\geq 1/2\varepsilon} \hat{u}_k^2 = \frac{\pi^2}{\varepsilon^2} \|Q_\varepsilon u\|_{L^2}^2 . \tag{61}$$

By (57), we now have

$$a[u,v] \leq \|g\|_{L^\infty} \|u\|_{H^1} \|v\|_{H^1} , \tag{62}$$

$$a[Q_\varepsilon u, Q_\varepsilon u] \geq \operatorname{ess\,inf}_{x\in\mathbb{T}} |g(x)|\ \|\partial_x Q_\varepsilon u\|_{L^2}^2 \geq C \operatorname{ess\,inf}_{x\in\mathbb{T}} |g(x)|\ \|Q_\varepsilon u\|_{H^1}^2 , \tag{63}$$

so Proposition 1 shows that A_ε is invertible and $\bar{L}_\varepsilon$ exists for all ε. We can then show

Theorem 1. *The projection generated homogenized operator $\bar{L}_\varepsilon$ corresponding to (47) satisfies*

$$\bar{L}_\varepsilon u = (D_\varepsilon - C_\varepsilon A_\varepsilon^{-1} B_\varepsilon) u = P_\varepsilon \bar{L} u, \qquad \forall u \in X_0^\varepsilon \tag{64}$$

if $\varepsilon = 1/n$ and $n \in \mathbb{Z}^+$.

Proof. By (54) and (55) we have for D_ε,

$$\begin{aligned}\langle D_\varepsilon u, v\rangle &= \langle g^\varepsilon \partial_x P_\varepsilon u, \partial_x P_\varepsilon v\rangle_{L^2} = \langle g^\varepsilon P_\varepsilon \partial_x u, P_\varepsilon \partial_x v\rangle_{L^2} = \langle P_\varepsilon g^\varepsilon P_\varepsilon \partial_x u, \partial_x P_\varepsilon v\rangle_{L^2} \\ &= \bar{g}\langle \partial_x P_\varepsilon u, \partial_x P_\varepsilon v\rangle_{L^2}.\end{aligned} \tag{65}$$

when $u, v \in H^1(\mathbb{T})$. For the second part of the operator, the equation $A_\varepsilon w = B_\varepsilon u$ needs to be solved. Then $C_\varepsilon A_\varepsilon^{-1} B_\varepsilon u$ is given by $C_\varepsilon w$. In this case we can directly write down the expression for w, namely

$$w(x) = \int_0^x \chi_y \left(\frac{x'}{\varepsilon}\right) \partial_x P_\varepsilon u(x') dx'. \tag{66}$$

where χ is the solution to the cell problem (11). It is well-known that $\chi \in H^1(\mathbb{T})$. By also using (54, 55, 56) and the fact that $P_\varepsilon u \in L^\infty(\mathbb{T})$, we have

$$\|w\|_{L^2} \leq |w|_\infty \leq \langle |\chi_y^\varepsilon|, |\partial_x P_\varepsilon u|\rangle_{L^2} \leq C \|\chi\|_{H^1} \|P_\varepsilon u\|_{H^1} , \tag{67}$$

$$\|\partial_x w\|_{L^2} \leq |\partial_x P_\varepsilon u|_\infty \|\chi_y^\varepsilon\|_{L^2} \leq C |\partial_x P_\varepsilon u|_\infty \|\chi\|_{H^1} , \tag{68}$$

$$w(1) = \overline{\chi_y^\varepsilon \partial_x P_\varepsilon u} = \overline{P_\varepsilon \chi_y^\varepsilon P_\varepsilon \partial_x u} = 0 = w(0). \tag{69}$$

Therefore $w \in H^1(\mathbb{T})$. Moreover, by (54) and (55)

$$\partial_x Q_\varepsilon w = \partial_x w - P_\varepsilon \partial_x w = \partial_x w - P_\varepsilon \chi_y^\varepsilon P_\varepsilon \partial_x u = \partial_x w. \tag{70}$$

Now, from (11) it follows that $g^\varepsilon \chi_y^\varepsilon = g^\varepsilon + c_0$ a.e., where c_0 is a constant. Together with (70) this shows that w indeed satisfies $A_\varepsilon w = B_\varepsilon u$,

$$\begin{aligned}\langle A_\varepsilon w, v\rangle &= \langle g^\varepsilon \partial_x Q_\varepsilon w, \partial_x Q_\varepsilon v\rangle_{L^2} = \langle g^\varepsilon \chi_y^\varepsilon \partial_x P_\varepsilon u, \partial_x Q_\varepsilon v\rangle_{L^2} \\ &= \langle g^\varepsilon \partial_x P_\varepsilon u, \partial_x Q_\varepsilon v\rangle_{L^2} + c_0 \langle \partial_x P_\varepsilon u, \partial_x Q_\varepsilon v\rangle_{L^2} \\ &= \langle B_\varepsilon u, v\rangle + c_0 \langle P_\varepsilon \partial_x u, Q_\varepsilon \partial_x v\rangle_{L^2} = \langle B_\varepsilon u, v\rangle. \end{aligned} \tag{71}$$

Then, finally,

$$\begin{aligned}\langle C_\varepsilon A_\varepsilon^{-1} B_\varepsilon u, v\rangle &= \langle C_\varepsilon w, v\rangle = \langle g^\varepsilon \partial_x Q_\varepsilon w, \partial_x P_\varepsilon v\rangle_{L^2} = \langle g^\varepsilon \chi_y^\varepsilon \partial_x P_\varepsilon u, \partial_x P_\varepsilon v\rangle_{L^2} \\ &= \langle P_\varepsilon g^\varepsilon \chi_y^\varepsilon P_\varepsilon \partial_x u, P_\varepsilon \partial_x v\rangle_{L^2} = \overline{g\chi_y} \langle \partial_x P_\varepsilon u, \partial_x P_\varepsilon v\rangle_{L^2}, \end{aligned} \tag{72}$$

which gives

$$\langle D_\varepsilon - C_\varepsilon A_\varepsilon^{-1} B_\varepsilon u, v\rangle = (\bar g - \overline{g\chi_y}) \langle \partial_x P_\varepsilon u, \partial_x P_\varepsilon v\rangle_{L^2} = \langle \bar L P_\varepsilon u, P_\varepsilon v\rangle. \tag{73}$$

Eventhough both $\bar L$ and $\bar L_\varepsilon$ are well-defined, neither $\bar L$ nor $\bar L_\varepsilon$ are one-to-one on the spaces we have chosen. We need to restrict the operators to smaller spaces on which they are invertible. Let $X' = H_0^1(0,1)$ and identify it as a closed subspace of $H^1(\mathbb{T})$ in the theorem below. On this space $\bar L$ is bounded and positive by Poincaré's inequality, $\langle \bar L u, u\rangle \geq \delta \|u\|_X^2$. Since $\bar L_\varepsilon = P_\varepsilon \bar L P_\varepsilon$ by Theorem 1, this implies (74) and (75). Hence, there are unique solutions $\bar u_\varepsilon$ and $\bar u$ satisfying (76) for $f \in X_0^\varepsilon$ by the Lax–Milgram lemma. Moreover, Parseval's equality ensures that (77) and (78) hold, since $P_\varepsilon' u = P_\varepsilon u - (P_\varepsilon u)(0)$ here. The theorem then shows that a solution of $\bar L_\varepsilon \bar u_\varepsilon = f$ converges strongly in $H_0^1(0,1)$ to the classically homogenized solution when $\varepsilon \to 0$.

Theorem 2. *Suppose X' is a closed subspace of X such that*

$$\langle \bar L_\varepsilon u, u\rangle \geq \delta \|u\|_X^2, \qquad \forall u \in X' \cap X_0^\varepsilon, \tag{74}$$

$$\langle \bar L_\varepsilon u, v\rangle \leq C \|u\|_X \|v\|_X, \qquad \forall u, v \in X_0^\varepsilon. \tag{75}$$

If

$$\bar L \bar u = f = \bar L_\varepsilon \bar u_\varepsilon, \qquad \bar u \in X', \quad \bar u_\varepsilon \in X' \cap X_0^\varepsilon, \tag{76}$$

$$\|(\bar L_\varepsilon P_\varepsilon - \bar L) u\|_X \to 0, \qquad u \in X', \tag{77}$$

and with P_ε' the orthogonal projection on $X' \cap X_0^\varepsilon$,

$$\|P_\varepsilon' u - u\|_X \to 0, \qquad \|P_\varepsilon u - u\|_X \to 0, \qquad \forall u \in X'. \tag{78}$$

Then

$$\|\bar u - \bar u_\varepsilon\|_X \to 0. \tag{79}$$

Proof. By (74), for all $u \in X' \cap X_0^\varepsilon$,

$$\|u\|_X^2 \leq \frac{1}{\delta} \langle \bar L_\varepsilon u, u\rangle \leq \frac{1}{\delta} \|\bar L_\varepsilon u\|_X \|u\|_X. \tag{80}$$

So,

$$\|u\|_X \leq \frac{1}{\delta} \|\bar L_\varepsilon u\|_X, \qquad \forall u \in X' \cap X_0^\varepsilon. \tag{81}$$

Also, by (75),

$$||\bar{L}_\varepsilon u||_X^2 = \langle \bar{L}_\varepsilon u, \bar{L}_\varepsilon u\rangle \le C||u||_X||\bar{L}_\varepsilon u||_X \qquad \Rightarrow \qquad ||\bar{L}_\varepsilon u||_X \le C||u||_X^2, \tag{82}$$

for all $u \in X_0^\varepsilon$. Now, since $\bar{u}_\varepsilon \in X' \cap X_0^\varepsilon$,

$$\begin{aligned} ||\bar{u} - \bar{u}_\varepsilon||_X &\le ||P'_\varepsilon \bar{u} - \bar{u}||_X + ||P'_\varepsilon(\bar{u} - \bar{u}_\varepsilon)||_X \\ &\le ||P'_\varepsilon \bar{u} - \bar{u}||_X + \frac{1}{\delta}||\bar{L}_\varepsilon P'_\varepsilon(\bar{u} - \bar{u}_\varepsilon)||_X \\ &\le ||P'_\varepsilon \bar{u} - \bar{u}||_X + \frac{1}{\delta}||(\bar{L}_\varepsilon P_\varepsilon - \bar{L})\bar{u}||_X + \frac{1}{\delta}||\bar{L}_\varepsilon(P'_\varepsilon - P_\varepsilon)\bar{u}||_X \\ &\le ||P'_\varepsilon \bar{u} - \bar{u}||_X + \frac{1}{\delta}||(\bar{L}_\varepsilon P_\varepsilon - \bar{L})\bar{u}||_X + \frac{C}{\delta}||(P'_\varepsilon - P_\varepsilon)\bar{u}||_X \\ &\le C||P'_\varepsilon \bar{u} - \bar{u}||_X + \frac{1}{\delta}||(\bar{L}_\varepsilon P_\varepsilon - \bar{L})\bar{u}||_X + \frac{C}{\delta}||P_\varepsilon \bar{u} - \bar{u}||_X, \end{aligned} \tag{83}$$

where (76) and (82) were also used. By (77) and (78) the remaining terms vanish.

Remark 1. The right hand side f in (47) does not agree with the right hand side $\tilde{f}$ of the standard weak formulation of the problem,

$$\langle g(x/\varepsilon)\partial_x u_\varepsilon, \partial_x v\rangle_{L^2} = \langle \tilde{f}, v\rangle_{L^2}, \qquad \forall v \in H^1(\mathbb{T}). \tag{84}$$

because of the differences in innerproducts. In fact, in the general standard formulation $\tilde{f} \in X^* = H^{-1}(\mathbb{T})$ but in order to stay in the framework of Sect. 2.3 we want the range of L_ε to be X. The right hand sides are related by the isomorphic Riesz mapping $\mathcal{A} : X^* \mapsto X$ such that $f = \mathcal{A}\tilde{f}$.

3 Wavelet Based Homogenization

In this section we show how the projection generated homogenization works when we choose wavelet spaces as our coarse and fine space decomposition. We start by reviewing the theory of multiresolution analysis. We then let $Lu = f$ be a finite-dimensional approximation of a partial differential equation and apply the machinery in Sect. 2.3 for this case. In particular, we demonstrate how the resulting $\bar{L}$ can be approximated by a sparse matrix and used for computing an approximate solution to the differential equation on a coarse grid.

3.1 Multiresolution Analysis

In this section we give a brief outline of the homogenization spaces that we will use. In general we want to homogenize to a subspace that represents the coarse scales of a function space. Therefore, it is close at hand to consider *multiresolution analysis* (MRA) and its related wavelet spaces. A wavelet representation lends itself naturally to analyzing the fine and coarse scales as well as the localization properties of a function. Here we restrict our attention to orthogonal MRA, to align with the theory in Sect. 2.3. The theory could easily be modified to allow also biorthogonal MRA. For a detailed description of MRA we refer the reader to e.g. the book by Daubechies, [13].

An MRA is made up of a ladder of closed function spaces,

$$\dots \subset V_j \subset V_{j+1} \subset \dots , \tag{85}$$

satisfying

$$\overline{\bigcup_{j\in\mathbb{Z}} V_j} = L^2(\mathbb{R}), \qquad \bigcap_{j\in\mathbb{Z}} V_j = \{0\}. \tag{86}$$

Moreover, the spaces V_j represent different scales in the sense that

$$f(x) \in V_j \Leftrightarrow f(2^{-j}x) \in V_0. \tag{87}$$

The zero space should also be invariant under integer translations,

$$f(x) \in V_0 \Rightarrow f(x-k) \in V_0, \qquad \forall k \in \mathbb{Z}, \tag{88}$$

and there should exist a *shape function*, $\varphi \in V_0$, such that

$$\{\varphi(x-k);\ k \in \mathbb{Z}\}, \tag{89}$$

is an orthonormal basis for V_0.

From the MRA definition above follows a number of important properties. Dilates and translates of the shape function form orthonormal bases for for the V_j spaces,

$$\{\varphi_{j,k}(x) = 2^{j/2}\varphi(2^j x - k);\ k \in \mathbb{Z}\} \text{ is an orthonormal basis for } V_j. \tag{90}$$

Denote the orthogonal complement of V_j in V_{j+1} by W_j,

$$V_{j+1} = V_j \oplus W_j, \qquad V_j \perp W_j. \tag{91}$$

Then, $L^2(\mathbb{R}) = \overline{\bigoplus_{j\in\mathbb{Z}} W_j}$ and there exists a *mother wavelet*, $\psi \in W_0$, with properties corresponding to that of φ,

$$\{\psi_{j,k}(x) = 2^{j/2}\psi(2^j x - k);\ k \in \mathbb{Z}\} \text{ is an orthonormal basis for } W_j. \tag{92}$$

This mother wavelet can be constructed explicitly from a given MRA. Note also that $\{\psi_{j,k};\ j,k \in \mathbb{Z}\}$ is an orthonormal basis for $L^2(\mathbb{R})$.

An important property of the wavelet ψ is its number of vanishing moments, i.e. the highest number M such that

$$\int_{\mathbb{R}} x^k \psi(x)dx = 0, \qquad k = 0, \dots , M-1. \tag{93}$$

This number is related to the wavelet system's approximation qualities. If W_j is orthogonal to higher degree polynomials, V_j better approximates smooth functions.

The simplest MRA is the one built on the Haar basis. It has $M = 1$ and the shape function and the mother wavelet are given by

$$\varphi(x) = \begin{cases} 1, & \text{if } 0 \le x \le 1, \\ 0, & \text{otherwise,} \end{cases} \qquad \psi(x) = \begin{cases} 1, & \text{if } 0 \le x \le 1/2, \\ -1, & \text{if } 1/2 < x \le 1, \\ 0, & \text{otherwise.} \end{cases} \tag{94}$$

In more than one dimension, the MRA is extended by considering tensor product spaces. Each dimension can be treated separately, using the one-dimensional analysis. The resulting multi-dimensional MRA is a tensor product of one-dimensional MRAs. In two dimensions, we define the tensor product spaces

$$\boldsymbol{V}_j = V_j \otimes V_j, \qquad \ldots \subset \boldsymbol{V}_j \subset \boldsymbol{V}_{j+1} \subset \ldots, \qquad j \in \mathbb{Z}, \tag{95}$$

with V_j being the one-dimensional spaces introduced above. The spaces $\boldsymbol{V}_j$ will satisfy the same conditions, (85)-(89), as the V_j spaces, with $\mathbb{R}$, $\mathbb{Z}$ replaced by $\mathbb{R}^2, \mathbb{Z}^2$. The shape function of this two-dimensional MRA can be written in terms of the one-dimensional shape function,

$$\boldsymbol{\varphi}(\boldsymbol{x}) = \varphi(x_1) \otimes \varphi(x_2) = \varphi(x_1)\varphi(x_2). \tag{96}$$

It follows, as in the one-dimensional case, that

$$\{\boldsymbol{\varphi}_{j,k,l}(\boldsymbol{x}) = \varphi_{j,k}(x_1) \otimes \varphi_{j,l}(x_2)\}, \qquad k, l \in \mathbb{Z}, \tag{97}$$

is an orthonormal basis for $\boldsymbol{V}_j$. Furthermore, denoting the orthogonal complement of $\boldsymbol{V}_j$ in $\boldsymbol{V}_{j+1}$ by $\boldsymbol{W}_j$, we have that

$$\boldsymbol{V}_{j+1} = \boldsymbol{V}_j \oplus \boldsymbol{W}_j, \qquad \boldsymbol{V}_j \perp \boldsymbol{W}_j, \qquad L^2(\mathbb{R}^2) = \overline{\bigoplus_{j \in \mathbb{Z}} \boldsymbol{W}_j}. \tag{98}$$

The wavelet space $\boldsymbol{W}_j$ is composed of three parts,

$$\boldsymbol{W}_j = (W_j \otimes W_j) \oplus (V_j \otimes W_j) \oplus (W_j \otimes V_j). \tag{99}$$

Here, the W_j spaces are those of the one-dimensional case, given by (91). Similar tensor product extensions can be made also for higher dimensions.

3.2 Wavelet Projections of Difference Operators

We will now discuss homogenization in a more concrete setting. The spaces we use are the finite dimensional wavelet spaces discussed in Sect. 3.1. The typical operators that we homogenize are discrete approximations of differential operators. As seen above in Sect. 2.3, given such an operator, defined on a fine grid, the idea of projection generated homogenization is to find an operator defined on a smaller space that extracts only the coarse scales of the solution. For a function in the wavelet space V_{j+1}, the coarse scale is represented by V_j, and we are in the first step thus interested in the homogenized operator given by $\bar{L}_{V_{j+1},V_j}$. We will mostly use the standard Haar basis.

Let us now consider the equation

$$L_{j+1}U = F, \qquad U, F \in V_{j+1}, \qquad L_{j+1} \in \mathcal{L}(V_{j+1}, V_{j+1}). \tag{100}$$

This equation may originate from a finite difference, finite element or finite volume discretization of a given differential equation. In the Haar case U can be identified as a piecewise constant approximation of $u(x)$, the solution to the continuous problem. Let $\mathcal{W}_j : V_{j+1} \mapsto V_j \times W_j$ be the unitary operator corresponding to (25). (By (91)

$V_j^\perp = W_j$.) As in the infinite dimensional case (31), we apply the transformation $\mathcal{W}_j$ on L_{j+1} from the left and get

$$\mathcal{W}_j L_{j+1} \mathcal{W}_j^* (\mathcal{W}_j U) = \mathcal{W}_j F \tag{101}$$

or if we decompose $\mathcal{W}_j L_{j+1} \mathcal{W}_j^*$ according to (26),

$$\begin{pmatrix} A_j & B_j \\ C_j & L_j \end{pmatrix} \begin{pmatrix} U_\mathrm{f} \\ U_\mathrm{c} \end{pmatrix} = \begin{pmatrix} F_\mathrm{f} \\ F_\mathrm{c} \end{pmatrix}, \qquad U_\mathrm{f}, F_\mathrm{f} \in W_j, \qquad U_\mathrm{c}, F_\mathrm{c} \in V_j. \tag{102}$$

Here the subindex "f" means projection onto the fine scale subspace, W_j in this case, and subindex "c" stands for projection onto the coarse scale subspace, V_j here. This is a convention we will use frequently. Also note that we changed the notation of the block previously called D_j to L_j, to indicate that $L_j = PL_{j+1}P$ actually corresponds to one type of direct discretization on the coarse scale.

The homogenized "coarse grid operator" is now given by block Gaussian elimination of (102),

$$(L_j - C_j A_j^{-1} B_j) U_\mathrm{c} = F_\mathrm{c} - C_j A_j^{-1} F_\mathrm{f}. \tag{103}$$

For simplicity we will use the notation $\overline{L_{j+1}}_{V_{j+1},V_j} \equiv \bar{L}_j$ and $\bar{F}_{V_{j+1},V_j} \equiv \bar{F}_j$. Hence,

$$\bar{L}_j = L_j - C_j A_j^{-1} B_j, \qquad \bar{F}_j = F_\mathrm{c} - C_j A_j^{-1} F_\mathrm{f}. \tag{104}$$

Since L_j corresponds to a direct coarse discretization, we can interpret $C_j A_j^{-1} B_j$ as a correction term, which includes subgrid phenomena in $\bar{L}_j$.

In the finite dimensional case, we will represent the operators using their matrix representations with respect to the bases given by (90) and (92). For instance in the Haar basis, we get the following matrix representation of $\mathcal{W}_j$,

$$\mathcal{W}_j = \begin{pmatrix} \tilde{Q}_j \\ \tilde{P}_j \end{pmatrix} = \frac{1}{\sqrt{2}} \begin{pmatrix} 1 & -1 & 0 & \cdots & & & \\ 0 & 0 & 1 & -1 & 0 & \cdots & \\ & & & & & & \\ \vdots & \vdots & & \ddots & \ddots & & \\ 0 & 0 & \cdots & 0 & 1 & -1 \\ 1 & 1 & 0 & \cdots & & \\ 0 & 0 & 1 & 1 & 0 & \cdots \\ \vdots & \vdots & & \ddots & \ddots & \\ 0 & 0 & \cdots & 0 & 1 & 1 \end{pmatrix}. \tag{105}$$

Note, in this representation $\tilde{P}_j^*$ and $\tilde{Q}_j^*$ are bases in V_{j+1} for V_j and W_j, respectively. The corresponding projections in V_{j+1} are $\tilde{P}_j^* \tilde{P}_j$ and $\tilde{Q}_j^* \tilde{Q}_j$. The matrix representation of (102) is a block matrix decomposition of the same form as the operator decomposition.

Before we go on we should also note that in general $\bar{L}_j$ will not be represented by a sparse matrix even if L_{j+1} is, because A_j^{-1} would typically be dense. For the homogenization method to be efficient we must be able to approximate $\bar{L}_j$ with a sparse matrix. If this sparse matrix is on banded form, it can be seen as a discretization of a local differential operator acting on the coarse space. In Sect. 3.3 we will get back to this issue and in subsequent sections we will show that this can indeed be done for many applications.

The procedure to obtain $\bar{L}_j$ described above can be applied recursively on $\bar{L}_j$ itself to get $\bar{L}_{j-1}$ and so on. This can easily be verified when L_{j+1} is symmetric positive definite. Then A_j is always invertible by Proposition 1 (with $a[u, v] = \langle Lu, v \rangle$) and Proposition 2 shows that also $\bar{L}_j$ is symmetric positive definite. By induction $\bar{L}_k$ exists for $k \le j$. Also note that Proposition 3 gives $\bar{L}_k = \overline{L_{j+1}}_{V_{j+1},V_k}$ and $\bar{F}_k = \bar{F}_{V_{j+1},V_k}$ for $k \le j$.

Moreover, an improvement in the condition number can often be estimated. Typically for standard discretizations

$$\max_{||v||=1} \langle L_k v, v\rangle < \max_{||u||=1} \langle L_{j+1} u, u\rangle, \tag{106}$$

when $k \le j$. Then, from Proposition 2,

$$\begin{aligned} \kappa(\bar{L}_k) &= \frac{\max_{||v||=1}\langle \bar{L}_k v, v\rangle}{\min_{||v||=1}\langle \bar{L}_k v, v\rangle} \le \frac{\max_{||v||=1}\langle L_k v, v\rangle}{\min_{||v||=1}\langle L_{j+1} v, v\rangle} \\ &< \frac{\max_{||v||=1}\langle L_{j+1} v, v\rangle}{\min_{||v||=1}\langle L_{j+1} v, v\rangle} = \kappa(L_{j+1}). \end{aligned} \tag{107}$$

The two-dimensional wavelet transform

$$\boldsymbol{\mathcal{W}}_j : \boldsymbol{V}_{j+1} \mapsto \boldsymbol{W}_j \times \boldsymbol{V}_j \tag{108}$$

can be written as a tensor product of one-dimensional transforms,

$$\boldsymbol{\mathcal{W}}_j = \mathcal{W}_j \otimes \mathcal{W}_j. \tag{109}$$

A linear operator L_{j+1} that acts on the space $\boldsymbol{V}_{j+1}$ can be decomposed in a way similar to the one-dimensional case. The equation

$$L_{j+1} U = F, \qquad U, F \in \boldsymbol{V}_{j+1}, \qquad L_{j+1} \in \mathcal{L}(\boldsymbol{V}_{j+1}, \boldsymbol{V}_{j+1}) \tag{110}$$

can then be transformed to

$$\begin{pmatrix} A_j & B_j \\ C_j & L_j \end{pmatrix} \begin{pmatrix} U_\mathrm{f} \\ U_\mathrm{c} \end{pmatrix} = \begin{pmatrix} F_\mathrm{f} \\ F_\mathrm{c} \end{pmatrix}, \qquad U_\mathrm{f}, F_\mathrm{f} \in \boldsymbol{W}_j, \qquad U_\mathrm{c}, F_\mathrm{c} \in \boldsymbol{V}_j, \tag{111}$$

and the coarse grid operator is again the Schur complement,

$$\overline{L}_{\boldsymbol{V}_{j+1},\boldsymbol{V}_j} \equiv \bar{L}_j = L_j - C_j A_j^{-1} B_j. \tag{112}$$

To get the matrix representation of (111), we take the one-dimensional matrix representations of $\mathcal{W}_j$ and, following (109), compute its Kronecker tensor product with itself. We must thereafter also apply a suitable permutation matrix to the product in order to get the same block structure of the matrix as in (111).

Note that the fine scale part of U can be decomposed as

$$U_\mathrm{f} = \begin{pmatrix} U_\mathrm{ff} \\ U_\mathrm{cf} \\ U_\mathrm{fc} \end{pmatrix}, \qquad U_\mathrm{ff} \in W_j \otimes W_j, \qquad U_\mathrm{cf} \in V_j \otimes W_j, \qquad U_\mathrm{fc} \in W_j \otimes V_j. \tag{113}$$

In some cases, the homogenized operator keeps important properties of the original operator. Let the forward and backward undivided differences be defined as

$$\Delta_+ u_i = u_{i+1} - u_i, \qquad \Delta_- u_i = u_i - u_{i-1}. \tag{114}$$

In [14] it was shown that the one-dimensional elliptic model equation $-(gu')' = f$ discretized with

$$-\Delta_+ \mathrm{diag}(g) \Delta_- U = h^2 F \tag{115}$$

will preserve its divergence form during homogenization. That is, we will get

$$\bar{L}_j = \frac{1}{h^2}\Delta_+ H_j \Delta_-, \tag{116}$$

where H_j is a strongly diagonal dominant matrix which can be interpreted as the effective material coefficient related to g. Analogously, for the first order differential operator $g(x)\frac{\partial}{\partial x}$ the discretized form $\mathrm{diag}(g)\Delta_-/h$ is preserved during homogenization,

$$\bar{L}_j = \frac{1}{h} H_j \Delta_-. \tag{117}$$

In two dimensions, the elliptic model equation $-\nabla(g(x,y)\nabla u) = f$ can be discretized as

$$L_{j+1} = -\frac{1}{h^2}\Delta_+^x G \Delta_-^x - \frac{1}{h^2}\Delta_+^y G \Delta_-^y, \qquad L_{j+1}U = F. \tag{118}$$

Then $\bar{L}_j$ is no longer on exactly the same form as L_{j+1}. The cross-derivatives must also be included. We get

$$\bar{L}_j = -\frac{1}{h^2}\Delta_+^x H^{xx} \Delta_-^x - \frac{1}{h^2}\Delta_+^y H^{yx} \Delta_-^x - \frac{1}{h^2}\Delta_+^x H^{xy} \Delta_-^y - \frac{1}{h^2}\Delta_+^y H^{yy} \Delta_-^y. \tag{119}$$

Remark 2. Spaces other than wavelet spaces could also be used. For instance, in the settings described in [29], [34] and [24] the coarse subspace is simply the grid function space given by removing every other grid point. In the classical homogenization setting the solution has the asymptotic expansion (24), as discussed in Sect. 2.2. The techniques mentioned above give homogenized operators with coarse grid solutions which directly samples (24). The oscillatory term $\varepsilon u_1(x, x/\varepsilon)$ is contained in the solution. In the wavelet homogenization the solution is a projection of u_ε onto a coarse scale V_j space. The influence of the u_1 term, which oscillates with mean zero, [5], is then significantly reduced.

3.3 Compact Representation of Projected Operators

When the operator L_{j+1} is derived from a finite difference, finite element or finite volume discretization, it is sparse and of a certain structure. In one dimension it might, for instance, be tridiagonal. However, as remarked in Sect. 3.2, the matrix $\bar{L}_j$ is in general not sparse since A_j^{-1} is usually dense. Computing all components of $\bar{L}_j$ would be inefficient. Fortunately, $\bar{L}_j$ will be diagonal dominant in many important cases and we should then be able to find a sparse matrix that is a close approximation of $\bar{L}_j$.

In some simple cases the matrix A_j is in fact diagonal. Examples include operators L of the form

$$Lu = \int_0^x a(t)u(t)dt + b(x)u(x). \tag{120}$$

Let P_j be the orthogonal projection on V_j. Suppose L is discretized in V_{j+1} as $L_{j+1} = P_{j+1}LP_{j+1}$ with a in (120) replaced by $P_{j+1}aP_{j+1}$, and similar for b. When the Haar system is used, A_j is diagonal and $\bar{L}_j$ is of the same form as L_{j+1}, cf. Chap. 7. Let $\bar{A}_k$ be related to $\bar{L}_{k+1}$ via (101, 102) in the same way as A_j relates to L_{j+1}. By induction, $\bar{A}_k$ is also diagonal for $k < j$. The operators in (120) turn up for instance in problems with systems of ordinary differential equations and one-dimensional elliptic equations, see [6,21]. In these cases, an explicit recurrence relation between scale levels can be established, which permits the computation of $\bar{L}_k$ on any fixed level k as the starting level, $j+1$, tends to infinity.

For more general problems one must instead rely on the rapid decay of elements in A_j and $\bar{L}_j$ off the diagonal, which is a consequence of the wavelet spaces' good approximation properties. The decay rate of A_j for Calderon–Zygmund and pseudo-differential operators were given by Beylkin, Coifman and Rokhlin in [8]. Letting $L_{j+1} = P_{j+1}LP_{j+1}$, $A_j = \{a^j_{k\ell}\}$, $B_j = \{b^j_{k\ell}\}$ and $C_j = \{c^j_{k\ell}\}$, they show that

$$|a^j_{k\ell}| + |b^j_{k\ell}| + |c^j_{k\ell}| \le \frac{2^{-\lambda j} C_M}{1+|k-\ell|^{M+1}}, \qquad |k-\ell| \ge \nu, \tag{121}$$

when the wavelet system has M vanishing moments. For Calderon–Zygmund operators $\lambda = 0$ and $\nu = 2M$. For a pseudo-differential operator $\nu = 0$ and if its symbol is $\sigma(x,\xi)$, the value of λ should be taken such that the estimates

$$|\partial^\alpha_\xi \partial^\beta_x \sigma(x,\xi)| \le C_{\alpha,\beta}(1+|\xi|)^{\lambda-\alpha+\beta}, \tag{122}$$

$$|\partial^\alpha_\xi \partial^\beta_x \sigma^*(x,\xi)| \le C_{\alpha,\beta}(1+|\xi|)^{\lambda-\alpha+\beta}, \tag{123}$$

are satisfied for some constants $C_{\alpha,\beta}$. For instance, in the second order elliptic case $\lambda = 2$. Moreover, Beylkin and Coult, [9], showed that if (121) holds with $\lambda = 0$ for A_j, B_j and C_j given by L_{j+1} in (101, 102), then the same estimate also holds for $\bar{A}_{j'}$, $\bar{B}_{j'}$ and $\bar{C}_{j'}$, here given by $P_{j'}\bar{L}_{k+1}P_{j'}$ for $j' \le k < j$ with $j+1$ being the starting homogenization level. Hence, the decay rate is preserved after homogenization.

The decay estimate in [9] for $\bar{A}_{j'}$ is uniform in k and may not be sharp for a fixed k. There is, for example, a general result by Concus, Golub and Meurant, [12], for diagonal dominant, symmetric and tridiagonal matrices. For those cases, which include A_j corresponding to the discretization in (115) of the one-dimensional elliptic operator, the inverse has exponential decay,

$$\left|\left(A_j^{-1}\right)_{k\ell}\right| \le C\varrho^{|k-\ell|}, \qquad 0 < \varrho < 1. \tag{124}$$

This holds also when the elliptic operator has a lower order term of type $b(x)\partial_x$ discretized with upwinding, [30].

One simple way to approximate $\bar{L}_j$ is to set all components outside a prescribed bandwidth ν equal to zero. Let us define

$$\operatorname{trunc}(M,\nu)_{ij} = \begin{cases} M_{ij}, & \text{if } 2\,|i-j| \le \nu - 1 \\ 0, & \text{otherwise.} \end{cases} \tag{125}$$

For $\nu = 1$ the matrix is diagonal. For $\nu = 3$ it is a tridiagonal and so on. We this procedure here in order to control the structure of $\bar{L}_j$ and we refer to it as *truncation*. For lower triangular matrices we define truncation to lower triangular form as

$$\operatorname{trunc}_{\mathrm{L}}(M,\nu)_{ij} = \begin{cases} M_{ij}, & \text{if } 0 \le i-j \le \nu - 1 \\ 0, & \text{otherwise.} \end{cases} \tag{126}$$

Another possible strategy is the basic truncation method used in [8],

$$\operatorname{trunc}_{\mathrm{BCR}}(M,\varepsilon)_{ij} = \begin{cases} M_{ij}, & \text{if } |M_{ij}| > \varepsilon, \\ 0, & \text{otherwise} \end{cases} \tag{127}$$

but it is not practical here since the location of the non-zero elements cannot be controlled. The matrix $\bar{L}_j$ can also be projected onto banded form in a more effective manner. The aim is that the projected matrix should give the same result as the original matrix on a given subspace, e.g. when applied to vectors representing smooth functions. Let $\{\boldsymbol{v}_j\}_{j=1}^{\nu}$ be a set of linearly independent vectors in $\mathbb{R}^N$. Denote the subspace of $\mathbb{R}^{N\times N}$ with matrices essentially[1] of bandwidth ν by $\mathcal{T}_\nu$. Moreover, let

$$\mathcal{L}_\nu = \{M \in \mathbb{R}^{N\times N} : \operatorname{span}\{\boldsymbol{v}_1, \boldsymbol{v}_2, \dots, \boldsymbol{v}_\nu\} \subset \mathsf{N}(M)\}, \tag{128}$$

where $\mathsf{N}(M)$ represents the null space of M. Then

$$\mathbb{R}^{N\times N} = \mathcal{T}_\nu \oplus \mathcal{L}_\nu \tag{129}$$

and we define the *band projection* of a matrix $M \in \mathbb{R}^{N\times N}$ as the projection of M onto $\mathcal{T}_\nu$ along $\mathcal{L}_\nu$, with the notation

$$\operatorname{band}(M,\nu) = \operatorname{Proj}_{\mathcal{T}_\nu} M. \tag{130}$$

As a consequence,

$$M\boldsymbol{x} = \operatorname{band}(M,\nu)\boldsymbol{x}, \qquad \forall \boldsymbol{x} \in \operatorname{span}\{\boldsymbol{v}_1, \boldsymbol{v}_2, \dots, \boldsymbol{v}_\nu\}. \tag{131}$$

In our setting M will usually operate on vectors representing smooth functions, for instance solutions to elliptic equations, and a natural choice for $\boldsymbol{v}_j$ vectors are thus the first ν polynomials,

$$\boldsymbol{v}_j = \{1^{j-1}, 2^{j-1}, \dots, N^{j-1}\}^*, \qquad j = 1, \dots, \nu. \tag{132}$$

Smooth solutions to the homogenized problem should be well approximated by these vectors. For the case $\nu = 1$ we get the standard "masslumping" of a matrix, often used in the context of finite element methods.

This technique is similar to the probing technique used by Chan et al., [11]. In that case the vectors $\boldsymbol{v}_j$ are sums of unit vectors. Other probing techniques have been suggested by Axelsson, Pohlman and Wittum, see Chapter 8 in [3]. The choice of $\boldsymbol{v}_j$ vectors could be optimized if there is some a priori knowledge of the homogenized solution. In some cases the band projection technique only gives improvements for small values of ν, see Fig. 2. The probing technique of [11] could be used for larger ν in our examples. Numerical evidence indicate that for small values of ν, the band projection technique is more efficient.

The focus here is on the principle and the truncation $\operatorname{trunc}(\bar{L}_j, \nu)$ is done after computing the full inverse of A_j, which is expensive. By capitalizing on the nearly sparse structure of the matrices involved, it was however shown in [9] that the cost can be reduced to $\mathcal{O}(N)$ operations for N unknowns and fixed accuracy. This method uses a multiresolution based LU decomposition procedure described in [10]. Moreover, the same homogenized operator will typically be reused multiple times, for instance with different right hand sides, or in different places of the geometry as a subgrid model. This aspect will be exemplifed below in Sect. 5.3 and Sect. 5.4. The

[1] We must require that each row of the matrices in $\mathcal{T}_\nu$ has the same number of elements. Therefore, the first and last $\nu - 1$ rows will have additional elements located immediately to the right and left of the band, respectively.

computation of band$(\bar{L}_j, \nu)$ can be based on trunc$(\bar{L}_j, \mu)$, $\mu > \nu$. The additional computational cost is proportional to $(\nu^3 + \mu\nu)N$. The $\nu^3 N$ term corresponds to solving N $\nu \times \nu$ systems and $\mu\nu N$ to computing the right hand sides, see also [3].

The truncation methods described above are even more efficient when applied to H_j instead of directly to the homogenized operator. The following proposition shows that when the solution to the homogenized problem belongs to the Sobolev space H^1, the accuracy is one order higher.

Proposition 4. *Suppose $L = \Delta_+ H \Delta_-$ and $LU = h^2 f$. Let $\kappa = ||L^{-1}|| \cdot ||H||$ and*

$$r = \frac{||\delta L||}{||L||} = \frac{||\delta H||}{||H||} < \frac{1}{4\kappa}, \tag{133}$$

where $||\cdot||$ denotes the l_2-norm. Furthermore, let u be any H^1-function such that $U_j = u(jh)$. Then, if

$$(L + \delta L)(U + \delta U_L) = h^2 f, \tag{134}$$

$$\Delta_+(H + \delta H)\Delta_-(U + \delta U_H) = h^2 f \tag{135}$$

the following bounds on the perturbed solutions hold $\forall f$

$$||\delta U_L|| \le \frac{4r\kappa}{1 - 4r\kappa}||U||, \tag{136}$$

$$||\delta U_H|| \le \frac{1}{2} \cdot \frac{4r\kappa}{1 - 4r\kappa}||\Delta_- U|| \le \frac{h}{2} \cdot \frac{4r\kappa}{1 - 4r\kappa}||u||_{H^1}. \tag{137}$$

The bound (136) is sharp.

Proof. Using $||\Delta_+|| = ||\Delta_-|| = 2$ and (133) we have

$$||L^{-1}\delta L|| \le \frac{||L^{-1}|| \cdot ||\delta H|| \cdot ||\Delta_+ H \Delta_-||}{||H||} \le 4r||L^{-1}|| \cdot ||H|| = 4r\kappa < 1. \tag{138}$$

Therefore,

$$||\delta U_L|| = ||(I + L^{-1}\delta L)^{-1} L^{-1} \delta L U|| \le \frac{||L^{-1}\delta L||}{1 - ||L^{-1}\delta L||}||U|| \le \frac{4r\kappa}{1 - 4r\kappa}||U||, \tag{139}$$

proving the first bound. In the second case,

$$||L^{-1}\Delta_+ \delta H \Delta_-|| \le 4||L^{-1}|| \cdot ||\delta H|| \le 4r\kappa < 1, \tag{140}$$

giving

$$\begin{aligned} ||\delta U_H|| &= ||(I + L^{-1}\Delta_+\delta H\Delta_-)^{-1} L^{-1}\Delta_+\delta H \Delta_- U|| \\ &\le \frac{||L^{-1}\Delta_+\delta H||}{1 - ||L^{-1}\Delta_+\delta H\Delta_-||}||\Delta_- U|| \le \frac{2r\kappa}{1 - 4r\kappa}||\Delta_- U|| \end{aligned} \tag{141}$$

which proves the second bound. The choice $L = -\Delta_+\Delta_-$, $\delta L = -rI||L||$ and f the eigenvector of L corresponding to its smallest eigenvalue shows the sharpness of (136). The final step, $||\Delta_- U|| \le h||u||_{H^1}$, follows from Cauchy-Schwarz's inequality. Note that (136) cannot be improved to $\mathcal{O}(h)$ by replacing $||U||$ with $||u||_{H^1}$.

In two dimensions truncation to simple banded form is in general not adequate, since the full operator will typically be block banded. However, both the crude truncation and the band projection generalizes easily to treat block banded form instead of just banded. Let M be the tensor product of two $N \times N$ matrices. Then we define truncation as

$$\operatorname{trunc}_2(M,\nu)_{ij} = \begin{cases} M_{ij}, & \text{if } 2\,|i-j-rN| \leq \nu - 1 - |2r|, \\ 0, & \text{otherwise}, \end{cases} \qquad |2r|+1 \leq \nu. \tag{142}$$

This mimics the typical block structure of a discretized differential operator. For the band projection, the space $\mathcal{T}_\nu$ of banded matrices in the one-dimensional definition, is simply replaced by the space of matrices with the block banded sparsity pattern defined in (142).

In two dimensions, untangling the various H components of (119) from $\bar{L}$ is more complicated than finding the H in (116) and (117) for one-dimensional problems. There is however no unsurmountable difficulty, and truncating H instead of $\bar{L}$ can be done also for two-dimensional problems.

4 Numerical Examples

In this section we present numerical results for the algorithms described above. Elliptic and hyperbolic model problems are studied. In one of the examples the wavelet based homogenization is applied to the Helmholtz equation for the generation of a course grid approximation. This could potentially be useful for sub grid scale modeling in computational electromagnetics.

4.1 Elliptic Problems

Consider first the one-dimensional elliptic model equation,

$$-\frac{d}{dx}\left(g(x)\frac{du}{dx}\right) = f(x), \qquad u(0) = u'(1) = 0, \tag{143}$$

with $g(x) > 0$ and then the Helmholtz equation,

$$-\sum_{i=1}^{d} \frac{\partial}{\partial x_i} g(x) \frac{\partial u}{\partial x_i} - \omega^2 u(x) = 0, \tag{144}$$

in one and two dimensions ($d = 1, 2$). The Helmholtz equation can be derived from the wave equation $v_{tt} = \nabla \cdot g(x) \nabla v$ under the assumption that $v(x,t) = u(x)e^{i\omega t}$. Hence $u(x)$ represents the amplitude and ω is the frequency.

Uniform finite volume discretization will be used in the numerical experiments. The computational domain is the unit interval (square in 2D). The grid size is denoted n, and the cell size $h = 1/n$. For a function $u(x)$ the corresponding grid function is written u_i (or u_{ij}). The grid function approximates cell averages,

$$u_i \sim \frac{1}{h}\int_{(i-1)h}^{ih} u(x)dx, \qquad i = 1, \ldots, n, \tag{145}$$

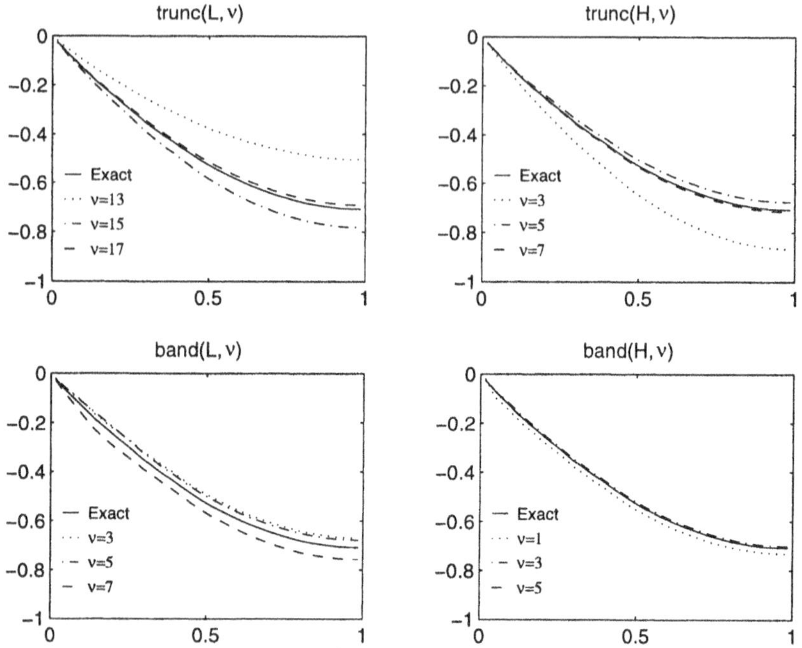

Fig. 2. Result for the elliptic model problem, $g(x)$ random, when the homogenized operator is approximated in different ways. The "exact" solution refers to the solution with the full 32×32 homogenized operator

in one dimension and

$$u_{ij} \sim \frac{1}{h^2} \int_{(i-1)h}^{ih} \int_{(j-1)h}^{jh} u(x_1, x_2) dx_1 dx_2, \qquad i, j = 1, \ldots, n, \tag{146}$$

in two dimensions. The whole vector $(u_1, \ldots, u_n)^T$ will be denoted by U. In two dimensions, $U = (u_{11}, \ldots, u_{1n}, u_{21}, \ldots, u_{2n}, \ldots, u_{nn})^T$.

The 1D Elliptic Model Equation

We approximate (143) with the discretization

$$-\frac{1}{h^2} \Delta_+ g_i \Delta_- u_i = f_i, \qquad i = 1, \ldots, n, \qquad u_0 = -u_1, \qquad u_{n+1} = u_n. \tag{147}$$

First, set $f = -1$ and let the coefficients of $g(x)$ have a uniform random distribution in the interval $[0.5, 1]$. Take $n = 256$ and make three homogenization steps. The coarsest level contains 32 grid points.

In Fig. 2 different truncation strategies are compared. The exact reference solution is given by the numerically homogenized operator at the coarsest level without any truncation. This is equivalent to the projection onto the coarse scale of the solution on the finest scale. In the top two subplots we use crude truncation, (125).

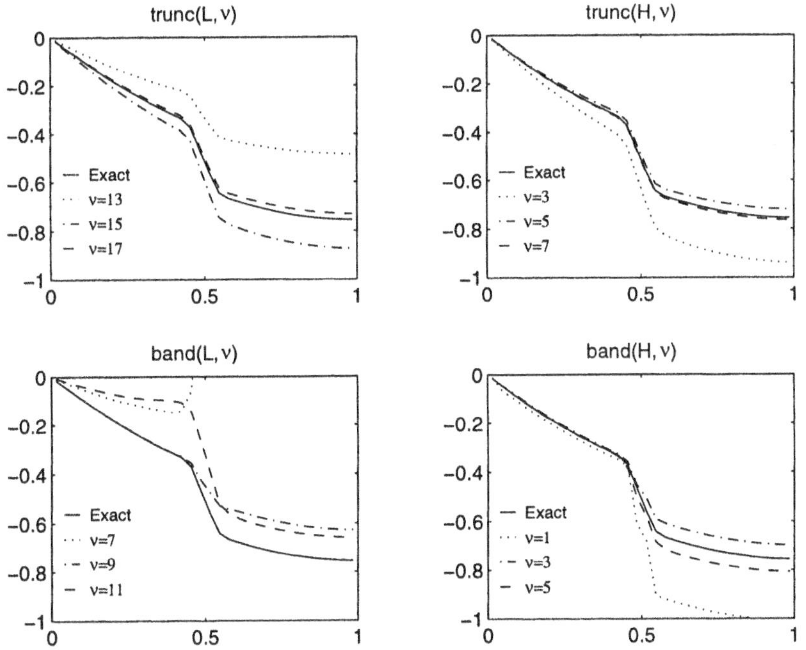

Fig. 3. Result for the elliptic model problem, $g(x)$ a slit, when the homogenized operator is approximated in different ways. The "exact" solution refers to the solution with the full 32×32 homogenized operator

In the bottom two subplots we use the band projection described in Sect. 3.3. The approximation is performed on H, see (116), and on $\bar{L}$ after all three homogenizations. We see that band projection gives a better approximation. We also see that it is more efficient to truncate H than to truncate $\bar{L}$.

Next, the coefficient in the differential equation is changed to

$$g(x) = \begin{cases} 1/6, & 0.45 < x < 0.55, \\ 1, & \text{otherwise.} \end{cases} \tag{148}$$

All other characteristics are kept. The result is given in Fig. 3 and it shows that the relative merits of the different methods are more or less the same. The structures of the untruncated $\bar{L}$ and H matrices are shown in Fig. 4. It should be noted that the local inhomogeneity of the full operator has spread out over a larger area, but it is still essentially local.

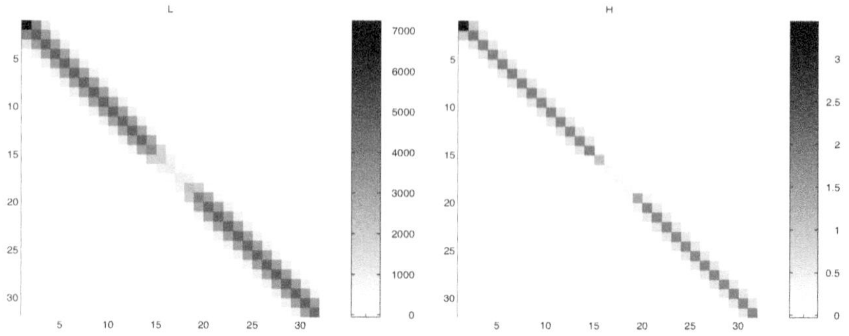

Fig. 4. Structure of the untruncated homogenized operators $\bar{L}$ (left) and H (right) for the elliptic model problem, $g(x)$ a slit. Gray level indicates absolute value of elements

The 1D Helmholtz Equation

The equation (144) is discretized in one dimension, with the boundary conditions $u(0) = 1$ and $u'(1) = 0$,

$$\begin{aligned} -\frac{1}{h^2}\Delta_+ g_i \Delta_- u_i - \omega^2 u_i &= 0, \quad i = 1, \dots, n, \\ u_0 &= 2u(0) - u_1, \\ u_{n+1} &= u_n. \end{aligned} \tag{149}$$

We use $\omega = 2\pi$ and the same $g(x)$ as in (148) and again we take $n = 256$ and use three homogenizations. We get

$$\bar{L}u = (\tilde{L} - \omega^2 I)u = 0. \tag{150}$$

Truncation is performed on $\tilde{L}$ (or $\tilde{H}$) and not on $\bar{L}$. The result is in Fig. 5. We see that Helmholtz equation gives results similar to those of the model equation. Again band projection is more efficient than truncation and approximating H_j is more efficient than approximating $\bar{L}$.

The 2D Helmholtz Equation

We consider the two-dimensional version of (144), with periodic boundary conditions in the y-direction, and at the left and right boundaries, $u(0, y) = 1$, $u_x(1, y) = 0$ respectively. This is a simple model of a plane time-harmonic wave of amplitude one entering the computational domain from the left and flowing out at the right. The discretization that we use is

$$\begin{gathered} -\frac{1}{h^2}\Delta_+^x g_{i\ell} \Delta_-^x u_{i\ell} - \frac{1}{h^2}\Delta_+^y g_{i\ell} \Delta_-^y u_{i\ell} - \omega^2 u_{i\ell} = 0, \qquad i, \ell = 1, \dots, n, \\ u_{i,0} = u_{i,n}, \quad u_{i,n+1} = u_{i,1}, \quad u_{n+1,\ell} = u_{n,\ell} \quad u_{0,\ell} = 2 - u_{1,\ell}. \end{gathered} \tag{151}$$

This leads to the matrix equation

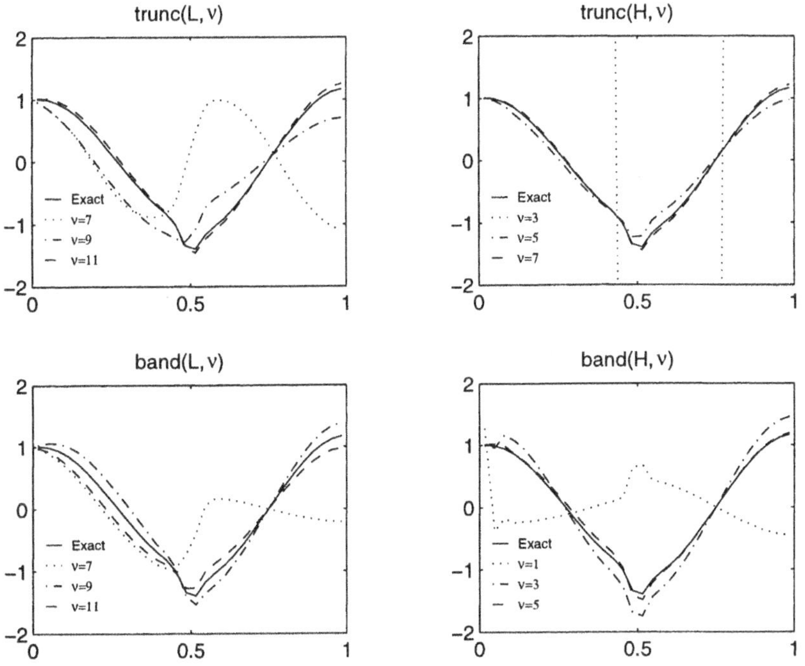

Fig. 5. Result for the Helmholtz equation, $g(x)$ a slit, when the homogenized operator is approximated in different ways. The "exact" solution refers to the solution with the full 32×32 homogenized operator

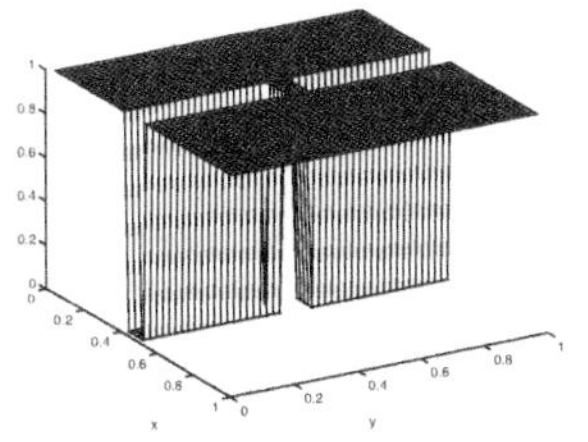

$$g(x,y) = \begin{cases} 10^{-4}, & 0.4 < x < 0.5 \text{ and } \\ & |y - 0.5| > 0.05, \\ 1, & \text{otherwise.} \end{cases}$$

Fig. 6. The variable coefficient $g(x, y)$ used in the 2D Helmholtz example

$$L_{j+1} U = F, \qquad U, F \in \boldsymbol{V}_{j+1}, \qquad n = m 2^{j+1}, \tag{152}$$

where m is a positive integer. L_{j+1} is homogenized following the theory for two-dimensional problems in Sect. 3.2.

As an example we choose the $g(x, y)$ shown in Fig. 6, which models a wall with a small slit where the incoming wave can pass through. With $\omega = 3\pi$ and $n = 48$, we obtained the results presented in Fig. 7.

The operator after one homogenization step is truncated according to (142). We show the results of truncation in Fig. 8, for various values of ν. The case $\nu = 9$ corresponds to a compression to approximately 7% of the original size. The structure of this operator is shown in Fig. 9.

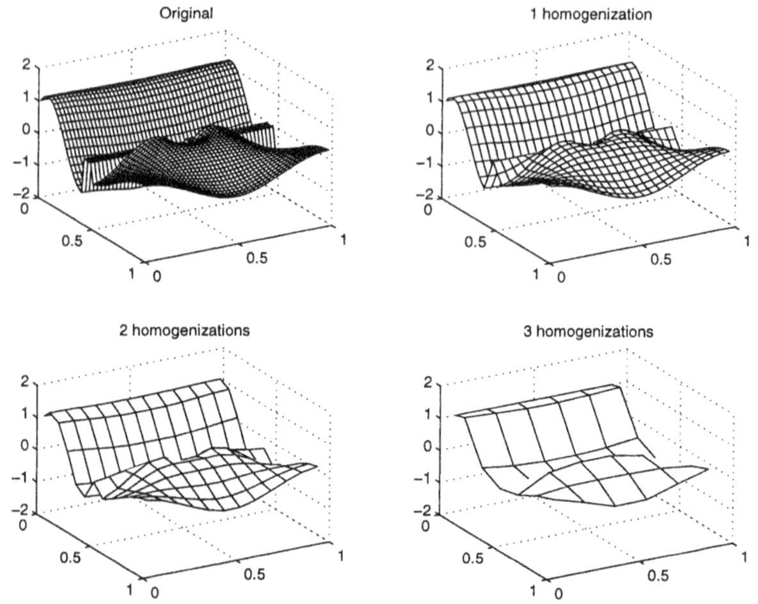

Fig. 7. Result for the 2D Helmholtz example. Solution shown for $0, \ldots, 3$ homogenization steps

4.2 Hyperbolic Problems

In this section we consider the time-dependent hyperbolic model equation in one space dimension,

$$\frac{\partial u}{\partial t} + g(x)\frac{\partial u}{\partial x} = f(x,t), \qquad u(0,t) = 0, \quad u(x,0) = b(x). \tag{153}$$

The variable coefficient $g(x)$ is positive and bounded.

We will use three different methods to homogenize (153). First, we consider a semi-discrete form of the problem, and only homogenize the spatial part of the operator. Second, we make a correction for the extra errors introduced when neglecting the time-derivative. Third, we homogenize the full two-dimensional operator.

The most straightforward way to homogenize (153) is by a semi-discrete approximation,

$$\frac{\partial u_i}{\partial t} + \frac{1}{h} g_i \Delta_- u_i = f_i(t), \qquad i = 1, \ldots, n, \qquad u_0 = -u_1, \quad u_i(0) = b_i, \tag{154}$$

with $g_i = g((i - 1/2)h)$. In matrix form we will have

$$U_t + L_{j+1}U = F, \qquad 2^{j+1} = n, \tag{155}$$

which can be expressed in a wavelet basis as

$$\frac{\partial}{\partial t}\begin{pmatrix} U_f \\ U_c \end{pmatrix} + \begin{pmatrix} A_j & B_j \\ C_j & D_j \end{pmatrix}\begin{pmatrix} U_f \\ U_c \end{pmatrix} = \begin{pmatrix} F_f \\ F_c \end{pmatrix}, \qquad U_f, F_f \in W_j, \qquad U_c, F_c \in V_j. \tag{156}$$

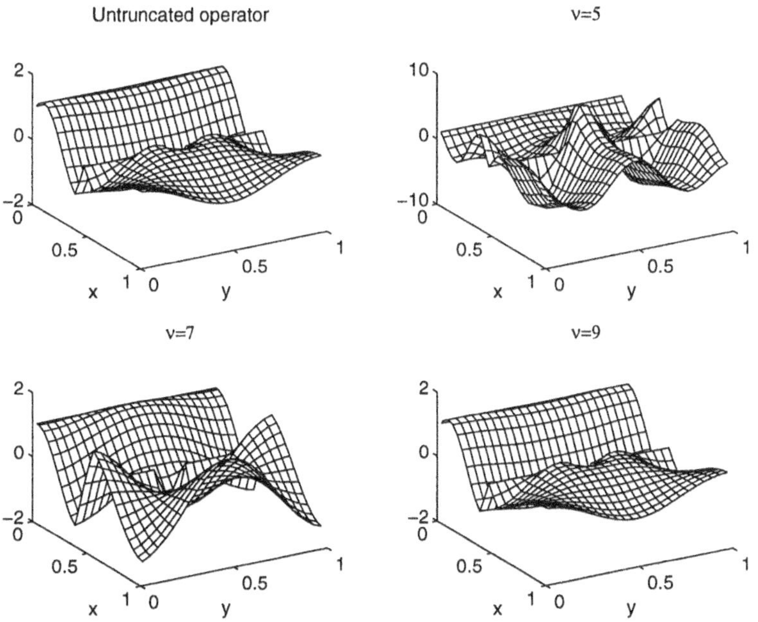

Fig. 8. Results for the 2D Helmholtz example, using the one step homogenized operator, truncated with different ν

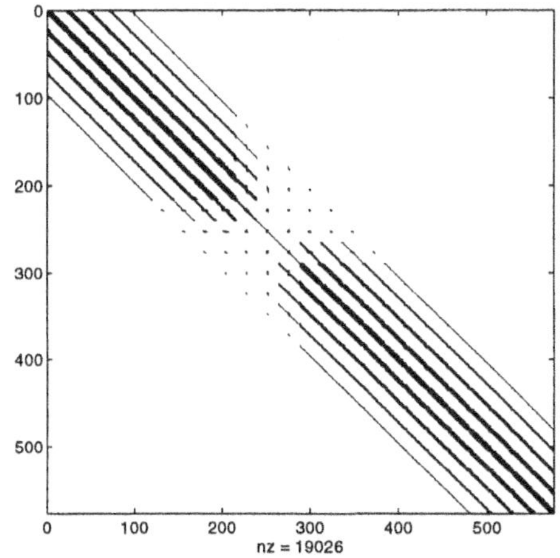

Fig. 9. Structure of the homogenized operator $\bar{L}$, after one homogenization step, for the 2D Helmholtz example. Elements larger than 0.1% of max value shown

In principle, L_{j+1} could be homogenized in the same way as in the elliptic case, using the Schur complement. The motivation for this is (9), which shows that when the scales of $g(x/\varepsilon)$ become very small, $\varepsilon \to 0$, the effect of the time-derivative vanishes, $\frac{\partial}{\partial t}U_{\mathrm{f}} \to 0$. As a limiting process in classical homogenization, this is fully justified, compare [5]. This argument implies that in the first set of equations of (156),

$$\frac{\partial U_{\mathrm{f}}}{\partial t} + A_j U_{\mathrm{f}} + B_j U_{\mathrm{c}} = F_{\mathrm{f}} \quad \Rightarrow \quad U_{\mathrm{f}} = A_j^{-1}(-B_j U_{\mathrm{c}} + F_{\mathrm{f}} - \frac{\partial U_{\mathrm{f}}}{\partial t}), \tag{157}$$

the last term, $-A_j^{-1}\frac{\partial}{\partial t}U_{\mathrm{f}}$ is eliminated. Substitution into the second set of equations of (156) yields

$$\frac{\partial U_{\mathrm{c}}}{\partial t} + \bar{L}_j U_{\mathrm{c}} = \bar{F}_j, \tag{158}$$

with $\bar{L}_j$ and $\bar{F}_j$ defined by (104).

We found experimentally that using an approximation of the term $-A_j^{-1}\frac{\partial}{\partial t}U_{\mathrm{f}}$ improved the homogenized operator. The correction is derived by taking the time-derivative of (157),

$$\frac{\partial U_{\mathrm{f}}}{\partial t} = A_j^{-1}\left(-B_j \frac{\partial U_{\mathrm{c}}}{\partial t} + \frac{\partial F_{\mathrm{f}}}{\partial t} - \frac{\partial^2 U_{\mathrm{f}}}{\partial t^2}\right). \tag{159}$$

After a transient mode we have for the model problem (153) with $g = \tilde{g}(x/\varepsilon)$ and $\tilde{g}(y)$ 1-periodic,

$$u(x,t) = \bar{u}(x,t) + \varepsilon u_1(x, x/\varepsilon, t) + \mathcal{O}(\varepsilon^2), \tag{160}$$

compare [5]. Hence, $\frac{\partial^2}{\partial t^2}U_{\mathrm{f}}$ is of the same order as $\frac{\partial}{\partial t}U_{\mathrm{f}}$ since the oscillations on the ε-scale are not functions of time and it is reasonable to neglect $-A_j^{-1}\frac{\partial^2}{\partial t^2}U_{\mathrm{f}}$ in (159) instead of $-A_j^{-1}\frac{\partial}{\partial t}U_{\mathrm{f}}$ in (157). Substituting (159) after elimination, into (156) gives

$$K_j \frac{\partial U_{\mathrm{c}}}{\partial t} + (D_j - C_j A_j^{-1} B_j) U_{\mathrm{c}} = F_{\mathrm{c}} - C_j A_j^{-1} F_{\mathrm{f}} + C_j A_j^{-2} \frac{\partial F_{\mathrm{f}}}{\partial t}, \tag{161}$$

with the correction matrix

$$K_j = I + C_j A_j^{-2} B_j. \tag{162}$$

Hence, the corrected homogenized operator $\tilde{L}_j$ and right hand side $\tilde{F}_j$ are

$$\tilde{L}_j = K_j^{-1}\bar{L}_j, \qquad \tilde{F}_j = K_j^{-1}\bar{F}_j + K_j^{-1}C_j A_j^{-2}\frac{\partial F_{\mathrm{f}}}{\partial t} \tag{163}$$

and we will solve (158), with $\tilde{L}_j$, $\tilde{F}_j$ substituted for $\bar{L}_j$, $\bar{F}_j$ respectively.

In principle, better approximations could be obtained by reiterating the steps above. After repeated differentiations of (159) we could choose to eliminate terms involving successively higher order time derivatives of U_{f}. However, the condition numbers of the correction matrices produced in this manner rapidly deteriorate.

Let $n = 128$, $f \equiv 0$ and

$$b(x) = \begin{cases} \sin^2(4\pi x), & 0 \le x \le 0.25 \\ 0, & 0.25 < x \le 1. \end{cases} \tag{164}$$

The result at time $t = 1$, is given in Fig. 10 for the case of $g(x)$ uniformly random distributed in $[0.1, 2]$ and $g(x)$ as in (148). The effect of the correction is shown to the right. In these calculations, the time-integration was replaced by a fourth-order Runge-Kutta method.

Because of causality the full operator must be lower triangular in the hyperbolic case. The homogenized operator will keep this property in the Haar case. After two

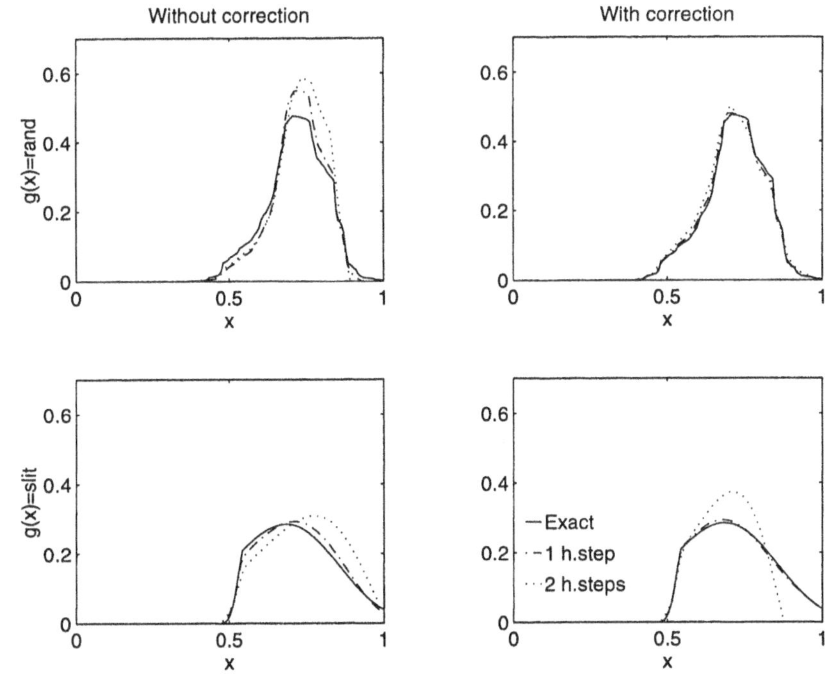

Fig. 10. The solution $u(x)$ at $t = 1$ for the hyperbolic case, using different $g(x)$, with and without first order correction. Solutions shown are the exact solution, computed with the full operator, and solutions computed with the operator homogenized one and two steps

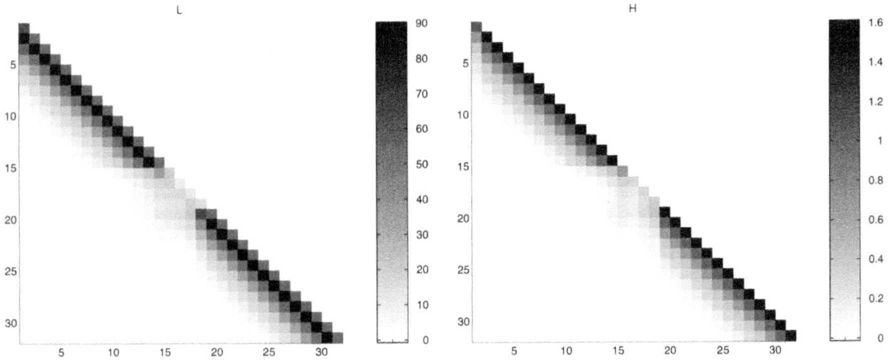

Fig. 11. Structure of the untruncated homogenized operators $\bar{L}$ (left) and H (right) for the hyperbolic problem, $g(x)$ a slit. Gray level indicates absolute value of elements

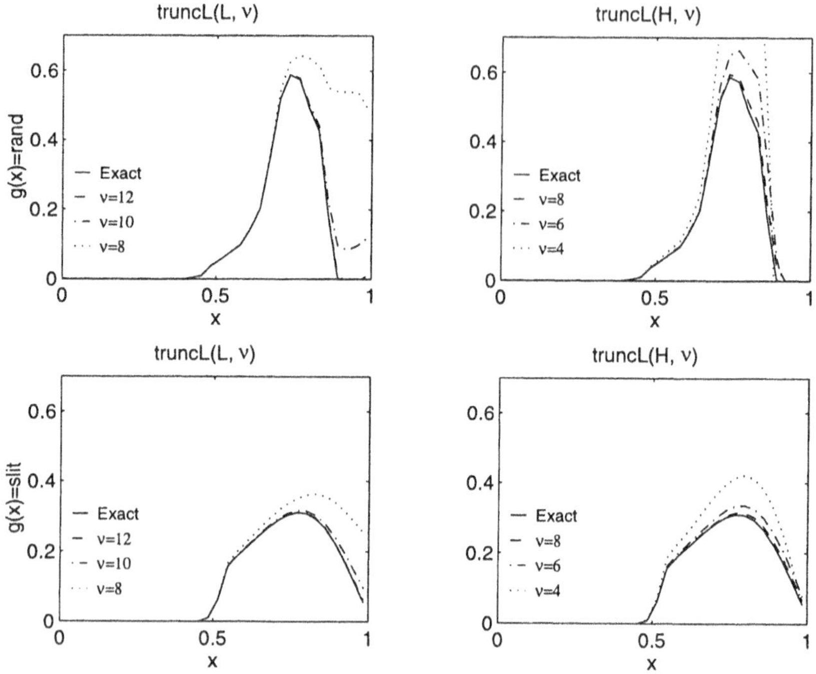

Fig. 12. Solution $u(x)$ for the hyperbolic case, using different $g(x)$ and different approximations of the homogenized operator. The "exact" solution refers to the solution with the full 32×32 homogenized operator

homogenization steps the operator, with $g(x)$ a slit, has the structure shown in Fig. 11. In view of this we approximate $\bar{L}$ and H, see (117), by truncation to lower triangular form. In Fig. 12 results using this truncation is displayed. The same two types of g_ε as in Fig. 10 were used. Like in the elliptic case, truncating the H matrix is more efficient than truncating $\bar{L}$. When the correction method is used it is harder to approximate the resulting homogenized operator, $\tilde{L}$, with a sparse matrix. It will not be as diagonal dominant as in the non-corrected case.

In our third approach we leave the semi-discrete form and instead make a full two-dimensional discretization of (153) in (x,t) space, using the implicit Euler approximation in time,

$$\begin{aligned} \frac{1}{h}\Delta_-^t u_{i\ell} + \frac{1}{h} g_i \Delta_-^x u_{i\ell} &= f_i, \qquad i,\ell = 1,\dots,n, \\ u_{0,\ell} &= -u_{1,\ell}, \\ u_{i,0} &= 2b_i - u_{i,1}. \end{aligned} \tag{165}$$

Like in the two-dimensional Helmholtz case, Sect. 4.1, we arrive at a linear system of equations,

$$L_{j+1}U = F, \qquad U, F \in \mathbf{V}_{j+1}, \qquad n = 2^{j+1}, \tag{166}$$

which is homogenized in the standard way, with the Schur complement.

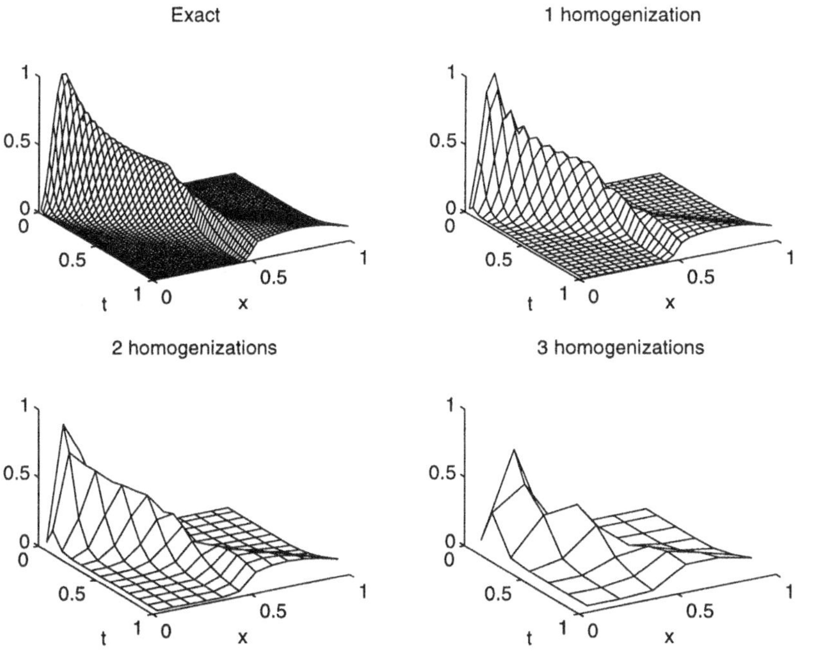

Fig. 13. Result for the 2D hyperbolic case, with $g(x)$ a slit

Using the same data as in the semi-discrete case, but with $n = 48$, we get for the case of g as in (148) the results in Fig. 13. We used the untruncated homogenized operator, but its structure indicates that, like in the 2D Helmholtz case, good sparse approximations can be found.

5 Waveguide

In this chapter we study a waveguide filter containing a fine scale structure, to illustrate some ways of using the numerically homogenized operator $\bar{L}$ in an application. The filter is shown in Fig. 14. It is composed of a straight waveguide with small gratings engraved on one of its sides at regular intervals. At one point the distance between two of the grating teeth is increased by 50%. This *quarter-wave step* gives a narrow-band filter effect and causes waves of one resonant frequency to pass through, while reflecting adjacent frequencies. The most important feature of the waveguide filter is the interplay between a small scale periodic structure (the gratings) and a localized inhomogeneity (the quarter-wave step). For this case, analytical homogenization techniques do not apply. The ability to separate frequencies makes the filter a useful component in many communication applications. For instance, it allows an optical communication link to be partitioned into many channels where signals can be selectively transmitted and detected. Applications can be found in [31] and [23]. The parameters used in the design of the filter determine many of its relevant properties, and it is of interest to simulate the filter numerically in order to anticipate the influence of these parameters.

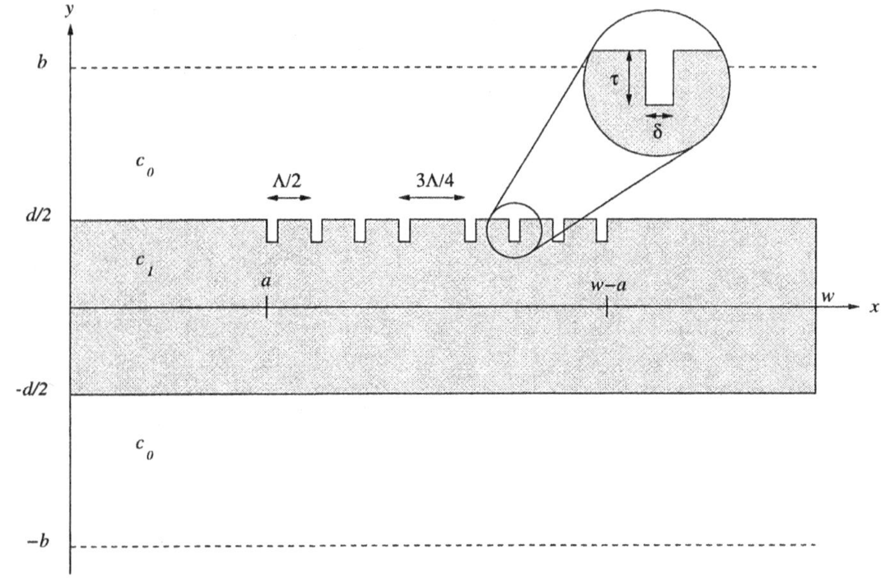

Fig. 14. Schematic picture of the waveguide with parameter specifications

We model the waveguide with the two-dimensional Helmholtz equation,

$$\nabla \cdot (c(x,y)^2 \nabla u) + \omega^2 u = 0, \tag{167}$$

where $c(x, y)$ represents the material dependent speed of propagation. In our case, c is piecewise constant and (167) should be interpreted in the distributional sense. See [36] for more details. In Fig. 14 we specify the parameters that define the waveguide (w, a, d, Λ, δ and τ). Related to these is S, denoting the number of gratings, which for a real filter would be a very large number. Further parameters are c_0 and c_1, the propagation speeds inside and outside the waveguide, which we fix to be $c_0 = 1$ (vacuum) and $c_1 = 1/3.3$ (GaAs).

5.1 Numerical Approximation

We approximate (167) in the rectangle $[0, w] \times [-b, b]$ on a uniform $N \times M$ size grid. With the same notation as in Chap. 4, we use a second order scheme given by

$$\frac{1}{h^2} \Delta_+^x c_{k-1/2,\ell}^2 \Delta_-^x u_{k\ell} + \frac{1}{h^2} \Delta_+^y c_{k,\ell-1/2}^2 \Delta_-^y u_{k\ell} + \omega^2 u_{k\ell} = 0 \tag{168}$$

for $k = 1, \dots, N$ and $\ell = 1, \dots, M$. This discretization corresponds to the same divergence form as the continuous problem (167), which ensures that the numerical solution satisfies the same interface conditions as the weak solution of (167) when c is discontinuous.

To complete the continuous and the discrete problems, (167, 168), extra conditions must be given at the boundary of the domain $[0, w] \times [-b, b]$. At $y = \pm b$

we use a Neumann boundary condition $u_y = 0$, discretized with the second order approximation

$$\frac{1}{h}\Delta_+^y u_{k,0} = 0, \qquad \frac{1}{h}\Delta_+^y u_{k,M} = 0, \qquad k = 1,\ldots,N. \tag{169}$$

This is motivated by the fact that the solution should be almost constant zero far away from the waveguide. For the boundaries at $x = 0$ and $x = w$ we need absorbing boundary conditions, since we are looking for a wave solution propagating along the x-axis. We use

$$\frac{1}{h}\Delta_+ u_{0,\ell} + a_1 u_{0,\ell} = a_2 u_{\rm in}(y_\ell), \qquad \frac{1}{h}\Delta_+ u_{N,\ell} + a_3 u_{N,\ell} = 0, \tag{170}$$

for $\ell = 1,\ldots,M$ with a_j, $j = 1,2,3$ chosen such that the a discrete solution of the waveguide problem without gratings satisfy both (168) and (170) exactly. Here $u_{\rm in}(y)$ is the wave we send into the waveguide at $x = 0$.

The resulting linear system of equations is then of the form

$$LU = F, \quad U, F \in V_n \otimes V_m, \quad \mathcal{L}(V_n \otimes V_m, V_n \otimes V_m), \tag{171}$$

where $M = 2^m$ and $N = 2^n$.

5.2 Direct Simulation

As an example of how the filter works, we simulate a test problem with the specifications

$$\begin{array}{lllll} w = 1, & a = 1/4, & S = 32, & d = 17/1024, & b = 4/128, \\ \Lambda = 32/1024, & \tau = 4/1024, & \delta = 4/1024, & M = 64, & N = 1024. \end{array} \tag{172}$$

The left part of Fig. 15 shows examples of solutions in the case of total reflection, $\omega/2\pi = 44.5$, and total transmission, $\omega/2\pi = 45.155$. In the right part of Fig. 15 the *frequency response* of the same filter is shown. This function measures how much power is transmitted through the filter at each frequency. For a fixed frequency the response is defined and computed as

$$T_{\rm pow} \equiv \frac{\int_{-\infty}^{\infty} |u(w,y)|^2 dy}{\int_{-\infty}^{\infty} |u_{\rm in}(y)|^2 dy} \approx \frac{\sum_{\ell=1}^{M} |u_{N,\ell}|^2}{\sum_{\ell=1}^{M} |u_{\rm in}(y_\ell)|^2}. \tag{173}$$

We note that within its operational range, the filter indeed just lets through waves in a narrow frequency band. The right figure also shows the stability of the response under perturbation of the parameters a and b. The case (i) is the original test problem (172). In (ii) we modified a to $1/8$ (and $N = 768$, $w = 3/4$). In (iii) we changed b to $24/1024$ (and $M = 48$). As can be seen, these perturbations only have a small effect on the response.

5.3 One-Dimensional Models

To reduce the complexity of solving the full Helmholtz equation (167), and to gain better understanding of the physical processes, one often wants to derive lower-dimensional models that capture the significant features of the full model. A typical model would be of the form

$$\frac{d}{dx} c_{\rm eff}(x,\omega)^2 \frac{du_{\rm e}}{dx} + \omega^2 u_{\rm e} = 0, \tag{174}$$

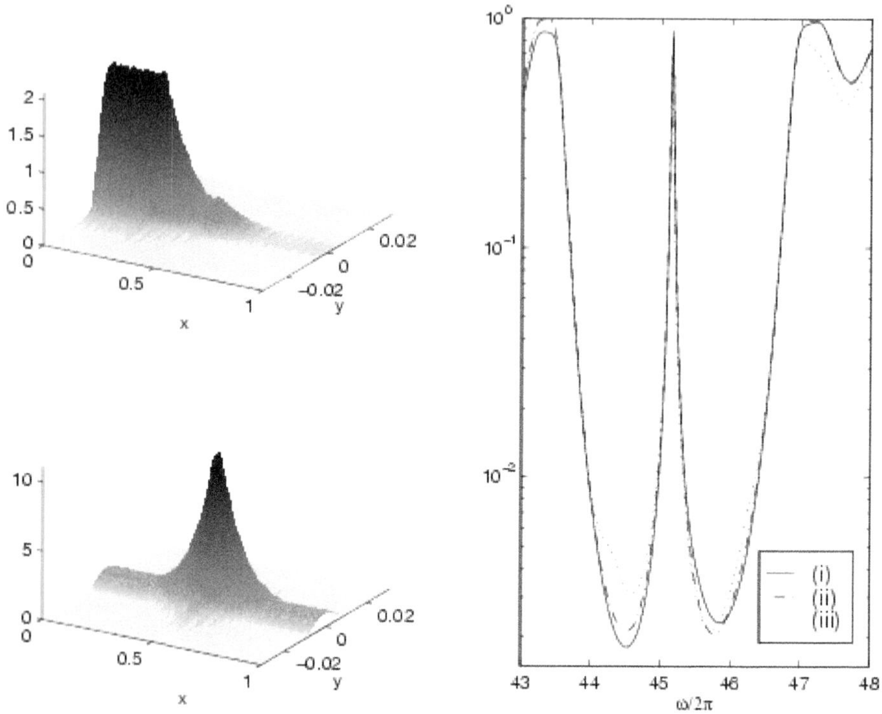

Fig. 15. Plots of solutions to test problem (172). Left figures show $|u|$ for frequencies in the stopband (top left) and at resonance (bottom left). Right figure shows the frequency response $T_{\text{pow}}(\omega)$ for the same test problem (i) and for problems with perturbed a (ii) and b (iii)

where u_e is related to u, the solution of (167), in a simple way, see e.g. [4,37]. We will here show how the homogenization technique described in Chap. 3 can be used to derive a family of one-dimensional models with different properties. In these one-dimensional models u_e is an approximation of the solution, projected in the y-direction on a space of low dimension.

The full two-dimensional operator is first homogenized multiple levels, but only in the y-direction. We hence use $X_0 = V_n \otimes V_{m'}$ with $m' < m$ as the small space. The operator is subsequently truncated and identified with a discretization of a one-dimensional differential operator acting on $V_n \times \cdots \times V_n$ ($2^{m'}$ factors). The resulting operator and right hand side approximate a system of one-dimensional partial differential equations of size $M' = 2^{m'}$, the number of grid points left in the y-direction. The order of the differential operator corresponds to the bandwidth of the truncated operator. In conclusion, three different parameters can be varied:

1. The size M' of the system of equations,
2. The order r of the differential operator,
3. The wavelet system used in the homogenization.

In general there is a trade-off between M' and r, such that a smaller M' requires a larger r. The reason is that the bandwidth after truncation usually needs to increase with the number of levels that an operator is homogenized in order to maintain good accuracy.

To exemplify the process described above, we derive a one-dimensional model for the test problem given by

$$\begin{array}{l} w=1, \quad a=3/16,\ S=10, \quad d=5/64,\ b=1/8, \\ \Lambda=1/8,\ \tau=1/64,\ \delta=1/64,\ M=16, \quad N=64. \end{array} \tag{175}$$

We use a medium size model, with $M'=4$, $r=2$, and X_0 given by the Daubechies wavelet system with four vanishing moments. Numerical experiments with different wavelet systems suggest that in general the higher order of the wavelets, the better the result.

Let us introduce the continuous 4×4 system model with the unknowns $u=(u_1,u_2,u_3,u_4)^*\in\mathbb{C}^4$,

$$\partial_x(B(x)\partial_x u)+C(x)u+\omega^2 u=0, \qquad B(x),C(x):\mathbb{R}\mapsto\mathbb{R}^{4\times 4}. \tag{176}$$

This ansatz differs from (174) in that it also has a variable coefficient in the lower order term. On the other hand, none of the coefficients are assumed to be frequency dependent. This is preferable here, since it will reduce the cost of computing the frequency response of the filter.

We now proceed as follows. The discrete two-dimensional problem is homogenized two levels in the y-direction and afterwards the unknowns in the equation $\bar{L}\bar{u}=\bar{f}$ are reordered to the blocked form

$$\begin{pmatrix} \bar{L}_{11} & \bar{L}_{12} & \bar{L}_{13} & \bar{L}_{14} \\ \bar{L}_{21} & \bar{L}_{22} & \bar{L}_{23} & \bar{L}_{24} \\ \bar{L}_{31} & \bar{L}_{32} & \bar{L}_{33} & \bar{L}_{34} \\ \bar{L}_{41} & \bar{L}_{42} & \bar{L}_{43} & \bar{L}_{44} \end{pmatrix}\begin{pmatrix}\bar{u}_1\\ \bar{u}_2\\ \bar{u}_3\\ \bar{u}_4\end{pmatrix}=\begin{pmatrix}\bar{f}_1\\ \bar{f}_2\\ \bar{f}_3\\ \bar{f}_4\end{pmatrix}, \qquad \bar{u}_i,\bar{f}_i\in V_n, \qquad \bar{L}_{ij}\in\mathcal{L}(V_n,V_n). \tag{177}$$

Each of the suboperators $\bar{L}_{ij}$ is truncated to tridiagonal form, which essentially corresponds to truncating the full operator $\bar{L}$ to 15-diagonal form (before reordering). We identify the unknowns $\bar{u}_i$ and $\bar{f}_i$ with one-dimensional grid functions, $\bar{u}_i=\{\bar{u}_{i,k}\}\approx\{u_i(x_k)\}$ and $\bar{f}_i=\{\bar{f}_{i,k}\}\approx\{f_i(x_k)\}$ for some u_i and f_i. Furthermore, the truncated operators $\tilde{L}_{ij}$ are identified with the discretization

$$\frac{1}{h^2}\tilde{L}_{ij}=\frac{1}{h^2}\Delta_+ b_{ij}(x_{k-1/2})\Delta_- +\tilde{c}_{ij}(x_k)+\frac{1}{2}\gamma_{ij}(x_k)(\mathcal{E}+\mathcal{E}^{-1})+\omega^2 \tag{178}$$

where $\mathcal{E}$ is the displacement operator defined by $\mathcal{E}x_k=x_{k+1}$. The γ_{ij} terms are needed since $\tilde{L}_{ij}$ is in general not symmetric. Each $\tilde{L}_{ij}$ can be seen as a discretization of a second order differential operator acting on the grid function $\bar{u}_i$. After assembling the suboperators by setting $B=\{b_{ij}\}$, $\tilde{C}=\{\tilde{c}_{ij}\}$ and $\Gamma=\{\gamma_{ij}\}$ we get

$$\frac{1}{h^2}\Delta_+ B(x_{k-1/2})\Delta_- U_k+\tilde{C}(x_k)U_k+\frac{1}{2}\Gamma(x_k)(U_{k-1}+U_{k+1})+\omega^2 U_k=F_k, \tag{179}$$

where we also assembled the grid functions $U=\{\bar{u}_i\}$ and $F=\{\bar{f}_i\}$. If $C(x_k)=\tilde{C}(x_k)+\Gamma(x_k)$ and $B(x_k)$ were real and independent of ω and if $F=0$ this could be identified with a second order discretization of (176). Thus, the matrices $B(x)$

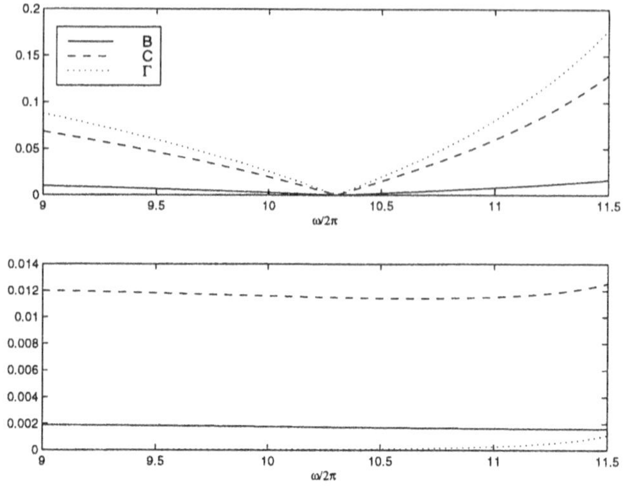

Fig. 16. Frequency dependency of coefficients. Relative L^2-difference between coefficients computed at ω and at $\omega/2\pi = 10.3$ (above). Relative L^2-difference between imaginary and real part of coefficients (below)

and $C(x)$ represent the effective material of the waveguide, similar to $c_{\text{eff}}(x,\omega)$ in (174).

In fact, the computed coefficients satisfy these conditions to a fairly good level of approximation, which is illustrated in Fig. 16. The coefficients are computed for a range of ω and the relative L^2-difference, compared to the coefficients computed at resonance, $\omega/2\pi = 10.3$, is plotted in the top picture. The $B(x)$ coefficients vary on the order of only 1% in the interval. In the bottom picture the L^2-norm of the imaginary part of the coefficients relative to the real part is plotted, showing that it is at worst a little more than 1%. This indicates that the model (176) is justified.

In Fig. 17 we show the frequency response computed in two different ways, and compare them with the response for the two-dimensional model. The first technique is to recompute the coefficients for each new frequency. This gives a very good agreement with the response of the full model. The second technique is to compute all coefficients, once and for all, at the resonance frequency, $\omega/2\pi = 10.3$. Even now the model captures the true response fairly well. We could hence calculate all parameters of the model for one single frequency and the model will remain approximately valid for the whole frequency band that we are interested in.

The computational gains can be large for the method described above. Since the bandwidth of the one-dimensional model is $M'r$, the cost of computing the solution for k frequencies is CkM'^2r^2N flops, plus the k-independent cost of constructing the homogenized operator. At least for large k, the work of computing the solutions dominates. Comparing this with a direct method, which requires $CkNM^3$ flops, we get a cost ratio of

$$\frac{M'^2r^2}{M^3}, \tag{180}$$

which is often small. For the test problem (175) it equals 1/64.

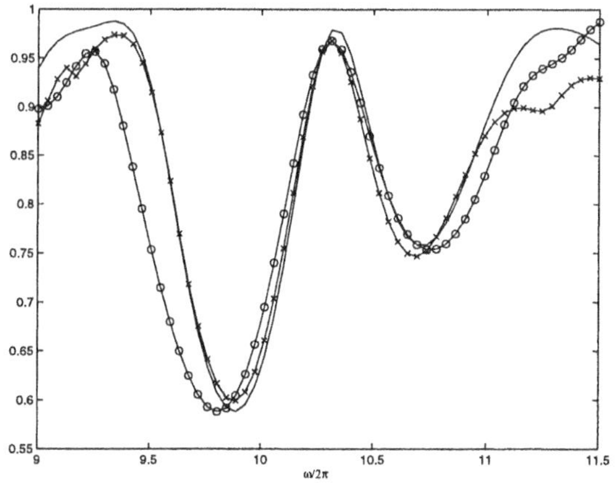

Fig. 17. Frequency response of test problem (175) computed using the one-dimensional model given in the text. Results shown when response was computed with the full two-dimensional model (solid line), with the one-dimensional model and coefficients recomputed for each new frequency (crosses), with the one-dimensional model and coefficients computed only once, for $\omega/2\pi = 10.3$, (circles)

5.4 The Homogenized Operator as a Subgrid Model

In this section we will use numerical homogenization to obtain linear subgrid models of the details of the waveguide and use them to solve a large problem on a coarse grid. Our target problem is given by the parameters

$$\begin{array}{lllll} a = 3/16, & d = 9/128, & b = 1/8, & \omega/2\pi = 11 & \\ \Lambda = 1/8, & \tau = 3/128, & \delta = 3/128, & M = 32, & N = 128, \end{array} \tag{181}$$

with varying length w and number of gratings S. In order to solve this problem on a coarse grid, we would need the homogenized operator $\bar{L}$ and right hand side $\bar{F}$. The key observation here is that a good approximation of these can be obtained *directly from a much smaller problem*, the analogue of the cell problem in classical homogenization.

We identify four distinct parts of the waveguide: the initial part, a grating tooth, the quarter-wave step in the middle and the end part. Let $\bar{u}^{\mathrm{i}}$, $\bar{u}^{\mathrm{m}}$, and $\bar{u}^{\mathrm{e}}$ denote the solution at grid points in the initial segment, middle segment and end segment. Moreover, let $\bar{u}^{\mathrm{t}}_k$ be the solution at grating tooth k. Then we can decompose $\bar{L}\bar{u} = \bar{F}$ as

$$(\tilde{L} + \delta L) \begin{pmatrix} \bar{u}^{\mathrm{i}} \\ \bar{u}^{\mathrm{t}}_1 \\ \vdots \\ \bar{u}^{\mathrm{t}}_{S/2} \\ \bar{u}^{\mathrm{m}} \\ \bar{u}^{\mathrm{t}}_{S/2+1} \\ \vdots \\ \bar{u}^{\mathrm{t}}_S \\ \bar{u}^{\mathrm{e}} \end{pmatrix} = \begin{pmatrix} \bar{F}^{\mathrm{i}} \\ \bar{F}^{\mathrm{t}}_1 \\ \vdots \\ \bar{F}^{\mathrm{t}}_{S/2} \\ \bar{F}^{\mathrm{m}} \\ \bar{F}^{\mathrm{t}}_{S/2+1} \\ \vdots \\ \bar{F}^{\mathrm{t}}_S \\ \bar{F}^{\mathrm{e}} \end{pmatrix} \tag{182}$$

where $\bar{F}$ follows the same partitioning as $\bar{u}$ and

$\tilde{L} =$

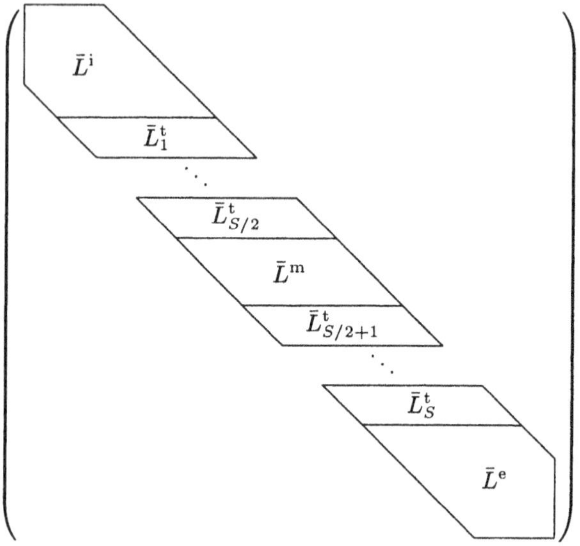

(183)

with elements being zero outside the delineated areas. The matrix δL is the residual between $\bar{L}$ and $\tilde{L}$, with all elements zero within the delineated areas.

With this notation we now observe that since $\bar{L}$ is diagonal dominant, δL will be small. More importantly,

$$\bar{L}^{\mathrm{t}}_k \approx \bar{L}^{\mathrm{t}}_\ell, \qquad \bar{F}^{\mathrm{t}}_k \approx \bar{F}^{\mathrm{t}}_\ell, \qquad \forall k, \ell. \tag{184}$$

We will interpret $\bar{L}^{\mathrm{i}}$, $\bar{L}^{\mathrm{t}}$, $\bar{L}^{\mathrm{m}}$, and $\bar{L}^{\mathrm{e}}$ together with $\bar{F}^{\mathrm{i}}$, $\bar{F}^{\mathrm{t}}$, $\bar{F}^{\mathrm{m}}$, and $\bar{F}^{\mathrm{e}}$ as the subgrid models for the corresponding parts of the geometry. (Since $\bar{L}^{\mathrm{t}}_k$ and $\bar{F}^{\mathrm{t}}_k$ are approximately the same for all k we will henceforth drop the subindices.)

The strategy for constructing an approximation of the homogenized operator of the full problem is then clear. First, we compute an approximation of $\bar{L}^{\mathrm{i}}$, $\bar{L}^{\mathrm{t}}$, $\bar{L}^{\mathrm{m}}$, and $\bar{L}^{\mathrm{e}}$ by homogenizing a much smaller problem. At the same time approximations of $\bar{F}^{\mathrm{i}}$, $\bar{F}^{\mathrm{t}}$, $\bar{F}^{\mathrm{m}}$, and $\bar{F}^{\mathrm{e}}$ are also obtained. Second, we assemble the subgrid models according to the block structures of (182) and (183) to obtain an approximation of $\tilde{L}$ and $\bar{F}$. Finally, δL is neglected.

In our example we use (181) with $w = 1$ and $S = 10$ as our small problem. This is a well resolved problem with eight gridpoints between the gratings and approximately 15 gridpoints per effective wavelength in the x-direction. The operator is homogenized three levels in the y-direction and one level in the x-direction, corresponding to a grid size of $M = 4$ and $N = 64$. This gives a very coarse resolution in the y-direction, in particular no resolution of the gratings, and a reasonable resolution in the x-direction. This grid size should be enough to represent the solution in a quantitatively correct way.

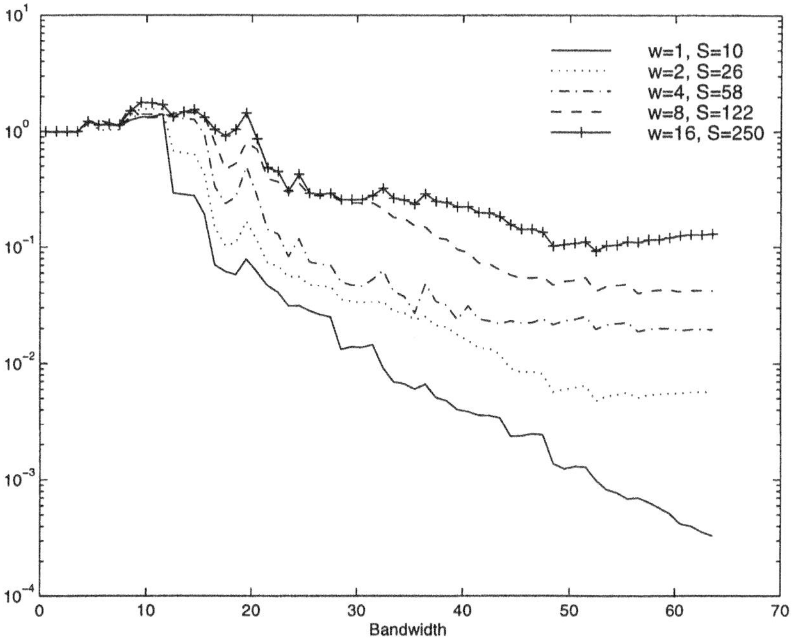

Fig. 18. Result of the subgrid technique. Relative L^2-error of solution as a function of the bandwidth

For the large problems denote the corresponding parameters with a prime sign. We will have $w' > 1$ and $S' > 10$ and we set $\bar{L}^{\mathrm{i}'} = \bar{L}^{\mathrm{i}}$, $\bar{L}^{\mathrm{m}'} = \bar{L}^{\mathrm{m}}$, $\bar{L}^{\mathrm{e}'} = \bar{L}^{\mathrm{e}}$, and

$$\bar{L}_k^{\mathrm{t}'} = \begin{cases} \bar{L}_k^{\mathrm{t}} & 1 \le k \le 2, \\ \bar{L}_3^{\mathrm{t}} & 3 \le k \le S'/2 - 2, \\ \bar{L}_{k+(S-S')/2}^{\mathrm{t}} & S'/2 - 1 \le k \le S'/2 + 2, \\ \bar{L}_{S/2+3}^{\mathrm{t}} & S'/2 + 3 \le k \le S' - 2, \\ \bar{L}_{k+S-S'}^{\mathrm{t}} & S' - 1 \le k \le S'. \end{cases} \tag{185}$$

So far we have deliberately been vague about the exact size of the areas in (183), and hence the size of the subgrid models. We now introduce a parameter ν, the band width of $\tilde{L}$, with which we all sizes can be defined at the same time. The value of ν will determine the accuracy of the solution.

The results of this computation is shown in Fig. 18, where the relative L^2-error between the solution computed using the subgrid technique and the exact solution, projected onto the coarse subspace, is plotted as a function of ν, for five different problem sizes. Our small problem, $w = 1, S = 10$ is also included, for comparison, in which case ν just indicates the bandwidth of the truncated operator. Even for a problem 16 times as large as the small problem, the relative L^2-error can be reduced to 10%, by choosing ν big enough. Note that even when the relative L^2-error is rather large, a good quantitative picture of the solution is obtained in the interior of the waveguide. This is exemplified for the case of $w = 16$ and $\nu = 22$

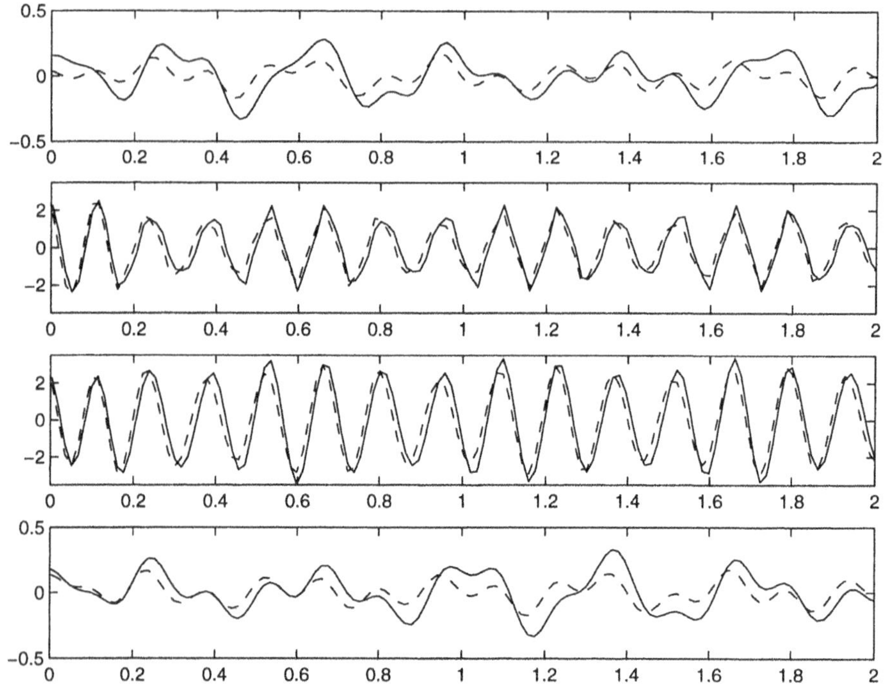

Fig. 19. Result of the large $w = 16$ problem approximated with $\nu = 22$. Plot of exact solution (solid) and approximate solution (dashed) as a function of x for four y-values corresponding to the four grid points in the y-direction of the coarse mesh. Real part of u shown in the x-interval $[0, 2]$, the region with the largest errors

in Fig. 19, where the part of the solution with the largest errors is plotted. The relative L^2-error in this case is of the order of 50%.

To estimate the gain in using the subgrid technique, let M and N denote the grid size of the large problem. The exact operator has a bandwidth of M. The cost to solve this problem is therefore CNM^3 flops with a direct solver. Suppose our coarse grid is of the size $N/2^n$ and $M/2^m$, corresponding to homogenization n times in the x-direction and m times in the y-direction. The direct solution using the subgrid model then costs $CNM\nu^2/2^{n+m}$ giving a ratio between the costs of

$$2^{-n-m}\left(\frac{\nu}{M}\right)^2, \tag{186}$$

In our largest case, $w = 16$, with $\nu = 45$, we get an approximate ratio of 1/8. We must also compute the subgrid model, although this only needs to be done once for all five computations above, and for a fixed grid size M', N' that can be considerably smaller than M, N. This cost is dominated by the other factors.

6 Multigrid

Many properties of the homogenized operator closely approximates those of the original operator. The lower eigenvalues and the corresponding eigenfunctions are, for example, well approximated in elliptic problems, see Santosa and Vogelius [40] for the classical case, and for the numerical case, Beylkin and Coult [9]. It is thus natural to try using an approximation of the homogenized operator as the coarse grid operator in a multigrid scheme. The largest potential advantage would be for problems where traditional multigrid is less efficient. Elliptic problems with highly oscillatory or strongly discontinuous coefficients are typical examples. Another class of such examples are advection dominated advection diffusion problems. For these equations the lower eigenmodes have large high derivatives and standard coarsening methods does not give satisfactory convergence. The coarser grids can not resolve the oscillations and it is therefore better to use an operator where these oscillations are not explicitly present but where there effect is taken into account.

In [19,20] this idea was explored and it was shown that using the numerical approximations of the analytically homogenized operator could be advantageous as a coarse grid operator for certain problems.

There are some factors, which make it more attractive to use the numerically homogenized operator, as developed in this paper, rather than the analytic form. A wider class of problems can be reached. Strict periodicity is not essential. Furthermore there are natural wavelet based prolongations and restrictions matching the definition of the coarse grid operator.

6.1 Algorithms

Let the system of linear equations,

$$L_h U_h = F_h, \tag{187}$$

come from the discretization of a differential equation, (1), on some grid Ω_h, where h represents the step size.

For notational purposes, we briefly describe the V-cycle method used in this context. Given an interpolation operator, I_{2h}^{h} where the superscript refers to the fine grid and the subscript refers to the coarse grid, and a restriction operator, I_h^{2h}, we can define a multigrid method recursively. The description of the two-level method is as follows. First, relax one or a few steps on the fine grid Ω_h to get an initial approximation U_h. Then, compute the residual $R_h = F_h - L_h U_h$, restrict the residual to the coarse grid Ω_{2h} : $R_{2h} = I_h^{2h} R_h$, and solve the residual equation,

$$L_{2h} E_{2h} = R_{2h}, \tag{188}$$

on the coarse grid. Set $U_h = U_h + I_{2h}^{h} E_{2h}$ and relax again one or a few steps on the fine grid. This describes the two-level method. Based on this, we define the V-cycle multigrid scheme recursively.

The first step in the wavelet homogenization of (187) can be written

$$\mathcal{W} L_h \mathcal{W}^* \mathcal{W} U = \begin{pmatrix} A & B \\ C & D \end{pmatrix} \begin{pmatrix} U_{\mathrm{f}} \\ U_{\mathrm{c}} \end{pmatrix} = \begin{pmatrix} F_{\mathrm{f}} \\ F_{\mathrm{c}} \end{pmatrix} = \mathcal{W} F_h . \tag{189}$$

Calculating the inverse of the factorization, and solving for U, it is natural that the interpolation and restriction operators should be defined by,

$$I_{2h}^{h} = \tilde{P}^* - \tilde{Q}^* A^{-1} B, \qquad I_{h}^{2h} = (I_{2h}^{h})^*, \qquad \mathcal{W} = \begin{pmatrix} \tilde{Q} \\ \tilde{P} \end{pmatrix} \tag{190}$$

where, as before, $\tilde{P}^*$ and $\tilde{Q}^*$ are bases for the scaling and wavelet spaces respectively. The coarse grid operator should be

$$\bar{L}_{2h} = D - CA^{-1}B, \tag{191}$$

which is the Schur complement. Although the matrices A, B, C and D are as sparse as the original operator L_h, A^{-1} is not. We observe that the fill-in that results from inversion decays exponentially as we move away from the original structure. This was discussed in Chap. 3.

The above procedure may be repeatedly applied until the desired coarseness is reached. Although the level of fill-in in the operator A^{-1} increases, the magnitude of the values decreases as we go away from the diagonal. This property of decay makes compression possible. Properties of wavelets play here an important role. In practice a compressed version $\tilde{L}_{2h}$ of $\bar{L}_{2h}$ should be used. One thing to keep in mind is that the number of gridpoints in each coordinate direction must be an even number.

One step towards improving efficiency is to avoid computing the inverse exactly. To this end, we use $ILU(0)$ to compute the incomplete LU factorization, and then use a sequence of forward and backward substitutions to compute the inverse. For the compression, we use a thresholding procedure. Any values that appear in the inverse in locations that hold zero values in D are set to zero, thus eliminating any fill-in over the original matrix, D. We will call this method the truncated wavelet multigrid method, and we will refer to the original method as the dense or full wavelet multigrid method.

6.2 Numerical Examples

We shall give a few examples in which the wavelet based multigrid method has been applied to two dimensional model problems. The examples are from De Leon, [30].

In the first set of examples the elliptic operator is,

$$L_\varepsilon = -\nabla \cdot g(x,y)\nabla \tag{192}$$

for a Dirichlet problem on the unit square. The standard 5-point numerical approximation is applied to 16×16 and 32×32 grids, resulting in system matrics of size 256×256 and 1024×1024 respectively.

The coefficient $g(x,y)$ is oscillatory and two examples are studied,

$$g(x,y) = 1 + 0.8\sin(10\sqrt{2}\pi x), \tag{193}$$

$$g(x,y) = 1 + 0.8\sin(10\sqrt{2}\pi(x-y)). \tag{194}$$

The analytically homogenized operators are,

$$\bar{L} = -\mu\frac{\partial^2}{\partial x^2} - \bar{a}\frac{\partial^2}{\partial y^2}, \tag{195}$$

$$\bar{L} = -\frac{\mu+\bar{a}}{2}\frac{\partial^2}{\partial x^2} + (\bar{a}-\mu)\frac{\partial^2}{\partial x \partial y} - \frac{\mu+\bar{a}}{2}\frac{\partial^2 U}{\partial y^2}, \tag{196}$$

respectively, with

$$\mu = \left(\int_0^1 (1 + 0.8 \sin(2\pi x))^{-1} dx \right)^{-1}, \tag{197}$$

$$\bar{a} = \int_0^1 1 + 0.8 \sin(2\pi x) dx. \tag{198}$$

Convergence results are given in Fig. 20 for the Haar case. Comparison is made with the standard finite difference multigrid, [22], and with standard finite difference multigrid with the coarse grid operator replaced by a difference approximation of $\bar{L}$ above. The effect of compression can be seen in Fig. 21 for booth Haar and Daubechies wavelets with four vanishing moments.

Consider also a problem with piecewise constant $g(x, y)$ in the form of a checkerboard pattern with 16 squares and the g-values 1 and 10^5. The standard multigrid method does not converge and we compare with an algebraic multigrid method, [38]. The wavelet technique works quite well but the convergence rate is reduced if the compression is too extreme, see Fig. 22.

Finally we approximate the advection diffusion problem,

$$-\varepsilon \Delta u + b \cdot u = 0, \quad x \in \Omega, \tag{199}$$

$$u = f(x, y), \quad x \in \partial\Omega, \tag{200}$$

where Ω is a unit square and $\|b\| \gg \varepsilon > 0$. The five point formula is used for the Laplacian and upwind differencing for the advection term,

$$b(x, y) = ((2y - 1)(1 - x^2), 2xy(y - 1)). \tag{201}$$

Figure 23 shows that the convergence rate is quite good and essentially independent of the stepsize. In this case the convergence does not deteriorate with $ILU(0)$ and compression.

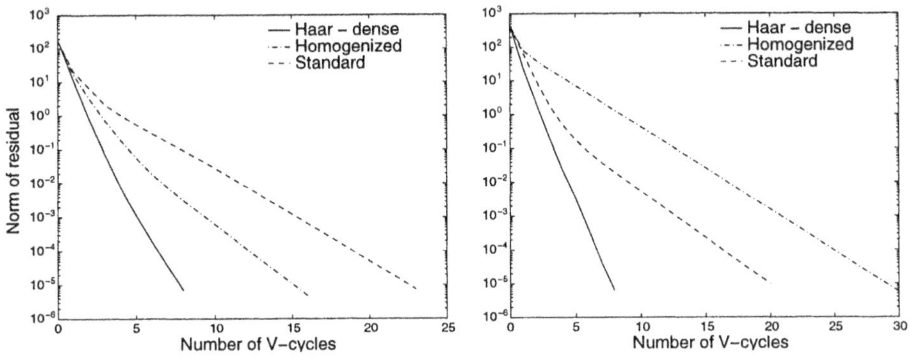

Fig. 20. Convergence result for $g(x, y)$ defined by (193). Three level V-cycle with one Gauss-Seidel smoothing iteration is used. Left frame 16×16 grid, right frame 32×32 grid

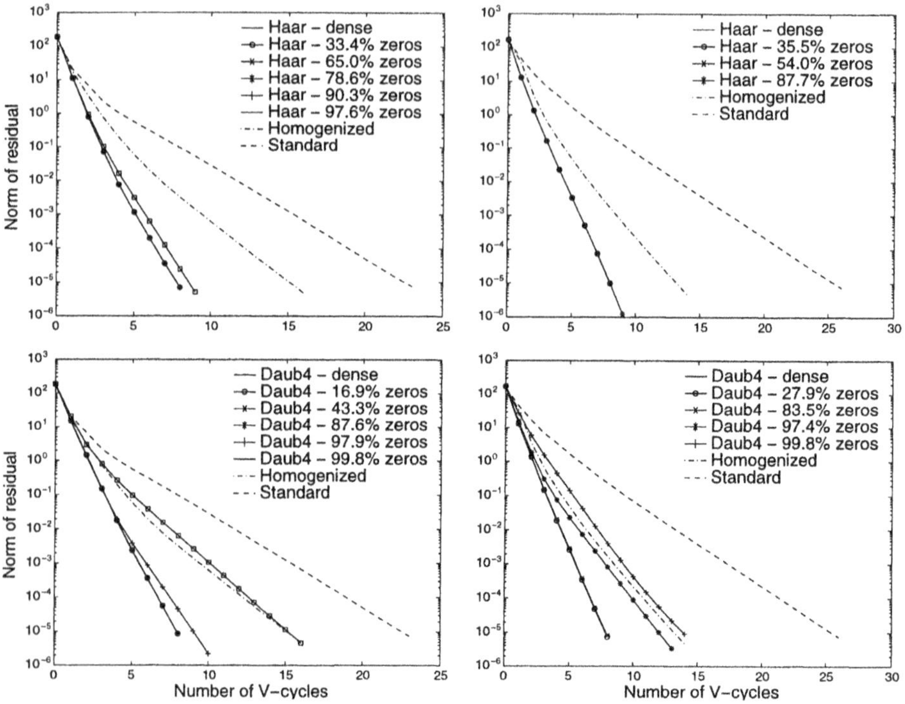

Fig. 21. Convergence result for $g(x, y)$ defined by (193) in left frames and by (194) in right frames. Three level V-cycle with one Gauss-Seidel smoothing iteration is used. Fine grid is 16×16. Percentage of zeros is for coarse grid operator, second level

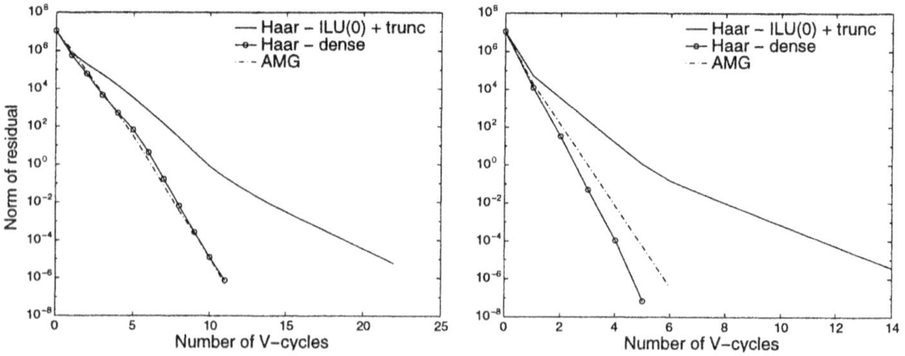

Fig. 22. Convergence result for 2-level iteration, $g(x, y)$ from checkerboard problem on 16×16 grid. Left frame 1 Gauss-Seidel smoothing iteration, right frame 5 Gauss-Seidel smoothing iterations

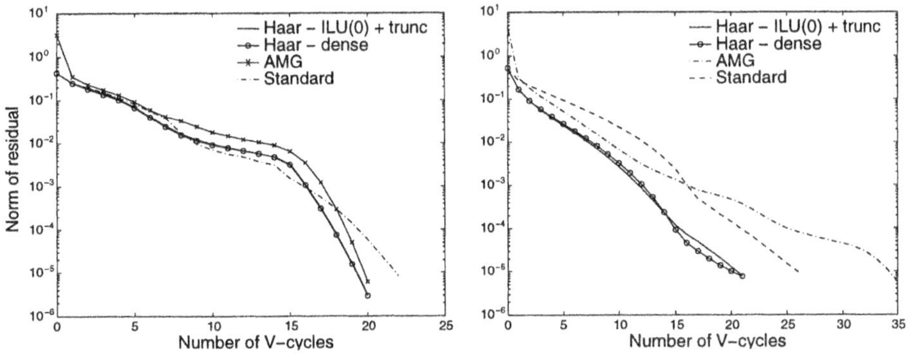

Fig. 23. Convergence result for advection diffusion problem. Left frame 16 × 16 grid, right frame 32 × 32 grid. Two level V-cycle with one Gauss-Seidel smoothing iteration is used

7 Nonlinear Equations

Homogenization of nonlinear equations are usually much more complicated than the linear equations we have considered so far. For some equations, however, the linear numerical homogenization methods described above generalize. The basic assumption needed is that the fine scale part of the solution is small so that we can linearize around the coarse scale. The ideas in this chapter were developed by Beylkin, Brewster and Gilbert in [7]. See also [27].

Suppose $L_{j+1} : V_{j+1} \mapsto V_{j+1}$ is a nonlinear operator. Using the same notation as in Sect. 3.2, the nonlinear decomposition of $L_{j+1}(U) = F$ can be written

$$\begin{aligned} &\mathcal{W}_j L_{j+1}(\mathcal{W}_j^* \mathcal{W}_j U) \\ &= \begin{pmatrix} \tilde{Q}_j \left(L_{j+1}\left(\tilde{Q}_j^* U_{\mathrm{f}} + \tilde{P}_j^* U_{\mathrm{c}}\right) - L_{j+1}\left(\tilde{P}_j^* U_{\mathrm{c}}\right)\right) + \tilde{Q}_j L_{j+1}\left(\tilde{P}_j^* U_{\mathrm{c}}\right) \\ \tilde{P}_j \left(L_{j+1}\left(\tilde{Q}_j^* U_{\mathrm{f}} + \tilde{P}_j^* U_{\mathrm{c}}\right) - L_{j+1}\left(\tilde{P}_j^* U_{\mathrm{c}}\right)\right) + \tilde{P}_j L_{j+1}\left(\tilde{P}_j^* U_{\mathrm{c}}\right) \end{pmatrix} \\ &\equiv \begin{pmatrix} A_j(U_{\mathrm{f}}, U_{\mathrm{c}}) + B_j(U_{\mathrm{c}}) \\ C_j(U_{\mathrm{f}}, U_{\mathrm{c}}) + D_j(U_{\mathrm{c}}) \end{pmatrix} = \mathcal{W}_j F = \begin{pmatrix} F_{\mathrm{f}} \\ F_{\mathrm{c}} \end{pmatrix}. \end{aligned} \tag{202}$$

Like in the linear case, U_{f} can be eliminated from the equations and the homogenized operator is

$$\bar{L}_j(U_{\mathrm{c}}) = D_j(U_{\mathrm{c}}) + C_j(A_j^{-1}(F_{\mathrm{f}} - B_j(U_{\mathrm{c}}), U_{\mathrm{c}})), \tag{203}$$

where $A_j^{-1}(u, v)$ is the solution w to $A_j(w, v) = u$.

Consider nonlinear operators of the type

$$L(u) = \int_0^x G(t, u(t))dt + H(x, u(x)), \tag{204}$$

The equation $L(u) = f$ could result for instance from a nonlinear ODE such as

$$\frac{d}{dx} H(x, u) = -G(x, u), \qquad H(0, u(0)) = f. \tag{205}$$

For simplicity we here let u be a scalar function, $G, H \in C^\infty$ and $f = \text{const}$. Similar derivations can be made also for systems of equations, non-smooth coefficients and variable f. One application that was studied in [27], is the steady state equations for

$$\begin{aligned} u_t &= u_{xx} - u^3 + u - v, \\ v_t &= \delta v_{xx} + \varepsilon(u + a_1(x)v + a_0(x)), \end{aligned} \tag{206}$$

which can be formulated as a system of nonlinear ODEs. These are the Fitzhugh Nagumo equations with diffusing inhibitor, widely used as a model for pattern formation in excitable reaction-diffusion systems.

For the discretization of (204) we use the Haar basis, but since G and H are nonlinear, it is more convenient to rescale and approximate cell averages of u instead of its Haar basis coefficients. Hence, we introduce $P_j = \tilde{P}_j/\sqrt{2}$ and $Q_j = \sqrt{2}\tilde{Q}_j$. Define the grid $\boldsymbol{x}_{j+1} = \{x_{j+1,k}\} = \{(k + 1/2)2^{-(j+1)}\}$ and the grid function $\boldsymbol{u}_{j+1} = \{u_{j+1,k}\} \approx 2^{(j+1)/2} P_{V_{j+1}} u(x)$, where $k = 0, \ldots, 2^{j+1} - 1$ and $P_{V_{j+1}}$ is the orthogonal projection in $L^2([0,1])$ to the Haar scaling space V_{j+1}. Furthermore, recursively define the succesively coarser grids $\boldsymbol{x}_j = P_j \boldsymbol{x}_{j+1}$, cell averages $\boldsymbol{u}_j = \boldsymbol{u}_{j+1}^{\mathrm{c}} = P_j \boldsymbol{u}_{j+1}$ and the fine scale parts $\boldsymbol{u}_{j+1}^{\mathrm{f}} = Q_j \boldsymbol{u}_{j+1}$. We have the following discretization of $L(u) = f$,

$$L_{j+1}(\boldsymbol{u}_{j+1}) = K_{j+1} G(\boldsymbol{x}_{j+1}, \boldsymbol{u}_{j+1}) + H(\boldsymbol{x}_{j+1}, \boldsymbol{u}_{j+1}) = f, \tag{207}$$

where

$$K_{j+1} = \delta_{j+1} \begin{pmatrix} \frac{1}{2} & 0 & \ldots & 0 \\ 1 & \ddots & \ddots & \vdots \\ \vdots & \ddots & \ddots & 0 \\ 1 & \ldots & 1 & \frac{1}{2} \end{pmatrix} \in \mathbb{R}^{2^{j+1} \times 2^{j+1}}, \qquad \delta_{j+1} = 2^{-(j+1)}, \tag{208}$$

and G, H being applied elementwise to its vector arguments, e.g. $G(\boldsymbol{x}_j, \boldsymbol{u}_j) = \{G(x_{j,k}, u_{j,k})\}$. The following easily verifiable formulae hold

$$Q_j K_{j+1} = -\frac{\delta_j}{2} P_j, \qquad P_j K_{j+1} = K_j P_j + \frac{\delta_j}{8} Q_j. \tag{209}$$

After also rescaling B_j and A_j we get

$$B_j(\boldsymbol{u}_j) = -\frac{\delta_j}{2} P_j G(\boldsymbol{x}_{j+1}, 2P_j^* \boldsymbol{u}_j) + Q_j H(\boldsymbol{x}_{j+1}, 2P_j^* \boldsymbol{u}_j), \tag{210}$$

$$\begin{aligned} A_j(\boldsymbol{u}_{j+1}^{\mathrm{f}}, \boldsymbol{u}_j) = {} & -\frac{\delta_j}{2} P_j \Big(G(\boldsymbol{x}_{j+1}, Q_j^* \boldsymbol{u}_{j+1}^{\mathrm{f}}/2 + 2P_j^* \boldsymbol{u}_j) - G(\boldsymbol{x}_{j+1}, 2P_j^* \boldsymbol{u}_j) \Big) \\ & + Q_j \Big(H(\boldsymbol{x}_{j+1}, Q_j^* \boldsymbol{u}_{j+1}^{\mathrm{f}}/2 + 2P_j^* \boldsymbol{u}_j) - H(\boldsymbol{x}_{j+1}, 2P_j^* \boldsymbol{u}_j) \Big). \end{aligned} \tag{211}$$

When G and H are linear functions of u, (204) reduces to (120), and $G(\boldsymbol{x}_{j+1}, \boldsymbol{u}_{j+1}) = \text{diag}(a(\boldsymbol{x}_{j+1}))\boldsymbol{u}_{j+1}$, $H(\boldsymbol{x}_{j+1}, \boldsymbol{u}_{j+1}) = \text{diag}(b(\boldsymbol{x}_{j+1}))\boldsymbol{u}_{j+1}$. Consequently, A_j is diagonal and the homogenization step is very easy, as was mentioned in Sect. 3.3. In the nonlinear case, $A_j(\boldsymbol{u}_{j+1}^{\mathrm{f}}, \boldsymbol{u}_j)$ is still local, and its jacobian with respect to $\boldsymbol{u}_{j+1}^{\mathrm{f}}$ for fixed $\boldsymbol{u}_j$ is diagonal.

In order to treat the nonlinear case, we make the simplifying assumption that $\delta_j \sim \boldsymbol{u}_{j+1}^{\mathrm{f}}$ is small. We then have the Taylor expansions,

$$\begin{aligned} P_j G(\boldsymbol{x}_{j+1}, \boldsymbol{u}_{j+1}) &= G(\boldsymbol{x}_j, \boldsymbol{u}_j) + \frac{\delta_j^2}{8} G_{xx}(\boldsymbol{x}_j, \boldsymbol{u}_j) + \frac{\delta_j \boldsymbol{u}_{j+1}^{\mathrm{f}}}{4} G_{xu}(\boldsymbol{x}_j, \boldsymbol{u}_j) \\ &+ \frac{(\boldsymbol{u}_{j+1}^{\mathrm{f}})^2}{8} G_{uu}(\boldsymbol{x}_j, \boldsymbol{u}_j) + \mathcal{O}(\delta_j^4), \qquad (212) \\ Q_j G(\boldsymbol{x}_{j+1}, \boldsymbol{u}_{j+1}) &= \delta_j G_x(\boldsymbol{x}_j, \boldsymbol{u}_j) + \boldsymbol{u}_{j+1}^{\mathrm{f}} G_u(\boldsymbol{x}_j, \boldsymbol{u}_j) + \mathcal{O}(\delta_j^3), \qquad (213) \end{aligned}$$

where, again, all functions are applied elementwise. This gives the simplified expressions

$$\begin{aligned} B_j(\boldsymbol{u}_j) &= -\frac{\delta_j}{2} G(\boldsymbol{x}_j, \boldsymbol{u}_j) + \delta_j H_x(\boldsymbol{x}_j, \boldsymbol{u}_j) + \mathcal{O}(\delta_j^3) \qquad (214) \\ A_j(\boldsymbol{u}_{j+1}^{\mathrm{f}}, \boldsymbol{u}_j) &= \boldsymbol{u}_{j+1}^{\mathrm{f}} H_u(\boldsymbol{x}_j, \boldsymbol{u}_j) + \mathcal{O}(\delta_j^3). \qquad (215) \end{aligned}$$

Hence, the solution to the cellproblem $A_j(\boldsymbol{u}_{j+1}^{\mathrm{f}}, \boldsymbol{u}_j) = -B_j(\boldsymbol{u}_j)$ satisfies

$$\boldsymbol{u}_{j+1}^{\mathrm{f}}(\boldsymbol{x}_j, \boldsymbol{u}_j) = \delta_j \frac{G(\boldsymbol{x}_j, \boldsymbol{u}_j) - 2H_x(\boldsymbol{x}_j, \boldsymbol{u}_j)}{2H_u(\boldsymbol{x}_j, \boldsymbol{u}_j)} + \mathcal{O}(\delta_j^3). \qquad (216)$$

We also compute approximations of C_j and D_j,

$$\begin{aligned} D_j(\boldsymbol{u}_j) &= P_j K_{j+1} G(\boldsymbol{x}_{j+1}, 2P_j^* \boldsymbol{u}_j) + P_j H(\boldsymbol{x}_{j+1}, 2P_j^* \boldsymbol{u}_j) \\ &= K_j \left(G(\boldsymbol{x}_j, \boldsymbol{u}_j) + \frac{\delta_j^2}{8} G_{xx}(\boldsymbol{x}_j, \boldsymbol{u}_j) \right) + H(\boldsymbol{x}_j, \boldsymbol{u}_j) \\ &+ \frac{\delta_j^2}{8} H_{xx}(\boldsymbol{x}_j, \boldsymbol{u}_j) + \frac{\delta_j^2}{8} G_x(\boldsymbol{x}_j, \boldsymbol{u}_j) + \mathcal{O}(\delta_j^4), \qquad (217) \\ C_j(\boldsymbol{u}_{j+1}^{\mathrm{f}}, \boldsymbol{u}_j) &= K_j P_j \left(G(\boldsymbol{x}_{j+1}, \boldsymbol{u}_{j+1}) - G(\boldsymbol{x}_{j+1}, 2P_j^* \boldsymbol{u}_j) \right) \\ &+ \frac{\delta_j}{8} Q_j \left(G(\boldsymbol{x}_{j+1}, \boldsymbol{u}_{j+1}) - G(\boldsymbol{x}_{j+1}, 2P_j^* \boldsymbol{u}_j) \right) \\ &+ P_j \left(H(\boldsymbol{x}_{j+1}, \boldsymbol{u}_{j+1}) - H(\boldsymbol{x}_{j+1}, 2P_j^* \boldsymbol{u}_j) \right) \\ &= K_j \left(\frac{\delta_j \boldsymbol{u}_{j+1}^{\mathrm{f}}}{4} G_{xu}(\boldsymbol{x}_j, \boldsymbol{u}_j) + \frac{(\boldsymbol{u}_{j+1}^{\mathrm{f}})^2}{8} G_{uu}(\boldsymbol{x}_j, \boldsymbol{u}_j) \right) \\ &+ \frac{\delta_j \boldsymbol{u}_{j+1}^{\mathrm{f}}}{8} G_u(\boldsymbol{x}_j, \boldsymbol{u}_j) + \frac{\delta_j \boldsymbol{u}_{j+1}^{\mathrm{f}}}{4} H_{xu}(\boldsymbol{x}_j, \boldsymbol{u}_j) \\ &+ \frac{(\boldsymbol{u}_{j+1}^{\mathrm{f}})^2}{8} H_{uu}(\boldsymbol{x}_j, \boldsymbol{u}_j) + \mathcal{O}(\delta_j^4). \qquad (218) \end{aligned}$$

Substituting (216) into the expressions for C_j and D_j gives a fourth order approximation of $\bar{L}_j$,

$$\bar{L}_j(\boldsymbol{u}_j) = K_j \bar{G}(\boldsymbol{x}_j, \boldsymbol{u}_j) + \bar{H}(\boldsymbol{x}_j, \boldsymbol{u}_j) + \mathcal{O}(\delta_j^4), \qquad (219)$$

where

$$\bar{G}(\boldsymbol{x}_j, \boldsymbol{u}_j) = G(\boldsymbol{x}_j, \boldsymbol{u}_j) + \frac{\delta_j^2}{8} G_{xx}(\boldsymbol{x}_j, \boldsymbol{u}_j) + \frac{\delta_j \boldsymbol{u}_{j+1}^{\mathrm{f}}}{4} G_{xu}(\boldsymbol{x}_j, \boldsymbol{u}_j) + \frac{(\boldsymbol{u}_{j+1}^{\mathrm{f}})^2}{8} G_{uu}(\boldsymbol{x}_j, \boldsymbol{u}_j), \tag{220}$$

$$\bar{H}(\boldsymbol{x}_j, \boldsymbol{u}_j) = H(\boldsymbol{x}_j, \boldsymbol{u}_j) + \frac{\delta_j^2}{8} H_{xx}(\boldsymbol{x}_j, \boldsymbol{u}_j) + \frac{\delta_j \boldsymbol{u}_{j+1}^{\mathrm{f}}}{8} G_u(\boldsymbol{x}_j, \boldsymbol{u}_j) + \frac{\delta_j \boldsymbol{u}_{j+1}^{\mathrm{f}}}{4} H_{xu}(\boldsymbol{x}_j, \boldsymbol{u}_j) + \frac{(\boldsymbol{u}_{j+1}^{\mathrm{f}})^2}{8} H_{uu}(\boldsymbol{x}_j, \boldsymbol{u}_j) + \frac{\delta_j^2}{8} G_x(\boldsymbol{x}_j, \boldsymbol{u}_j), \tag{221}$$

and $\boldsymbol{u}_{j+1}^{\mathrm{f}}$ given by the first term in (216). We hence get explicit formulae for $\bar{G}$ and $\bar{H}$ given G and H. As for linear equations, the homogenization can the be repeated. In certain cases, the form of G and H will be preserved by $\bar{G}$ and $\bar{H}$, such as when

$$G(x, u) = g_0(x, u) + \delta_j^2 g_1(x, u), \quad H(x, u) = h_0(x, u) + \delta_j^2 h_1(x, u), \tag{222}$$

where g_j and h_j are $\mathcal{O}(1)$.

Another approach is based on solving the nonlinear cellproblem (216) numerically, instead of using Taylor expansion, see [7]. When there is an apriori bound on $\boldsymbol{u}_j$ one can discretize also in this variable. The functions $\bar{G}$ and $\bar{H}$ are then represented and computed on a finite grid in both arguments. This is a more expensive method, but also more accurate when $\boldsymbol{u}_{j+1}^{\mathrm{f}}$ is not small.

References

1. J. Anderson. *Computational Fluid Dynamics, The Basics with Applications.* McGraw-Hill, 1995.
2. U. Andersson, B. Engquist, G. Ledfelt, and O. Runborg. A contribution to wavelet-based subgrid modeling. *Appl. Comput. Harmon. Anal.*, 7:151–164, 1999.
3. O. Axelsson. *Iterative Solution Methods.* Cambridge University Press, 1994.
4. T. M. Benson, R. J. Bozeat, and P. C. Kendall. Rigorous effective index method for semiconductor Rib waveguides. *IEE Proc. J*, 139(1):67–70, 1992.
5. A. Bensoussan, J.-L. Lions, and G. Papanicolau. *Asymptotic Analysis for Periodic Structures.* North-Holland Publ. Comp., The Netherlands, 1978.
6. G. Beylkin and M. Brewster. A multiresolution strategy for numerical homogenization. *Appl. Comput. Harmon. Anal.*, 2:327–349, 1995.
7. G. Beylkin, M. E. Brewster, and A. C. Gilbert. A multiresolution strategy for numerical homogenization of nonlinear ODEs. *Appl. Comput. Harmon. Anal.*, 5:450–486, 1998.
8. G. Beylkin, R. Coifman, and V. Rokhlin. Fast wavelet transforms and numerical algorithms I. *Comm. Pure Appl. Math.*, 44:141–183, 1991.
9. G. Beylkin and N. Coult. A multiresolution strategy for reduction of elliptic PDEs and eigenvalue problems. *Appl. Comput. Harmon. Anal.*, 5:129–155, 1998.

10. G. Beylkin, J. Dunn, and D. L. Gines. LU factorization of non-standard forms and direct multiresolution solvers. *Appl. Comput. Harmon. Anal.*, 5:156–201, 1998.
11. T. Chan and T. Mathew. The interface probing technique in domain decomposition. *SIAM J. Matrix Anal. Appl.*, 13(1):212–238, January 1992.
12. C. Concus, G. H. Golub, and G. Meurant. Block preconditioning for the conjugate gradient method. *SIAM J. Sci. Stat. Comp.*, 6:220–252, 1985.
13. I. Daubechies. *Ten Lectures on Wavelets.* SIAM, 1991.
14. M. Dorobantu and B. Engquist. Wavelet-based numerical homogenization. *SIAM J. Numer. Anal.*, 35(2):540–559, April 1998.
15. L. Durlofsky. Numerical calculation of equivalent grid block permeability tensors for heterogeneous porous media. *Water Resour. Res.*, 27:699–708, 1991.
16. Y. R. Efendiev, T. Y. Hou, and X. H. Wu. Convergence of a nonconformal multiscale finite element method. *SIAM J. Numer. Anal.*, 37:888–910, 2000.
17. B. Engquist and T. Y. Hou. Particle method approximation of oscillatory solutions to hyperbolic differential equations. *SIAM J. Num. Analysis*, 26:289–319, 1989.
18. B. Engquist and J.-G. Liu. Numerical methods for oscillatory solutions to hyperbolic problems. *Comm. Pure Appl. Math.*, 46:1–36, 1993.
19. B. Engquist and E. Luo. New coarse grid operators for highly oscillatory coefficient elliptic problems. *J. Comput. Phys.*, 129(2):296–306, 1996.
20. B. Engquist and E. Luo. Convergence of a multigrid method for elliptic equations with highly oscillatory coefficients. *SIAM J. Numer. Anal.*, 34(6):2254–2273, 1997.
21. A. C. Gilbert. A comparison of multiresolution and classical one-dimensional homogenization schemes. *Appl. Comput. Harmon. Anal.*, 5(1):1–35, 1998.
22. W. Hackbusch. *Multigrid Methods and applications.* Springer-Verlag, 1985.
23. H. A. Haus and Y. Lai. Narrow-band optical channel-dropping filter. *J. Lightwave Technol.*, 10(1):57–61, 1992.
24. T. Y. Hou and X. H. Wu. A multiscale finite element method for elliptic problems in composite materials and porous media. *J. Comput. Phys.*, 134(1):169–189, 1997.
25. T. J. R. Hughes. Multiscale phenomena: Green's functions, the dirichlet-to neumann formulation, subgrid, scale models, bubbles and the origins of stabilized methods. *Comput. Methods Appl. Mech. Engrg.*, 127:387–401, 1995.
26. T. J. R. Hughes, G. R. Feijóo, L. Mazzei, and J.-B. Quicy. The variational multiscale method – a paradigm for computational mechanics. *Comput. Methods Appl. Mech. Engrg.*, 166:3–24, 1998.
27. O. Runborg J. Krishnan and I.G. Kevrekidis. Application of wavelet-based reduction methods to models of transport and reaction in heterogeneous media. 2001. To appear.
28. J. Keller. Geometrical theory of diffraction. *J. Opt. Soc. Amer.*, 52, 1962.
29. S. Knapek. Matrix-dependent multigrid-homogenization for diffusion problems. *SIAM J. Sci. Stat. Comp.*, 20(2):515–533, 1999.
30. D. D. Leon. *Wavelet Operators Applied to Multigrid Methods.* PhD thesis, Department of Mathematics, UCLA, 2000.
31. M. Levy, L. Eldata, R. Scarmozzino, R. M. Osgood Jr, P. S. D. Lin, and F. Tong. Fabrication of narrow-band channel-dropping filters. *IEEE Photonic. Tech. L.*, 4(12):1378–1381, 1992.

32. Z. Li. *The Immersed Interface Method—A Numerical Approach for Partial Differential Equations with Interfaces.* PhD thesis, Department of Applied Mathematics, Univ. of Washington, 1994.
33. A.-M. Matache and C. Schwab. Homogenization via p-FEM for problems with microstructure. *Appl. Numer. Math.*, 33:43–59, 2000.
34. N. Neuss. *Homogenisierung und Mehrgitter.* PhD thesis, Fakultät Mathematik Universität Heidelberg, 1995.
35. N. Neuss, W. Jäger, and G. Wittum. Homogenization and multigrid. Preprint 98-04 (SFB 359), 1998.
36. P.-O. Persson and O. Runborg. Simulation of a waveguide filter using wavelet-based numerical homogenization. *J. Comput. Phys.*, 166:361–382, 2001.
37. V. Ramaswamy. Strip-loaded film waveguide. *Bell Syst. Tech. J.*, 53:697–705, 1974.
38. J. W. Ruge and K. Stüben. Algebraic multigrid. In *Multigrid Methods*, Frontiers in Applied Mathematics, pages 73–170. SIAM, 1987.
39. O. Runborg. *Multiscale and Multiphase Methods for Wave Propagation.* PhD thesis, Department of Numerical Analysis and Computing Science, KTH, Stockholm, 1998.
40. F. Santosa and M. Vogelius. First order corrections to the homogenized eigenvalues of a periodic composite medium. *SIAM J. Appl. Math.*, 53:1636–1668, 1993.
41. B. Z. Steinberg, J. J. McCoy, and M. Mirotznik. A multiresolution approach to homogenization and effective modal analysis of complex boundary value problems. *SIAM J. Appl. Math.*, 60(3):939–966, 2000.
42. A. Taflove. *Computational Electromagnetics, The Finite-Difference Time-Domain Method*, chapter 10. Artech House, 1995.
43. L. Tartar. *Etudes des oscillations dans les equations aux derivées partielle non linéaires*, volume 195 of *Lecture Notes in Physics*, pages 384–412. Springer-Verlag, 1984.
44. D. Wilcox. *Turbulence Modeling for CFD.* DCW Industries, Inc., 1993.

Beamlets and Multiscale Image Analysis

David L. Donoho[1] and Xiaoming Huo[2]

[1] Statistics Department, Stanford University;
[2] School of Industrial and Systems Engineering, Georgia Institute of Technology

Abstract. We describe a framework for multiscale image analysis in which *line segments* play a role analogous to the role played by *points* in wavelet analysis.

The framework has five key components. The *beamlet dictionary* is a dyadically-organized collection of line segments, occupying a range of dyadic locations and scales, and occurring at a range of orientations. The *beamlet transform* of an image $f(x, y)$ is the collection of integrals of f over each segment in the beamlet dictionary; the resulting information is stored in a *beamlet pyramid*. The *beamlet graph* is the graph structure with pixel corners as vertices and beamlets as edges; a path through this graph corresponds to a polygon in the original image. By exploiting the first four components of the beamlet framework, we can formulate *beamlet-based algorithms* which are able to identify and extract beamlets and chains of beamlets with special properties.

In this paper we describe a four-level hierarchy of beamlet algorithms. The first level consists of simple procedures which ignore the structure of the beamlet pyramid and beamlet graph; the second level exploits only the parent-child dependence between scales; the third level incorporates collinearity and co-curvity relationships; and the fourth level allows global optimization over the full space of polygons in an image.

These algorithms can be shown in practice to have suprisingly powerful and apparently unprecedented capabilities, for example in detection of very faint curves in very noisy data.

We compare this framework with important antecedents in image processing (Brandt and Dym; Horn and collaborators; Götze and Druckenmiller) and in geometric measure theory (Jones; David and Semmes; and Lerman).

Key Words and Phrases. Multiscale Line Segments. Multiscale Radon Transform. Line Segment Extraction. Curve Extraction. Object Extraction.

Acknowledgements. This work has been partially supported by AFOSR MURI 95–P49620–96–1–0028, by National Science Foundation grants DMS 98–72890 (KDI), DMS 95–05151, and ECS-97-07111, and by DARPA ACMP BAA 98-04. The authors would like to thank Achi Brandt, Emmanuel Candès, Scott S. Chen, Raphy Coifman, Davi Geiger, David Horn, Hiroshi Ishikawa, Ian Jermyn, Peter Jones, Gilad Lerman, Ofer Levi, Frank Natterer, Michael Saunders, Stephen Semmes, and Jean-Luc Starck for helpful comments, preprints, and references.

1 Introduction

In the last ten years, multiscale thinking in general and wavelet analysis in particular have become quite popular. The journal *Applied and Computational Harmonic Analysis*, founded only in 1993, has quickly become one of the most-cited journals in the mathematical sciences, and leading figures on wavelet research became among the most cited authors in the mathematical sciences as well.

If one searches for factors that may have contributed to the influence of wavelets, one notes that the twin ideas of multiscale representation and of localized representation appeal to the imagination, with phenomena 'at a certain scale' represented by waveforms at that scale localized at a certain positions in space.

1.1 Beyond Wavelets?

Wavelets barely scratch the surface of what is possible with multiscale representation; in this paper, we will describe a different multiscale system which allows us to explore and expose very different phenomena.

To see the need for such an alternate path to multiscale thinking, consider some of the principal phenomena arising in one of the key fields — Vision — that inspired much of the early work on wavelets [41,40].

Vision, of course, is a rapidly developing field [38,50], approached from many viewpoints, from 'hard' techniques such as sophisticated brain imaging experiments on humans, and direct surgical implantation of electrodes in cats and monkeys, to 'soft' techniques based on theoretical speculation.

It is well-accepted by now that vision is an inherently multiscale phenomenon, both because of the projective invariance required of the visual interpretation task (a face must be recognized as the same, whether seen from close up or far away) and because of the scale-invariance of natural scene statistics [25,45]. Moreover, the visual task is about representing and interpreting intensity fields replete with singularities — principally discontinuities. For years now, it has been suggested that a mathematical multiscale representation of vision might be the key to understanding how early human vision works [38,25]; and efforts have been made to compare response properties of neurons in region V1 of the visual cortex with wavelets [44,26]. Three specific phenomena seem suggestive.

First, that individual neurons responded best to stimuli definitely possessing a particular corresponding location and scale; second, that organized families of neurons spanning ranges of locations and scales were identified in the V1 region; and third, that the visual system very definitely works to apply nonlinear thresholding linearities to the outputs of V1 cells. These facts all supported the image of the V1 as computing a wavelet transform and thresholding the transform coefficients with the aim of obtaining a sparse representation of the underlying scene.

However, as research progressed, evidence arose supporting two sorts of counter-arguments which seem to contradict this naive view of the visual system as performing a kind of wavelet thresholding.

The first family of arguments runs against the idea that a visual system should be built from wavelets. First, it is known that the response properties of cortical cells are highly sensitive not only to the location and scale of the stimulus, but also the orientation and elongation of the stimulus. Second, empirical studies of databases of natural imagery [43,8,49] have found that the kinds of basis elements which best sparsify natural scenes are highly direction-specific, unlike wavelets. Third, since images are dominated by edges, we can consider mathematical studies pursuing optimally sparse representation in mathematical models of edge-dominated imagery [15], which likewise suggest a need for analyzing elements based on very highly oriented elongated features. Some of this work was described at Yosemite by Emmanuel Candès. It seems clear at the time of this writing that if a multiscale system is used to analyze images, it is likely to contain multiscale families of elements with 'needle'-like waveforms, assuming a wide range of locations, scales, orientations and elongations; wavelets, on the other hand, exhibit a small, fixed number of preferred orientations, and are better described as roughly isotropic.

The second family of arguments runs against the idea that a visual system can be based simply on thresholding of the outputs of a single layer of analyzing elements. Recent brain imaging experiments have shown subjects images containing long lines and curves, and have viewed the neural response by fMRI [39,50]. The results point to the V3 region (as well as other areas) as exhibiting substantial neural activity in response to such stimuli, with activity distributed throughout that region, as if a complex integrative task were taking place there. The sort of integrative task that may be taking place is illustrated by the the Gestalt principle of 'Good Continuation'. For example, Field and collaborators [27] and Julesz and collaborators [35] each conducted experiments in which a subject was shown, for a brief period of time, one of two images: either a confusing image containing many small randomly-oriented sausage-like features, or a similar image in which, in a addition, a small number of the same sausage-like features are organized as well-separated links which could with effort be recognized as part of a chain. The consistent finding was that when the second stimulus was presented, the chain fairly 'jumped out' at the observer, who perceived its existence 'preattentively', i.e. essentially reflexively. The argument is made that this 'integrative' or 'global' perceptual behavior runs counter to what one would expect in a system where isolated analyzing elements generate an output which is then processed separately from one element to the next.

These two streams of arguments suggest that any multiscale process underlying assimilation of visual information must have two properties:

[a] Analysis by a family of multiscale analyzing elements which are 'needle-like' at fine scales;

[b] A process of interconnecting evidence which considers the alignments and spatial interrelationships of these elements, for example, whether they belong to a common straight line, connected curve, or even a closed curve.

They also suggest that systems exhibiting these properties will require notions beyond those commonly encountered in wavelet analysis (and indeed in traditional multiscale analysis).

1.2 Beamlet Analysis

This paper will describe an approach to multiscale image analysis which we call *beamlet analysis*, which offers an interesting contrast to wavelet analysis, and which may broaden the reader's view of the possible kinds of multiscale thinking. While wavelets offer localized scale/location representation near fixed regions of space with specified scale and location, beamlets have localized scale/location/orientation based on dyadically-organized *line segments.*

The Beamlet framework involves 5 central components, which will be described in detail in Section 2 below.

1. The *beamlet dictionary* is a dyadically-organized library of line segments at a range of locations, orientations, and scales, which gives a multiscale approximation to the collection of all line segments.
2. The *beamlet transform* is the collection of all line integrals of the image along beamlets in the beamlet dictionary.
3. The *beamlet pyramid* is the collection of all beamlet transform coefficients, arranged in a data structure with a hierarchical multiscale nature.
4. The *beamlet graph* is the graph structure in which vertices correspond to pixel corners in the underlying image and the edges correspond to beamlets joining pairs of such pixel corners.
5. *Beamlet Algorithms* extract data from the beamlet pyramid in a way driven by the structure of the the beamlet graph, for example network flow algorithms, or recursive dyadic partitioning algorithms.

Clearly, a lot is going on here; it will take time and effort to understand this machinery and to appreciate the opportunities it offers.

Note first of all that this system certainly exhibits properties [a] and [b] proposed above as desired ingredients of a model for the visual system: analysis by 'needle-like' elements and methods for incorporating connectivity, collinearity and co-curvity. This pair of properties alone is enough to motivate study of this system. However, our personal motivation is based on certain intrinsic attractions of the system, as a scheme for organizing image information, and our perception that it has applications spanning a wide range of important scientific fields.

As far as the intrinsic attractions go, we think particularly of three:

- Beamlet analysis is clearly a fundamental multiscale idea, different from wavelet notions; by learning about it we will expand our notion of multiscale analysis and what it can accomplish.
  The consciousness-raising aspect of the beamlet framework is one we will repeatedly emphasize below; they can help us get 'out of the wavelet rut' in our habits of multiscale thinking.
- Beamlets play a fundamental approximation-theoretic role; chains of beamlets offer sparse approximate representations of nice curves in the plane — in a certain sense optimally sparse.
  A comparison with the approximation theory for wavelets is instructive. While wavelets can offer optimally sparse representations of smooth functions, beamlets offer optimally sparse representations of smooth curves embedded in an image.
- Beamlet methods provide a fundamentally correct data structure for dealing with noisy filament detection and boundary finding problems.
  The beamlet pyramid contains integrals of the image over line segments at all scales and locations. In certain signal detection problems, the usual detectors based on pixel-level filtering can have very poor signal-to-noise ratios, and hence low detection probabilities, while hiding somewhere in that pyramid can be integrals with high signal-to-noise ratio, allowing for signal detection which would not be possible using standard filtering or standard edge detection approaches.

We also believe that beamlet analysis offers a fascinating range of possible applications. There are several important scientific areas where a data structure oriented around line segments seems just what is called for. We mention two examples.

- *Particle Physics.* David Horn and co-workers Halina Abramowicz, Gideon Dror, Ury Naftali, Carmit Sahar-Pikielny at Tel-Aviv University have been developing tools for working with data from particle detectors associated with current state-of-the-art particle physics experiments. In such experiments, certain watched-for events involve excessive densities of particles in linear and curvilinear tracks. This team has had good results in identifying such structures in a highly cluttered and noisy environment using a kind of proto-beamlet analysis, combined with neural network postprocessing. See [3,4,23].
- *Large-Scale Structure of the Universe.* The availability of massive 3-d catalogs of coordinates of galaxies, such as the Las Campanas redshift survey and the forthcoming Sloan Digital Sky Survey, makes it possible to ask fundamental questions about the large-scale distribution of matter in the universe and to attempt to match simulations of galaxy clustering based on currently favored physical theories ('inflation' + 'cold dark matter') with data. Since the large-scale structure is largely dominated by the clustering of matter in sheets and filaments [24,5,47], the development of sensitive measures of such clustering behavior is paramount. Existing

studies extract subcatalogs lying in so-called 'pencil beams' (thin tubes) and study the properties of galaxy distribution in such beams. This is again a kind of proto-beamlet analysis, this time in 3-d.

We believe that such examples represent only the 'tip of the iceberg': many more such detection and clustering problems will be identified, over a range of fields, both in the analysis of image data and in the analysis of 3-d data. We have been developing a systematic infrastructure for such line segment based analysis, which we plan to describe in several publications.

We also observe that there are currently several groups of researchers who seem to be attracted as we are to the possibilities of data analysis based on local linear structures. In addition to David Horn's group mentioned above, we mention

- Achi Brandt at the Weitzman Institute of Science, and his former Ph.D. student J. Dym, who proposed fast algorithms for evaluating families of line integrals of an image [11], and who suggested that they might be used in vision data analysis [12]; and
- Peter Jones at Yale, and his former Ph.D. student Gilad Lerman, now at N.Y.U., who considered the evaluation of Jones' β's in the analysis of multivariate statistical data [36].

Finally, we mention our own work [22,30] making use of beamlet ideas, using a pre-beamlet terminology.

We find all this proto-beamlet work pioneering in various ways, and wish to encourage others to study and cite these earlier efforts. However, we also expect readers to find this prior body of work uneven and lacking a common language. For example, it is not even clear to us that all the above cited authors would agree among themselves that they were working on similar problems, though we claim they are.

In this paper we identify and formalize 5 key components of beamlet analysis, as mentioned above, and we call attention to key principles and heuristics. While a few of these were already evident in earlier work, it seems that the full package offers many possibilities not evident at all in earlier work, and which we hope to explain.

1.3 A Hierarchy of Beamlet Algorithms

In our view, one of the most important differences between traditional multiscale thinking based on wavelets and multiscale thinking based on beamlets lies in the kind of algorithms that can be built. Wavelet analysis offers basically two kinds of algorithms:

1. Very simple algorithms based on thresholding of wavelet coefficients, or simply calculating levelwise summaries;

2. Moderately simple algorithms based on the tree structure (parent-child relationships) obeyed by the dyadic squares hosting the wavelet coefficients.

In contrast, we can discern *four* levels of structure in beamlet-based algorithms. The first two are very analogous to wavelet-style thinking:

1. Very simple algorithms based on thresholding of beamlet coefficients, or simply calculating levelwise summaries;
2. Moderately simple algorithms based on the tree structure (parent-child relationships) of dyadic squares hosting the beamlet coefficients.

The analogy is due to the fact that, like wavelets, beamlets are indexed by scale and location, the scale/location indexing providing 'generation' relationships between parents and children at the same location but different scales. The next two levels go considerably beyond the patterns familiar from wavelet-style thinking; they involve exploitation of *neighboringness* relations among beamlets. Here neighboringness can include simple *connectedness* of pairs of beamlets – one ends where the next begins – as well as *collinearity* – two beamlets belong to the same beam – and *co-curvity* – two beamlets belong to the same curve. The two levels of structure are, in order of complexity:

3. Moderately sophisticated algorithms based on *local* uses of the neighboringness of beamlet coefficients according to the beamlet graph.
4. More sophisticated algorithms based on *globally* optimal uses of the neighboringness of beamlet coefficients according to the beamlet graph.

We find the third and fourth levels of structure in this hierarchy unprecedented and quite interesting. Our goal in this paper is to help the reader begin to think in terms of these new relationships and begin to perceive the new algorithmic possibilities they offer.

As a brief example, we will see below that the beamlet framework allows us to explore the *space of curves in an image* in a multi-resolution fashion, efficiently extracting a filament or an object boundary by solving a global optimization criterion. While it might seem rather daunting to optimize over all curves in an image, this is possible by exploiting several facts about the beamlet framework.

- Polygonal curves in the $x - y$ plane can be built up by chaining together a very few connected beamlets.
- Each such chain is a path in the beamlet graph.
- Certain network flow algorithms, when deployed in the beamlet graph, allow us to efficiently explore the space of paths in the graph, looking for paths which make a sum of associated terms optimally large.
- The integral of an image along a polygonal curve made of beamlets is simply the sum of a few terms from the beamlet transform.

Combining these ingredients allows one to formulate algorithms which can efficiently search the space of all curves in an image for a globally optimal curve – provided the notion of optimality involves an integral computed along the curve.

We plan in this paper to systematically build up the reader's understanding of the many kinds of algorithms which can be built based on the beamlet pyramid and beamlet graph. Starting from a discussion in Section 2 of the main components of the beamlet framework we develop in sections 3-6 a discussion of the hierarchy of levels of structure in beamlet-based algorithms. This begins in Section 3 with elementary algorithms which make no use of the interrelatedness of beamlets (Level 1); progresses in Section 4 to tree-based algorithms (Level 2) which exploit parent-child relations; and then finally, in Scetions 5 and 6, to discuss graph-based algorithms (at Levels 3 and 4) which impose the collinearity, co-curvity, and spatial continuity by exploiting the beamlet graph.

In Section 7 we review our software environment, *Beamlab*, which computed the many examples described here; and in Section 8, we review important literature on antecedents of beamlets, in work of Götze & Druckenmiller on Fast Hough Transforms [29], Brandt & Dym on calculation of multiple line integrals [11], and Peter Jones, Guy David, Stephen Semmes, and Gilad Lerman on geometric analysis of point sets [32].

2 Beamlet Analysis

We begin with some terminology and notation. We are interested in image data as n by n arrays of numbers, but we choose to view the image domain as the continuum square $[0,1]^2$ and the pixels as an array of $1/n$-by-$1/n$ squares arranged in a grid in $[0,1]^2$.

A dyadic square S is the collection of points $\{(x_1, x_2) : [k_1/2^j, k_1 + 1/2^j] \times [k_2/2^j, k_2 + 1/2^j]\}$ where $0 \le k_1, k_2 < 2^j$ for an integer $j \ge 0$. For clarity, we will sometimes write $S(k_1, k_2, j)$, so that for example $S(0,0,0)$ is the unit square $[0,1]^2$, and so that if we have an n-by-n grid with $n = 2^J$ dyadic, then the individual pixels are the n^2 cells $S(k_1, k_2, J)$, $0 \le k_1, k_2 < n$.

2.1 Beamlet Dictionary

Suppose we take vertices $v_1, v_2 \in [0,1]^2$ and consider the line segment $b = \overline{v_1 v_2}$. We call such a segment a *beam.* If we consider only beams connecting vertices $(k_1/n, k_2/n)$ at pixel corners, there are order $O(n^4)$ such beams. We note that there are $N = n^2$ pixels, and that typical fast image processing algorithms take order $O(n^2) = O(N)$ flops or at most $O(n^2 \log(n^2)) = O(N \log(N))$; we are seeking algorithms of order $O(n^2)$ or as near to that as we can get. Relying on collections of cardinality $O(n^4)$ beams as an organizational device will lead to unworkable algorithms, and so we seek a *reduced-cardinality substitute.*

Take the collection of all dyadic squares at scales $0 \le j \le J$. Fix a quantum of resolution $\delta = 2^{-J-K}$ for $K \ge 0$. On each dyadic square, traverse the boundary in clockwise fashion starting at the upper right corner and mark off equispaced vertices a distance δ apart. As δ is dyadic and it divides the perimeter length of every dyadic square with sidelength $\ge 1/n$, there are precisely $M_j = 4 \cdot 2^K \cdot 2^{J-j}$ vertices marked out in this fashion on a dyadic square S with side $2^{-j} \ge 1/n$. Call this collection of vertices $V(S)$; label the vertices according to the order they are encountered in the clockwise boundary traverse, so that $V(S) = \{v_{i,S} : 0 \le i < M_j\}$. If we consider any two dyadic squares which have intersecting boundaries, along the intersection of the boundaries the two squares have the same vertices in common; under our labeling system we might have $v_{i,S} = v_{i',S'}$ even though $i \ne i'$. For later use, we let $\mathcal{V}(n, \delta)$ denote the collection of all vertices in all $V(S)$ where S is dyadic of sidelength $\ge 1/n$, and we let $\mathcal{L}(n)$ denote the latticework of all horizontal and vertical lines in the square at spacing $1/n$.

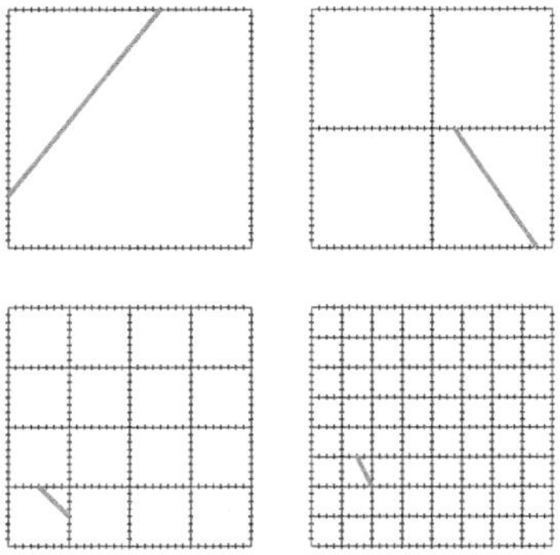

Fig. 1. Beamlets at different scales (squares with different sidelength).

Within each dyadic square S, consider the collection of all beams connecting vertices on the boundary of S:

$$B_\delta(S) = \{b = \overline{v_{i,S}v_{i',S}} : 0 \le i_1, i_2 < M_j\}.$$

There are $\binom{M_j}{2}$ such beams in total.

Definition 1. For given dyadic n and δ, the set of **beamlets** is the collection $\mathcal{B}_{n,\delta}$ of all beams belonging to some $B_\delta(S)$ for some dyadic square S of sidelength 2^{-j}, $0 \le j \le J$.

See Figure 1 for some examples of beamlets.

We note that beamlets only connect vertices on the boundary of a dyadic square, so that although the family of beamlets is built from $O(n^2)$ vertices, it contains many fewer than $O(n^4)$ beams. In fact, as $\binom{M_j}{2} \approx M_j^2/2$, we have

$$\#\mathcal{B}_{n,\delta} \approx (J+1) \cdot 8 \cdot 4^K \cdot n^2.$$

For example, suppose that $\delta = 1/n$. Then there are just 4 vertices $v_{i,S}$ associated to any dyadic square with sidelength $1/n$; these are of course the corners of the squares, and we have

$$\#\mathcal{B}_{n,1/n} \approx 8 \cdot (\log_2(n) + 1) \cdot n^2.$$

Although there are order $O(n^2)$ pixels, and although order $O(n^4)$ beams can be defined based on pixel corners, the collection of beamlets has a cardinality only logarithmically larger than $O(n^2)$. It follows that exhaustive searches through the collection of beamlets can run much faster than exhaustive searches through the collection of beams.

Despite reduced cardinality, the dictionary of beamlets is expressive. It consists of beams at a variety of scales, locations, and orientations. A relatively small number of beamlets can be used as a substitute for any single beam. In [22] we prove

Lemma 1. *Any beam with endpoints anywhere in* $[0,1]^2$ *can be approximated within Hausdorff distance* $\delta/2 + 1/n$ *by a continuous chain* $(b_1, b_2, \dots, b_m)$ *of beamlets* $b_i \in \mathcal{B}_{n,\delta}$ *where the number* m *of beamlets required is bounded by* $8\log_2(n)$ *for* $n > 2$.

The idea is shown in Fig. 2. The rationale is much like the logic of modern airline traffic. There are very many long connections between a few hubs and relatively fewer short connections between provincial centers.

2.2 Beamlet Transform

Let $f(x_1, x_2)$ be a continuous function on $[0,1]^2$. The *beamlet transform* of f is the collection of all line integrals

$$T_f(b) = \int_b f(x(l))dl, \quad b \in \mathcal{B}_{n,\delta},$$

the integrals being taken along line segments $b \in \mathcal{B}_{n,\delta}$; here $x(l)$ traces out the beamlet b along a unit speed path. The digital beamlet transform of an $n \times n$ array (f_{i_1,i_2}) is understood to be the beamlet transform of the function f defined on the continuum by interpolation of the values (f_{i_1,i_2}):

$$f(x_1, x_2) = \sum_{i_1,i_2} f_{i_1,i_2} \phi_{i_1,i_2}(x_1, x_2),$$

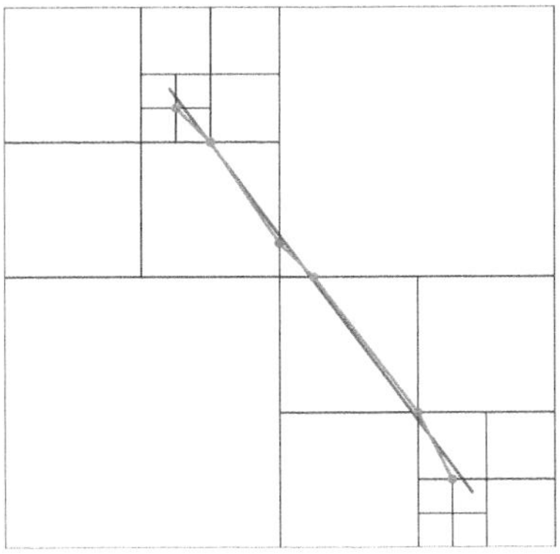

Fig. 2. A line segment approximated by a chain of beamlets. The total number of beamlets (in the chain) is no more than $O(\log_2(n))$. The two endpoints of the beam are $(0.3, 0.85)$ and $(0.85, 0.1)$.

where (ϕ_{i_1,i_2}) is a specified family of continuous interpolation functions. There are several ways the functions ϕ_{i_1,i_2} may be chosen. In this paper we use only *Average-Interpolating Functions.* Let $Pixel(i_1, i_2)$ denote the square $[i_1/n, (i_1 + 1)/n) \times [i_2/n, (i_2 + 1)/n)$. The functions ϕ_{i_1,i_2} may be chosen to obey the conditions

$$n^2 \int_{Pixel(i_1',i_2')} \phi_{i_1,i_2}(x_1, x_2) dx_1 dx_2 = \delta_{i_1,i_1'} \cdot \delta_{i_2,i_2'}$$

where $\delta_{i,i'}$ is the Kronecker symbol. Then f is a function that obeys

$$f_{i_1,i_2} = Ave\{f|Pixel(i_1, i_2)\};$$

in other words, the values of f_{i_1,i_2} are viewed as pixel-level averages of the continuous function f. The ensuing definition is illustrated in Fig. 3.

2.3 Beamlet Pyramid

In traditional usage [6], a pyramid is a data structure exhibiting a notion of *scale* and a notion of *causality.* The data stored at coarser scales should be derivable from data at any finer scale.

In our definition, the beamlet pyramid is the collection of all integrals $T_f[b]$ for $b \in \mathcal{B}_{n,\delta}$. This structure exhibits the multiscale nature assumed in a pyramid, but exhibits the causality only in an approximate sense.

If we consider, for the moment, an ideal continuum beamlet dictionary, obtained by letting the spacing $\delta \to 0$ in the definition of beamlet, then we arrive at a system obeying a *two-scale relation:* each continuum beamlet can be decomposed into a union of at most three beamlets at the next finer scale.

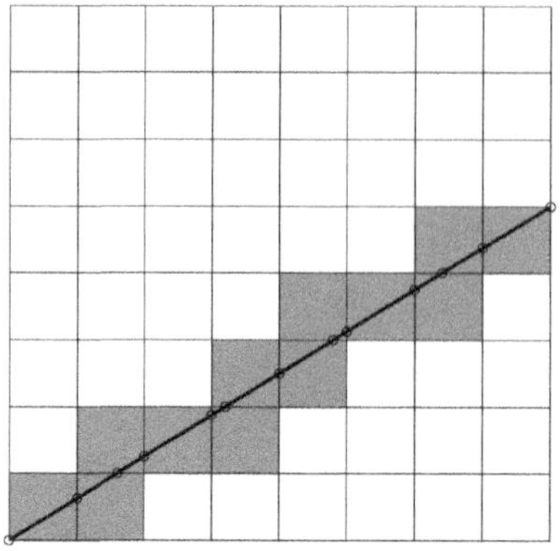

Fig. 3. An illustration of piecewise constant interpolation of a digital image, and its associated beamlet transform. Beamlet transform is a weighted sum of pixel values whose associated squares (shaded) the beamlet traverses.

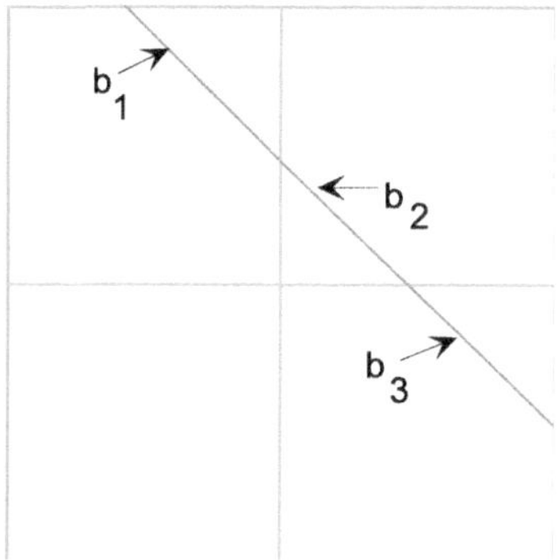

Fig. 4. Decomposition of a beamlet into three beamlets at next finer scale.

We demonstrate this point in Fig. 4, which illustrates the decomposition of a beamlet b into the union of three beamlets b_1, b_2, and b_3 at a finer scale.

The key analytic property we get from this is that the decomposition

$$b = \cup_i b_i$$

of a coarse-scale beamlet into finer-scale beamlets gives

$$T_f[b] = \sum_i T_f[b_i].$$

In practice, we work with $\delta > 0$, of course, and so this two-scale relation only holds in an approximate sense: there is a chain of at most three beamlets at the next finer scale which deviates at most δ from the coarse-scale beamlet.

2.4 Beamlet Graph

Associated with the beamlet pyramid is the *beamlet graph.* This graph has $(n+1)^2$ vertices — corresponding to pixels of an $n \times n$ image — and $16n^2(\log(n)+1)$ edges — corresponding to the beamlets from $\mathcal{B}_{n,1/n}$. Two vertices of the beamlet graph are connected by an edge if and only if they correspond to pixels whose lower left corners are connected by beamlets.

In the beamlet graph, some vertices have only nearest-neighbor connections, but many others have connections to other vertices as well, vertices corresponding to pixels which might be say 2, 4, 8, 16 etc. pixels distant. An illustration of different connectivities for different vertices is in Fig. 5.

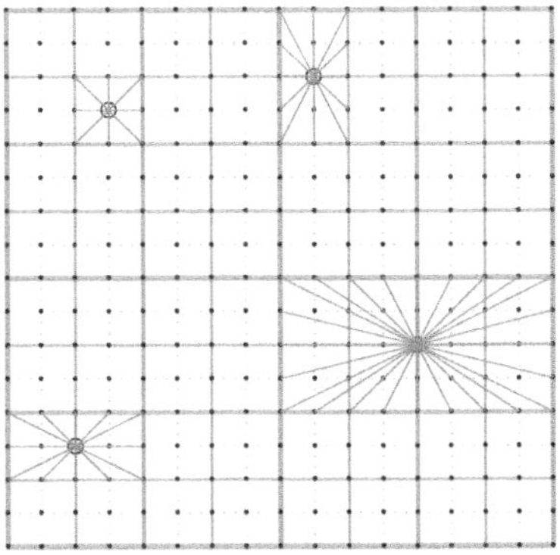

Fig. 5. An illustration of different connectivity of different vertices in a beamlet graph.

The beamlet graph should be compared to another graph used in image analysis: the *nearest-neighbor graph* [42,28,31]. In the nearest neighbor graph, each vertex again corresponds to a pixel in the image, but is connected only to vertices corresponding to pixels which are its nearest neighbors in the image (actually there are two variants – the 4-neighbor graph (cross connectivity) and the 8-neighbor graph (star connectivity) – for our purposes the two are similar).

It is interesting to contrast the two structures.

- *Diameter.* While two vertices in the nearest-neighbor graph corresponding to pixels at opposite ends of the image can only be connected by a path of $O(n)$ edges, the corresponding vertices in the beamlet graph can be connected by a path of many fewer edges: no two vertices in the beamlet graph are ever more than $4\log_2(n)+1$ edges apart.
- *Path/Polygon Isomorphism.* In each graphical structure, every path through the graph corresponds to a polygonal curve in the image. The polygonal curves generated from paths in the nearest-neighbor graph can have

a strikingly different appearance from those generated in the beamlet graph. This is detailed in our next three observations:

- *Multiscale nature.* While a polygonal curve deriving from the nearest-neighbor graph has many pieces of about the same size and so in general is monoscale, a polygonal curve deriving from the beamlet graph may have edges of various lengths, some of which can be global scale, others pixel scale, and others in between.
- *Integrated Curvature.* While a path in the nearest-neighbor graph in general corresponds to a polygonal curve in the image which makes many small 'turns', it is possible for a path in the beamlet graph to correspond to a polygonal curve in the image which is very nearly straight, i.e. to consist of relatively few exactly straight segments with relatively little turning between the pieces.
  Define the total curvature $K(\pi)$ of a polygonal curve in an image, by summing the angular changes at path vertices. Consider vertices v_0 and v_1 in the graph corresponding to pixels chosen at random on opposite borders of the image, and consider the shortest path between them in the nearest-neighbor graph. This path in the nearest-neighbor graph corresponds to a polygonal curve π across the image, and the integrated curvature $K(\pi)$ is proportional to the length n of the path. In contrast, we can construct a path through the beamlet graph, so that the corresponding polygonal curve in the image has integrated curvature $O(\log(n))$.
- *Directional Accuracy.* While a polygonal curve deriving from the nearest-neighbor graph must make many small turns away from its ultimate goal, a curve deriving from the beamlet graph may exhibit a high degree of directional faithfulness.
  Define the overall direction of a polygonal curve connecting two points to be the direction vector between endpoints. Define the *average directional deviation* of a polygonal curve to be the average, along the arc length of the curve, of the angular difference between the direction of the particular segment and the overall direction of the polygonal curve. For a monoscale polygonal curve between typical distant pixels at with average direction $\theta \in [0, \pi]$ derived from the shortest path in the nearest-neighbor graph, the average angular difference is typically a nonzero constant, roughly of magnitude $\min(|\theta|, |\theta - \pi/2|, |\theta - \pi|)$ in the 4-neighbor graph, and of magnitude $\min(|\theta|, |\theta - \pi/2|, |\theta - \pi/4|, |\theta - 3\pi/4|, |\theta - \pi|)$ for the 8-neighbor graph. In contrast, if the endpoints are n pixels apart, the shortest multiscale polygonal curve derived from the beamlet graph has average angular difference of order $1/n$.

In Fig. 6, we compare shortest paths with the same endpoints in both the near-neighbor graph and the beamlet graph. In Table 1, for two fixed points $(0.3, 0.85)$ and $(0.85, 0.1)$, we give its *integrated curvature* and *average directional deviation* (which is to measure the directional accuracy) at a range of resolutions (see details in the table).

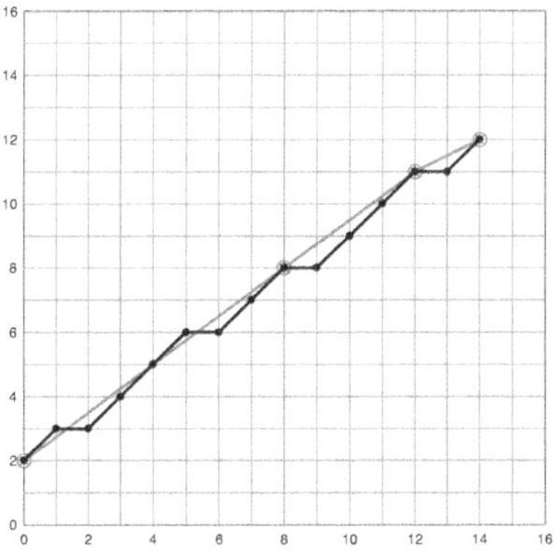

Fig. 6. A line segment approximated by paths in the beamlet graph and the nearest neighbor graph; image size $n = 16$. The two endpoints of the line segment are $(0, \frac{2}{16})$ and $(\frac{14}{16}, \frac{12}{16})$. The vertices on the path in the beamlet graph are marked by the symbol $\otimes$. The edges on the path in the nearest neighbor graph are marked by thick line. The path in the beamlet graph contains 3 edges, while the path in the nearest neighbor graph has 14 edges.

Table 1. Integrated Curvature and Average Directional Deviation for a line segment at various resolutions. (For an n by n image, the pixel-level resolution is $1/n$.) NNG = nearest-neighbor graph; BG = beamlet graph.

Resolution	Integrated Curvature		Average Directional Deviation	
(n)	NNG	BG	NNG	BG
16	0.8π	0.23π	0.08π	0.0217π
32	1.6π	0.63π	0.08π	0.0252π
64	3.2π	0.23π	0.08π	0.0092π
128	6.4π	0.25π	0.08π	0.0067π
256	12.8π	0.27π	0.08π	0.0039π
512	25.6π	0.58π	0.08π	0.0024π
1024	51.2π	0.13π	0.08π	0.0007π
2048	102.4π	0.15π	0.08π	0.0006π
4096	204.8π	0.17π	0.08π	0.0003π

These differences between the two graph structures are, we hope to show, responsible for major differences in the statistical effectiveness of otherwise similar algorithms for detecting filaments and objects.

3 Level One: Structureless Algorithms

In order to develop our theme — the range of interesting new algorithms made available by the beamlet framework — we begin with a description of three algorithms treating beamlet coefficients without regard to the interconnections which might exist between the beamlets at different scales and locations. These algorithms bear strong analogies to algorithms used routinely in wavelet analysis.

3.1 Detection of a Line Segment or a Curve

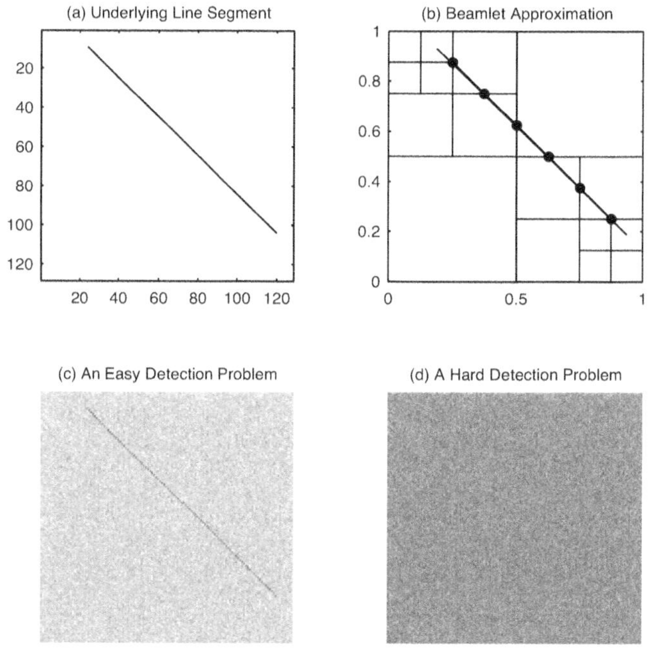

Fig. 7. The problem of detecting the existence of a line segment in a noisy picture. (a) An underlying line segment; (b) Its approximation by beamlets. Note there always a beamlet overlapping a substantial proportion of any such line segment; (c) An easy detection case; (d) A hard detection case. The only difference between (c) and (d) is that the noise levels are different.

Suppose we have a noisy n-by-n image, perhaps containing somewhere within it a faint image of a line segment of unknown length, orientation and position. See Figure 7. In this model, we have noisy pixel-level data (y_{i_1,i_2}), $0 \leq i_1, i_2, < n$, and we model these data as

$$y_{i_1,i_2} = A \cdot \tilde{\Phi}_{i_1,i_2} + \varepsilon z_{i_1,i_2}, \quad 0 \leq i_1, i_2, < n, \tag{1}$$

where ε is a noise level, z_{i_1,i_2} is a white Gaussian noise, A is an unknown amplitude parameter and $\tilde{\Phi}_{i_1,i_2} = \tilde{\Phi}(i_1, i_2; \overline{v_0 v_1})$ is the observed effect at the sensor array of an unknown beam $\overline{v_0 v_1}$. Formalizing matters, we may say that our problem is to test the simple null hypothesis

$$H_0 : A = 0 \tag{2}$$

against the composite alternative

$$H_1 : A > 0, \quad v_0, v_1 \in [0,1]^2. \tag{3}$$

This is highly composite because of the wide range of possible endpoint pairs being considered. Standard ideas in hypothesis testing (so-called 'matched filtering' for generalized likelihood ratio testing) suggest that we consider the random field

$$Y[v_0, v_1] = \langle \psi_{\overline{v_0 v_1}}, y \rangle.$$

where $\psi_{\overline{v_0 v_1}}$ is the so called filter matched to $H_{1,\overline{v_0 v_1}}$:

$$\psi_{\overline{v_0 v_1}}(i_1, i_2) = \tilde{\Phi}(i_1, i_2; \overline{v_0 v_1}) / \|\tilde{\Phi}(\cdot, \cdot; \overline{v_0 v_1})\|_2.$$

and consider the *maximum beam statistic*

$$Y^* = \max\{Y[v_0, v_1] : v_0, v_1 \in [0,1]^2\}.$$

Here the maximum is taken over all beams $\overline{v_0 v_1}$ with endpoints in $[0,1]^2$. We then reject H_0 if Y^* exceeds a certain threshold. This is often called the Generalized Likelihood Ratio Test, because it takes the standard Likelihood Ratio Test (deriving from the above H_0 and a simple alternative hypothesis $H_{1,\overline{v_0 v_1}}$) and, then optimizes over all choices of $H_{1,\overline{v_0 v_1}}$ in order to test the composite alternative.

There is a large number of beams to consider; even if we discretize to work at the scale of pixels, we have at least $O(n^4)$ beams. Naive evaluation of this maximum requires $O(n^4)$ evaluations costing $O(n)$ flops each, which is $O(n^5)$ flops total; prohibitively expensive for large n.

Consider instead testing based on the *maximum beamlet statistic.* We simply compute a maximum over all beamlets rather than over all beams. The comparable statistic is

$$Y^+ = \max\{T_y(b)/\sqrt{L(b)} : b \in \mathcal{B}_{1/n,n}\},$$

where $T_y(b)$ is the beamlet transform of data y and $L(b)$ is the euclidean length of beamlet b. We reject H_0 whenever this exceeds a certain threshold.

Once the beamlet pyramid has been computed, this statistic can be calculated in order $O(N \log(N))$ additional flops, where $N = n^2$ is the number of pixels in the image.

We remark that this procedure is analogous to a wavelet domain test for white noise. In such a test one evaluates the largest wavelet coefficient and asks if it exceeds a threshold deriving from the white noise assumption.

We also remark that this procedure uses beamlets *structurelessly*; the simple maximum is taken, without regard for the location or scale or direction of the maximum, or for interactions among maxima.

As it turns out, although the beamlet-based statistic is much cheaper to evaluate than the GLRT it has rather substantial power. In Figure 8, we compare performance of tests based on maximum beamlet statistics with maximum beam statistics. Evidently, the beamlet approach based on maximizing over a small, strategically chosen subset of beams provides nearly identical sensitivity. A theoretical analysis proving this will be published elsewhere.

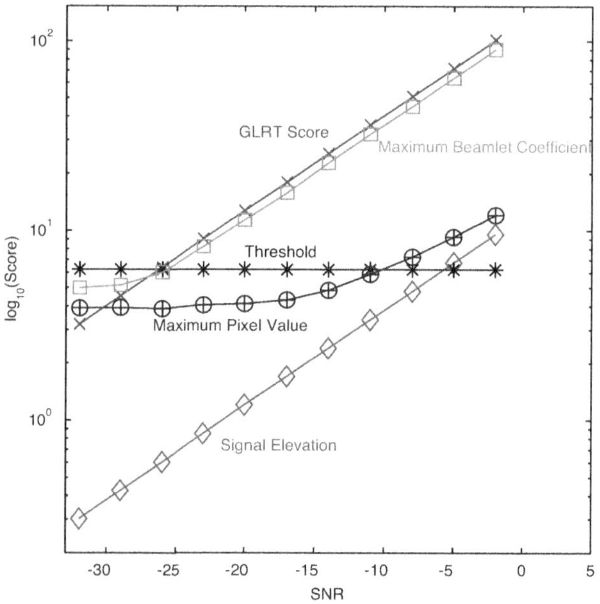

Fig. 8. Illustrating the nearly identical sensitivity of the maximum beam statistic the maximum beamlet statistics. We test at eleven different *signal to noise ratios* (SNR). The lines with $\Diamond$, $\oplus$, $*$, $\Box$ and $\times$ indicate: the pixel SNR of the underlying line object; the median of maximum raw pixel intensities; the threshold of detectability for .05 false-alarm rate; the median of the maximum standardized beamlet coefficients; and the median GLRT score. All the medians refer to results of 100 Monte Carlo simulations. The size of the test image is $n = 128$. The underlying line segment is the same as in Fig 7.

3.2 Thresholding and Visualization

Consider the following visualization procedure.

- *Transform.* Calculate the entries in the beamlet pyramid;
- *Threshold.* Set to zero each beamlet coefficient which does not exceed a (possibly scale-dependent) threshold;
- *Visualize.* For each beamlet coefficient surviving the thresholding operation, draw a line segment depicting that beamlet.

This is analogous to thresholding reconstruction in wavelet analysis, where an object is transformed, its small coefficients discarded, and then the transform is inverted.

Note that this procedure is indeed structureless, since each beamlet coefficient lives or dies based solely on its own value, and not those of others.

An example is given in Fig. 9 (using a digitization of a line drawing by Picasso from the collection of the Metropolitan Museum of Art in New York, kindly provided by Ruth Kozodoy in blessed memory of Revira Singer).

While the procedure is graphically interesting, it generally lacks the degree of informativeness we might have expected. Indeed, owing to the lack of 'inter-scale inhibition' many beamlets at the same location exceed threshold, giving an undesirable degree of 'over-excitation'. In particular, many larger scale beamlets exceed threshold without providing an adequate reflection of features in the underlying object.

We expect that a better method would impose 'spatial inhibitions', keeping two beamlets from both being present in the visualization if they were spatially almost in the same place and 'inter-scale inhibitions', keeping two beamlets from both being present in the visualization if they were spatially in the same location but at adjacent scales. More sophisticated algorithms at Levels 2-4 are able to impose just such inhibitions.

3.3 Multiscale Thickness of Sets

Suppose now we have a *binary* image which is 'mostly white' (represented by the value zero) with a small fraction of black pixels (represented by the value one). The collection of black pixels can be thought of as a 'set' whose geometrical properties we wish to characterize. For example it could be a scattered set of points or it could be points on a smooth curve. Consider the following multiscale summarization procedure.

- *Transform.* Calculate the entries in the beamlet pyramid;
- *Measure Thickness.* For each dyadic square, use the beamlet coefficients associated with that square to measure the relative 'thickness' β_S of the set of nonzero pixels inside that square, i.e. the width of the narrowest strip containing the nonzero data. An illustration of calculation of the thickness is given in Fig. 10.

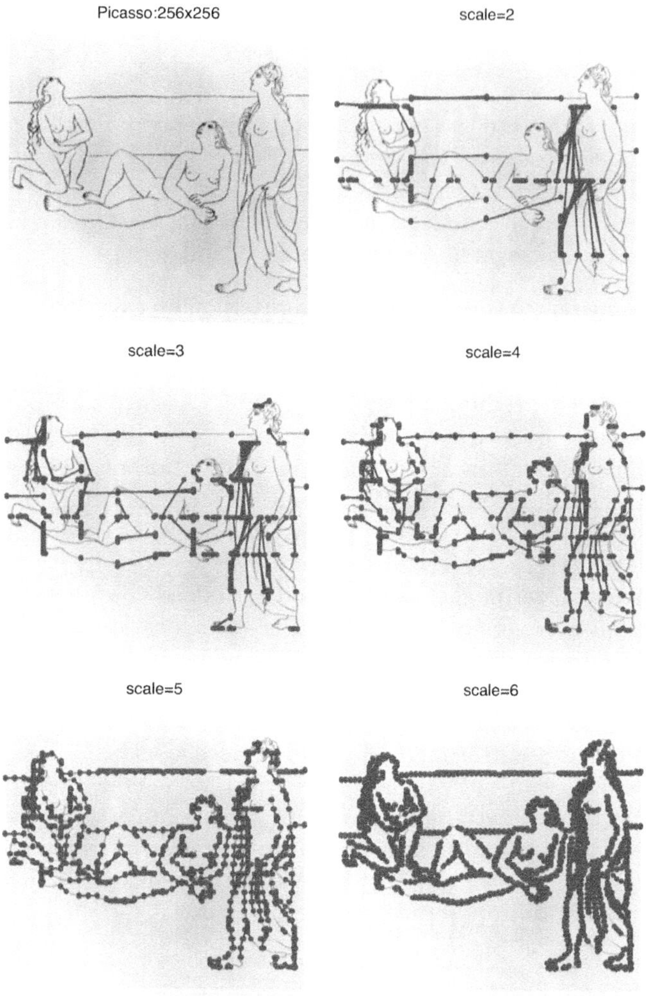

Fig. 9. For the Picasso image, we plot the beamlets at each of several scales whose coefficients exceed a predetermined threshold.

- *Summarize by Scale.* Calculate the average thickness across all dyadic squares at each individual scale;
- *Visualize.* Plot the average thickness at scale j versus j.

Note that this procedure is indeed structureless in the sense that the thickness of a dyadic square is computed without regard to any other squares. The procedure is beamlet driven because we can use the beamlet analysis of the image to compute it – we simply analyze the pattern of nonzero beamlet coefficients associated to a dyadic square in order to identify the narrowest strip.

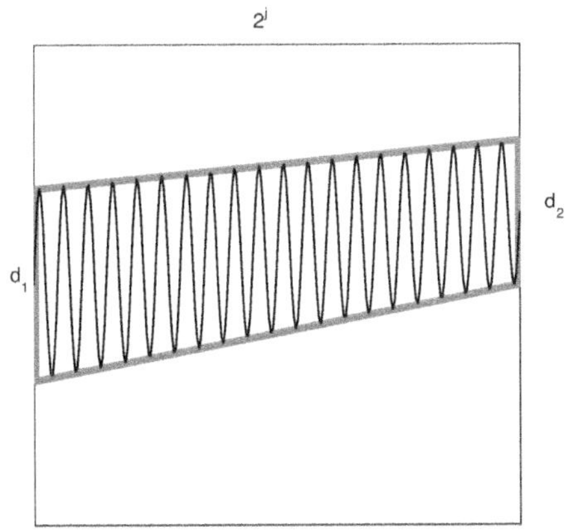

Fig. 10. Calculating the thickness of data in a dyadic square. We find the narrowest strip that is bounded by two beamlets and contains the target curve. The thickness is $\max(d_1, d_2)$ and the normalized thickness or *beta-thickness* is defined as in $\beta_S = \max(d_1, d_2)/2^{-j}$.

Since the procedure identifies a pair of beamlets associated with each square – namely the pair bounding the narrowest strip containing the 'set' – the procedure can be seen as imposing a degree of local spatial inhibition. Only two out of a large number of beamlets are isolated as being 'important' for each square.

This average thickness measures the extent to which the 'set' can be locally approximated as lying on a line. If the set is given by a smooth curve then at fine scales this will of course be the case. If the set is very complex, this will not be the case.

This operation – comparing a set with its approximation by linear sets – is analogous to measuring certain norms in wavelet analysis. Wavelet coefficients measure the deviation of the function from a simple linear approximation, and if the function being analyzed is smooth, they are expected to decay at a certain rate as a function of scale. Comparing the size of the coefficients with scale allows to measure the regularity of the object.

In the mathematical literature, an analysis tool of this kind was introduced by Peter Jones and developed to a great extent by G. David and S. Semmes [32,19]. See also the Yale Ph.D. Thesis of Gilad Lerman. While this earlier work made no direct connection to the pixelated image setting, nor to use of specific discrete data structures, like the beamlet pyramid, it turns out that beamlet tools indeed allow to translate their ideas into a digital image setting very conveniently.

Examples are given in Figure 11. As the Figure shows, the behavior of Thickness versus Scale for a smooth curve (in this case the circle) is quite distinct from the behavior for a fractal curve (in this case random walk). In

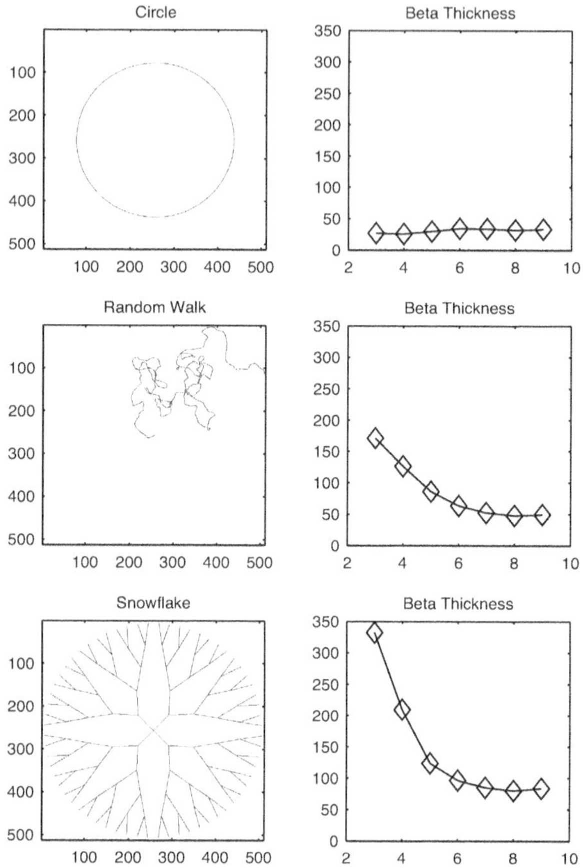

Fig. 11. The "scale-thickness graph" of three different 'sets' at various scales. The left column gives the original curve(s). The right column gives the corresponding thickness–scale plots, which portray the sum of squares of all thicknesses at each given scale.

effect, the beamlet pyramid contains information sufficient to measure the fractal dimension of curves embedded in $\mathbf{R}^2$.

4 Level Two: Tree-Structured Algorithms

We now turn to beamlet-driven algorithms exploiting the tree-structure of the dyadic squares underlying the beamlet pyramid.

We first remind the reader of what we mean by tree structure. Fig. 12 depicts a series of dyadic squares in the plane stratified into different floating panels according to the size of the squares. Parent-Child relationships are illustrated by lines connecting centers of parent squares to centers of child squares.

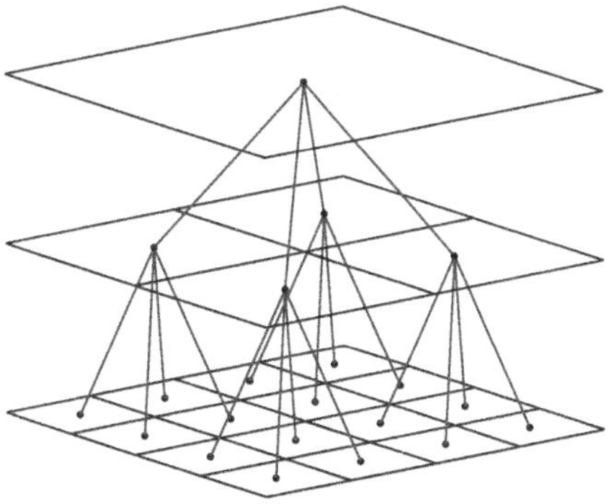

Fig. 12. A complete Recursive Dyadic Partition (RDP) and its tree.

We also recall for the reader the definition of a recursive dyadic partition. This is any partition of the square reachable by application of arbitrary combinations of these two rules:

- $P = \{[0,1]^2\}$ is an RDP.
- If $P = \{S_1, \ldots, S_{\ell-1}, S_\ell, S_{\ell+1}, \ldots, S_n\}$ is an RDP, and if S_ℓ can be decomposed into four dyadic squares $S_{\ell,00}$, $S_{\ell,01}$, $S_{\ell,10}$, and $S_{\ell,11}$, then the new partition

$$P' = \{S_1, \ldots, S_{\ell-1}, S_{\ell,00}, S_{\ell,01}, S_{\ell,10}, S_{\ell,11}, S_{\ell+1}, \ldots S_n\}$$

is again an RDP.

Fig. 13 gives an example of an incomplete RDP.

We also note that each RDP corresponds to a tree; as illustrated in Figure 12, a complete RDP consisting of all dyadic squares of side 1/4 corresponds to a complete tree of depth 2. Fig. 14 shows that an incomplete partition as drawn in Fig. 13 with squares having finest dyadic scale 1/4 . corresponds to an incomplete tree of depth 2.

Finally, we introduce the notion of a *beamlet-decorated* RDP, or BD-RDP. This is an RDP in which the pieces of the partition are (optionally) decorated by associated beamlets. Fig. 15 gives a picture of a BD-RDP and its associated tree structure — note how a fairly general curve can be depicted by this device.

A key property offered by the BD-RDP is the *inter-scale inhibition* between beamlets in such a decoration. While extensive collections of beamlets

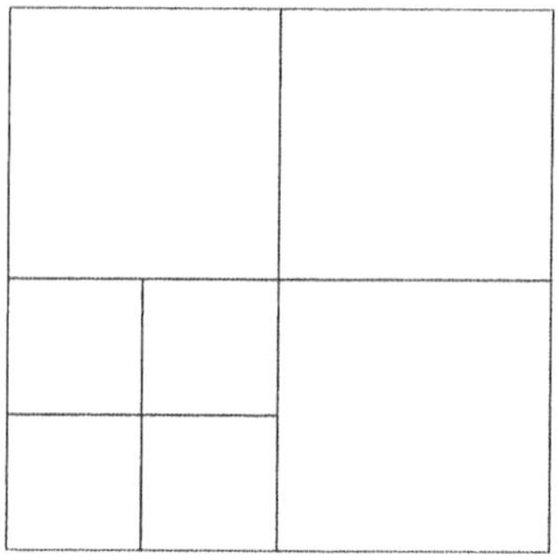

Fig. 13. An incomplete RDP.

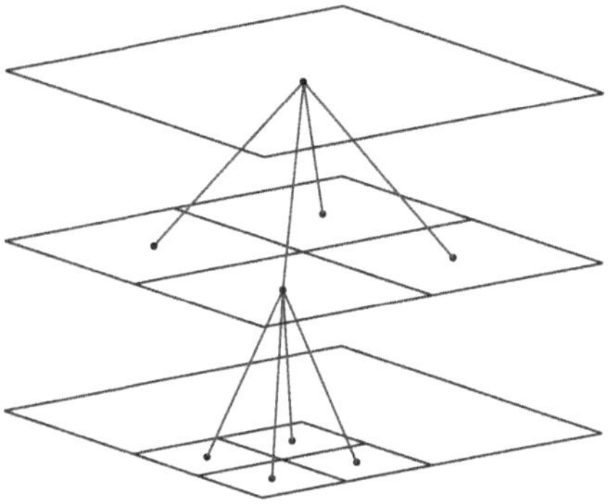

Fig. 14. A tree associated with an incomplete RDP.

can arise as decorations in a BD-RDP, they are always spatially disjoint — they can never contain two beamlets at different scales crossing over each other.

4.1 Extraction of Multiple Line Fragments

Consider now an algorithm for extracting multiple line fragments from an image using tree concepts. As in Section 3.2, we wish to visualize an image using a sequence of beamlets. To do so, we search for a BD-RDP solving an optimization problem. Because of the inter-scale inhibition of beamlets obtained in this way, we will avoid the overlap properties seen in Fig. 9.

We associate to each dyadic square S a quantity C_S defined as

$$C_S = \max_{b \sim S} T(b)/\ell(b)^{1/2}$$

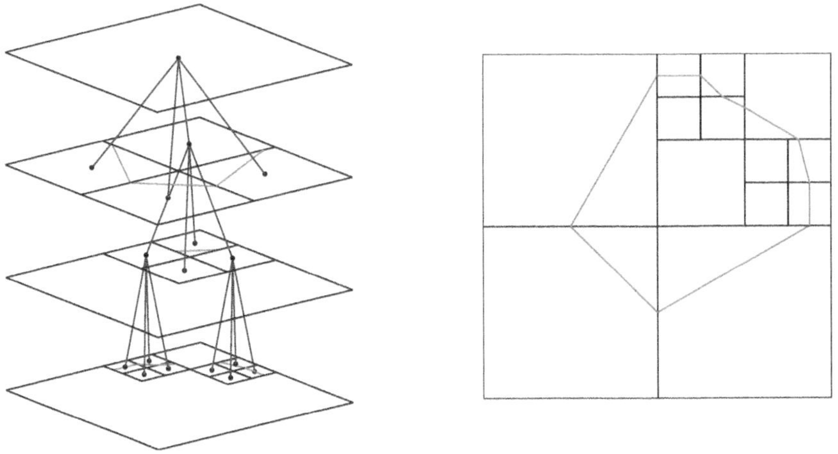

Fig. 15. A beamlet decorated recursive dyadic partition (BD-RDP) (right) and its associated tree structure (left).

where $T(b)$ is the beamlet coefficient associated to beamlet b, $\ell(b)$ is the length of the associated beamlet, and $b \sim S$ means that the beamlet b is associated to S.

We then consider the partition that maximizes, over all recursive dyadic partitions $P = \{S_1, \ldots, S_n\}$ of $[0,1]^2$, the complexity-penalized energy

$$J(P) = \sum_{S \sim P} C_S^2 - \lambda \# P.$$

In effect, C_S measures the energy of the 'model' which says that inside S there is a single beamlet – i.e. the image is zero except along the trace of the beamlet within the square. Indeed, if we model the data falling inside a square as proportional to the indicator 1_b of a certain beamlet, then the energy of the fitted model is the square of the coefficient $\langle f, 1_b \rangle / \|1_b\|^2$ times the norm $\|1_b\|^2$. Now as $T_f(b) \approx \langle f, 1_b \rangle$ and $\|1_b\|^2 \approx L(b)$, this is essentially the square of C_S.

Hence we are searching for a collection of non-overlapping beamlets which provides a substantially accurate description of the image.

An example is given in Fig. 16; note how the description is truly multiscale, containing line segments of all different lengths, positions, and orientations. Significantly, the description avoids the messiness of simple thresholding in Fig. 9. This illustrates the importance of inter-scale inhibition, and the value of tree structures.

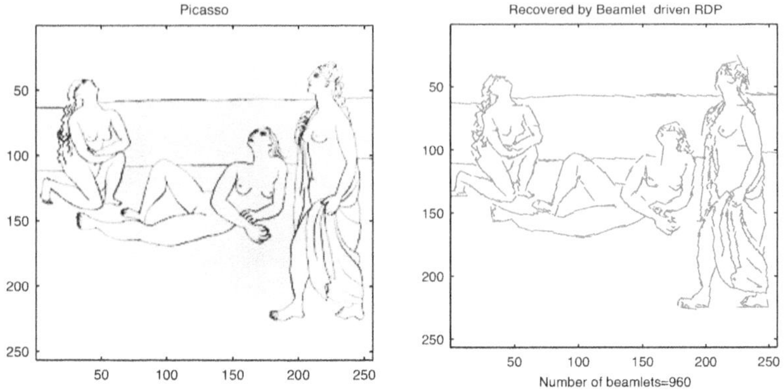

Fig. 16. Result of Beamlet-Decorated Recursive Dyadic Partitioning to extract multiple line segments from an image with abundant linear and curvilinear features.

4.2 Optimization over BD-RDP's

An important point about optimization over RDP's is that, for appropriate objective functions J, the globally optimal partition solving

$$\max_{P \in RDP} J(P). \tag{4}$$

can be obtained rapidly. As a result, in the setting of the previous section, the optimal BD-RDP can be obtained in $O(N \log(N))$ flops, where $N = n^2$ is the number of pixels.

The essential point is that the objective function J should be an additive functional of RDP's, taking the form:

$$J(P) = \sum_{S \sim P} \Gamma_S$$

The fast algorithm for (4) relies on a bottom-up tree pruning process, recursively deciding whether it is better to leave a dyadic square unsubdivided or to divide it into four pieces. Starting from a complete quadtree refined out all the way to pixel leve, we consider each dyadic square S of 2-by-2 pixels and consider whether it is better to treat this as a single, undivided block, or to divide it into four pieces. To make the decision we compare the objective value Γ_S for the unsubdivided square with sum of the four objective values Γ_{S_i} for the four dyadic subsquares going to make up S. Call the maximum of these two numbers $\Gamma^*(S)$, and record the identity of the maximizer in a label μ_S for later use. After completing the process at the 2-by-2 scale, we move up one level, to consider 4-by-4 squares S and consider for each one the question of whether it should be split into four pieces or not, according as Γ_S exceeds the sum of the four optimized $\Gamma^*_{S_i}$ or not; we define Γ^*_S as the indicated maximum and record the identity of the maximizer in μ_S. Continuing to move up level by level, we reach the root, and after processing the root, the value $\Gamma^*_{[0,1]^2}$ is the value of (4) and the various labels μ_S reveal the optimal partition. For details, see [22] and references there.

To apply this in the situation of Section 4.1, we simply set

$$\Gamma_S = (C_S^2 - \lambda)_+$$

where C_S is as defined there, and, if $\Gamma_S > 0$, we record in κ_S the identity of the beamlet maximizing C_S. Then the optimal RDP is given by the variables μ_S set by the pruning algorithm; and the optimal decoration is given by the κ_S variables of terminal nodes in the optimal RDP. The whole algorithm is fast in this case. Once we have the beamlet transform in hand, it costs order $O(N/4^j)$ flops to evaluate each of the 4^j C_S at level j. Then the bottom-up tree pruning algorithm can run in $O(N\log(N))$ flops.

4.3 Extraction of Multiple Regions

The same framework can work for extracting regions rather than filaments. Suppose that we have data y which can be modeled piecewise constant with curved boundaries between pieces. We can obtain an approximation which finds an optimal recursive-dyadic partition with linear splits by the principle of minimizing a complexity-penalized residual sum of squares. This allows us to find piecewise constant approximations with simple polgonal boundaries.

In a BD-RDP, a beamlet associated with a dyadic square splits the square into two regions which we call *wedgelets*; compare [22]. Using beamlet ideas, we can represent noisy data in terms of wedgelets.

Consider the model $M(S,b)$ which says that in dyadic square S, the underlying noiseless signal is piecewise constant, split into two pieces along beamlet b – i.e. a linear superposition of the two wedgelets associated with b. Let y_S be the restriction of the data vector y to this square, and denote the least-squares projection of this data onto the span of model $M(S,b)$, by $Proj\{y_S|M(S,b)\}$.

Define now, for each beamlet b associated with square S, the residual sum of squares associated with model $M(S,b)$

$$RSS(S,b) = \|y_S - Proj\{y_S|M(S,b)\}\|_2^2$$

and define, for each dyadic square S

$$PRSS'_S = \min_{b\sim S} RSS(S,b) + 2\cdot\lambda^2$$

where λ is a penalization factor. Since it does not always make sense to split a square into pieces along a beamlet, we put this in competition with a model that is simply constant in the square, taking the value $m_S = Ave\{y_S\}$:

$$PRSS''_S = \min\left(PRSS'_S, \|y_S - m_S\|_2^2 + \lambda^2\right).$$

For each RDP P define its complexity-penalized residual sum of squares

$$J(P) = \sum_{S \sim P} PRSS''_S.$$

Using the ideas of the last section, the optimum over all RDP's is rapidly computible, once the $PRSS''_S$ are given. As it turns out the numbers $RSS(S, b)$ are easily derivable, from the beamlet transform of the cumulative sums of the data (y_{i_1,i_2}) in i_1 and i_2. Compare [22].

Obviously this approach can be used on non-noisy data for a form of compression; an example is given in Figure 17, where the piecewise constant model has actually been replaced (using similar algorithms) by a piecewise linear model, and the penalization adjusted to reflect the increased complexity of piecewise linear over piecewise constant reconstruction. We exhibit several such approximations obtained by complexity-penalized sum of squares approximations using different values of the penalty parameter λ^2. Through control of this parameter one is able to approximate fine detail or only coarse structure.

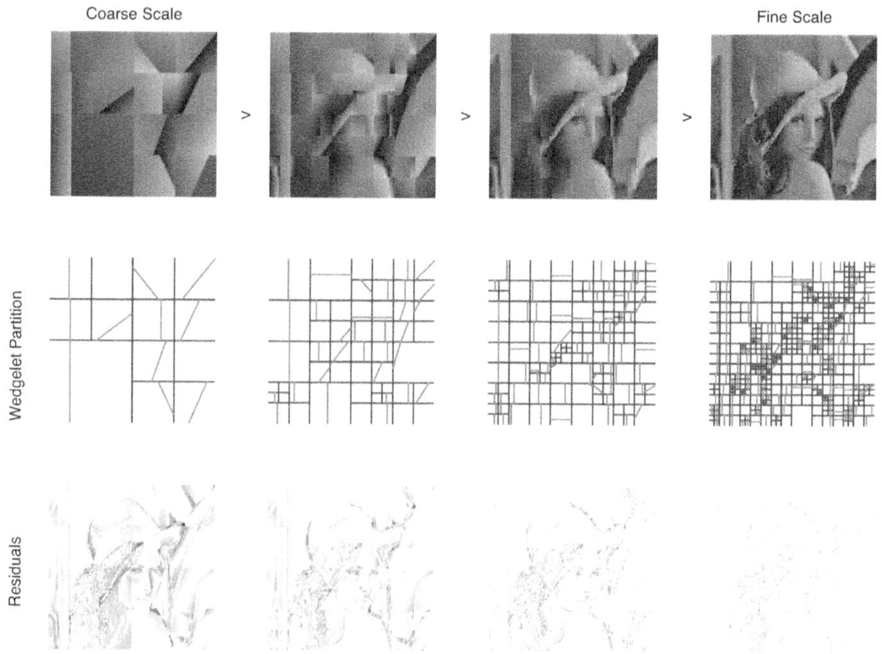

Fig. 17. Optimal piecewise linear approximations to Lenna derived from wedges. First row: series of approximations with increasing complexity; Second row, correpsonding series of partitions, notice beamlets splitting squares into wedges; third row: corresponding approximation errors.

This tool can also be used for noise removal. Suppose we have noisy data

$$y_{i_1,i_2} = f(i_1,i_2) + \sigma z_{i_1,i_2}, \quad 0 \leq i_1, i_2 < n$$

where Z is a random white noise and again the object of interest f can be modelled piecewise constant with curved boundaries between pieces. Now λ should be calibrated to the noise level; obvious choices include 3σ and also $\sigma \cdot 2 \cdot \sqrt{\log(n)}$.

A numerical example is presented in Figure 18. The quality of approximation is good – it produces reasonable fidelity in matching boundaries – but notice that the boundaries given by the wedgelet approximation consist simply of disconnected linear fragements – no continuity or directional coherence is imposed. As the third column of Figure 18 illustrates, the decorations in the optimal partition do not cohere globally.

5 Level Three: Local Chaining of Line Segments

It can be argued that the tree-based algorithms of Section 4 bring up issues already familiar from tree-based algorithms used in the wavelets community [46,13] and, more broadly, in the multiscale community, where quadtree partitioning has been in use for many years.

In this and following sections we will use the structure of the beamlet graph in a way not available within the wavelet lexicon, exploiting the neighborhood properties of the beamlet graph to express *connectedness* and *directional continuation.* As suggested in the introduction, understanding the expression of continuation is vital in several fields, including vision.

5.1 Fast Approximate GLRT for Line Segment Detection

Return to the line segment detection problem discussed in Section 3.1, and now consider the problem of computing the GLRT – maximizing over $N(N-1)/2$ beams (here $N = (n+1)^2$ is the number of pixel corners in the underlying image). As discussed earlier, this is an $O(N^2 \cdot n) = O(n^5)$ computation.

Earlier, we proposed a *substitute problem*: maximizing over beamlets rather than beams. The cardinality of the set of beamlets being only $O(N \log(N))$ rather than $O(N^2)$, the resulting computations become much more manageable. We saw that this simple idea had substantial power in detecting line segments, while nearly, but not quite attaining the performance level of the GLRT (Figure 8).

To get even better performance, we might seek an even better approximation to the maximum beam statistic Y^* with the same order of computattional complexity as the maximum beamlet statistic Y^+. Our idea is to adaptively chain together several beamlets to make better non-dyadic approximations to beams. Ignoring the cost of the beamlet transform itself,

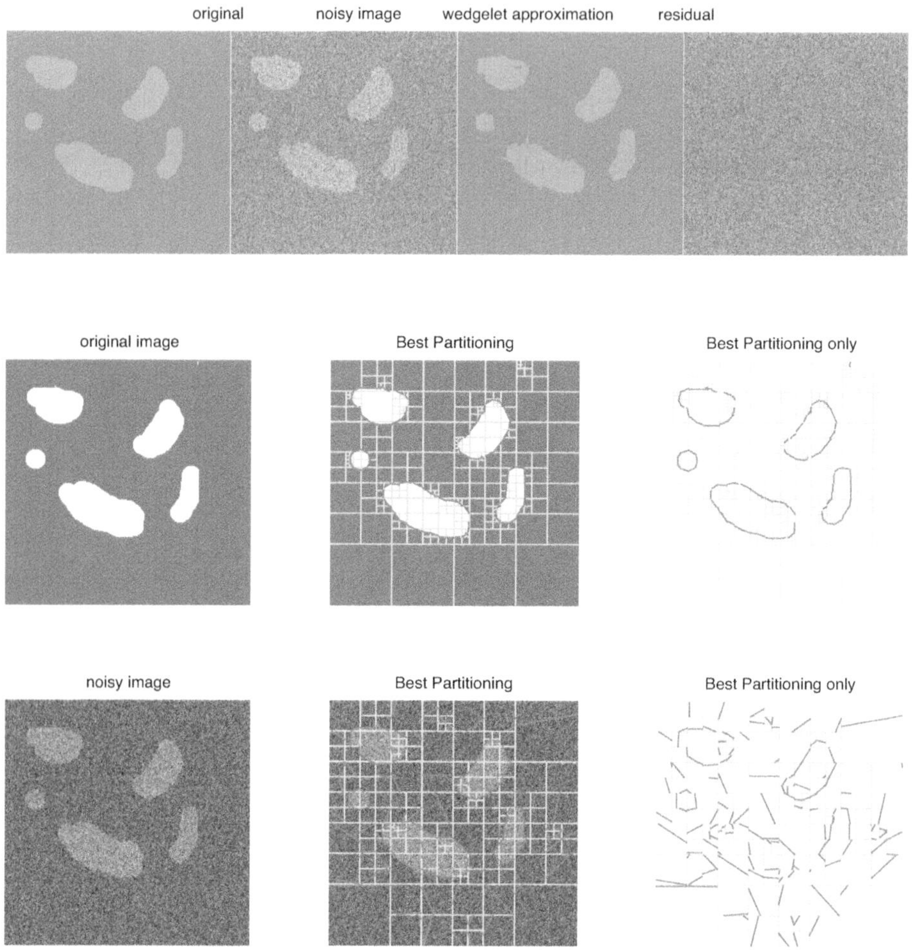

Fig. 18. Extracting multiple blobs. First row, from left to right: the noiseless object; object in white noise; wedgelet approximation; residual deviation between estimator and original object. The residual deviation contains no obvious pattern. The second and third rows illustrate the associated RDPs for the noiseless object and the noisy image respectively. Note that in the clean image, the RDP decorations occur at boundaries of blobs. The two RDPs are in crude qualitative agreement, although the one for noisy images is noisier.

the computational strategy can operate in order $O(N \log(N))$ flops and can reject H_0 under approximately the same circumstances as would Y^*.

The algorithm is driven by the idea of using the largest beamlet coefficients to identify promising beamlets as 'seeds'; we then explore many non-dyadic beams in the neighborhood of each such promising beamlet – each seed 'sprouts' many chains, among which we search for an optimum. Because of geometric properties of lines and dyadic subintervals, the overall maximum among the subset we explore is very likely to be very close to the maximum beam statistic Y^*.

An example suggesting the feasibility of this approach is given in Figure 19. A faint line segment is embedded in a very noisy image, at a level not far from the absolute threshold of detectibility. The figure illustrates the beamlets passing the screening threshold t_1. Out of many tens of thousands of beamlets in the pyramid, only several dozen pass the screening phase. Sprouting chains starting from this limited set is an eminently practical idea.

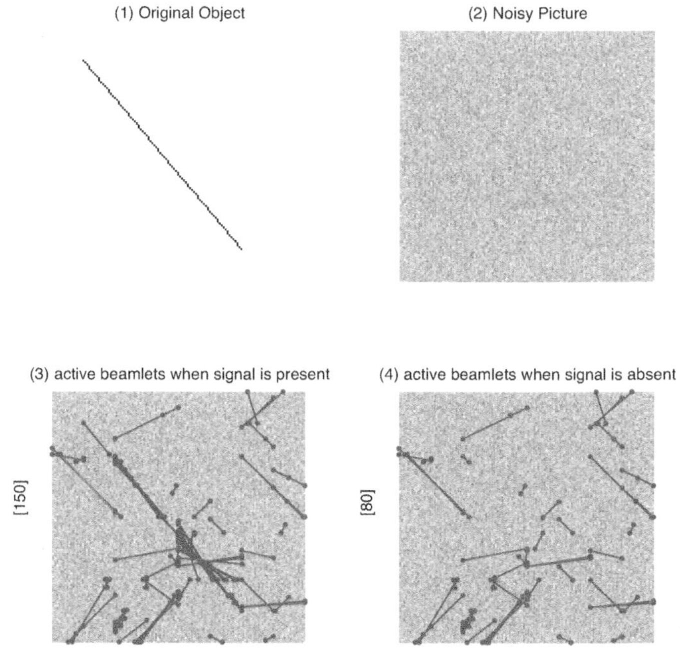

Fig. 19. Illustrating the idea of 'seed' beamlets in detecting a line segment. Beamlets passing the initial screening threshold are depicted. The number in brackets ([,]) gives the number of beamlets at each scale passing the screening threshold.

To give a formal algorithmic description, we recall that the beamlet system has a defining parameter δ denoting the spacing between adjacent beamlet endpoints, typically one pixel wide $\delta = 1/n$, but also possibly smaller in certain applications. Let us say that a collection of beams (which can include, naturally beamlets, or a mixture of beams and beamlets, etc.) is *δ-collinear* if there exists a single straight line which furnishes a δ-approximation to all members of the collection – i.e. if each beam lies within δ Hausdorff distance of the common line. Also, let us say that a beamlet is a *δ-subinterval* of a beam if the beamlet lies within a Hausdorff distance δ of the beam.

Algorithm: *Approximate Beam Maximization by Beamlet Chaining.* This algorithm depends on parameters t_1, t_2, m_n and ℓ which must be appropriately specified.

1. *Screening the Beamlet Pyramid.*
   - Obtain beamlet transform of image.
   - Select all beamlets achieving $Y[b] > t_1$. Place them in a list $\mathcal{P}$ of *promising* intervals.
   - If $\mathcal{P}$ contains more than m_n such intervals, Print `Reject` H_0 and stop.
2. *Searching Neighborhoods of Promising Beamlets.*
   - For each beamlet $b \in \mathcal{P}$,
     - Construct a list $\mathcal{C}_\ell[b]$ of δ-collinear chains extending b through depth ℓ.
     - For each chain $c \in \mathcal{C}_\ell[b]$
       — Calculate $Y[c] \equiv \sum_{b' \sim c} Y[b']$.
       — If $Y[c]$ exceeds t_2, Print `Reject` H_0 and stop.
     - End
   - End
3. *Termination: No calculated $Y[c]$ exceeds t_2*
   - Print `Accept` H_0 and stop.

In the above description we have referred to $\mathcal{C}_\ell(b)$. This is the set of *depth ℓ chains* rooted at b, i.e. the collection of all beamlet chains which

(a) have b as a maximal δ-subinterval;
(b) involve only beamlets generated from squares at most ℓ levels finer than the square defining b;
(c) are δ-collinear.

Such chains may be viewed as continuations of b which have b as their most substantial piece. Figure 20 illustrates a few chains constructible from a fixed beamlet b.

Our claim that this algorithm can obey an $O(N\log(N))$ complexity estimate depends on the choice of parameters for the algorithm. Recalling that ε is the per-pixel noise level, we suggest setting $t_1 = (\varepsilon/2)\sqrt{\log n}$, $t_2 = 2\varepsilon\sqrt{\log n}$, $m_n = N^{1/2}$ and $\ell = \ell_n$ tending to infinity slowly, but in practice equal to 3 or 4. We ignore the cost of calculating the beamlet transform. The cost of screening the normalized beamlet coefficients $Y[b]$ for values exceeding t_1 is of moderate computational complexity $O(N\log(N))$. So, to get

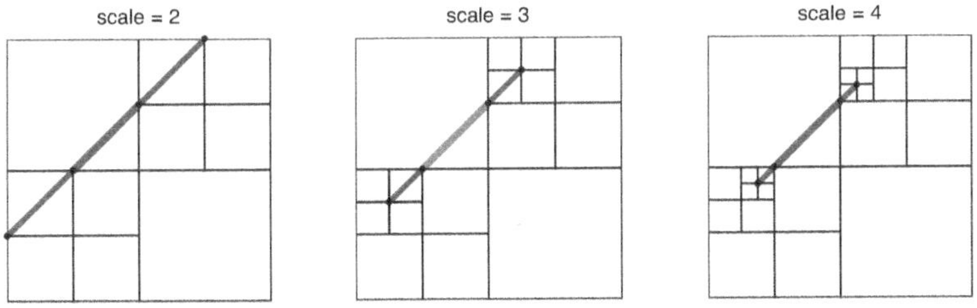

Fig. 20. Collinear chains sprouting from a fixed beamlet. Illustrates some chains sprouted using squares at the same scale and squares as much as two scales fines.

a fast algorithm, we need that the follow-up phase of exploring chains which extend each promising beamlet not require more than order $O(N \log(N))$ work. In that stage we must therefore have not too many promising seeds and not too much work per seed. Now as we have chosen $m_n = o(N)$ dyadic intervals, at most this many dyadic intervals are allowed to exceed threshold in the preliminary dyadic screening phase; the number of chains searched is $O(2^{2\ell})$, which we can arrange to increase slowly; and it is easy to see that the effort in computing the corresponding chain sums $Y[c]$ is likewise $O(2^{2\ell})$. Thus if $m_n \cdot O(2^{2\ell_n}) = O(N)$, we have established the desired control of complexity.

The question this leaves open, naturally, is whether this procedure can really come close to achieving Y^*. This being not an algorithmic question, – our focus here – but a question about empirical processes, we leave it to a forthcoming paper.

We comment on an important geometric idea underlying the above algorithm. In a forthcoming paper we show that, for every beam b, there is a beamlet b' which is a δ-subinterval of b and has a length $\ell(b')$ at least $1/7$ the length $\ell(b)$ of the beam b. In other words, associated to each beam there is a beamlet almost coinciding along at least $1/7$ of its length. We anchor our search around beamlets playing this role; such beamlets can be expected to be the ones with highest signal-to-noise ratio. The idea of viewing promising beamlets as maximal δ-subintervals, and of exploring only chains sprouting from seeds which maintain this property, is an important simplification, because it significantly limits the number of chains which have to be explored.

5.2 Detection of Filaments by Local Chaining

Consider now the problem of detecting a faint filament in noisy data; this is similar to model (1), but now the term Φ models the presence of a curve embedded in the image, rather than a line segment.

An obvious approach is to compare the maximum beamlet coefficient with a threshold, and this will work well if the curve is not very faint and not very curvy. However, according to statistical theory we will not review here, the limit of detectibility when employing the maximum beamlet score is very far from the absolute limit of detectibility according to the best test.

We suggest instead a method which constructs potentially many chains and which compares the observed typical length of the chains with the typical length under the signal-absent hypothesis. If there are many more long chains than we would expect under the signal-absent hypothesis, we reject it in favor of signal-present. Our current theoretical understanding suggests that this gives a near-optimal test; we hope to explore this in a future publication.

We again consider a two-stage algorithm which constructs a new graph, in some sense 'dual' to the beamlet graph. The first stage defines the vertices of this graph: they are beamlets which are 'promising', having survived a first level of screening. The second stage defines edges connecting beamlets.

For the first stage screening, we simply define a threshold t_1 and select all beamlets with beamlet coefficients exceeding t_1 in amplitude, producing our list of 'promising' beamlets, and so vertices of our graph. For the second stage, we conduct a follow-up to screening, adding an edge connecting two vertices of the new graph iff the corresponding beamlets have two properties:

- They fall in dyadic squares whose separation is no more than 3 times the radius of the smaller square; and
- They are δ-co-curvilinear.

The resulting graph we call the *co-curvilinearity graph.*

Here δ-co-curvilinearity of a collection of beamlets means that there is an underlying curve with curvature $\leq C$ (a specified constant, fixed in advance) such that all the beamlets in the collection lie within δ of the curve.

Figure 21 illustrates the idea of beamlets eligible for chaining under co-curvilinearity.

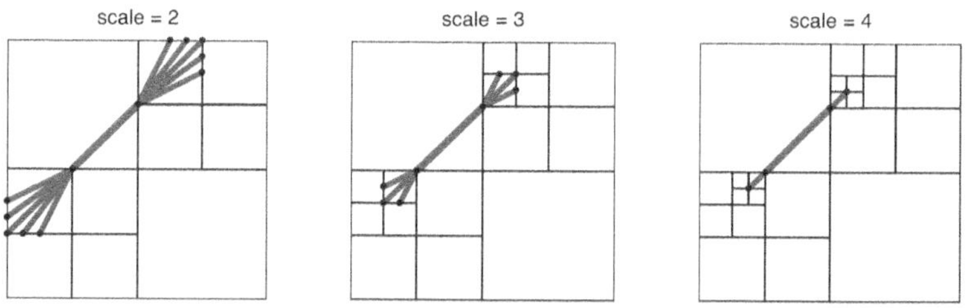

Fig. 21. Chaining of Co-curvilinear beamlets.

In Figure 22, we consider a faint curve in a very noisy image, and illustrate some of the 'promising' elements according to the above recipe.

Finally, the test statistic for detection of a filament measures the extent to which there are more lengthy chains present in this image than one would expect in a signal-absent image.

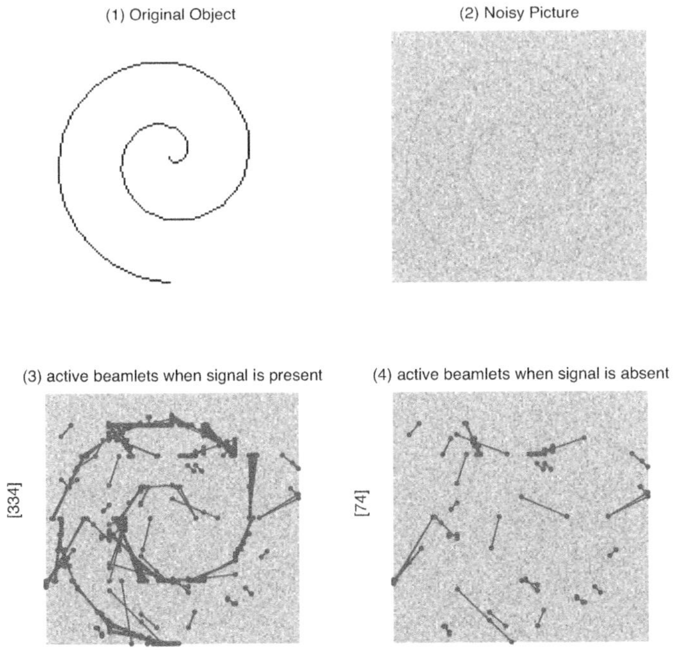

Fig. 22. Illustrating the beamlets passing a preliminary threshold test in the case where the image contains and does not contain a filament. In brackets we display the number of promising beamlets.

Evidently, the threshold t_1 for being declared promising needs to be carefully set in order to obtain useful results; this is a topic in empirical process theory we will explore in a forthcoming paper.

6 Level Four: Global Chaining of Line Segments

There is ample evidence that the human visual system is able to effortlessly integrate local information into coherently identified global objects [27,35,39], and considerable interest in developing vision algorithms with such properties [37,12]. This, and some of the other application opportunities mentioned in the introduction, make the identification of globally coherent objects a compelling topic.

So far we have considered methods for detecting objects and filaments which are entirely local, and so tend to recover disconnected fragments of whole filaments and objects. We now turn to methods which obtain globally coherent objects, and not just fragments, using global optimization techniques in the beamlet graph.

6.1 Extraction of an Anchored Filament

Consider now the problem of locating a curve which connects two *given* pixels in a noisy picture; see Figure 23.

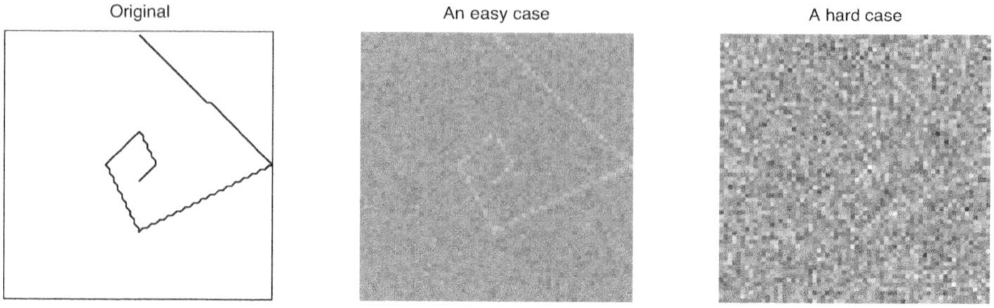

Fig. 23. Filament Detection problem. The three figures are (a) the noiseless underlying spiral filament, (b) an easy low-noise detection problem, and (c) a hard high-noise detection problem.

As in Section 3.1, the array of data (y_{i_1,i_2}) contains embedded in it the indicator function of an underlying curve γ buried in a Gaussian noise with standard deviation ϵ, so that

$$y_{i_1,i_2} = A \cdot \tilde{\Phi}_\gamma(i_1, i_2) + \epsilon z_{i_1,i_2}, \qquad 0 \le i_1, i_2 < n,$$

where A is an unknown constant, $\tilde{\Phi}_\gamma$ is the indicator function of a curve γ, and the random variables z_{i_1,i_2} are i.i.d. $N(0,1)$. We are particularly interested in the case where ε is large compared to A, in which case we say that the Pixel-level signal-to-noise-level ratio is small.

We have considered several different approaches to recovering γ based on global optimization. In all of these, we use *ratios of additive criteria* defined as follows. Given a beamlet polygon p we consider optimization criteria of the form

$$J(p) = \sum_{b\sim p} \Psi_1(b) / \sum_{b\sim p} \Psi_2(b),$$

where Ψ_1 and Ψ_2 are two specified criterion functions. We have developed software which can compute the solution using ideas from network flow.

In one family of computational examples, we use

$$\Psi_1(b) = T_y(b) - \lambda\sqrt{\ell(b)},$$
$$\Psi_2(b) = \sqrt{\ell(b)},$$

where $T_y(b)$ is the beamlet transform of the noisy image y, and λ is a penalization parameter. In theoretical investigations, we have studied

$$\Psi_1(b) = T_y(b) - \lambda\sqrt{\ell(b)},$$
$$\Psi_2(b) = \ell(b).$$

Each criterion has interesting mathematical properties, which we hope to describe elsewhere.

To solve such optimization problems, we deploy dynamic programming (DP) in the beamlet graph. While the use of dynamic programming in filament extraction in the nearest-neighbor graph is rather firmly established [42,28], our approach is qualitatively different, both because of the multiscale nature of the beamlet graph, and because of the ratio nature of our objective. (In fact it turns out that the beamlet graph can give a considerable advantage in computational effort as compared to the nearest-neighbor graph, as we will show below).

We recall the concept of *propagation* methods, which provide a generic framework for dynamic programming on a graph; compare for example [7,9]. To solve the optimization problem whose solution is a path from one fixed point v_0 to another v_1 in a graph, we 'propagate outward from v_0.' We form a set called **OPEN** consisting of v_0 alone. We then consider its immediately adjacent vertices; these are merged into **OPEN**. At each newly added vertex, we keep track of information on the pairs $\left(\sum_{b\in p}\Psi_1(b), \sum_{b\in p}\Psi_2(b)\right)$ for all paths leading from v_0 to this vertex. (More properly we keep information only on the convex hull of the set of such pairs. By doing this, we reduce the storage needed at each vertex, and hence reduce the algorithmic complexity.)

We continue propagation, considering at the next stage the next 'layer' of connected vertices. At the k-th stage, the data stored at a given vertex v_1 are either uninitialized (if no k-step path will reach v_1) or else they record information about the pairs $\left(\sum_{b\in p}\Psi_1(b), \sum_{b\in p}\Psi_2(b)\right)$ for all k-step paths leading from v_0 to v_1. If the true optimal path has fewer than k steps then these data are sufficient to determine the optimal value of $J(p)$.

An important distinction between the beamlet graph and the nearest-neighbor graph is that *simple paths in the beamlet graph have few links.* Heuristically, we can use this to eliminate a lot of unnecessary popagation, because if the true solution is simple, we will reach the optimal solution at an early stage. As discussed earlier, any polygonal curve has a simpler representation in the beamlet graph than it does in the near-neighbor graph, and the degree of simplification can be extensive. For example, to represent a long line segment in an n by n image always takes $O(n)$ edges in the near-neighbor graph, but never more than $O(\log_2(n))$ edges in the beamlet graph. So for an identical objective function and propagation method, running the problem the beamlet graph tends to require many fewer stages than running it on the nearest-neighbor graph.

In "real" applications, due to computing resource limitations, we might *not* be able to afford to run a propagation algorithm to a guaranteed optimal solution. We may terminate the process after a certain number of stages, and take the best solution up-to-then. Apparently because it is easy to reach the optimal solution in the beamlet graph, we are safer with an early termination than we would be in the nearest-neighbor graph.

Results of numerical experiments comparing algorithms based on the beamlet graph and the nearest-neighbor graph are given in Figure 24. The key point is that the results in the beamlet graph are dramatically closer to the desired object, a spiral curve.

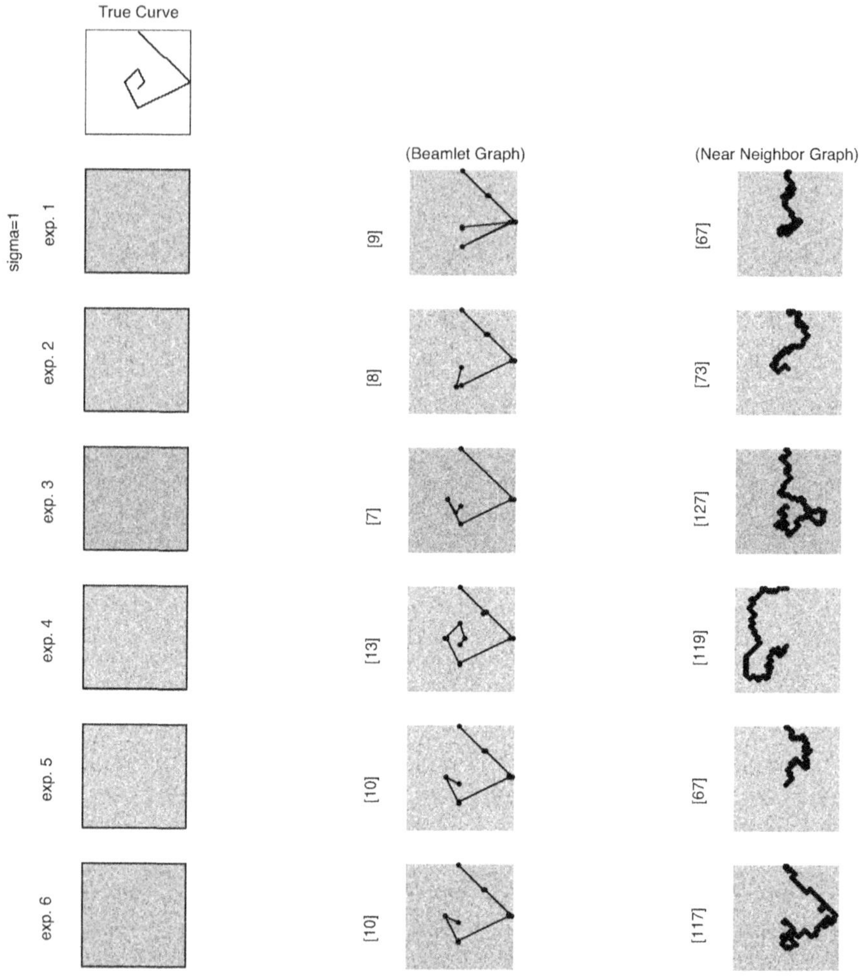

Fig. 24. Illustrating advantages of running a propagation algorithm on the beamlet graph. The original image (in the first row) contains a spiral. The first column contains six noisy realizations. We terminate the propagation after 20 steps. The middle column illustrates the optimal path in the beamlet graph. The right column illustrates the optimal path in the near-neighbor graph.

This method can also be used to automatically extract curvilinear patterns from data which are not literally filamentary. For example, Figure 25 presents an astronomical image, which clearly exhibits a spiral galaxy pattern; the figure illustrates the arm extracted by the beamlet/MCTTRC technique.

Fig. 25. A galaxy image (left) and the result of applying a dynamic programming algorithm on the beamlet graph to estimate its trace (right).

6.2 Extraction of a Globally Optimal Region

In this section, suppose that the underlying object of interest is a region of the plane bounded by a simple closed curve, having a constant but unknown amplitude which is greater inside than outside the region. The pixel values of this noisy image are still denoted by y_{i_1,i_2}, and we let $y(x_1, x_2)$ denote a continuous function obtained by average-interpolation.

Motivated by the idea of matched filtering, we might seek to recover the region by solving the problem:

$$\max_R \qquad S(R)/\sqrt{Area(R)}, \tag{5}$$

where

$$S(R) = \int_R y(x_1, x_2)dx_1 dx_2$$

is the integral of the interpolant over region R.

This is not an easy problem to solve, because of the huge dimensionality of the space of 'all regions'. Surprisingly, an efficient algorithm to find an approximate solution can be built on the following two observations.

- *Reduction to Beamlet Transforms.* Consider the numerator $S(R)$. Using integration by parts, in the form of the Gauss-Green theorem, we have

$$S(R) = \int_{\partial R} \mathbf{Y} \cdot \boldsymbol{n}\, ds,$$

where $\mathbf{Y}$ is the 2-D vector field of partial integrals

$$\mathbf{Y}(x_1, x_2) = \left[\frac{1}{2} \int_0^{x_1} y(t, x_2) dt, \frac{1}{2} \int_0^{x_2} y(x_1, t) dt \right],$$

$\boldsymbol{n}$ is the unit normal to the boundary ∂R. Note that on a line segment b, the normal direction $\boldsymbol{n}$ is constant vector $\boldsymbol{n}(b)$. Also note that if we consider regions R bounded by beamlet polygons, then we have

$$S(R) = \sum_{b \sim \partial R} (\int_b \mathbf{Y} ds) \cdot \boldsymbol{n}(b). \tag{6}$$

It follows that

$$S(R) = \sum_{b \sim \partial R} T_1(b)c(b) + T_2(b)s(b)$$

where the unit normal $\boldsymbol{n}(b)$ to a beamlet b has components $(c(b), s(b))$, T_1 is the beamlet transform of the horizontal primitive $\int_0^{x_1} y(t, x_2) dt$, and T_2 is the beamlet transform of the vertical primitive $\int_0^{x_2} y(x_1, t) dt$. Hence, $S(R)$ can be computed from simple operations on a pair of beamlet transforms.

- *Simplification by Isoperimetry.* We consider the denominator in (5). Working with the square root of the area causes numerous dificulties, so approximate it by isoperimetry:

$$\sqrt{Area(R)} \approx \ell(\partial R),$$

where $\ell(\partial R)$ denotes the length of the boundary ∂R. Assuming this is a beamlet polygon,

$$\ell(\partial R) = \sum_{b \sim \partial R} \ell(b).$$

Combining these observations, the original problem (5) can be dropped in favor of the following surrogate optimization problem:

$$\max_{p \text{ closed}} \quad \frac{\sum_{b \sim p} \int_b \mathbf{Y} \cdot \boldsymbol{n} ds}{\sum_{b \sim p} \ell(b)}. \tag{7}$$

This problem (7) turns out to be an instance of a well known problem in the field of Network Flows, the *optimum cost-to-time ratio cycle* problem. We have a directed graph in which edges have a pair of weights $\Psi_1(b)$ and $\Psi_2(b)$ and are seeking a cycle p optimizing a ratio, as follows:

$$\min_{p \text{ closed}} \quad \frac{\sum_{b \sim p} \Psi_1(b)}{\sum_{b \sim p} \Psi_2(b)}; \tag{8}$$

see, for example, [7]. It can be elegantly solved by linear programming [18,33].

To apply this machinery in the present case, we define the *beamlet digraph*, making a distinction between a beamlet going from v_0 to v_1 and a beamlet going in the opposite direction. We then of course define Ψ_1 and Ψ_2 consistent with (7). Finally, we employ a fast linear programming code based on primal-dual interior point methods for its solution.

In Figure 26, we provide a numerical example. In fact, we need one further detail: in this example we do not actually try to solve the problem (7), but instead a regularized problem, where Ψ_1 reflects both $\int_b \boldsymbol{Y} \cdot \boldsymbol{n}\, ds$ and a penalty term $-\lambda|b|^{1/2}$. In this modified objective function, longer beamlets are favored. In Figure 26, the estimates are regularized based on this. Note that in one of the examples the signal/noise ratio is very poor, and it is hard to convince oneself that an object is even present based solely on visual impressions. However, our method still detects it with satisfactory accuracy.

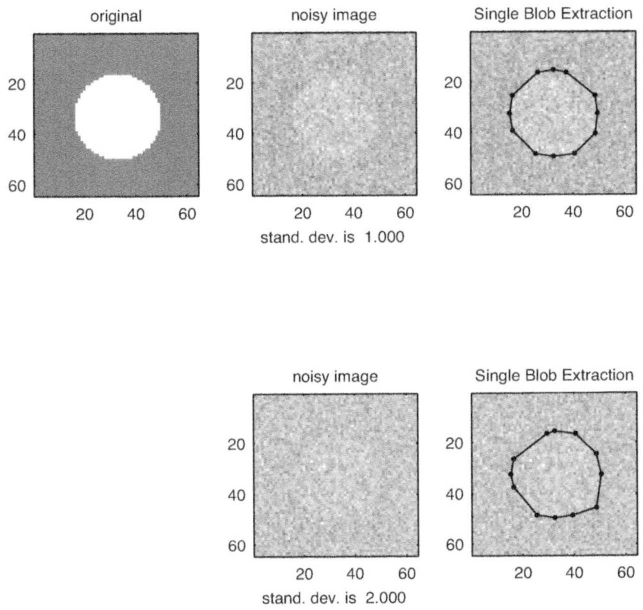

Fig. 26. Estimating a region at two different signal to noise ratios. The upper left panel displays the original object. The middle column displays noisy images at two different noise levels. The right column displays results of beamlet-based object recovery.

7 Beamlab

The examples computed in this paper were obtained using *Beamlab*, a Matlab toolbox we have developed. It contains about 250 files, 130 `.m` (Matlab) files and 65 `.c` sources for accelerated `.mex` files. Online information is available at URL's [1,2].

Beamlab has twelve subdirectories, which can be grouped into four categories:

1. Functions for exact beamlet and wedgelet transforms, and their supporting subroutines.
2. Functions that implement network flow algorithms on the beamlet graph and the near-neighbor graph, and their supporting subroutines.
3. Functions and scripts for applications, such as the maximum beamlet test for detecting existence of a faint linear object in noisy data, the tree algorithms associated with beamlet-decorated recursive dyadic partition, and the beta thickness tools for measuring the regularity of a curve.
4. Precompiled `.mex` functions. Use of these binaries will significantly speed up execution of central components of the library.

These main categories are then further subdivided into their own subdirectories:

1. Exact Beamlet and Wedgelet Transforms.
   - **ExactBeamletTrans**—Functions implementing basic beamlet transforms.
   - **ExactWedgeletTrans**—Functions implementing basic wedgelet transforms.
   - **Utility**—Functions playing supporting roles in the implementation of the beamlet and wedgelet transform, and functions to visualized these transforms.

   The subdirectories **ExactBeamletTrans** and **ExactWedgeletTrans**, only contain `.m`-files. For large problem sizes, these implementations will be prohibitively slow.
2. Network Algorithms.
   - **MCTTRC**—Functions and scripts that apply linear programming approaches to solve the *minimum cost to time ratio cycle* problem.
   - **MCTTRP**—Functions and scripts that implement a dynamic-programming-style propagation algorithm to find a *minimum cost to time ratio path* on a graph.
   - **MinDistPath**—Functions that find the minimum distance path on a graph. We implemented a variant of Dijkstra's algorithm [9,7].
   - **BeamletGraph**— Supporting functions used to manipulate structures in the beamlet graph.
3. Applications.

- **MaxBeamletCoeff**—Scripts to implement the maximum beamlet coefficient test for detecting existence of a line segment.
- **BDRDP**—Functions and scripts to implement beamlet-decorated recursive dyadic partitioning to extract multiple line segments and wedgelets.
- **Regularity**—Functions and scripts that calculate beta-thickness.

4. Binary sources.
   - **BWMEX**—CMEX sources for functions related to the exact beamlet and wedgelet transforms.
   - **GraphMEX**—CMEX sources for functions that carry out network flow algorithms on either the beamlet graph, or the near-neighbor graph.

8 Relation to Previous Work

The multiscale analysis tools we have developed here have several antecedents, already mentioned in the introduction. Having come near the end of our story, it is easy to explain the relationship of these ideas with previous work. Also, we can explore the relationship of the ideas with previous uses of network-flow based algorithms.

8.1 Fast Radon Transforms

A pyramid data structure identical to $\mathcal{B}_{n,1/n}$ has been used before by Götze and Druckenmiller [29], though not by that name, and for different purposes. Their goal was to develop a fast Hough transform for digital data. They recursively applied the pseudo-two-scale relation mentioned in section 2.3 to calculate approximations to coarse-scale integrals from approximations to finer-scale integrals. For them, the tool is just an intermediate one; the real goal is to approximate Radon transformation. Brady [10] has independently developed a fast approximate Radon transform based on approximate-two-scale concepts, but with less direct connection to the beamlets system.

Brandt and Dym [11] have also considered the use of approximate two-scale relations to rapidly calculate approximations to line integrals over large families of multiscale line segments, and not just the global-scale line segments appearing in Radon transformation.

Our philosophy is in some sense the reverse of these efforts at fast Radon transformation. We take as given in this paper (and also in [22]), that one is calculating the multiscale Radon transform *exactly*, i.e. that the objects of interest are precise integrals over a specific systematic family of line segments at all different scales and locations.

We are specifically *not* interested in fast approximate calculation of the multiscale Radon transform, and we have not used such a procedure in computing any of the examples above. We instead view such exact integrals as a starting point or given, and our goal is to explore algorithms which can make use of such integrals, rather than to treat approximations to them as an end in themselves.

8.2 Image Analysis

The articles of David Horn and collaborators at Tel-Aviv University [3,4,23] we have cited in the introduction show that for analyzing imagery produced by certain particle physics experiments, certain line-segment integrals calculations provide valuable feature detectors. This pioneering work is the first we know of actually working out real applications using beamlet-like ideas. Our methods differ in emphasizing a formal dyadic structure, as opposed to a collection of beams arrived at intuitively and non-dyadically, and in emphasizing a range of algorithms which can be developed using this structure, rather than specific applications attacked with line segment features and neural network classifiers.

A. Brandt, in oral communication at the Yosemite Workshop, informed us that his former Ph.D. student J. Dym, in a thesis at the Weitzmann Institute, has proposed that integrals of an image over all rectangles at all orientations, scales, and locations would be an important tool for image analysis. Brandt and co-authors mentioned in passing [12] that a tool of this kind could be useful for detecting curves in images. Evidently, Brandt had the idea some years ago to consider features sets based on multiscale line integrals, and so his work strongly anticipates much of ours. It is often said by scientists working in multiscale methods that one has 'arrived' as a mature contributor to the subject when one discovers footprints of Achi Brandt in what one thought originally to be virgin, unexplored territory. So we have arrived!

8.3 Geometric Measure Theory

The beamlet system we have described here is, in our view, closely related to important recent work in harmonic analysis [32,19,36]. Peter Jones started off an extensive line of research by showing that one could gather information about approximations to the pieces of a curve defined by intersections with dyadic boxes — recording the error of approximation of such pieces by line segments — and used that information to characterize curves of finite arclength (Traveling-Salesman Problem). David and Semmes have extended such dyadic-organization ideas to R^d (where one dissects a surface into dyadically-organized pieces and studies approximation by k-planes); they used such tools to understand a number of important questions in analysis. Gilad Lerman, in a Yale Ph.D. thesis, has used such tools to look for structure in point clouds in high-dimensional space.

In this paper, we have focused on image data, which are conceptually very different from point-cloud data. The connection with point-cloud data is illustrated in Section 3.3 above; we can define an image which corresponds to binning points in the point-cloud and placing counts in cells based on the number of members.

The difference in setting creates a very different emphasis. We have defined a very special discrete set of beams and showed how to use them to extract curves from noisy data.

A particular comment of some importance the goal of finding smooth curves is conceptually very different from finding merely rectifiable ones. The tool we have used for this complexity-penalization is considerably different than minimal arclength; for example, arclength is unable to distinguish between a chain of beamlets and a single beamlet having the same image in the plane, whereas our complexity penalty prefers a single beamlet over an equivalent chain of shorter ones.

8.4 Other Applications of Beamlets

The companion paper [22], gave a thorough theoretical analysis of the method of Section 4.3. It was shown that, when applied to the problem of recovering an object which is black or white with a smooth boundary between colors, when that object is embedded in noisy data, this method could achieve near-minimax mean squared error.

The thesis [30] used the beamlet pyramid to decompose images into superpositions of beamlet features and wavelets. See Fig. 27, which shows that the beamlets objectively 'explain' more of the interesting structure than do wavelets.

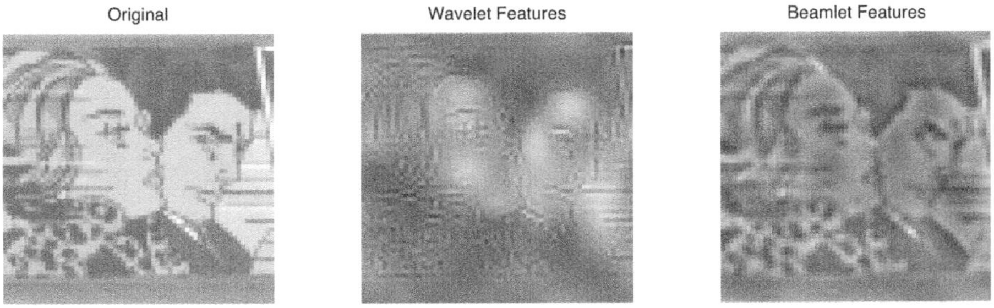

Fig. 27. An example of using beamlets together with wavelets as building blocks to decompose an image. From left to right, they are an original image, its wavelet components, and its beamlet components. This decomposition is based on minimizing the ℓ^1 norm of its coefficients. (Like the work in [16].)

8.5 Other Applications of Network Flow Ideas

Applications of network flow ideas to image processing have been around for three decades, since the work of Montanari [42]. Recent applications include

[28,31,17] and many other authors in the field of perceptual grouping. In all this work, methods are developed for networks based on the nearest-neighbor graph with either 4 or 8 neighbors.

The work of Jermyn and Ishikawa [31] and of Cox, Rao, and Zhong are particularly notable for the ability to find globally optimal regions and boundaries.

In this paper we have deployed network flows in the beamlet graph; as a result of the variable scale of beamlets, the algorithm of Jermyn and Ishikawa is not applicable; as a result of the nonplanarity of the beamlet graph, the algorithm of Cox, Rao, and Zhong is not applicable. The application of interior point methods of linear programming to region recovery seems genuinely novel.

The multiscale nature of the beamlet graph seems an important feature for region and filament recovery, with important implications when the boundary of the object is not detectable at the pixel level, but only by integrating over many pixels. A review of the figures given throughout this paper will show that in many of our examples, the solution happens to use multiscale beamlets rather than the fine-scale connections used by nearest-neighbor graph methods. We believe that this property is responsible for the fact that these methods can recover the object at dramatically smaller signal-to-noise ratios than those possible using monoscale pixel-level thinking such as in the nearest-neighbor graph.

References

1. http://www.isye.gatech.edu/~xiaoming/beamlab.
2. http://www-stat.stanford.edu/~wavelab.
3. H. Abramowicz, D. Horn,U. Naftali, C. Sahar-Pikielny. (1996) An Orientation Selective Neural Network and its Application to Cosmic Muon Identification *Nucl. Instr. Meth. Phys. Res.* A 378 (1996) 305-311.
4. H. Abramowicz, D. Horn,U. Naftali, C. Sahar-Pikielny. (1997) An Orientation Selective Neural Network for Pattern Identification in Particle Detectors in *Advances in Neural Information Processing Systems* 9 eds. M. C. Mozer, M. J. Jordan and T. Petsche, MIT Press 1997, pp. 925–931.
5. J.R. Bond, L. Kofman and D. Pogosyan. How filaments of galaxies are woven into the cosmic web. *Nature*, 380(6575):603–606, April 1996.
6. P. J. Burt, and E. H. Adelson, "The Laplacian pyramid as a compact image code", *IEEE Transactions on Communications*, 9:(4), pp. 532-540, 1983.
7. R. K. Ahuja, T. L. Magnanti, and J. B. Orlin. *Network Flows: Theory, Algorithms, and Applications.* Prentice Hall, 1993.
8. A.J. Bell and T.J. Sejnowski. An information-maximization approach to blind separation and blind deconvolution. *Neural Computation*, 7:1129–1159, 1995.
9. D. Bertsekas. *Dynamic Programming and Optimal Control*, volume 1. Athena Scientific, 1995.
10. M. L. Brady. A fast discrete approximation algorithm for the Radon transform. *SIAM J. Computing*, 27(1):107–19, February 1998.

11. A. Brandt and J. Dym. Fast calculation of multiple line integrals. *SIAM J. Sci. Comput.*, 20(4):1417–1429, 1999.
12. E. Sharon, A. Brandt, and R Basri. Fast multiscale image segmentation. In *Proceedings IEEE Conference on Computer Vision and Pattern Recognition*, volume 1, pages 70–7, 2000.
13. R.W. Buccigrossi and E.P. Simoncelli. Image compression via joint statistical characterization in the wavelet domain. *IEEE Transactions on Image Processing*, 8(12):1688–1701, 1999.
14. E. Candès and D. Donoho. Ridgelets: the key to high-dimensional intermittency? *Phil. Trans. R. Soc. Lond. A.*, 357:2495–2509, 1999.
15. E. J. Candès and D. L. Donoho. Curvelets: a surprisingly effective nonadaptive representation of objects with edges. In Albert Cohen, Christophe Rabut, and Larry L. Schumaker, editors, *Curve and Surface Fitting: Saint-Malo 1999.* Vanderbilt University Press, Nashville, TN.,ISBN 0-8265-1357-3, 2000.
16. S. S. Chen, D. L. Donoho, and M. A. Saunders. Atomic decomposition by basis pursuit. *SIAM J. Sci. Comput.*, 20(1):33–61, 1999.
17. I.J. Cox, S.B. Rao, and Y. Zhong. "Ratio Regions": A Technique for Image Segmentation, *Proc. ICPR '96*, 557-564.
18. G.B. Dantzig, W.O. Blattner, and M.R. Rao. Finding a cycle in a graph with minimum cost to time ratio with application to a ship routing problem. *Theory of Graphs*, pages 77–83, 1966.
19. G. David and S. Semmes. *Analysis of and on uniformly rectifiable sets*, volume 38 of *Math. Surveys and Monographs.* Amer. Math. Soc., 1993.
20. D. L. Donoho. Sparse components analysis and optimal atomic decomposition. Technical report, Department of Statistics, Stanford University, To appear, *Constructive Approximation*, 1998.
21. D. Donoho and X. Huo. Beamlet pyramids: A new form of multiresolution analysis, suited for extracting lines, curves, and objects from very noisy image data. In *Proceedings of SPIE*, volume 4119, July 2000.
22. D.L. Donoho. Wedgelets: Nearly minimax estimation of edges. *Annals of Statistics*, 27(3):859–897, 1999.
23. G. Dror, H. Abramowicz and D. Horn (1998) Vertex identification in High Energy Physics Experiments. NIPS*98.
24. A. Fairall. *Large-Scale Structures in the Universe.* Chichester, West Sussex, 1998.
25. D.J. Field. Relations between the statistics of natural images and the response properties of cortical cells. *J. Opt. Soc. Am.*, 4:2379–2394, 1987.
26. D.J. Field. Scale-invariance and self-similar 'wavelet' transforms: an analysis of natural scenes and mammalian visual systems. In M. Farge, et al. eds., *Wavelets, Fractals and Fourier Transforms.* Oxford Univ. Press, 1993.
27. D.J. Field, A. Hayes, and R.F. Hess. Contour integration by the human visual system: evidence for a local "association field". *Vision Research*, 33(2):173–93, Jan. 1993.
28. D. Geiger, A. Gupta, L.A. Costa, and J. Vlontzos. Dynamic programming for detecting, tracking and matching deformable contours. *IEEE Trans. on Pattern Analysis and Machine Intelligence*, 17(3):294–302, 1995.
29. W.A. Götze and H.J. Druckmüller. A fast digital radon transform – an efficient means for evaluating the hough transform. *Pattern Recognition*, 28(12):1985–1992, 1995.

30. X. Huo. *Sparse Image Representation via Combined Transforms.* PhD thesis, Stanford, August 1999.
31. I. Jermyn and H. Ishikawa. Globally optimal regions and boundaries. In *7th ICCV*, Kerkyra, Greece, September 1999.
32. P. W. Jones. Rectifiable sets and the traveling salesman problem. *Inventiones Mathematicae*, 102:1–15, 1990.
33. R. M. Karp. A characterization of the minimum cycle mean in a digraph. *Discrete Mathematics*, 23:309–311, 1978.
34. R.M. Karp and J.B. Orlin. Parametric shortest path algorithms with an application to cyclic staffing. *Discrete Applied Mathematics*, 3:37–45, 1981.
35. I. Kovacs and B. Julesz. A closed curve is much more than an incomplete one: Effect of closure in figure-ground segmentation. *Proc. Natl. Acad. Sci. USA*, 90:7495–7497, August 1993.
36. G. Lerman (2000) *Geometric Transcriptions of Sets and Their Applications to Data Analysis.* Ph.D. Thesis, Yale University Department of Mathematics.
37. T. Leung and J. Malik. Contour continuity in region based image segmentation. In *5th. Euro. Conf. Computer Vision*, Frieburg, Germany, June 1998.
38. D. Marr. *Vision : a computational investigation into the human representation and processing of visual information.* W.H. Freeman, San Francisco, 1982.
39. J. D. Mendola, A. Dale, B. Fischl, A.K. Liu, and R. G.H. Tootell. The representation of illusory and real contours in human cortical visual areas revealed by functional mri. *J. of Neuroscience*, 19(19):8560–8572, Oct. 1999.
40. Y. Meyer. Review of "An Introduction to Wavelets" and "Ten Lectures on Wavelets". *Bulletin Amer Math Soc*, 28(2), April 1993.
41. Y. Meyer. *Wavelets. Algorithms & Applications.* Society for Industrial and Applied Mathematics (SIAM), 1993
42. U. Montanari. On the optimal detection of curves in noisy pictures. *Communications of the ACM*, 14(5):335–45, 1971.
43. B.A. Olshausen and D.J. Field. Emergence of simple-cell receptive field properties by learning a sparse code for natural images. *Nature*, 381:607–609, 1996.
44. D.A. Pollen and S.F. Ronner. Phase relationships between adjacent cells in the visual cortex. *Science*, vol. 212, pp. 1409-1411, 1981.
45. D.L. Ruderman. The statistics of natural images. *Network*, 5(4):517–548, 1993.
46. A. Said and W.A. Pearlman, "A new, fast and efficient image codec based on set partitioning in hierachical trees," IEEE Tr. on Circuits and Systems for Video Techn., Vol. 6, pp.243-250, June 1996.
47. B.S. Sathyaprakash, V. Sahni, and S.F. Shandarin. Emergence of filamentary structure in cosmological gravitational clustering. *Astrophysical Journal, Letters*, 462(1):L5–8, 1996.
48. J.-L. Starck, F. Murtagh, and A. Bijaoui. *Image Processing and Data Analysis.* Cambridge, 1998.
49. J.H. van Hateren and A. van der Schaaf. Independent component filters of natural images compared with simple cells in the primary visual cortex. *Proc. R. Soc. Lond. B*, 265:359–366, 1998.
50. B. Wandell. *Foundations of Vision.* Sinauer Assoc., 1995.

Generalized FEM for Homogenization Problems*

Christoph Schwab and Ana-Maria Matache**

Seminar for Applied Mathematics, ETH-Zentrum, CH-8092 Zürich, Switzerland

Abstract. We introduce the concept of generalized Finite Element Method (gFEM) for the numerical treatment of homogenization problems. These problems are characterized by highly oscillatory periodic (or patchwise periodic) pattern in the coefficients of the differential equation and their solutions exhibit a multiple scale behavior: a macroscopic behavior superposed with local characteristics at micro length scales. The gFEM is based on two-scale FE spaces that are obtained by augmenting the standard polynomial FE spaces with problem dependent, non-polynomial micro shape functions that reflect the oscillatory behavior of the solution. Our choice of micro-scale shape functions is motivated by a new class of representations formulas for the solutions on unbounded domains, which are based on an assumption of scale separation and generalized Fourier inversion integrals. Under analiticity assumptions on the input data, gFEM converges robustly and exponentially. The micro-shape functions obtained from the representation formula are solutions to suitable unit-cell problems. In the gFEM, they are obtained in the start-up phase of the calculations by solving numerically these unit-cell problems.
Different choices of the micro shape functions are possible within the framework of the gFEM and we also analyze gFEM based on micro shape functions obtained from the theory of Bloch waves. Numerical results for one-dimensional and two-dimensional problems corroborate our theoretical results.

1 Introduction

We consider the Finite Element approximation of the classical elliptic homogenization problem:

$$L^{\varepsilon}(x,\partial_x)u^{\varepsilon} := -\nabla\cdot(a^{\varepsilon}(x)\nabla u^{\varepsilon}) + a_0^{\varepsilon}(x)u^{\varepsilon} = f \quad \text{in } \Omega^{\varepsilon}, \tag{1}$$

$$Bu^{\varepsilon} = 0 \quad \text{on } \partial\Omega^{\varepsilon}. \tag{2}$$

Here, the set $\Omega^{\varepsilon} = \Omega^{\varepsilon}_{\infty} \cap \Omega$ is defined as the intersection of a bounded Lipschitz domain Ω and the infinite periodic lattice of width ε given by

$$\Omega^{\varepsilon}_{\infty} := \varepsilon\Omega_{\infty}, \quad \Omega_{\infty} := \bigcup_{k\in\mathbb{Z}^n} (2\pi k + \widehat{Q}),$$

* Research supported by the Swiss National Science Foundation under Project Number BBW 21-58754.99

** Member of the EC-Network"Homogenization and Multiple Scales" HMS 2000 and $\widehat{Q} \subset \widehat{Y} := [-\pi, \pi]^n$, $\partial\widehat{Q}$ Lipschitz, is the unit-cell. We assume that $\Omega^\varepsilon_\infty$ is connected. The case when $\widehat{Q} \neq [-\pi, \pi]^n$ and the interior boundary of $\widehat{Q}$ is merely piecewise smooth models problems where the material has holes or rigid inclusions. If $\widehat{Q} = [-\pi, \pi]^n$, then $\Omega^\varepsilon = \Omega$. Under these assumptions, the domain Ω^ε is a bounded, connected subset of $\mathbb{R}^n$ with boundary $\partial\Omega^\varepsilon$ and associated boundary operator B which may be either the trace operator or the conormal derivative operator. In electrostatics, for example, u^ε is the electrostatic potential, a^ε represents the dielectric permittivity matrix and the parameter ε describes the size of the material heterogeneities. Since ε is small compared to the size of Ω, we have a problem with multiple scales: a macroscopic one (the size of the sample Ω) and a microscopic one (the size of the heterogeneities, given by ε). The heterogeneities are assumed to be periodically distributed, i.e., $a^\varepsilon(x) = a(x/\varepsilon)$, $a_0^\varepsilon(x) = a_0(x/\varepsilon)$ and the coefficients $a(\cdot) := (a_{pq}(\cdot))_{pq} \in L^\infty(\mathbb{R}^n)^{n\times n}_{\mathrm{sym}}$, $a_0(\cdot) \in L^\infty(\mathbb{R}^n)$ are 2π-periodic functions in each variable. It is also assumed that there exist positive constants $\gamma_0, \gamma_1 > 0$ such that

$$\gamma_0\, \eta^\top \eta \leq \sum_{p,q=1}^{n} a_{pq}(\xi)\eta_p\eta_q \leq \gamma_1\, \eta^\top \eta,\ \forall \eta \in \mathbb{R}^n\,,\ \text{for a.e. } \xi \in \mathbb{R}^n\,, \tag{3}$$

$$\gamma_0 \leq a_0(\xi) \leq \gamma_1,\ \text{for a.e. } \xi \in \mathbb{R}^n\,. \tag{4}$$

Under these assumptions, problem (1), (2) admits a unique solution.

In this work, we consider numerical solution of (1), (2) by the Finite Element Method (FEM). After discussing the scale-resolution requirement for standard FEM, we design in Section 2 generalized FEM which allow to resolve the ε scale of the solution without mesh refinement down to scale ε. These generalized FEM are built from a macroscopic mesh together with a 'microscopic' FE-space $\mathcal{M}^\mu_\varepsilon$ of ε-periodic functions attached to each macroscopic degree of freedom. Several possible choices of $\mathcal{M}^\mu_\varepsilon$ are conceivable and the performance of the generalized FEM strongly depends on the choice of the micro-space. The specific choice of $\mathcal{M}^\mu_\varepsilon$ will be determined by the regularity of the exact solution u^ε. From classical homogenization theory, rather complete asymptotic expansions as $\varepsilon \to 0$ of the exact solution u^ε are available. The profiles in these expansions could, in principle, be used in the construction of $\mathcal{M}^\mu_\varepsilon$. Here, we shall use a representation of the solution in terms of a (generalized) Fourier integral. This is presented in Section 3 together with a methodology for deriving ε-dependent FE-spaces which allow robust (i.e. uniform in the scale ε) exponential convergence to the exact solution. The Fourier representation allows for new two-scale regularity estimates for the solution u^ε which are robust and explicit in ε (see [28]) and which, in turn, lead to the design of generalized FEM which allow for robust and optimal (in terms of the regularity of the data) convergence rates. We present here theoretical results only in an expository fashion, and refer to [31,32,26,28,29] for a detailed analysis.

2 Generalized FEM

In this section we present the abstract formulation of the two-scale FEM with generic micro space $\mathcal{M}_\varepsilon^\mu \subset H^1_{\mathrm{per}}(\varepsilon\widehat{Q})$ and discuss implementational aspects of the two-scale FEM. In particular, for macroscopic FE triangulations that are aligned with the periodic structure for which the computation of the element stiffnes matrices can be done with work independent of ε. Specific microspaces $\mathcal{M}_\varepsilon^\mu$ will be discussed below.

2.1 The homogenization problem on bounded domains

We present the ideas for a Dirichlet problem, but emphasize that other boundary conditions may be considered as well. As before, Ω is a bounded, Lipschitz subdomain of $\mathbb{R}^n$. Let $f(x) \in L^2(\Omega)$ and let $\varepsilon > 0$ fixed. Denote by $u^\varepsilon(x) \in H^1_0(\Omega)$ the solution of the following boundary value problem

$$\begin{aligned} L^\varepsilon(x, \partial/\partial_x)u^\varepsilon(x) := -\nabla\cdot\left(a\left(\frac{x}{\varepsilon}\right)\nabla u^\varepsilon(x)\right) + a_0\left(\frac{x}{\varepsilon}\right)u^\varepsilon(x) &= f(x) \text{ in } \Omega \\ u^\varepsilon|_{\partial\Omega} &= 0, \end{aligned} \tag{5}$$

with coefficients $a(\cdot) \in L^\infty(\mathbb{R}^n)^{n\times n}_{\mathrm{sym}}$, $a_0(\cdot) \in L^\infty(\mathbb{R}^n)$ being 2π-periodic functions and satisfying the strong ellipticity conditions (3)–(4). The finite element method is based on the usual variational formulation of (5):

$$\text{Find } u^\varepsilon(x) \in H^1_0(\Omega) \text{ such that } B^\varepsilon(u^\varepsilon, v) = (f, v)_{L^2(\Omega)} \quad \forall v \in H^1_0(\Omega), \tag{6}$$

where the bilinear form $B^\varepsilon(\cdot,\cdot) : H^1_0(\Omega) \times H^1_0(\Omega) \to \mathbb{R}$ is given by

$$B^\varepsilon(u, v) := \int_\Omega a\left(\frac{x}{\varepsilon}\right)\nabla u(x)\cdot\nabla v(x) + a_0\left(\frac{x}{\varepsilon}\right)u(x)v(x)\,dx \quad \forall u, v \in H^1_0(\Omega).$$

The finite element method [12,10] is then defined as follows: for a subspace $V_N \subset H^1_0(\Omega)$ of finite dimension $N = \dim(V_N)$, find $u_N^\varepsilon \in V_N$ such that

$$B^\varepsilon(u_N^\varepsilon, v) = (f, v)_{L^2(\Omega)} \quad \forall v \in V_N. \tag{7}$$

(7) is a linear system of N equations. Selecting a basis $\{\varphi_j^N : j = 1, \ldots, N\}$ for V_N, we can expand u_N^ε and $v \in V_N$ in terms of the shape functions $\{\varphi_j^N\}$ as

$$u_N^\varepsilon = \sum_{i=1}^N u_{i,N}^\varepsilon \varphi_i^N, \quad v = \sum_{j=1}^N v_{j,N}\varphi_j^N,$$

with coefficients $u_{i,N}^\varepsilon, v_{j,N} \in \mathbb{R}$. Inserting this into (7) we obtain the linear system for the coefficient vector $\underline{u_N^\varepsilon} = \left\{u_{i,N}^\varepsilon\right\}_{i=1}^N$

$$\underline{\underline{A^\varepsilon}}\,\underline{u_N^\varepsilon} = \underline{f}, \tag{8}$$

in which the stiffness matrix $\underline{\underline{A}}^\varepsilon$ and the load vector $\underline{f}$ are given by

$$\underline{\underline{A}}^\varepsilon_{ij} = B^\varepsilon(\varphi_j^N, \varphi_i^N), \quad \underline{f}_i = (f, \varphi_i^N)_{L^2(\Omega)}, \quad \forall i,j = 1,\ldots,N. \tag{9}$$

By the Lax-Milgram Theorem, there exists a unique weak solution of (7) and the error $e_N^\varepsilon := u^\varepsilon - u_N^\varepsilon$ is quasioptimal in the sense that for some $C > 0$ independent of ε it holds [11]

$$\|e_N^\varepsilon\|_{H^1(\Omega)} \leq C \inf_{v \in V_N} \|u^\varepsilon - v\|_{H^1(\Omega)}. \tag{10}$$

Thus, the error is completely determined by the approximation properties of the space V_N.

In view of (10) we design the generalized FEM to satisfy the following two criteria: a) the spaces V_N should have *robust* (i.e. independent of ε) approximation properties, and b) the linear system (8) can be set up efficiently.
For standard FE spaces, *scale resolution*, i.e., the discretization of the small-scale problem features with finite elements, is not feasible, even with advanced computing hardware. Even if the right-hand side f, the unit-cell $\widehat{Q}$, the domain Ω and the coefficients a and a_0 are smooth (i.e., C^∞), for $\varepsilon/\mathrm{diam}(\Omega) \ll 1$ the solution u^ε exhibits oscillations on the ε-scale which spoil the FE-convergence. To see this assume that $\widehat{Q} = [-\pi, \pi]^n$. Then, $\Omega^\varepsilon = \Omega$ and there exist positive constants $C = C(\Omega)$ and $C(\alpha) = C(\alpha, \Omega)$, $\alpha \in \mathbb{N}^n$, such that

$$\|u^\varepsilon\|_{L^2(\Omega)} \leq C, \quad \|D^\alpha u^\varepsilon\|_{L^2(\Omega)} \leq C(\alpha)\varepsilon^{1-|\alpha|}, \quad \forall \alpha \in \mathbb{N}^n,\ |\alpha| > 0.$$

Then, denoting by $\mathcal{V}_\varepsilon^N = \mathcal{V}_N = S^{p,1}(\Omega, \mathcal{T}_H) \subset H^1(\Omega)$ the FE space of piecewise polynomials of degree $p \geq 1$ on a quasiuniform mesh $\mathcal{T}_H$ of meshwidth H, it holds that

$$\min_{v \in S^{p,1}(\Omega, \mathcal{T}_H)} \|u^\varepsilon - v\|_E \leq C H^p \|D^{p+1} u\|_{L^2(\Omega)} \leq C(H/\varepsilon)^p.$$

Since the energy projection is bounded, we also have

$$\min_{v \in S^{p,1}(\Omega, \mathcal{T}_H)} \|u^\varepsilon - v\|_E \leq \|u^\varepsilon\|_E \leq C\|f\|_{L^2(\Omega)}.$$

It follows therefore that the FE error with respect to the FE space $\mathcal{V}_N = S^{p,1}(\Omega, \mathcal{T}_H)$ satisfies the following *a-priori* bounds

$$\frac{\|u^\varepsilon - u_N^\varepsilon\|_E}{\|u^\varepsilon\|_E} \leq \min(1, C(H/\varepsilon)^p),$$

with $C = C(p, \Omega, f, a, a_0) > 0$ being a constant independent of ε and H. Standard FEM such as piecewise linears on a quasiuniform mesh $\mathcal{T}_H$ of size H converge only if $H < \varepsilon$, i.e., if $N = \dim \mathcal{V}_\varepsilon^N = O(\varepsilon^{-n})$ so that the ε-scale

is resolved. Moreover, high order FEM ($p > 1$) will essentially not improve the scale resolution.
To overcome this nonrobustness of the standard FEM, we have developed *generalized FEM* where the size H of the element is larger than the scale ε of the oscillatory coefficient. The main idea is to replace the polynomials by function spaces that are, in a sense, adapted to the coefficients of the elliptic operator, i.e., information at scale ε is built into the shape functions.

2.2 Generalized two-scale FEM

We consider (1) with homogeneous Dirichlet boundary conditions on $\Gamma_D^\varepsilon := \partial\Omega^\varepsilon \cap \partial\Omega$ and homogeneous Neumann boundary conditions on the (if non-empty) interior cavity' boundaries $\Gamma_N^\varepsilon := \partial\Omega^\varepsilon \backslash \Gamma_D^\varepsilon$

$$\begin{aligned} u^\varepsilon|_{\Gamma_D^\varepsilon} &= 0 \\ n \cdot \left(a\left(\frac{x}{\varepsilon}\right) \nabla u^\varepsilon \right)|_{\Gamma_N^\varepsilon} &= 0. \end{aligned} \tag{11}$$

The solution u^ε of (1), (11) is an element of the space $H^1_{\Gamma_D^\varepsilon}(\Omega^\varepsilon)$ defined as

$$H^1_{\Gamma_D^\varepsilon}(\Omega^\varepsilon) := \left\{ u \in H^1(\Omega^\varepsilon) \mid \gamma_0 u|_{\Gamma_D^\varepsilon} = 0 \right\},$$

where γ_0 here denotes the Dirichlet trace operator. Two-scale FEMs are built from a macro and a micro FE-space. Let $\mathcal{T}$ be a regular mesh in Ω. The macro FE-space $S_0^{\underline{p},1}(\Omega, \mathcal{T}) \subset H_0^1(\Omega)$ is the usual FE space given by

$$S_0^{\underline{p},1}(\Omega, \mathcal{T}) := \{ u \in H_0^1(\Omega) \; : \; u|_K \in S^{p_K}(K), \forall K \in \mathcal{T} \}. \tag{12}$$

Here $\underline{p} = \{p_K\}_{K \in \mathcal{T}}$ is a degree vector and $S^{p_K}(K)$ denotes the elemental polynomial space in each element. We will refer to $\underline{p}$ as the *macro degree* and to $\mathcal{T}$ as the *macro mesh* of the two-scale FEM.
The micro FE-space $\mathcal{M}_\varepsilon^\mu$, $\mu \geq 0$, is a generic finite dimensional subspace of $H^1_{\text{per}}(\varepsilon\widehat{Q})$ of ε-periodic, oscillatory shape functions

$$H^1_{\text{per}}(\varepsilon\widehat{Q}) \supset \mathcal{M}_\varepsilon^\mu := \text{Span}\left\{ \phi_j\left(\frac{x}{\varepsilon}\right) \; : \; j = 1, \ldots, \mu + 1 \right\}, \quad \mathcal{M}^\mu := \mathcal{M}_1^\mu, \tag{13}$$

with $\phi_j(y) \in H^1_{\text{per}}(\widehat{Q})$, $\forall j$. A basic assumption on the space $\mathcal{M}_\varepsilon^\mu$ is that $\mathcal{M}_\varepsilon^0 = \mathcal{M}^0 = \text{Span}\{1\}$. As in the degree vector $\underline{p}$, we associate with each macroelement $K \in \mathcal{T}$ a micro degree μ_K and we denote by $\underline{\mu}$ the micro degree vector $\underline{\mu} := \{\mu_K\}_{K \in \mathcal{T}}$. We define the generalized two-scale FE space

$S_0^{\underline{p}}(\Omega,\mathcal{T};\mathcal{M}_\varepsilon^{\underline{\mu}}) \subset H^1_{\Gamma_D^\varepsilon}(\Omega^\varepsilon)$ as

$$\begin{aligned} S_0^{\underline{p}}(\Omega,\mathcal{T};\mathcal{M}_\varepsilon^{\underline{\mu}}) &= S_0^{\underline{p},1}(\Omega,\mathcal{T}) \otimes_{\mathcal{T}} \mathcal{M}_\varepsilon^{\underline{\mu}} \\ &:= \left\{ u \in H^1_{\Gamma_D^\varepsilon}(\Omega^\varepsilon) \mid u|_K \in S^{p_K}(K) \otimes \mathcal{M}_\varepsilon^{\mu_K} \ \forall K \in \mathcal{T} \right\}. \end{aligned} \tag{14}$$

If $\mathcal{M}_\varepsilon^\mu$ itself is a standard FE-space, we will use the notation $S^{\underline{p},\underline{\mu}}(\Omega,\mathcal{T};\varepsilon\widehat{Q},\varepsilon\widehat{\mathcal{T}})$ for the two grid FE space corresponding to

$$\mathcal{M}^{\underline{\mu}} = S^{\underline{\mu}}_{\mathrm{per}}(\widehat{Q},\widehat{\mathcal{T}}),$$

the polynomial FE space on the unit-cell of degree $\underline{\mu}$ on the unit-cell mesh $\widehat{\mathcal{T}}$.

Due to the assumption that Span $\{1\} \subset \mathcal{M}_\varepsilon^\mu$ for all $\mu \geq 0$, the macroscopic polynomial space is always a subspace of $S_0^{\underline{p}}(\Omega,\mathcal{T};\mathcal{M}_\varepsilon^{\underline{\mu}})$. We can refer to $S_0^{\underline{p}}(\Omega,\mathcal{T};\mathcal{M}_\varepsilon^{\underline{\mu}})$ as the *generalized* FE space and to the FEM based on $S_0^{\underline{p}}(\Omega,\mathcal{T};\mathcal{M}_\varepsilon^{\underline{\mu}})$ as *generalized* two-scale FEM.
We emphasize the flexibility provided by the tensor product construction of (14). Using a macro-mesh $\mathcal{T}$ allows us to represent the geometry as is customary in standard FEM. If we use elements $K \in \mathcal{T}$ of size $O(\varepsilon)$ in a small neighborhood of $\partial\Omega$ and choose there the micro degree $\mu = 0$, then the gFEM space reduces to a standard FEM with scale resolution at the boundary. This allows us to enforce boundary conditions directly, i.e. without recourse to the homogenized equation. In the interior of the domain, we choose large elements K and choose there $\mu > 0$, i.e. we couple in effect the generalized FEM in the interior to a standard FEM near the boundary.

Remark 1. For domains Ω with corners, we could also refine the macro mesh $\mathcal{T}$ geometrically toward the corners as is done in the *hp*-FEM for problems with piecewise analytic input data, [37,23]. However, for macro meshes $\mathcal{T}$ with strong local refinements, care has to be taken in the selection of μ: once for $K \in \mathcal{T}$ diam$(K) = O(\varepsilon)$, the macro mesh locally resolves the fine scale of u^ε and in such elements the micro degree μ may be reduced (even to $\mu = 0$) in order to avoid linear dependence.
It might be of interest to use a macro mesh $\mathcal{T}$ with anisotropic elements at the boundary: boundary macro elements abutting at a corner or singularity with size $< \varepsilon$ where the micro shape functions are omitted ($\mu_K = 0$) or elongated elements abutting at the boundary where $\mu_K > 0$ are necessary to resolve the tangential fine structure of the solution. In that case, the micro space $\mathcal{M}_\varepsilon^\mu$ should be chosen so as to reflect highly oscillatory character of the solution in the tangential direction. These micro shape functions would have to be derived from appropriate problems on the unit-cell. The interior elements are of size $\gg \varepsilon$ and $\mu_K > 0$ there, i.e., the micro scales are accounted by the oscillatory shape functions $\phi_j(x/\varepsilon)$, $0 < j \leq \mu_k$.

2.3 Implementation of two-scale FEM

In order to obtain an efficient algorithm it is essential that the element stiffness and mass matrices can be computed in a complexity independent of ε and to an accuracy which will not compromise the asymptotic convergence rates. Due to the rapid oscillations of the coefficients and of the micro-shapefunctions, the elemental stiffness matrices on the macro mesh can *not* be evaluated robustly, i.e., with work independent of ε, by standard quadratures. A remedy is either to resort to massively parallel computation as in [20,21], or to exploit periodicity. It turns out that the macro stiffness and mass matrices can be developed from *moments*, i.e., integrals with respect to monomial weighted coefficients in the fast variable of the micro scale shape functions on the unit-cell combined with certain lattice summation formulas.

Proposition 1. *Let $\varepsilon > 0$ and let $\mathcal{M}^{\mu}(\widehat{Q}) = Span\{\phi_i\}_{i=0}^{\mu}$ be any conforming FE discretization of $H^1_{\text{per}}(\widehat{Q})$ of dimension $\mu+1$ independent of ε, and assume that the macro mesh $\mathcal{T}_H$ is aligned with the periodic microstructure. Then the FEM with respect to the two-scale discretization is*

$$S^p(\Omega, \mathcal{T}_H; \mathcal{M}^{\mu}_{\varepsilon}(\varepsilon\widehat{Q})),$$

where $\mathcal{M}^{\mu}_{\varepsilon}(\varepsilon\widehat{Q}) = Span\{\phi_i(x/\varepsilon)\}_{i=0}^{\mu}$, can be implemented with computational work independent of ε.

The ε-independence is achieved by judiciously exploiting the periodicity in the fast variable. The stiffness matrix of the generalized FEM can be computed with a fixed number of operations (independent of ε) by exploiting the periodicity of the coefficients $a(\cdot), a_0(\cdot)$ and that of the special shape functions $\phi_j(y)$. We must compute only once integrals of $\phi_j(y)$ and their derivatives multiplied by monomials on the unit cell. In order to see this we state the following lemma:

Lemma 1. *Assume that $j \in \mathbb{N}$, $F(\cdot)$ is 1-periodic and $\varepsilon = 1/M$ with $M \in \mathbb{N}$. Then*

$$\int_0^1 x^j F\left(\frac{x}{\varepsilon}\right) dx = \sum_{k=0}^{j} \varepsilon^{j-k} \binom{j}{k} S(M,k) \int_0^1 y^{j-k} F(y) dy, \tag{15}$$

where

$$S(M,k) = \frac{1}{M^{k+1}} \sum_{l=0}^{M-1} l^k.$$

Remark 2. The sum of powers of natural numbers can be expressed in terms of Bernoulli numbers B_n and can be easily tabulated [36,17]

$$\begin{aligned}\textstyle\sum_{l=1}^m l^k &= \frac{m^{k+1}}{k+1} + \frac{m^k}{2} + \frac{1}{2}\binom{k}{1} B_2 m^{k-1} + \frac{1}{4}\binom{k}{3} B_4 m^{k-3} \\ &\quad + \frac{1}{6}\binom{k}{5} B_6 m^{k-5} + \dots\end{aligned}$$

the last term containing either m or m^2. It should be remarked that

$$\frac{1}{m^{k+1}} \sum_{l=1}^{m} l^k = O\left(\frac{1}{k+1}\right), \quad \text{as } m \to \infty.$$

In this way, computing integrals of monomials x^j multiplying $F(x/\varepsilon)$ over the reference domain $(0,1)$ reduces to computing the integrals of all monomials of degree k, with $0 \le k \le j$, multiplying $F(y)$ over the main period of $F(\cdot)$ and insert this in (15).

We comment on the development of the macroelement stiffness matrix for the case when the macro mesh $\mathcal{T}_H$ is aligned with the periodic lattice. If the shape functions of the micro space $\mathcal{M}^\mu$ are piecewise polynomials with respect to some unit-cell triangulation $\widehat{\mathcal{T}}$, then the elemental stiffness matrices on the macro mesh can be evaluated *exactly*. For convenience, we consider only the 1-d case. For the calculation of the two-scale element stiffness matrices, the basic integrals

$$\underline{\underline{\widehat{K}}}_\mu^{\gamma\delta\tau} = \left(\int_{\widehat{Q}} a(\hat{y}) \phi_i^{(\gamma)}(\hat{y}) \phi_j^{(\delta)}(\hat{y}) \hat{y}^\tau \, d\hat{y} \right)_{i,j=0,\ldots,\mu} \tag{16}$$

are needed. For the case when $\mathcal{M}^\mu = S^\mu_{\text{per}}(\widehat{Q}, \widehat{\mathcal{T}})$, i.e., the space of micro shape functions is the FE space corresponding to the FE discretization of the unit-cell, the macro stiffness and mass matrices can be developed from *moments*, i.e., integrals in the fast variable corresponding to discretization of the unit-cell problem with monomial weighted coefficients, combined with certain lattice summation formulas.
Let $K \in \mathcal{T}$ be a macro element of size H with 'macroscopic' polynomial space $S^p(K) = \text{Span}\{\nu_I^{[K]}\}_I$. The entries $\underline{\underline{A}}^{\varepsilon,[K]}_{(Ii)\,(Jj)}$ of the element stiffness matrix corresponding to K are of the form

$$\underline{\underline{A}}^{\varepsilon,[K]}_{(Ii)\,(Jj)} = \sum_{\gamma,\delta \le 1} \sum_{\alpha=\alpha(p)} \sum_{\tau \le \alpha} \left(\underline{\underline{\widehat{K}}}_\mu^{\gamma\delta\tau} \right)_{ij} \sum_{m=0}^{M-1} S^{IJ}_{\gamma\delta\alpha\tau}(m, H, \varepsilon),$$

with $M := H/\varepsilon$ and with $\sum_{m=0}^{M-1} S^{IJ}_{\gamma\delta\alpha\tau}(m, H, \varepsilon)$ being directly computable. The idea (see Lemma 1) is to compute sums of powers of natural numbers appearing in $\sum_{m=0}^{M} S^{IJ}_{\gamma\delta\alpha\tau}(m, H, \varepsilon)$ in terms of Bernoulli numbers B_i. Thus, such sums can be computed with work independent of $M = H/\varepsilon$. More precisely, one exploits the fact that for $N \in \mathbb{N}$, $\sum_{k=1}^{N} k^q$ is given by

$$\begin{aligned} \textstyle\sum_{k=1}^{N} k^q = {} & \frac{N^{q+1}}{q+1} + \frac{N^q}{2} + \frac{1}{2}\binom{q}{1} B_2 N^{q-1} + \frac{1}{4}\binom{q}{3} B_4 N^{q-3} \\ & + \frac{1}{6}\binom{q}{5} B_6 N^{q-5} + \ldots, \end{aligned} \tag{17}$$

the last term containing either N or N^2. It should be remarked that

$$\frac{1}{N^{q+1}} \sum_{k=1}^{N} k^q = O\left(\frac{1}{q+1}\right), \quad \text{as } N \to \infty,$$

so that (17) could also be used as asymptotic expansion for small $\varepsilon = H/M$. The amount of work for computing the element stiffness matrix in the two-scale FEM is therefore independent of ε. If $n > 1$, the same arguments apply if all indices are changed to suitable multi-indices.

In the remainder of this paper, we investigate the performance of the generalized FEM for homogenization problems. We start by investigating the regularity of the solutions u^ε of (1) in dependence on ε in the following section. Rather than using the classical asymptotic expansion technique (e.g. [7,8,14]), we use Fourier representations of the solution on the unbounded domain $\Omega^\varepsilon_\infty$ which are valid uniformly for all $\varepsilon > 0$ to deduce uniform two-scale regularity of u^ε. Moreover, we will give finite dimensional approximations of u^ε together with robust error estimates for the approximation error which are based on the Fourier representation. The finite dimensional approximations of u^ε allow directly to deduce proper choices of $\mathcal{M}^\mu_\varepsilon$ in (14) so that corresponding error bounds are also valid for the two scale FEM.

3 Regularity and Approximation of u^ε

In this section we address problem (1) on unbounded domains, i.e., when $\Omega = \mathbb{R}^n$. Based on a mixed variational formulation we show existence and uniqueness of the solution u^ε in exponentially weighted Sobolev spaces $H^1_{-\nu}(\Omega^\varepsilon_\infty)$, $\nu \in (0, \nu_0)$, for general right hand sides f in the dual space of $H^1_\nu(\Omega^\varepsilon_\infty)$ [31,26]. We also represent u^ε as a Fourier-Bochner integral with respect to the Banach spaces $H^1_{-\nu}(\Omega^\varepsilon_\infty)$. The starting point is the analysis of the Fourier-Bochner integral kernel introduced as the inverse Fourier symbol of the elliptic operator (1). In this symbol, *scale separation* occurs: it is the product of the standard Fourier waves $e^{it\cdot x}$ and periodic functions $\phi(x/\varepsilon, \varepsilon, t)$ which encode the fast scale of the solution. More importantly, the functions $\phi(y, \varepsilon, t)$ arise as solutions of parameter dependent problems on the unit-cell $\widehat{Q}$ and are therefore amenable to FE-approximations. Next, we present the regularity properties of the integral kernel $e^{it\cdot x}\phi(x/\varepsilon, \varepsilon, t)$ in dependence on the parameters $\varepsilon > 0$, $t \in \mathbb{R}^n$. It turns out that the dependence on these parameters is *analytic*, even if the unit-cell problem admits only very low elliptic regularity. This analyticity is exploited in the construction of various finite-dimensional approximations of the solution as well as in proving robust error estimates for these approximations.

3.1 Variational setting

Let $\varepsilon > 0$ be a length scale and let $\widehat{Q} \subset \widehat{Y} := [-\pi, \pi]^n$ be a unit cell with Lipschitz boundary. We consider the homogenization problem on the unbounded domain

$$L^\varepsilon \left(\frac{x}{\varepsilon}, \partial_x\right) u^\varepsilon(x) = f(x) \quad \text{in} \quad \Omega^\varepsilon_\infty \tag{18}$$

$$n \cdot \left(a\left(\frac{x}{\varepsilon}\right)\nabla u^\varepsilon\right) = 0 \quad \text{on} \quad \partial\Omega^\varepsilon_\infty \tag{19}$$

in which the second order elliptic operator $L^\varepsilon(x/\varepsilon, \partial_x)$ is given by

$$L^\varepsilon \left(\frac{x}{\varepsilon}, \partial_x\right) u(x) := -\nabla \cdot \left(a\left(\frac{x}{\varepsilon}\right)\nabla u\right) + a_0\left(\frac{x}{\varepsilon}\right) u,$$

where n is the unit outward normal to $\partial\Omega^\varepsilon_\infty$, $\varepsilon > 0$ and $f \in L^2(\Omega^\varepsilon_\infty)$.
We considered homogenous Neumann boundary conditions on the interior cell boundaries, but emphasize that the case when $\widehat{Q} = \widehat{Y}$ or the case when homogeneous Dirichlet boundary conditions on the interior boundary $\partial\widehat{Q}\backslash\partial\widehat{Y}$ are prescribed can be treated completely analogously.
Then, by the Lax-Milgram Theorem [25], for every $f \in L^2(\Omega^\varepsilon_\infty)$ and for every $\varepsilon > 0$, (18)–(19) admits a unique weak solution $u^\varepsilon \in H^1(\Omega^\varepsilon_\infty)$, i.e.,

$$B^\varepsilon(u^\varepsilon, v) = (f, v)_{L^2(\Omega^\varepsilon_\infty)} \quad \forall v \in H^1(\Omega^\varepsilon_\infty), \tag{20}$$

where the bilinear form $B^\varepsilon(\cdot,\cdot)$ is given by

$$B^\varepsilon(u,v) := \left(\nabla v, a\left(\frac{x}{\varepsilon}\right)\nabla u\right)_{L^2(\Omega^\varepsilon_\infty)} + \left(v, a_0\left(\frac{x}{\varepsilon}\right)u\right)_{L^2(\Omega^\varepsilon_\infty)}, \quad \forall u, v \in H^1(\Omega^\varepsilon_\infty).$$

3.2 Fourier representation of the solution

Let $f = f(x)$ be independent of ε and $f \in L^2(\mathbb{R}^n)$ (in particular, f has to be defined everywhere on $\mathbb{R}^n$, not just on $\Omega^\varepsilon_\infty$). Then, the following Fourier inversion formula in $L^2(\mathbb{R}^n)$ holds

$$f(x) = \frac{1}{(2\pi)^{n/2}} \int_{\mathbb{R}^n} \hat{f}(t) e^{it\cdot x} dt,$$

where

$$\hat{f}(t) := \frac{1}{(2\pi)^{n/2}} \int_{\mathbb{R}^n} e^{-it\cdot x} f(x) dx \in L^2(\mathbb{R}^n)$$

denotes the standard Fourier transform of f.

We might therefore think of $u^\varepsilon(x)$ as a superposition of elementary solutions $\hat{f}(t)\psi^\varepsilon(x,t)$ of (18)–(19) with right hand side $\hat{f}(t)e^{it\cdot x}$, where the kernel $\psi^\varepsilon(x,t)$ satisfies

$$\begin{aligned} L^\varepsilon(x/\varepsilon, \partial_x)\psi^\varepsilon(x,t) &= e^{it\cdot x} \quad \text{in } \Omega^\varepsilon_\infty, \\ n\cdot a\left(\frac{x}{\varepsilon}\right)\nabla_x\psi^\varepsilon(x,t) &= 0 \quad \text{on} \quad \partial\Omega^\varepsilon_\infty. \end{aligned} \tag{21}$$

To this end, we have to give a rigorous mathematical meaning to the integral

$$\int_{\mathbb{R}^n} \hat{f}(t)\psi^\varepsilon(x,t)\,dt.$$

Since $f(x) = e^{it\cdot x}$ in (21) is not in $L^2(\Omega^\varepsilon_\infty)$, (21) has no solution in $H^1(\Omega^\varepsilon_\infty)$ in the sense of (20). We have therefore to weaken the notion of solution of (18)–(19) and to allow for input data such as $f(x) = e^{it\cdot x}$.
To characterize precisely this notion of solution of (18)–(19) in $\Omega^\varepsilon_\infty$, we introduce weighted Sobolev spaces $H^j_\nu(\Omega^\varepsilon_\infty)$ of complexed-valued functions with exponential weights depending on a real parameter ν.

Definition 1. For $j = 0,1$ and for any $\nu \in \mathbb{R}$ the weighted Sobolev spaces $H^j_\nu(\Omega^\varepsilon_\infty)$ equipped with the norm $\|\cdot\|_{j,\nu}$ are defined to be

$$H^j_\nu(\Omega^\varepsilon_\infty) = \overline{C^\infty_0(\mathbb{R}^n;\mathbb{C})|_{\Omega^\varepsilon_\infty}}^{\|\cdot\|_{j,\nu}}, \tag{22}$$

where

$$\|u\|^2_{j,\nu} = \int_{\Omega^\varepsilon_\infty}\left(\sum_{|\alpha|\le j}|D^\alpha_x u|^2\right)e^{2\nu\|x\|}\,dx \quad (D^\alpha_x u = \partial^{\alpha_1}_{x_1}\ldots\partial^{\alpha_n}_{x_n}u,\ \forall\alpha\in\mathbb{N}^n_0). \tag{23}$$

Note that for $\nu > 0$ holds $H^1_\nu \subset H^1_0 = H^1 \subset H^1_{-\nu}$. The variational solution $u^\varepsilon \in H^1(\Omega^\varepsilon_\infty)$ of (20) can also be viewed as an element of the exponentially weighted Sobolev spaces $H^1_{-\nu}(\Omega^\varepsilon_\infty)$ for some $\nu > 0$. Moreover, it has the following representation: for $x \in \Omega^\varepsilon_\infty$, there holds

$$u^\varepsilon(x) = \frac{1}{(2\pi)^{n/2}}\int^{(B)}_{\mathbb{R}^n}\hat{f}(t)\psi^\varepsilon(x,t)dt. \tag{24}$$

Remark 3. Here and in the following we write $\int^{(B)}$ to designate a Bochner integral of $H^1_{-\nu}$-valued functions.

To render (21), (24) meaningful, we must generalize (20) for right hand sides $f(x)$ which are not decaying at ∞. To do so, we extend $B^\varepsilon(.,.)$ to a sesquilinear form $\Psi(\varepsilon)[\cdot,\cdot] : H^1_{-\nu}(\Omega^\varepsilon_\infty) \times H^1_\nu(\Omega^\varepsilon_\infty) \to \mathbb{C}$:

$$\Psi(\varepsilon)[u,v] = \int_{\Omega^\varepsilon_\infty} \left\{ \sum_{p,q=1}^{n} a_{pq}\left(\frac{x}{\varepsilon}\right) \frac{\partial u}{\partial x_p}(x) \overline{\frac{\partial v}{\partial x_q}(x)} + a_0\left(\frac{x}{\varepsilon}\right) u(x)\overline{v(x)} \right\} dx. \tag{25}$$

For all $\varepsilon > 0$ and for $\nu > 0$ sufficiently small, $\Psi(\varepsilon)$ is bounded and 'coercive' with respect to $H^1_{-\nu}(\Omega^\varepsilon_\infty) \times H^1_\nu(\Omega^\varepsilon_\infty)$, in the sense that the *inf-sup* stability condition holds (note that for $\nu = 0$, $\Psi(\varepsilon)$ coincides with B^ε in (20)). More precisely, there holds [31,32,26]:

Proposition 2. *There exist positive constants ν_0, C and γ such that for all $\nu \in (0, \nu_0)$ and all $\varepsilon > 0$*

1. $|\Psi(\varepsilon)[u,v]| \le C\|u\|_{1,-\nu}\|v\|_{1,\nu}$,
2. $\inf_{\|u\|_{1,-\nu}=1} \sup_{\|v\|_{1,\nu}=1} |\Psi(\varepsilon)[u,v]| \ge \gamma > 0$
3. $\sup_{u\in H^1_{-\nu}(\Omega^\varepsilon_\infty)} |\Psi(\varepsilon)[u,v]| > 0$ *for all* $v \in H^1_\nu(\Omega^\varepsilon_\infty)$ *and* $v \ne 0$.

The continuity of the sesquilinear form $\Psi(\varepsilon)$ stated in *1.* is obvious. The *inf-sup* condition *2.* and the injectivity property *3.* can be verified in the following way: based on the coercivity of the bilinear form B^ε for the case $\nu = 0$ a perturbation argument can be employed to prove the existence of a positive $\nu_0 > 0$ such that *2.* and *3.* hold. We emphasize that ν_0 is independent of ε and depends only on the upper and lower bounds of the matrix a and of the zero order coefficient a_0. Proposition 2 implies immediately the existence of $\psi^\varepsilon(x,t)$:

Proposition 3. *The properties 1, 2 and 3 of $\Psi(\varepsilon)$ imply that the variational problem*

$$\begin{gathered}\text{Given } f \in \left(H^1_\nu(\Omega^\varepsilon_\infty)\right)^*, \text{ find} \\ u \in H^1_{-\nu}(\Omega^\varepsilon_\infty) \ : \ \Psi(\varepsilon)[u,v] = \langle f, v\rangle_{(H^1_\nu(\Omega^\varepsilon_\infty))^* \times H^1_\nu(\Omega^\varepsilon_\infty)}, \quad \forall v \in H^1_\nu(\Omega^\varepsilon_\infty),\end{gathered} \tag{26}$$

admits a unique weak solution $u \in H^1_{-\nu}(\Omega^\varepsilon_\infty)$ and the a-priori estimate

$$\|u\|_{H^1_{-\nu}(\Omega^\varepsilon_\infty)} \le (1/\gamma)\|f\|_{(H^1_\nu(\Omega^\varepsilon_\infty))^*}$$

holds.

Since $e^{it\cdot x} \in \left(H^1_\nu(\Omega^\varepsilon_\infty)\right)^*$, this result implies the existence of a solution $\psi^\varepsilon(x,t)$ of (21). The representation (24), however, is yet of limited use for applications since the kernel $\psi^\varepsilon(x,t)$ requires the solution of an unbounded domain problem and is not explicitly available. However, in the problem (18),

(19) holds *scale-separation* which is reflected in the structure of the kernel $\psi^\varepsilon(x,t)$, as will now be explained.

Let $\phi(\cdot,\varepsilon,t)$ be the 2π-periodic weak solution of the so-called unit-cell problem

$$\begin{aligned}\mathcal{L}(\varepsilon,t;y,\partial_y)\phi &:= e^{-i\varepsilon t3\cdot y}L^\varepsilon(y,\varepsilon^{-1}\partial_y)\,e^{i\varepsilon t\cdot y}\phi\\ &= -e^{-i\varepsilon t\cdot y}\sum_{p,q=1}^{n}\varepsilon^{-2}\frac{\partial}{\partial y_p}\left(a_{pq}(y)\frac{\partial}{\partial y_q}\left(\phi(y,\varepsilon,t)e^{it\cdot\varepsilon y}\right)\right)\\ &+\ a_0(y)\phi(y,\varepsilon,t)=1,\quad \text{in } \widehat{Q},\end{aligned} \tag{27}$$

$$n\cdot a(y)(it\varepsilon+\nabla_y)\phi=0 \text{ on } \widehat{\Gamma}_N.$$

We denoted here by $\widehat{\Gamma}_N$ the Neumann boundary $\widehat{\Gamma}_N := \partial\widehat{Q}\backslash\widehat{\Gamma}_{\text{per}}$, where $\widehat{\Gamma}_{\text{per}}=\partial\widehat{Q}\cap\partial[-\pi,\pi]^n$.
We will refer to $x\in\mathbb{R}^n$ as the *slow* variable and $y=x/\varepsilon$ will denote the *fast* variable of the problem.
The Sobolev spaces of periodic functions on Ω_∞ for which $\widehat{Y}$ is the fundamental period will be denoted by $H^j_{\text{per}}(\widehat{Q})$, and are defined to be the completion with respect to the standard Sobolev norm $\|\cdot\|_j$ of the complex-valued, C^∞ functions on Ω_∞ that are 2π-periodic in each coordinate. Next, we define

$$\psi(y,\varepsilon,t):=\phi(y,\varepsilon,t)e^{it\cdot\varepsilon y},\quad y\in\Omega_\infty. \tag{28}$$

The function ψ is the generalized fundamental solution of L^ε (resp. the two-scale Fourier symbol of $(L^\varepsilon)^{-1}$ on $\Omega^\varepsilon_\infty$). The basic tool for the proof of the Fourier-Bochner integral representation in (24) is given in the following proposition:

Proposition 4. *For each $\varepsilon>0$, $t\in\mathbb{R}^n$ and $\nu\in(0,\nu_0)$, the kernel ψ defined in (28) is the unique weak solution in $H^1_{-\nu}(\Omega^\varepsilon_\infty)$ of*

$$\Psi(\varepsilon)\left[\psi\left(\frac{\cdot}{\varepsilon},\varepsilon,t\right),v\right]=\langle e^{it\cdot x},v\rangle_{(H^1_\nu)^*\times H^1_\nu},\quad \forall v\in H^1_\nu(\Omega^\varepsilon_\infty). \tag{29}$$

In particular therefore $\psi^\varepsilon(x,t)=\psi(x/\varepsilon,\varepsilon,t)$ in $H^1_{-\nu}(\mathbb{R}^n)$ and

$$\left\|\psi\left(\frac{x}{\varepsilon},\varepsilon,t\right)\right\|_{1,-\nu}\le\frac{1}{\gamma\nu^{n/2}}.$$

We state now the Fourier-Bochner integral respresentation formula (24) in weighted Sobolev spaces for the solution $u^\varepsilon(x)$ of (18) on $\Omega^\varepsilon_\infty$. The proof can be found in [31].

Theorem 1. *Let $f(x)\in L^2(\mathbb{R}^n)$, $\nu\in(0,\nu_0)$ and $\varepsilon>0$. Further let $u^\varepsilon(x)\in H^1(\Omega^\varepsilon_\infty)$ be the solution of (18) on $\Omega^\varepsilon_\infty$. Then, as an element of the weighted Sobolev space $H^1_{-\nu}(\Omega^\varepsilon_\infty)$, u^ε admits the Bochner integral representation (24).*

We will show how the kernels $\phi(y,\varepsilon,t),\psi(y,\varepsilon,t)$ can be used to design FE subspaces of (1), (2) which encode the microstructure and coefficient regularity. A crucial role for obtaining error estimates will be played by the kernels' analyticity with respect to t. As a byproduct of the analysis, we obtain specific choices of the micro-subspace $\mathcal{M}^{\mu}_{\varepsilon}$ in the generalized FE-space (14).

3.3 Regularity of the kernels ψ, ϕ

The kernel $\phi(\cdot,\varepsilon,t)$ can be continued analytically with respect to (ε,t) in a complex neighborhood $\widehat{G}\subset\mathbb{C}^{n+1}$ of $\mathbb{R}^{n+1}$, with values in $H^1_{\mathrm{per}}(\widehat{Q})$. More precisely, for every fixed $\varepsilon>0$, the mappings

$$\mathbb{R}^n \ni t \to \phi\left(\frac{\cdot}{\varepsilon},\varepsilon,t\right),\ \psi\left(\frac{\cdot}{\varepsilon},\varepsilon,t\right) \in H^1_{-\nu}(\Omega^{\varepsilon}_{\infty})$$

can be continued analytically with respect to t in a strip neighbourhood of $\mathbb{R}^n$, and the width of the strip is independent of ε.
Let $\nu\in(0,\nu_0)$, $\nu_n=\nu/(2\sqrt{n})$ and denote by $\mathcal{D}_{\nu_n}$ the strip region of width ν_n

$$\mathcal{D}_{\nu_n} := \{t\in\mathbb{C}^n \ :\ \max_{j=1,\dots,n}|\mathrm{Im}\,t_j|<\nu_n\}. \tag{30}$$

It can be verified that the mapping

$$\mathcal{D}_{\nu_n}\ni t\to G(t):=e^{it\cdot x}\in\left(H^1_{\nu}(\Omega^{\varepsilon}_{\infty})\right)^* \tag{31}$$

is holomorphic in $\mathcal{D}_{\nu_n}$ with values in the Banach space $\left(H^1_{\nu}(\Omega^{\varepsilon}_{\infty})\right)^*$. Moreover, $G_k(t):=(ix)^k e^{it\cdot x}\in\left(H^1_{\nu}(\Omega^{\varepsilon}_{\infty})\right)^*$ is the $k\in\mathbb{N}^n$–th partial derivative with respect to t of the $\left(H^1_{\nu}(\Omega^{\varepsilon}_{\infty})\right)^*$-valued mapping $G(t)$ and its norm has the following bound

$$\forall t\in\mathcal{D}_{\nu_n}:\quad \|G_k(t)\|_{(H^1_{\nu}(\Omega^{\varepsilon}_{\infty}))^*}\le C_{\nu,n}\frac{2^{|k|}k!}{\nu^{|k|}}n^{|k|/2},\quad k\in\mathbb{N}^n, \tag{32}$$

where $k!:=k_1!\dots k_n!$. For fixed $t\in\mathcal{D}_{\nu_n}$ and $k\in\mathbb{N}^n$, let $\psi^{\varepsilon}_k(t)$ be the weak solution in $H^1_{-\nu}(\Omega^{\varepsilon}_{\infty})$ of the following variational problem:

$$\begin{aligned}\psi^{\varepsilon}_k\in H^1_{-\nu}(\Omega^{\varepsilon}_{\infty})\ :\ &\Psi(\varepsilon)[\psi^{\varepsilon}_k(t),v]=\\ &\langle G_k(t),v\rangle_{(H^1_{\nu}(\Omega^{\varepsilon}_{\infty}))^*\times H^1_{\nu}(\Omega^{\varepsilon}_{\infty})},\quad \forall v\in H^1_{\nu}(\Omega^{\varepsilon}_{\infty}).\end{aligned} \tag{33}$$

By Proposition 3, ψ^{ε}_k exists and is unique. It has the following properties (see [26] for a proof):

1) The mapping $\mathcal{D}_{\nu_n}\ni t\to\psi(\cdot/\varepsilon,\varepsilon,t)\in H^1_{-\nu}(\Omega^{\varepsilon}_{\infty})$ is holomorphic in $\mathcal{D}_{\nu_n}$ with values in the Banach space $H^1_{-\nu}(\Omega^{\varepsilon}_{\infty})$. Moreover, $\psi^{\varepsilon}_k(t)$ defined by (33) is given by

$$\psi^{\varepsilon}_k(t)=D^k_t\psi\left(\frac{\cdot}{\varepsilon},\varepsilon,t\right)=\partial^{k_1}_{t_1}\dots\partial^{k_n}_{t_n}\psi\left(\frac{\cdot}{\varepsilon},\varepsilon,t\right) \tag{34}$$

and the $\|\cdot\|_{1,-\nu}$-norm of $\psi_k^\varepsilon(t)$ is uniformly bounded with respect to $t \in \mathcal{D}_{\nu_n}$ by

$$\|\psi_k^\varepsilon(t)\|_{1,-\nu} \leq C_{\nu,n} \frac{2^{|k|} k!}{\gamma \nu^{|k|}} n^{|k|/2}. \tag{35}$$

2) For a given $\nu \in (0, \nu_0)$ the mapping $\mathcal{D}_{\nu_n} \ni t \to \phi(\cdot/\varepsilon, \varepsilon, t) \in H^1_{-2\nu}(\Omega_\infty^\varepsilon)$ is a holomorphic function of $t \in \mathcal{D}_{\nu_n}$ with values in the Banach space $H^1_{-2\nu}(\Omega_\infty^\varepsilon)$. Moreover,

$$\left\| D_t^k \phi\left(\frac{\cdot}{\varepsilon}, \varepsilon, t\right)\right\|_{H^1_{-2\nu}(\Omega_\infty^\varepsilon)} \leq C_{\nu,n,\gamma}(1 + \|t\|) \frac{(k+\mathbf{1})!}{(\nu/2)^{|k|}} n^{|k|/2}, \ \forall t \in \mathcal{D}_d, \tag{36}$$

where $\mathbf{1} := (1, 1, \ldots, 1)^\top \in \mathbb{R}^n$.

3.4 Approximation of u^ε

The analyticity of the kernel $\psi(y, \varepsilon, t)$ with respect to ε can be used to obtain from (24) asymptotic expansions with respect to ε which essentially reproduce the classical asymptotic homogenizations of (1), (2). Here, we derive local approximation spaces $\mathcal{V}_\varepsilon^N$ from approximations of u^ε based on a Poisson summation formula applied to the Bochner representation (24). This spectral approach gives robust, exponentially convergent approximations of u^ε under certain analyticity properties of the data.

Sets of analytic data

Assume that X is a Banach space and let $\mathcal{D}_d$ denote the strip region of width d given by

$$\mathcal{D}_d = \{z \in \mathbb{C}^n \ : \ |\mathrm{Im}\,(z)| \leq d\}.$$

Then we define

$$\begin{aligned}&\mathbf{H}^p(\mathcal{D}_d; X) = \\ &\{g : \mathcal{D}_d \to X \,|\, g \text{ is analytic in } \mathcal{D}_d \text{ and } \mathbf{N}_p(g, \mathcal{D}_d; X) < \infty\},\end{aligned} \tag{37}$$

where

$$\mathbf{N}_p(g, \mathcal{D}_d; X) = \begin{cases} \lim\limits_{\delta \to 0+} \left(\displaystyle\int_{\partial \mathcal{D}_d(\delta)} \|g(z)\|_X^p |dz| \right)^{1/p} & , \quad \text{if } 1 \leq p < \infty \\ \lim\limits_{\delta \to 0+} \sup\limits_{z \in \mathcal{D}_d(\delta)} \|g(z)\|_X & , \quad \text{if } p = \infty, \end{cases}$$

and for $0 < \delta < 1$, $\mathcal{D}_d(\delta)$ is defined by

$$\mathcal{D}_d(\delta) = \{z \in \mathbb{C}^n \ : \ |\mathrm{Re}\,(z)| < 1/\delta, \ |\mathrm{Im}\,(z)| < d(1 - \delta)\}.$$

Definition 2. We say, a function f fulfils the 'usual assumptions', if $f \in L^2(\mathbb{R}^n)$, and its Fourier transformation can be extended to a holomorphic function $\hat{f}$ in the strip $\mathcal{D}_d$, with $d = \nu_n$ as in (30), that satisfies the following growth condition:

$$|\hat{f}(z)| \leq C(f)e^{-\alpha\|z\|}, \quad \forall\, z \in \mathcal{D}_d, \tag{38}$$

for some positive constants $C(f), \alpha > 0$.

Remark 4. A typical example of functions that satisfy the 'usual assumptions' is that of functions arising through convolution with a Gauss kernel. Let $f \in L^2(\mathbb{R}^n)$ be compactly supported. Without loss of generality we may assume that $\operatorname{supp} f \subset \{x \;:\; \|x\| \leq 1\}$. Further let $\delta > 0$ and denote by ρ_δ the following Dirac sequence

$$\rho_\delta(y) := \delta^{-n/2}e^{-\|y\|^2/\delta}, \quad y \in \mathbb{R}^n .$$

Define $f_\delta := f * \rho_\delta$, i.e.,

$$f_\delta(x) = \delta^{-n/2} \int_{\mathbb{R}^n} f(y)e^{-\|x-y\|^2/\delta} dy, \quad x \in \mathbb{R}^n . \tag{39}$$

Then, f_δ defined by (39) is an analytic, exponentially decaying function that satisfies the 'usual assumptions'. Indeed, the Fourier transform of f_δ is given by the product of the Fourier transforms of f and ρ_δ

$$\hat{f}_\delta(\xi) = \hat{f}(\xi)\hat{\rho}_\delta(\xi) = e^{-\delta\|\xi\|^2/4}\hat{f}(\xi)$$

which is an analytic function that decays exponentially in a complex strip neighborhood of $\mathbb{R}^n$.

Under the 'usual assumptions' on f the mapping

$$\mathcal{D}_d \ni t \to g(t,\cdot) = g_\varepsilon(t,\cdot) := \frac{1}{(2\pi)^{n/2}}\hat{f}(t)\phi\left(\frac{\cdot}{\varepsilon},\varepsilon,t\right) \in H^1_{-2\nu}(\Omega^\varepsilon_\infty) \tag{40}$$

is in $\mathbf{H}^p(\mathcal{D}_d; H^1_{-2\nu}(\Omega^\varepsilon_\infty))$. Moreover, there exists a $C(\gamma,\nu,n)$ such that $g_\varepsilon(t,\cdot)$ satisfies the growth condition:

$$\|g_\varepsilon(t,\cdot)\|_{1,-2\nu} \leq C(\gamma,\nu,n)C(f)\left(1+\frac{1}{\alpha}\right)e^{-\frac{\alpha}{2}\|t\|}, \quad \forall\, t \in \mathcal{D}_d, \tag{41}$$

where α and $C(f)$ are as in (38).

Trapezoidal approximation of (24)

We shall now derive finite dimensional approximations by truncating a generalized Poisson summation formula applied to the Bochner representation

$$\begin{aligned} u^{\varepsilon}(x) &= \frac{1}{(2\pi)^{n/2}} \overset{(B)}{\underset{\mathbb{R}^n}{\int}} \hat{f}(t)\psi(x/\varepsilon,\varepsilon,t)\,dt \\ &= \frac{1}{(2\pi)^{n/2}} \overset{(B)}{\underset{\mathbb{R}^n}{\int}} \hat{f}(t)e^{it\cdot x}\phi\left(\frac{x}{\varepsilon},\varepsilon,t\right)\,dt. \end{aligned} \tag{42}$$

Let $L \geq 1$ and assume in what follows that $\pi/h \geq 2\,L$, i.e., $h \leq \pi/(2\,L)$. We define

$$\mathbb{Z}^n_{(N)} := \{k \in \mathbb{Z}^n \ : \ |k_p| \leq N, \, \forall p = 1, \cdots, n\}$$

and the trapezoidal sum approximations of the Bochner integral (42) is

$$\begin{aligned} u^{\varepsilon}_{N,h}(x) &= \mathbf{1}_{[-\frac{\pi}{h},\frac{\pi}{h}]^n}(x)\frac{1}{(2\pi)^{n/2}}h^n \sum_{k\in\mathbb{Z}^n_{(N)}} \phi\left(\frac{x}{\varepsilon},\varepsilon,kh\right)\hat{f}(kh)e^{ikh\cdot x} \\ &= \mathbf{1}_{[-\frac{\pi}{h},\frac{\pi}{h}]^n}(x)\frac{1}{(2\pi)^{n/2}}h^n \sum_{k\in\mathbb{Z}^n_{(N)}} \psi\left(\frac{x}{\varepsilon},\varepsilon,kh\right)\hat{f}(kh). \end{aligned} \tag{43}$$

Then, u^{ε} in the integral (24) can be split as

$$u^{\varepsilon}(\cdot) = u^{\varepsilon}_{N,h}(\cdot) + \delta_N(f,h)(\cdot).$$

Remark 5. Since we have real-valued input data we have a real valued solution u^{ε}. The same holds for the finite approximation $u^{\varepsilon}_{N,h}$. Indeed, since

$$\hat{f}(-\xi) = \overline{\hat{f}(\xi)}, \quad \psi\left(\frac{\cdot}{\varepsilon},\varepsilon,-\xi\right) = \overline{\psi\left(\frac{\cdot}{\varepsilon},\varepsilon,\xi\right)},$$

it follows that

$$\begin{aligned} &u^{\varepsilon}_{N,h}(x) = \\ &\mathbf{1}_{[-\frac{\pi}{h},\frac{\pi}{h}]^n}(x)\frac{1}{(2\pi)^{n/2}}h^n\psi\left(\frac{x}{\varepsilon},\varepsilon,0\right)\hat{f}(0) + 2\,\mathbf{1}_{[-\frac{\pi}{h},\frac{\pi}{h}]^n}(x)\frac{1}{(2\pi)^{n/2}}h^n \\ &\times \sum_{k\in\mathbb{Z}^n_{(N)}\setminus\{0\},\,k_1\geq 0} \left\{\operatorname{Re}\psi\left(\frac{x}{\varepsilon},\varepsilon,kh\right)\operatorname{Re}\hat{f}(kh) - \operatorname{Im}\psi\left(\frac{x}{\varepsilon},\varepsilon,kh\right)\operatorname{Im}\hat{f}(kh)\right\}, \end{aligned}$$

which is real valued.

We next present two types of robust (in ε) convergence estimates which are based on different assumptions on the data f.

Exponential convergence

Under the 'usual assumptions' on the data f, the sum (43) approximates u^{ε} in (24) at a robust exponential rate, i.e., independent of ε, (see [26,29]).

Proposition 5. *Let f satisfy the 'usual assumptions' in Definition 2 with some $\alpha, d > 0$ and let $L > 0$ be arbitrary. Define*

$$h = \left(\frac{\pi d}{\alpha N}\right)^{1/2} \tag{44}$$

and assume $N \geq 4dL^2/(\alpha\pi)$, i.e., $\pi/h \geq 2L$. Then, the error $\delta_N(f,h)(\cdot) = u^\varepsilon(\cdot) - u^\varepsilon_{N,h}(\cdot)$, with $u^\varepsilon_{N,h}(\cdot)$ as in (43), decays exponentially with respect to N and uniformly with respect to ε in the $\|\cdot\|_{0,-2\nu}$, $\|\cdot\|_{H^1_{-2\nu}((-L,L)^n)}$-norms:

$$\begin{aligned} &\|\delta_N(f,h)(\cdot)\|_{0,-2\nu} + \|\delta_N(f,h)(\cdot)\|_{H^1_{-2\nu}((-L,L)^n)} \\ &\qquad \leq C(\gamma,\nu,n)C(f)\left(1+\frac{1}{\alpha}\right)^2 \frac{1}{\alpha^n} e^{-(\pi\alpha d N/n)^{1/2}}. \end{aligned} \tag{45}$$

The constants $C(\gamma,\nu,n), C(f)$ are independent of ε, N, L.

As a corollary, the following approximation result holds

Corollary 1. *Assume that f satisfies the 'usual assumptions' and that*

$$h = \left(\frac{\pi d}{\alpha N}\right)^{1/2}, \quad N \geq \frac{4dL^2}{\alpha\pi}. \tag{46}$$

Define the finite-dimensional subspace

$$\mathcal{V}_\varepsilon^N := \operatorname{Span}\left\{\operatorname{Re}\psi\left(\frac{\cdot}{\varepsilon},\varepsilon,kh\right),\ \operatorname{Im}\psi\left(\frac{\cdot}{\varepsilon},\varepsilon,kh\right)\ :\ k \in \mathbb{Z}^n_{(N)}\right\}. \tag{47}$$

Then

$$\begin{aligned} &\inf_{v\in\mathcal{V}_\varepsilon^N} \|u^\varepsilon - v\|_{H^1_{-2\nu}((-L,L)^n)} + \inf_{v\in\mathcal{V}_\varepsilon^N} \|u^\varepsilon - v\|_{0,-2\nu} \\ &\qquad \leq C(\gamma,\nu,n)C(f)\left(1+\frac{1}{\alpha}\right)^2 \frac{1}{\alpha^n} e^{-(\pi\alpha d N/n)^{1/2}}, \end{aligned} \tag{48}$$

where $C(f)$ and $\alpha = \alpha(f)$ are those from Definition 2, in particular independent of ε, N.

Spectral convergence

The usual assumptions on the data f entail essentially boundedness of all derivatives. Now we assume that $f(x) \in H^s(\mathbb{R}^n)$ in (18) has finite regularity ($s > n/2$) and is compactly supported, i.e., $f \in H^s_{comp}(\mathbb{R}^n)$. Then for any $\varepsilon > 0$ the solution $u^\varepsilon(x)$ of (24) can be approximated by

$$u^\varepsilon_N \in \mathcal{V}_\varepsilon^N = \operatorname{Span}\{\operatorname{Re}\psi(\cdot/\varepsilon,\varepsilon,kh),\ \operatorname{Im}\psi(\cdot/\varepsilon,\varepsilon,kh)\}_{k\in\mathbb{Z}^n_{(N)}}$$

with respect to $\|\cdot\|_{1,-\nu}$ at a robust algebraic convergence rate.

Proposition 6. *Assume that f in (18) is $H^s(\mathbb{R}^n)$ with $s > n/2$ and $\operatorname{supp} f \subset \Omega_M := (-M, M)^n$, with $M > 0$. Let $d := \min\{1/M, \nu_n\}$, $\nu_n = 1/(2\sqrt{n})$, and $N \geq 1$. Then, for any $\varepsilon > 0$ and for all N*

$$\inf_{v \in \mathcal{V}_\varepsilon^N} \|u^\varepsilon - v\|_{1,-\nu} \leq C_{\nu,s,n} N^{-(s-n/2)/2} \|f\|_{H^s(\mathbb{R}^n)}, \tag{49}$$

where $C_{\nu,s,n} > 0$ is independent of ε, N, and M.

4 Approximation Properties on the Bounded Domain

On the unbounded domain, the spaces (47) have robust (with respect to ε) approximation properties. Thus we obtained rates of convergence that were independent of the regularity of the coefficients and of the unit-cell problem (Proposition 5, Corollary 1). We base a generalized FEM in Ω^ε on the spaces $\mathcal{V}_\varepsilon^N$ from the unbounded domain. However, the spaces $\mathcal{V}_\varepsilon^N$ are not $H^1_{\Gamma_D^\varepsilon}(\Omega^\varepsilon)$-conforming, i.e., $\mathcal{V}_\varepsilon^N \not\subset H^1_{\Gamma_D^\varepsilon}(\Omega^\varepsilon)$ in general. Thus, $\mathcal{V}_\varepsilon^N$ cannot be used as FE-spaces directly and have to be modified. Our approch for enforcing $H^1_{\Gamma_D^\varepsilon}(\Omega^\varepsilon)$-conformity is to use a standard FEM in a neighborhood of the boundary and the spaces $\mathcal{V}_\varepsilon^N$ in the interior of the domain.

4.1 Density of the approximation spaces

A minimal requirement for the generalized FEM (7) is that $\cup_N V_N$ be dense (in the $H^1(\Omega^\varepsilon)$-norm) in the set of all solutions. Since we will base our spaces V_N on the spaces $\mathcal{V}_\varepsilon^N$ in (47), we first ascertain in Proposition 7 that $\cup_N \mathcal{V}_\varepsilon^N$ is dense in the set of solutions of (1), (2) with boundary conditions as in (11).

Proposition 7. *Let $f \in L^2(\Omega)$ and let $u^\varepsilon \in H^1_{\Gamma_D^\varepsilon}(\Omega^\varepsilon)$ be the solution of (1), (11). Then, for all bounded Lipschitz domains $\Omega_0 \subset\subset \Omega$, with $\Omega_0^\varepsilon := \Omega_0 \cap \Omega_\infty^\varepsilon$*

$$\lim_{N \to \infty} \inf_{v \in \cup_{\mu=1}^N \mathcal{V}_\varepsilon^\mu} \|u^\varepsilon - v\|_{H^1(\Omega_0^\varepsilon)} = 0,$$

i.e., $\cup_{N>0} \mathcal{V}_\varepsilon^N|_\Omega$ is dense with respect to the $H^1(\Omega_0^\varepsilon)$-topology in the space of solutions to (5) corresponding to right-hand sides $f \in L^2(\Omega)$.

The proof of this result for the case when the domain has no holes, i.e., $\Omega^\varepsilon = \Omega$ can be found in [26]. In the general case the proof follows analogously.

4.2 Conforming approximation spaces via scale resolution at the boundary

The approximation spaces $\mathcal{V}_\varepsilon^N$ defined in (47) are, when restricted to Ω^ε, in general not $H^1_{\Gamma_D^\varepsilon}(\Omega^\varepsilon)$-conforming. We propose scale resolution at the boundary with standard FEM and use $\mathcal{V}_\varepsilon^N$ in the interior only. In Proposition 8, we show that such composite FE-spaces V_N are still dense in the set of all $H^1_{\Gamma_D^\varepsilon}(\Omega^\varepsilon)$ solutions. Enforcement of the essential boundary conditions is only one reason for using scale resolution near the boundary. For problems posed on the unbounded domain the solution $u^\varepsilon(x)$ has a two-scale behavior and these scales are separated, but the situation is different for problems on bounded domains, since near the boundary, scale separation occurs only for special situations. For example, if the boundary is flat and aligned with the periodicity as in the case of an infinite, perforated layer, then two-scale separation occurs in the tangential direction at the boundary. In general, however, scale separation does not take place at curved or nonaligned parts of the boundary (see, e.g., [34]).

Different approaches to enforcing Dirichlet boundary conditions are possible. In the asymptotic homogenization approach ($\varepsilon \to 0$) one constructs so-called *correctors*. The explicit construction of so-called "boundary correctors" (which compensate for boundary conditions violated by outer asymptotic expansions, but which are, generally, of boundary layer type) is tedious and limited to special cases, in which some geometry simplifications are assumed, as in the case when the periodic structure is aligned with the boundaries. For the general case (e.g., when the boundaries are curved and intersect the cavities of the microscopic periodic lattice) only little is known about the detailed structure of the boundary correctors (see also [34]). Such situations can be dealt with by scale resolution at the boundary. The geometry approximation comes as an additional argument for scale resolution at the boundary.

Let $\widetilde{V}_N \subset H^1(\Omega^\varepsilon)$ be any finite dimensional subspace (it need not satisfy the homogeneous Dirichlet boundary conditions) such that

$$\inf_{v\in\widetilde{V}_N} \|u_\varepsilon - v\|_{H^1(\Omega^\varepsilon_{(\varepsilon)})} \leq \delta,$$

where $\Omega^\varepsilon_{(\varepsilon)} := \{x \in \Omega^\varepsilon \,|\, \mathrm{dist}(x, \partial\Omega) > \varepsilon\}$. Let $\chi \in H^1_0(\Omega) \cap W^{1,\infty}(\Omega)$ be a cut-off function such that

$$\begin{aligned}\chi \equiv 1 \ \text{ on } \Omega_{(2\varepsilon)} &:= \{x \in \Omega \,|\, \mathrm{dist}(x, \partial\Omega) > 2\varepsilon\},\\ \chi \equiv 0 \ \text{ on } \Omega\backslash\Omega_{(\varepsilon)} &= \{x \in \Omega \,|\, \mathrm{dist}(x, \partial\Omega) \leq \varepsilon\}.\end{aligned}$$

Assume further that $V_{0,\Gamma_D^\varepsilon}(\Omega^\varepsilon\backslash\Omega_{(2\varepsilon)}) \subset H^1_{0,\Gamma_D^\varepsilon}(\Omega^\varepsilon\backslash\Omega_{(2\varepsilon)})$ is a finite dimensional subspace of $H^1_{0,\Gamma_D^\varepsilon}(\Omega^\varepsilon\backslash\Omega_{(2\varepsilon)}) := \{v \in H^1(\Omega^\varepsilon\backslash\Omega_{(2\varepsilon)}) \ : \ v|_{\Gamma_D^\varepsilon} = 0\}$. Then we define the $H^1_{\Gamma_D^\varepsilon}(\Omega)$-conforming FE-space V_N as

$$V_N := \chi\widetilde{V}_N + (1-\chi)V_{0,\Gamma_D^\varepsilon}(\Omega^\varepsilon\backslash\Omega_{(2\varepsilon)}).$$

Proposition 8. *Assume that the cut-off function χ introduced above satisfies:*

$$\|\chi\|_{W^{1,\infty}(\Omega)} \leq C_\chi \varepsilon^{-1}.$$

Then, for each $s \in [0, 1/2)$ there exist $C(s) = C(s, \Omega, C_\chi) > 0$ and $C > 0$ such that there holds

$$\begin{aligned} \inf_{v \in V_N} \|u^\varepsilon - v\|_{H^1(\Omega^\varepsilon)} &\le C(s)\varepsilon^{-(1-s)}\delta \\ &+ C \inf_{w \in V_{0,\Gamma_D^\varepsilon}(\Omega^\varepsilon \setminus \Omega_{(2\varepsilon)})} \Big(\|\nabla(u^\varepsilon - w)\|_{L^2(\Omega^\varepsilon \setminus \Omega_{(2\varepsilon)})} \\ &\qquad + \varepsilon^{-1}\|(u^\varepsilon - w)\|_{L^2(\Omega^\varepsilon \setminus \Omega_{(2\varepsilon)})}\Big). \end{aligned} \tag{50}$$

The proof of this result for domains without holes ($\widehat{Q} = (-\pi, \pi)^n$) can be found in [26], in the general case the result can be proved analogously.

Remark 6. Let $V_{0,\Gamma_D^\varepsilon}(\Omega^\varepsilon \setminus \Omega_{(2\varepsilon)}) = S^{1,1}_{0,\Gamma_D^\varepsilon}(\Omega^\varepsilon \setminus \Omega_{(2\varepsilon)}, \mathcal{T}_h)$ be the space of piecewise linears with respect to a quasiuniform triangulation $\mathcal{T}_h$ of $\Omega^\varepsilon \setminus \Omega_{(2\varepsilon)}$ that are zero on the boundary Γ_D^ε. Note that $\cup_{h>0} S^{1,1}_{0,\Gamma_D^\varepsilon}(\Omega^\varepsilon \setminus \Omega_{(2\varepsilon)}, \mathcal{T}_h)$ is dense in $H^1_{0,\Gamma_D^\varepsilon}(\Omega^\varepsilon \setminus \Omega_{(2\varepsilon)})$ so that the infimum in the right hand side of the error estimate (50) tends to zero as $h \to 0$. If u^ε is piecewise regular then the size of this infimum can be quantified. Assuming that $u^\varepsilon(x)|_K \in H^2(K)$ for all $K \in \mathcal{T}_h$ we have the standard estimate [12] for the piecewise linear interpolant $u^\varepsilon_{\text{Int}}$ of u^ε on $\mathcal{T}_h$

$$\begin{aligned} &h\|\nabla(u^\varepsilon - u^\varepsilon_{\text{Int}})\|_{L^2(\Omega^\varepsilon \setminus \Omega_{(2\varepsilon)})} + \|u^\varepsilon - u^\varepsilon_{\text{Int}}\|_{L^2(\Omega^\varepsilon \setminus \Omega_{(2\varepsilon)})} \le \\ &Ch^2 \left\{\textstyle\sum_{K \in \mathcal{T}_h} |u^\varepsilon|^2_{H^2(K)}\right\}^{1/2}. \end{aligned}$$

For smooth coefficients $a(y)$, $a_0(y)$ in (5) we have $\left\{\sum_{K \in \mathcal{T}_h} |u^\varepsilon|^2_{H^2(K)}\right\}^{1/2} \le C\varepsilon^{-1}$. Hence, we obtain with $w = u^\varepsilon_{\text{Int}}$ in (50) the following error estimate with respect to the subspace $V_N = \chi\widetilde{V}_N + (1-\chi)S^{1,1}_{0,\Gamma_D^\varepsilon}(\Omega^\varepsilon \setminus \Omega_{(2\varepsilon)}, \mathcal{T}_h)$

$$\inf_{v \in V_N} \|u_\varepsilon - v\|_{H^1(\Omega^\varepsilon)} \le C_1(s)\varepsilon^{-(1-s)}\delta + C_2 h/\varepsilon,$$

with C_i, $i = 1, 2$ independent of h and ε. Note that here $h < \varepsilon$. The space V_N is essentially one in which *scale resolution* is achieved at the boundary and in which the functions from $\widetilde{V}_N$ are used in the interior.

5 Numerical experiments

We turn now to the computational investigation of the generalized FEM on bounded domains. To this end, we consider homogenization problems in one and two dimensions. The goal of the experiments is twofold. First, we would like to confirm the theoretical asymptotic convergence estimates which were presented in the previous sections, in particular the robustness of the methods with respect to ε. Second, we will investigate the performance of several possible choices of micro spaces $\mathcal{M}^\mu_\varepsilon$. We consider in particular the microspaces based on oversampling, on the asymptotic theory, and on Bloch-wave eigenfunctions.

5.1 Microscale shape functions space $\mathcal{M}^{\mu}_{\varepsilon}(\varepsilon\widehat{Q})$ based on oversampling

We have seen so far that collocation of the Fourier-Bochner integral kernel $\psi(x/\varepsilon, \varepsilon, t)$ at various sets of collocation points $\mathcal{N} = \{t_j\}_j$ gives dense systems of shape functions with very favorable approximation properties for elliptic homogenization problems. These shape functions are products of smooth functions in x and periodic, oscillatory micro shape functions that arise as solutions of parameter dependent unit-cell problems. We present a FEM for the solution of the unit cell problem and a methodology to derive a well conditioned set of shape functions from the collocated kernels $\psi(x/\varepsilon, \varepsilon, t_j)$, $t_j \in \mathcal{N}$. This will be based on the SVD of the matrix of coefficient vectors of the Finite Element approximations to the $\phi(y, \varepsilon, t_j)$, $t_j \in \mathcal{N}$.
To cast (27) in weak form, we define for each $\varepsilon > 0$ and $t \in \mathbb{C}^n$ the sesquilinear form $\Phi(\varepsilon, t)[\cdot, \cdot] : H^1_{\mathrm{per}}(\widehat{Q}) \times H^1_{\mathrm{per}}(\widehat{Q}) \to \mathbb{C}$ by

$$\Phi(\varepsilon, t)[\phi, v] :=$$

$$\int_{\widehat{Q}} a(y)\nabla_y \left(\phi(y)e^{it\cdot\varepsilon y}\right) \cdot \overline{\nabla_y \left(v(y)e^{it\cdot\varepsilon y}\right)} + \varepsilon^2 a_0(y)\phi(y)\overline{v(y)}\, dy. \tag{51}$$

With this notation, the integral kernel $\phi(\cdot, \varepsilon, t) \in H^1_{\mathrm{per}}(\widehat{Q})$ in the representation (24) is the weak solution of the following variational problem

$$\text{Find } \phi(\cdot) \in H^1_{\mathrm{per}}(\widehat{Q}) \text{ such that}$$

$$\Phi(\varepsilon, t)[\phi, v] = \varepsilon^2 \int_{\widehat{Q}} \overline{v(y)}\, dy \quad \forall v \in H^1_{\mathrm{per}}(\widehat{Q}). \tag{52}$$

Let $\mathcal{N} = \{t_j : j = 1, \ldots, \hat{\mu}\}$ be any set of collocation points in $\mathbb{C}^n$. Given a partition $\widehat{\mathcal{T}}$ of the unit cell $\widehat{Q}$ into intervals K, for an arbitrary $t_j \in \mathcal{N}$, compute the FE approximations

$$\phi_{FE}(y, \varepsilon, t_j) \in S^{\hat{p},1}_{\mathrm{per}}(\widehat{Q}, \widehat{\mathcal{T}}) :$$

$$\Phi(\varepsilon, t_j)[\phi_{FE}, v] = \varepsilon^2 \int_{\widehat{Q}} \overline{v(y)}\, dy, \quad \forall v \in S^{\hat{p},1}_{\mathrm{per}}(\widehat{Q}, \widehat{\mathcal{T}}), \tag{53}$$

where $\Phi(\varepsilon, t)[\cdot, \cdot]$ is as in (51) and the FE space $S^{\hat{p},1}_{\mathrm{per}}(\widehat{Q}, \widehat{\mathcal{T}})$ is defined by

$$S^{\hat{p},1}_{\mathrm{per}}(\widehat{Q}, \widehat{\mathcal{T}}) := \left\{ u \in H^1_{\mathrm{per}}(\widehat{Q}) \; : \; u\Big|_K \in \mathcal{S}^{\hat{p}}(K), \forall K \in \widehat{\mathcal{T}} \right\}, \tag{54}$$

and $\mathcal{S}^{\hat{p}}(K)$ is a space of polynomials of degree at most $\hat{p}$ on K. The design of $\widehat{\mathcal{T}}$, $\hat{p}$ must take into account regularity of the solution. For example, if the unit cell problem has a crack, then the solution $\phi(y,\varepsilon,t) \in B^2_\beta(\hat{Q})$, the countably normed space [5], and *hp*-FEM [37,23] with a geometrical mesh refinement towards the crack-tips must be employed for the solution of the unit-cell problem.

Remark 7. In the proof of the error estimate (45) the oscillating functions $\psi(x/\varepsilon,\varepsilon,t) = e^{it\cdot x}\phi(x/\varepsilon,\varepsilon,t)$ are exact, i.e., the unit-cell solutions $\phi(y,\varepsilon,t)$ are assumed exact. The micro-scale shape functions in the definition of $\mathcal{M}^\mu_\varepsilon$ are derived from the FE approximations $\phi_{FE}(y,\varepsilon,t)$ obtained by energy projection through a FE discretization of the unit-cell problem. The stiffness matrix computations based on the generalized FE spaces $S^{\underline{p}}(\Omega,\mathcal{T};\mathcal{M}^\mu_\varepsilon)$ require only *moments*, i.e., weighted integrals (16) of the finite element solutions $\phi_{FE}(y,\varepsilon,t)$ over the unit-cell. Careful inspection of these moments reveals that they converge quasioptimally and at the same rate as the energy norm.

Since the sesquilinear form $\Phi(\varepsilon,t)[\cdot,\cdot]$ is coercive (in the sense that a Gårding inequality holds and the unit cell solution operator is injective), there exists a unique solution $\phi_{FE}(y,\varepsilon,t_j) \in S^{\hat{p},1}_{\text{per}}(\widehat{Q},\widehat{\mathcal{T}})$ of (53). In agreement with Proposition 5, we choose $t_j(\mu) = jh$ where $h = 1/\sqrt{\mu}$ with $j \in \mathbb{Z}^n_{(\mu)}$. Notice that the values of d and α in (38) are generally not available. Therefore, the choice of t_j is to some extent heuristic. By Section 3.3, $\phi(y,\varepsilon,t)$ is analytic in t. As μ increases, the collocation points t_j will cluster, resulting in almost linear dependence of the shape functions $\phi_{FE}(y,\varepsilon,t_j)$; these functions are hence not well-suited as basis for a generalized p-FEM. Some orthogonalization is needed to obtain a well-conditioned basis. In addition, the points $t_j(\mu)$ depend on μ meaning that the shape functions $\phi_{FE}(y,\varepsilon,t_j(\mu))$ are not hierarchical.

We propose an *oversampling*, i.e., to select $\hat{\mu} \gg \mu$ sufficiently large and

$$\mathcal{N} = \{t_j(\hat{\mu}) \,:\, j = 1,\ldots,\hat{\mu}\},$$

and to perform an *orthogonalization* as follows:

Algorithm 2 Let $\underline{N}(y)$ be a basis of $S^{\hat{p},1}_{\text{per}}(\widehat{Q},\widehat{\mathcal{T}})$. Then

$$\phi_{FE}(y,\varepsilon,t_j) = \Phi_j(\varepsilon)^\top \underline{N}(y), \quad j = 1,\ldots,\hat{\mu}.$$

Compute the SVD

$$[\Phi_1(\varepsilon),\ldots,\Phi_{\hat{\mu}}(\varepsilon)] = \underline{U}\, diag(\sigma_1,\ldots,\sigma_{\hat{\mu}})\underline{V}^\top$$

with $\sigma_1 \geq \sigma_2 \geq \sigma_3 \geq \cdots \geq \sigma_{\hat{\mu}} \geq 0$ and set

$$\mathcal{M}^\mu_\varepsilon = \text{Span}\left\{\phi_j\left(\frac{x}{\varepsilon},\varepsilon\right) := U_j^\top \underline{N}\left(\frac{x}{\varepsilon}\right) \,:\, j = 1,\ldots,\mu+1\right\}, \tag{55}$$

with U_j being the j-th column of $\underline{U}$.

It is clear that this orthogonalization changes only the basis, but not the span of the shape functions if $\mu = \hat{\mu}$. If $\mu < \hat{\mu}$, however, the definition (55) will change the span. Nevertheless, if $\sigma_j < eps$ for $\mu < j \leq \hat{\mu}$ with eps of the order of machine precision, this change will be negligible. The list of significant singular values for a one dimensional model problem with $a_0 \equiv 1$, $a(\cdot)$ as in (56) and different ε scales is presented in Table 1. The largest singular values appear to be proportional to some powers of ε. Span $\{\psi(\cdot/\varepsilon, \varepsilon, t_j)\}$ is

$\varepsilon = 1.0e-01$	$\varepsilon = 1.0e-02$	$\varepsilon = 1.0e-03$
$\sigma_1 = 1.017400157e+01$	$\sigma_1 = 8.708065522e+00$	$\sigma_1 = 9.651065322e+00$
$\sigma_2 = 3.602041416e+00$	$\sigma_2 = 1.716876891e+00$	$\sigma_2 = 4.686737612e-01$
$\sigma_3 = 3.499684680e+00$	$\sigma_3 = 1.517277336e+00$	$\sigma_3 = 3.516838658e-02$
$\sigma_4 = 1.653439563e+00$	$\sigma_4 = 3.158726721e-01$	$\sigma_4 = 4.904764017e-05$
$\sigma_5 = 1.118879591e+00$	$\sigma_5 = 1.890039200e-02$	$\sigma_5 = 8.322681798e-08$
$\sigma_6 = 2.052151441e-01$	$\sigma_6 = 2.906576569e-05$	$\sigma_6 = 1.003629032e-11$
$\sigma_7 = 4.486387147e-04$	$\sigma_7 = 2.709646458e-09$	$\sigma_7 = 7.670974077e-15$
$\sigma_8 = 7.805541440e-07$	$\sigma_8 = 2.782052676e-13$	$\sigma_8 = 7.669801103e-18$
$\sigma_9 = 2.429807212e-11$	$\sigma_9 = 1.612941949e-17$	$\sigma_9 = 5.661242396e-18$
$\varepsilon = 1.0e-04$	$\varepsilon = 1.0e-05$	$\varepsilon = 1.0e-06$
$\sigma_1 = 9.651024285e+00$	$\sigma_1 = 9.650620918e+00$	$\sigma_1 = 9.856848581e+00$
$\sigma_2 = 4.641386272e-02$	$\sigma_2 = 4.640937839e-03$	$\sigma_2 = 4.641062782e-04$
$\sigma_3 = 3.481845540e-04$	$\sigma_3 = 3.481393137e-06$	$\sigma_3 = 3.533916622e-08$
$\sigma_4 = 4.898472869e-08$	$\sigma_4 = 4.898386017e-11$	$\sigma_4 = 4.907798256e-14$
$\sigma_5 = 8.316984046e-12$	$\sigma_5 = 4.464187517e-14$	$\sigma_5 = 2.457293272e-14$

Table 1. The largest singular values σ_j of the coefficient matrix with respect to the shape functions $\{\phi_{FE}(\cdot, \varepsilon, t_j),\ j = 0, 1, \ldots, \hat{\mu}\}_{t_j = j/\sqrt{\hat{\mu}}}$, $\hat{\mu} = 64$

practically independent on the choice of the collocation points. Therefore the precise choice of t_j will not matter much, as long as with increasing μ they cover the interval $[-\sqrt{\mu}, \sqrt{\mu}]^n$ and are spaced as $1/\sqrt{\mu}$ by Proposition 5.

Figure 2 presents the shape functions $\{\phi_j(y, \varepsilon)\}_{j=1}^{\mu}$ obtained with Algorithm 2 for the one-dimensional problem with $\widehat{Q} = (-\pi, \pi)$, $a_0 \equiv 1$, $a(\cdot)$ as in (56), $\varepsilon = 0.001$ and, based on Proposition 5, the set of collocation points $\mathcal{N} = \{t_j(\hat{\mu}) = j/\sqrt{\hat{\mu}} : j = 0, \ldots, \hat{\mu},\ \hat{\mu} = 64\}$. In this case the number of j such that the corresponding singular values $\sigma_j > eps = 10^{-10}$ is $\mu = 5$. Hence the orthogonalization has also reduced the number of micro shape functions substantially. We clearly see the substantial reduction of degrees of freedom due to the almost linear dependence of the $\phi_{FE}(y, \varepsilon, t_j)$ and the low regularity of these shape functions at the jumps of $a(\cdot)$ at $y = \pm\pi/2$. Note also that, unlike the kernels $\phi(y, \varepsilon, t_j)$, $\phi_j(y, \varepsilon)$ are piecewise polynomials.

$$a(y) = \begin{cases} 10 & \text{if } |y| \leq \frac{\pi}{2}, \\ 1 & \text{else }, \end{cases} \tag{56}$$

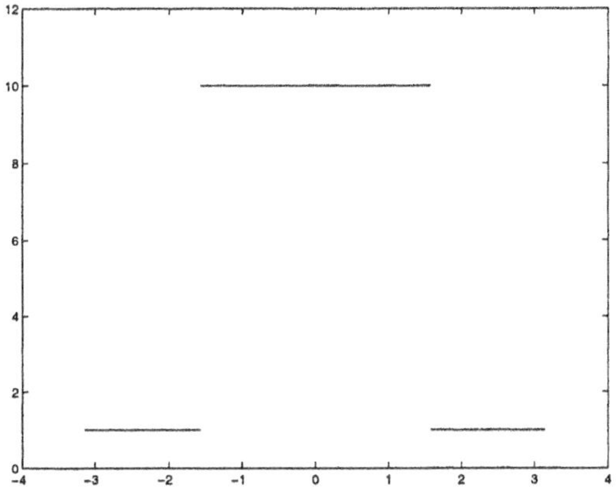

Fig. 1. The coefficient $a(\cdot)$.

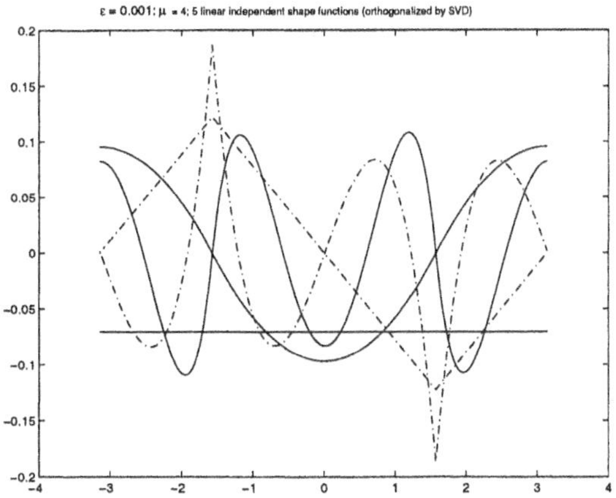

Fig. 2. $\phi_j(\cdot, \varepsilon)$, $j = 1, \ldots, \mu$.

Remark 8. In numerical experiments we found that Algorithm 21 is very robust with respect to the choice of collocation points. After the SVD the first shape functions associated with the largest singular values are practically independent of the number and the choice of t_j. The shape functions $\phi_j(\cdot, \varepsilon)$ resulting from Algorithm 21 are therefore, at least numerically, *hierarchical,* and enable *hierarchic modeling* of problems with microstructure.

5.2 gFEM based on oversampling in the one-dimensional case

We consider the following model problem in $\Omega := (-1,1)$: let $f(\cdot) \in L^2(\Omega)$ and $\varepsilon > 0$ fixed. Denote by $u^\varepsilon(\cdot) \in H^1_0(\Omega)$ the weak solution of the following boundary value problem

$$-\frac{d}{dx}\left(a\left(\frac{x}{\varepsilon}\right)\frac{du^\varepsilon}{dx}(x)\right) + a_0\left(\frac{x}{\varepsilon}\right)u^\varepsilon(x) = f(x) \quad \text{in} \quad \Omega, \qquad u^\varepsilon(\pm 1) = 0, \tag{57}$$

with

$$a(y) = \begin{cases} 10 & \text{if } |y| \leq \frac{\pi}{2}, \\ 1 & \text{else }, \end{cases} \tag{58}$$

and $a_0(y) \equiv 1$. We implemented the generalized p-FEM described in the previous section for (57) with $a(\cdot)$ as in (56) and absolute terms $a_0 \equiv 1$, $a_0 \equiv 0$, respectively. The exact solution $u^\varepsilon(x)$ is in general piecewise analytic but non-polynomial on the microscale. We used in all our computations the orthonormalized micro shape functions $\phi_j(y,\varepsilon)$ in (55) from the unit cell problem with absolute term $a_0 \equiv 1$.

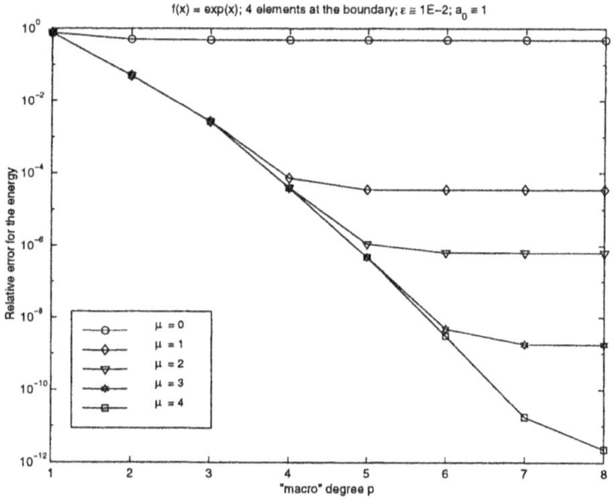

Fig. 3. Exponential rate of convergence for the FE energy with $\mathcal{M}^\mu_\varepsilon$ based on oversampling. $f(x) = exp(x)$.

Figure 3 shows the convergence of the generalized p-FEM for $a_0 \equiv 1$, $f(x) = exp(x)$ and $\varepsilon \cong 10^{-2}$ when p is increased on a fixed mesh $\mathcal{T} = \mathcal{T}_0 \cup \mathcal{T}_b$ with $\mathcal{T}_b$

covering 4 periods of length $2\pi\varepsilon$ at each boundary point for various values of μ. The curves corresponding to $\mu = 0$ show the error when only macroscopic shape functions, i.e. global polynomials, are used.
We see that for fixed $\mu > 0$ and increasing p, first exponential convergence is apparent, however a saturation occurs at a p-level which depends on the micro degree μ. Exponential convergence requires therefore the joint increase of the micro degree μ with the macro degree p.

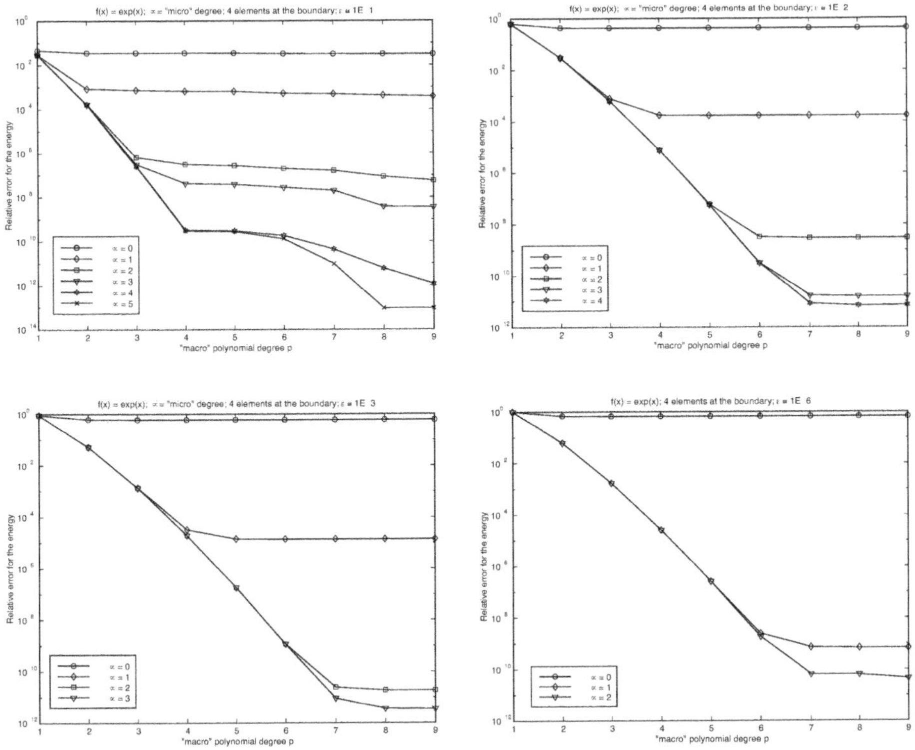

Fig. 4. Convergence rate for fixed micro degree μ and increasing macro polynomial degree p. The system of micro shape functions $\mathcal{M}_\varepsilon^\mu$ is based on the oversampling procedure in Algorithm 21 and $f(x) = exp(x)$.

In Figure 4 we show analogous results for $a_0 \equiv 0$, $f(x) = exp(x)$ (with respect to the same mesh) and different microscales ε, varying from $\cong 10^{-1}$ down to $\cong 10^{-6}$. We note that for $\mu = 1$ and for $\varepsilon \cong 10^{-1}$ a very slow convergence is apparent - here the scales are resolved, but the low solution regularity stalls the spectral convergence. As before, one can see from the results in Figure 5 that keeping p fixed and increasing μ does not lead to exponential convergence. Rather, Figures 3, 4, 5 show again that μ must be increased together with p.

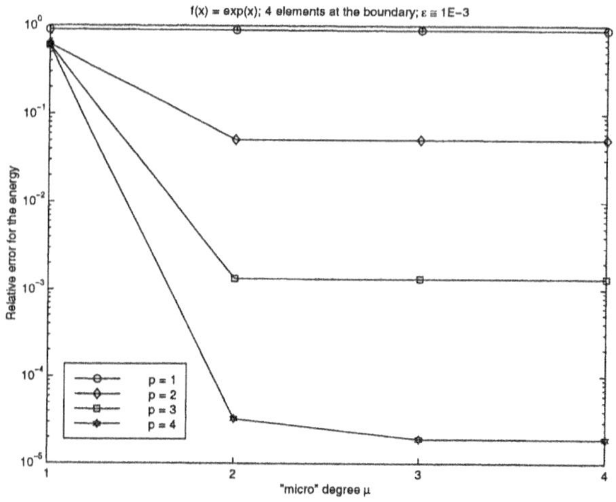

Fig. 5. Convergence rate for fixed macro polynomial degree p and increasing micro degree μ ($\mathcal{M}_\varepsilon^\mu$ based on oversampling). $f(x) = exp(x)$.

5.3 Quasi 1-D Case

The generalized FEM with micro scale shape functions $\mathcal{M}_\varepsilon^\mu$ derived by oversampling the Fourier-Bochner integral kernel has been implemented for two dimensional problems. Our gFEM implementation is based on Concepts-1.4, an object oriented programming framework for general elliptic problems in C++ [24]. We illustrate here the performance of the gFEM considering first the (quasi 1-d) model problem

$$-\nabla \cdot \left(a\left(\frac{x}{\varepsilon}\right) \nabla u^\varepsilon(x) \right) = f(x) \quad \text{in } \Omega := (0,1)^2$$

$$u^\varepsilon|_{\Gamma_D} = 0 \quad \text{on } \Gamma_D := \{x_1 = 0\} \cup \{x_1 = 1\}$$

$$n \cdot a\left(\frac{x}{\varepsilon}\right) u^\varepsilon(x)|_{\Gamma_N} = 0 \quad \text{on } \Gamma_N := \partial\Omega \backslash \Gamma_D,$$

where $a(y) = \cos(y_1) + 2$, $y = (y_1, y_2) \in \widehat{Q} := (0, 2\pi)^2$ and $f(x) \equiv 1$. The implementation is completely general, but since we know for this problem the exact solution

$$u^\varepsilon(x) = -\int_0^{x_1} \frac{t}{a(t/\varepsilon)} dt + \frac{\int_0^1 \tau/a(\tau/\varepsilon) d\tau}{\int_0^1 1/a(\tau/\varepsilon) d\tau} \int_0^{x_1} \frac{1}{a(t/\varepsilon)} dt$$

it is reasonable to investigate the numerical performance for such solutions.

The macro mesh $\mathcal{T}$ used in all our experiments is shown in Figure 6. It consists of four interior elements of size $(1-4\pi\varepsilon)/\varepsilon$ and two layers of elements in a tubular neighborhood of $\partial\Omega$ covering two periods in the normal direction to $\partial\Omega$. The boundary elements are either elements near corners of the size of one micro cell or elongated elements along edges that are of size $O(\varepsilon)$ in the normal direction to the boundary. We recall that the macro polynomial degree has to be increased together with the number of micro shape functions in order to achieve convergence, thus we have successively and simultaneously increased p and the micro degree μ in all elements $K \in \mathcal{T}$ such that $p = \mu+1$. The relative error in the energy for increasing macro polynomial degree p and micro degree μ such that $p = \mu + 1$, $\mu = 0, 1, 2, 3$, and for different ε scales is presented in Table 2.

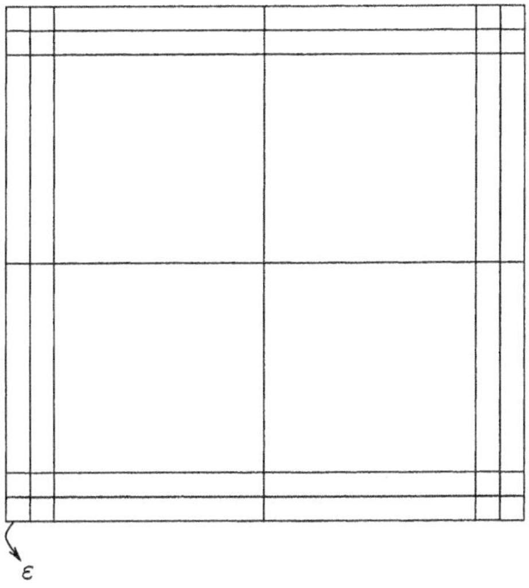

Fig. 6. The macro mesh $\mathcal{T}$ (not at scale) for 2-D calculations.

p	$\varepsilon \cong 10^{-2}$	$\varepsilon \cong 10^{-3}$	$\varepsilon \cong 10^{-4}$	$\varepsilon \cong 10^{-5}$	$\varepsilon \cong 10^{-6}$
1	0.150836	0.309959	0.346179	0.350048	0.350437
2	0.079917	0.133956	0.133974	1.714673e-05	2.229294e-06
3	1.202447e-05	2.320999e-06	7.504589e-07	5.895692e-07	5.734377e-07
4	1.874951e-07	4.898619e-07	5.629935e-07	5.710097e-07	5.715827e-07

Table 2. Relative error in the energy for gFEM with $p = \mu + 1$.

5.4 2-D Problem

In this section we analyse the performance of the gFEM for a genuinely two-dimensional model problem

$$-\nabla \cdot \left(a\left(\frac{x}{\varepsilon}\right) \nabla u^{\varepsilon}(x) \right) + a_0\left(\frac{x}{\varepsilon}\right) u^{\varepsilon}(x) = f(x) \quad \text{in } \Omega := (0,1)^2$$
$$u^{\varepsilon}|\partial\Omega = 0 \quad \text{on } \partial\Omega,$$

with

$$f(x) = 10x_1 + 10x_2, \quad a(y) = 6 + \cos(y_1) + \cos(y_2), \quad \text{and } a_0(y) \equiv 1.$$

Since the exact solution is not available, we compare our numerical results with the results obtained by doing scale resolution via standard FE discretizations. The unit-cell problem has been solved numerically by standard FEM. The unit-cell triangulation $\widehat{\mathcal{T}}$ was obtained by a three level uniform refinement of the unit-cell domain and the polynomial degree in each element was $\hat{p} = 2$. The collocation set in our computations here was $\mathcal{N} = \{(10^i, 10^j) : -4 \leq i, j \leq 2\}$. After the SVD of the matrix of coefficient vectors of the discrete unit-cell solutions we select $\mathcal{M}^{\mu}_{\varepsilon} := \text{Span}\{\phi_i(x/\varepsilon;\varepsilon) \;:\; i = 0, 1, \ldots, \mu\}$ as the space of resulting shape functions corresponding to the first μ largest singular values. The unit-cell computations represent an independent subroutine of our gFEM implementation. For the setting of the elemental stiffness matrices only the monomial weighted integrals of the form

$$\int_{\widehat{Q}} a(y) y_1^r y_2^s D_y^{\alpha} \phi_m(y;\varepsilon) D_y^{\beta} \phi_n(y;\varepsilon) dy, \quad |\alpha|, |\beta| \leq 1$$

are needed, thus the gFEM can handle other choices of micro scale spaces $\mathcal{M}^{\mu}_{\varepsilon}$ as well. The micro length scale ε in our computations here is $\varepsilon = 1/(40\pi)$, that means that there are 20 periods of length $2\pi\varepsilon$ along one side of $\partial\Omega$. The relatively coarse macro mesh $\mathcal{T}$ for the gFEM is as illustrated in Figure 6. We used a constant macro polynomial degree $p_K = 4$ for all $K \in \mathcal{T}$ and the micro degree was also constant $\mu_K = 4$ in all elements. The number of degrees of freedom for the gFEM computations was $\#DOF \cong 1000$, whereas for the standard FEM the number of degrees of freedom was $\#DOF \cong 40000$.
We emphasize that the number of degrees of freedom and the CPU time by the gFEM is independent of the micro length scale ε. Computation for $\varepsilon \cong 10^{-6}$, for example, costs the same work as $\varepsilon = 10^{-2}$.
We present at the top of Figure 7 the first component of the energy flux $a(x/\varepsilon)\nabla u^{\varepsilon}$ obtained by the generalized FEM based on oversampling and compare it with the results obtained by the standard FEM (bottom). Figure 8 displays the second component of $a(x/\varepsilon)\nabla u^{\varepsilon}$ obtained with gFEM and the standard FEM and finally, in Figure 9 we compare the solution plots. We see that the gFEM is able to accurately capture all the solution characteristics with significantly less degrees of freedom than the standard FEM.

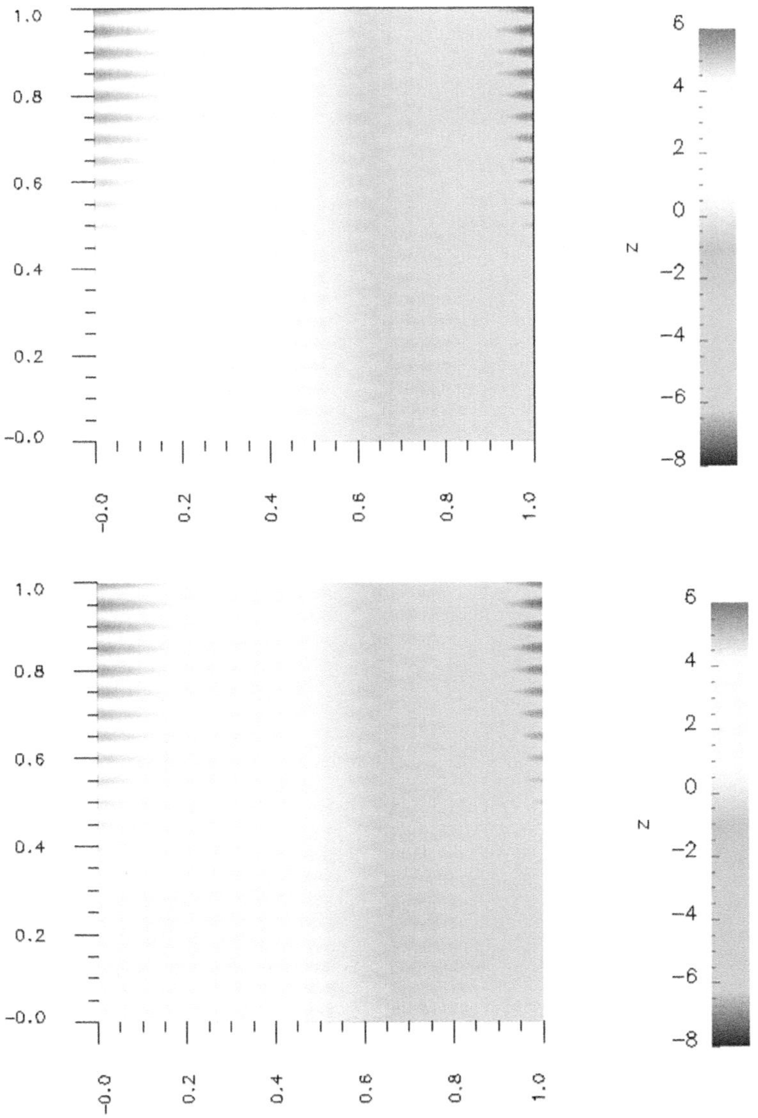

Fig. 7. The first component of the flux $a(x/\varepsilon)\nabla u^\varepsilon$ obtained by the generalized FEM based on oversampling (top) and by scale resolution (bottom); $\widehat{Q} = (0, 2\pi)^2$, $\varepsilon = 1/(40\pi)$. See also Color Plate 1.

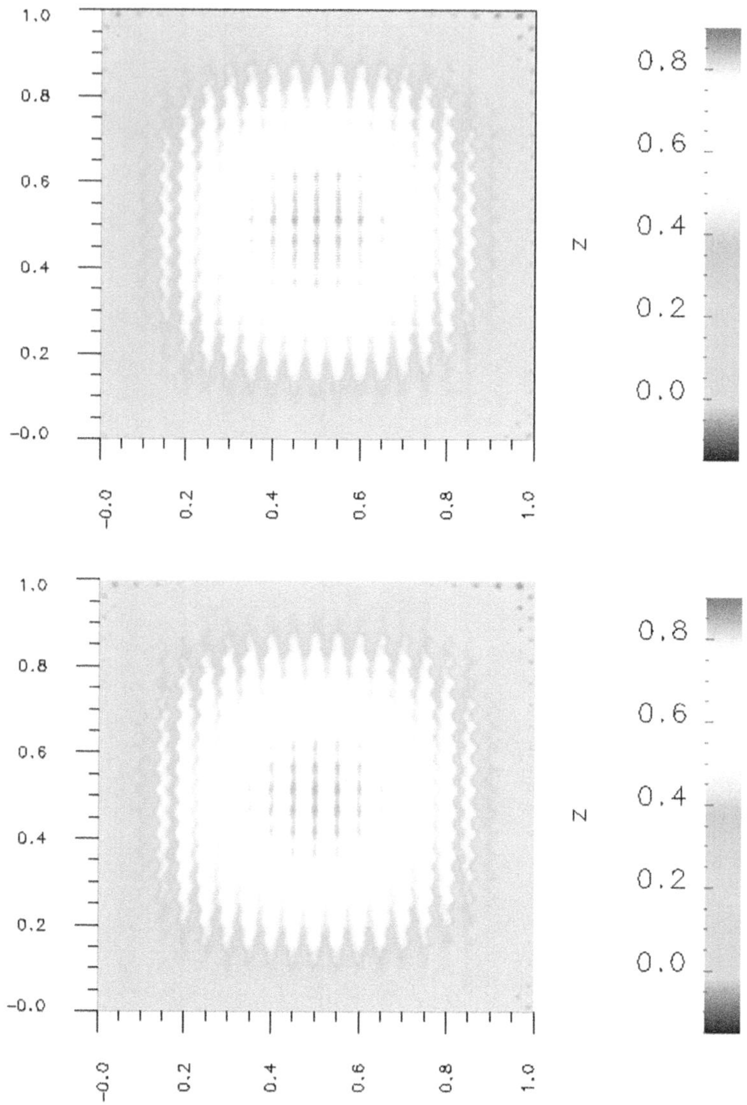

Fig. 8. The second component of the flux $a(x/\varepsilon)\nabla u^{\varepsilon}$ obtained by the generalized FEM based on oversampling (top) and by direct FEM computation (bottom); $\widehat{Q} = (0, 2\pi)^2$, $\varepsilon = 1/(40\pi)$. See also Color Plate 2.

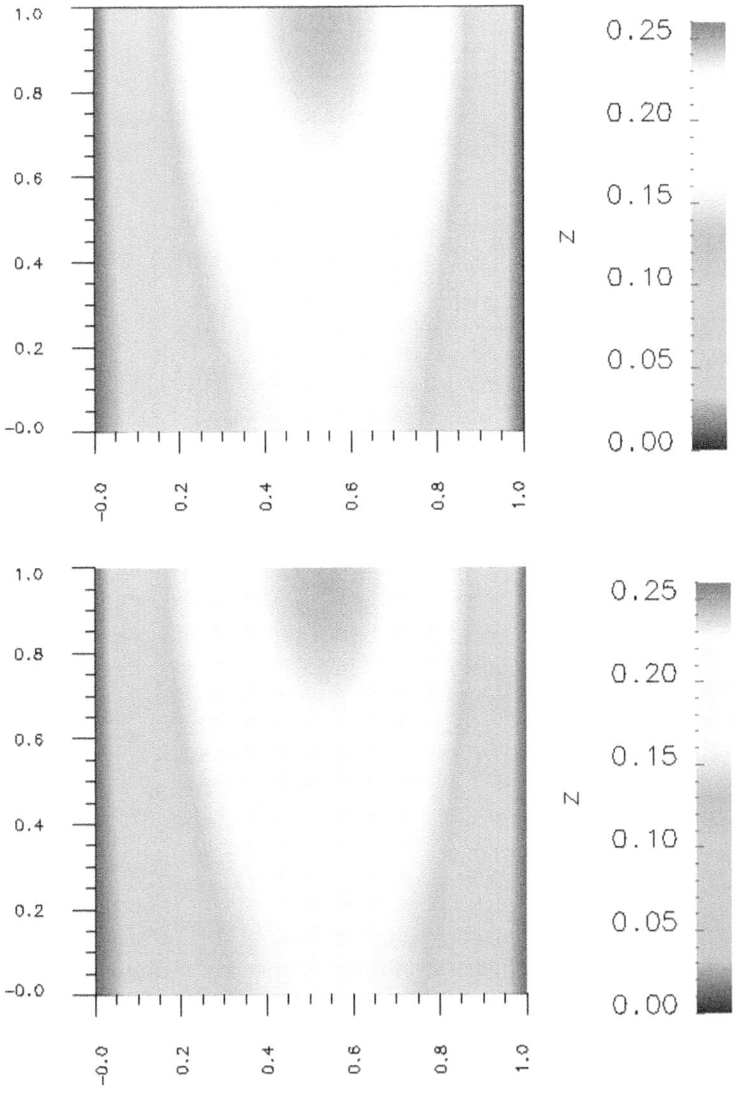

Fig. 9. The numerical solutions obtained by the generalized FEM based on oversampling (top) and by direct FEM computation (bottom); $\widehat{Q} = (0, 2\pi)^2$, $\varepsilon = 1/(40\pi)$. See also Color Plate 3.

5.5 Microscale space based on Bloch Wave eigenmodes

Bloch waves

One possible way to generalize the ideas of Fourier analysis to the study of non-homogeneous periodic media is the Bloch transform. The generalized

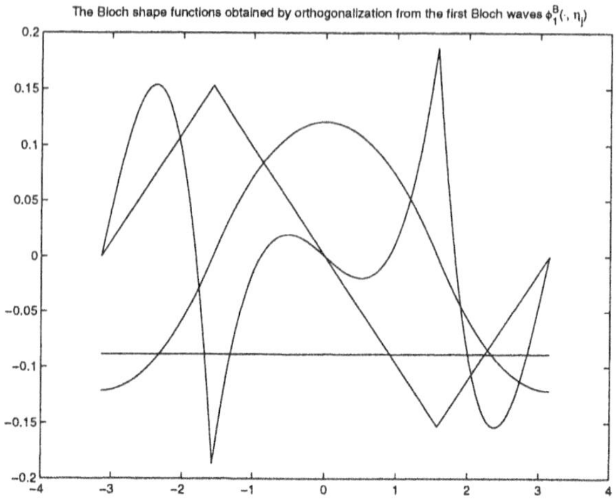

Fig. 10. The shape functions $\phi_i^B(\cdot)$ obtained by orthogonalization from the first Bloch eigenmode $\phi_1(\cdot;\eta_j)$ sampled at $\eta_j = 10^{-j}$, $j = 1, 2, \ldots, 10$

eigenfunctions are known as classical Bloch waves and were introduced in solid state physics in the context of propagation of electrons in a crystal, [9]. The starting point is the spectral analysis of the operator

$$L(\partial_y) := -\nabla_y \cdot (a(y)\nabla_y).$$

We define the shifted operator $L(\partial_y;\eta)$ associated with $L(\partial_y)$ by

$$L(\partial_y;\eta) := -(\nabla_y + i\eta) \cdot [a(y)(\nabla_y + i\eta)]. \tag{59}$$

Note that $L(\partial_y;\eta)$ represents the principal part of the unit cell operator (27). We attach to the shifted operator a family of spectral problems parametrized by $\eta \in \mathbb{R}^n$: Find $\lambda(\eta) \in \mathbb{R}$ and $\phi(\cdot;\eta) \in H^1_{\mathrm{per}}(\widehat{Q})$ not identically zero such that

$$L(\partial_y;\eta)\,\phi(\cdot;\eta) = \lambda(\eta)\phi(\cdot;\eta). \tag{60}$$

With these notations, the unit-cell operator in (27) can be written as $L(\partial_y;\varepsilon t)+\varepsilon^2 a_0(\cdot)$. Then $\psi(\cdot;\eta) := e^{i\eta\cdot y}\phi(\cdot;\eta)$ solve the parameter dependent spectral problems

$$\begin{aligned} L(\partial_y)\psi(y;\eta) &= \lambda(\eta)\psi(y;\eta) \quad \text{in } \mathbb{R}^n \\ \psi(y + 2\pi k;\eta) &= e^{2\pi i k\cdot\eta}\psi(y), \quad \forall\, k \in \mathbb{Z}^n,\ y \in \mathbb{R}^n. \end{aligned}$$

The eigenfunctions $\psi(y;\eta)$ are the so called Bloch waves and represent the natural generalization of the Fourier waves, the eigenfunctions in the case

of constant coefficient, to the case of periodic media. It is also easy to see that the Bloch eigenvalues in (60) are $\mathbb{Z}^n$-periodic and the eigenmodes are shifted-periodic in the second variable in the sense that for all $k \in \mathbb{Z}^n$ there holds holds $\phi(\cdot;\eta+k) = e^{-ik\cdot y}\phi(\cdot;\eta)$. It is therefore enough to study them for $\eta \in [0,1)^n$, the first *Brillouin* zone.
The property of the shifted operator to depend polynomially on η does not imply regularity of the Bloch eigenvalues and eigenvectors with respect to this parameter. In fact, it can be shown [15] that the eigenvalues are Lipschitz functions of η, and in general exhibit singularities. The first eigenvalue $\lambda_1(\cdot)$ is analytic and geometrically simple near $\eta = 0$. In addition, the first eigenfunction can be chosen so that it depends analytically on η in a small neighborhood of the origin. More detailed regularity properties of the first Bloch eigenvalue and eigenmode are listed below:

Proposition 9. *The first Bloch eigenvalue is stationary at $\eta = 0$*

$$\nabla_\eta \lambda_1(0) = 0. \tag{61}$$

Further, the second derivative (or in higher dimensions the Hessian matrix) of $\lambda_1(\cdot)$ at $\eta = 0$ is the homogenized coefficient (matrix) [13]

$$\frac{1}{2}\frac{\partial^2 \lambda_1}{\partial \eta_i \partial \eta_j}(0) = \frac{1}{(2\pi)^n}\Phi_0(\chi_{1;i} + y_i, \chi_{1;j} + y_j) \tag{62}$$

whereas the first order derivatives of the first Bloch mode at $\eta = 0$ give the correction terms $\chi_{1;j}(\cdot)$

$$\frac{\partial \phi_1}{\partial y_j}(\cdot;0) = i\frac{1}{(2\pi)^{n/2}}\chi_{1;j}(\cdot), \tag{63}$$

where $\chi_{1;j}(\cdot) \in H^1_{\mathrm{per}}(\widehat{Q})$ are solutions to

$$\chi_{1;j}(\cdot) \in H^1_{\mathrm{per}}(\widehat{Q}), \int_{\widehat{Q}} \chi_{1;j}(y)dy = 0 : \quad L(\partial_y)(\chi_{1;j} + y_j) = 0 \quad \text{in } \widehat{Q}. \tag{64}$$

A proof of this proposition can be found in [15].
The next step is to introduce the Bloch eigenvalues $\{\lambda^\varepsilon_m(t)\}_{m=1}^\infty$ and the eigenvectors $\{\phi^\varepsilon_m(x;t)\}_{m=1}^\infty$ at ε scale. By diagonalizing the principal part of the second order elliptic operator (18)

$$-\nabla_x \cdot \left(a\left(\frac{x}{\varepsilon}\right)\nabla_x\left(e^{it\cdot x}\phi^\varepsilon_m(x;t)\right)\right) = \lambda^\varepsilon_m(t)\phi^\varepsilon_m(x;t),$$

the differential equation $L(\varepsilon^{-1}\partial_x)u^\varepsilon = f$ can be transformed into a set of algebraic equations for the Bloch transforms. The energy of u^ε contained in all Bloch modes except the first one goes to zero when $\varepsilon \to 0$. Therefore, the first Bloch mode is essential for the asymptotic limit. It turns out [1] that the first Bloch mode transform tends to the usual Fourier transform in the $L^2(\mathbb{R}^n)$ topology as $\varepsilon \to 0$. Thus, the Bloch waves representing periodic medium tend to Fourier waves representing the homogenized medium. By a scaling argument, for $t \in \varepsilon^{-1}(-1/2, 1/2)^n$, the Bloch eigenvalues and eigenfunctions at ε scale are found to be related to $\lambda_m(\eta)$, $\phi_m(\cdot;\eta)$ in (60) as follows:

$$\lambda_m^\varepsilon(t) = \varepsilon^{-2}\lambda_m(\varepsilon t), \quad \phi_m^\varepsilon(x;t) = \phi_m\left(\frac{x}{\varepsilon};\varepsilon t\right).$$

The Bloch waves $e^{it\cdot x}\phi_m^\varepsilon(x;t)$ at ε scale provide now the spectral resolution of $L(\varepsilon^{-1}\partial_x)$. For $g \in L^2(\mathbb{R}^n)$ arbitrary, the m-th Bloch coefficient of g is defined as

$$\hat{g}_m^\varepsilon(t) := \varepsilon^{-n/2} \int_{\mathbb{R}^n} g(x) e^{-it\cdot x}\overline{\phi_m^\varepsilon(x;t)}\, dx.$$

Analogously to Fourier theory, inversion formula and a generalized Plancherel identity hold:

$$g(x) = \varepsilon^{n/2} \int_{\varepsilon^{-1}\left(-\frac{1}{2},\frac{1}{2}\right)^n} \sum_{m=1}^{\infty} \hat{g}_m^\varepsilon(t) e^{it\cdot x}\phi_m^\varepsilon(x;t)\, dt,$$

$$\varepsilon^{-n}\int_{\mathbb{R}^n} g(x)\overline{h(x)}\, dx = \int_{\varepsilon^{-1}\left(-\frac{1}{2},\frac{1}{2}\right)^n} \sum_{m=1}^{\infty} \hat{g}_m^\varepsilon(t)\overline{\hat{h}_m^\varepsilon(t)}\, dt, \quad \forall g, h \in L^2(\mathbb{R}^n).$$

Due to these results the differential equation $L(\varepsilon^{-1}\partial_x)u^\varepsilon(x) = f(x)$ in $\mathbb{R}^n$ transforms into an infinite algebraic system for the Bloch coefficients of u^ε

$$\hat{f}_m^\varepsilon(\xi) = \lambda_m^\varepsilon(\xi)\hat{u}_m^\varepsilon(\xi), \quad \forall m \geq 1, \forall \xi \in \varepsilon^{-1}(-1/2, 1/2)^n.$$

Furthermore, when $\varepsilon \to 0$, all Bloch transforms except the first one converge to 0 [15].

Proposition 10. *Let*

$$w^\varepsilon(x) = \varepsilon^{n/2} \int_{\varepsilon^{-1}(-1/2,1/2)^n} \sum_{m\geq 2} \hat{u}_m^\varepsilon(\xi) e^{i\xi\cdot x}\phi_m^\varepsilon(x;\xi) d\xi.$$

Then, there exists a constant $C > 0$ such that $\|w^\varepsilon\|_{L^2(\mathbb{R}^n)} \leq C\varepsilon$.

Generalized FEM based on Bloch waves

From the classical theory on homogenization [7,8,35,13] based on asymptotic analysis as $\varepsilon \to 0$ it is known that the solutions $u^\varepsilon \in H_0^1(\Omega)$ to

$$L(\varepsilon^{-1}\partial_x)u^\varepsilon = f \quad \text{in } \Omega$$

$$u^\varepsilon|_{\partial\Omega} = 0 \quad \text{on } \partial\Omega$$

converge, as $\varepsilon \to 0$, to a limit u^0 that solves the so-called homogenized problem

$$\begin{aligned} -\nabla \cdot (A \nabla u^0) &= f \quad \text{in } \Omega \\ u^0|_{\partial\Omega} &= 0 \quad \text{on } \partial\Omega, \end{aligned} \tag{65}$$

where the homogenized coefficients $A = \{A_{ij}\}_{i,j}$ are given by $A_{ij} = \Phi_0(\chi_{1;i} + y_i, \chi_{1;j} + y_j)$, see also (62). We see that the homogenized coefficients are not only means over the initial ones, but contain an additional term coming from the microstructure. This term is computed using the solutions $\chi_{1;i}$ in (64) on the unit cell.

More precisely, $\|u^\varepsilon - u^0\|_{L^2(\Omega)} \to 0$ as $\varepsilon \to 0$, however $\|u^\varepsilon - u^0\|_{H^1(\Omega)} \not\to 0$, as $\varepsilon \to 0$, since the small scale features of the solutions are lost by this averaging process. To improve the convergence, one needs the correctors

$$\left\| u^\varepsilon(x) - u^0(x) - \varepsilon \sum_i \frac{\partial u^0}{\partial x_i}(x) \chi_{1;i}\left(\frac{x}{\varepsilon}\right) \right\|_{H^1(\Omega)} \to 0 \quad \text{as } \varepsilon \to 0. \tag{66}$$

The homogenized limit u^0 is independent of ε and can be well approximated by standard polynomial FE spaces. Due to (66) it would be of interest to consider for the choice of the micro-scale space $\mathcal{M}^\mu$ the system of correctors $\{\chi_{1;i}(y)\}_i$ in (64) or, due to the fact that the first Bloch eigenmode $\phi_1(\cdot;\eta)$ is analytic at 0 and its first order derivatives with respect to η at $\eta = 0$ are precisely the correctors $\chi_{1;i}(\cdot)$, the system of micro-scale shape functions derived by sampling the first Bloch eigenmode in a collocation set refined geometrically at $\eta = 0$.

We studied the performance of the generalized FEM when the system of microscale shape functions was derived by sampling the first Bloch eigenmode in a collocation set refined geometrically at $\eta = 0$. We discretize the unit-cell by the FE space $S^{\hat{p},1}_{\text{per}}(\widehat{Q}, \widehat{\mathcal{T}})$ and then solve a parameter dependent class of generalized eigenvalue problems parametrized with respect to η in the collocation set $\mathcal{N}_B = \{\eta_j = 10^{-j},\ j = 0, \ldots, \hat{\mu}\}$.
Let us denote by a superscript B (in order to distinguish them from the generalized shape functions $\phi_j(\cdot, \varepsilon)$ introduced in (55)) the Bloch shape functions $\phi^B_i(y)$, $i = 0, 1, \ldots$, which are obtained by orthogonalizing the coefficient matrix of the first Bloch eigenfunctions with respect to η in a collocation set $\mathcal{N}_B$. We define the generalized FE space

$$S^{\underline{p}}_0(\Omega, \mathcal{T}; \mathcal{M}^{\underline{\mu}}_{\varepsilon,B}) = \left(S^{\underline{p}}(\Omega, \mathcal{T}) \otimes_{\mathcal{T}} \mathcal{M}^{\underline{\mu}}_{\varepsilon,B}(\Omega, \mathcal{T}) \right) \cap H^1_0(\Omega),$$

where

$$\mathcal{M}^\mu_{\varepsilon,B} = \text{Span}\left\{ \phi^B_i\left(\frac{x}{\varepsilon}\right) \ :\ i = 0, \ldots, \mu \ll \hat{\mu} \right\}.$$

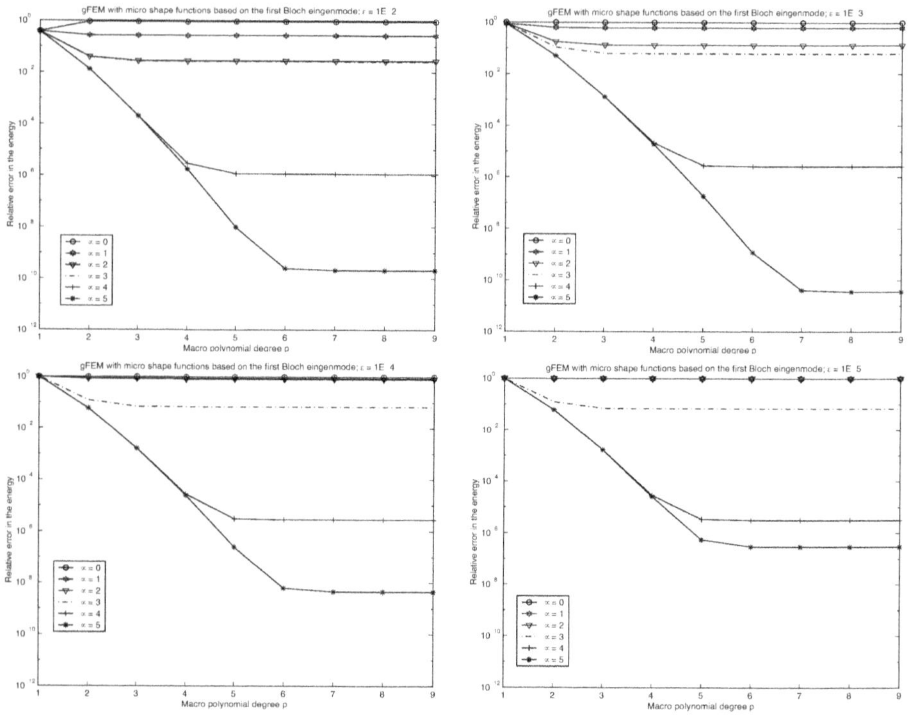

Fig. 11. Relative error for the energy versus p with $\mathcal{M}^{\mu}_{\varepsilon} = \mathcal{M}^{\mu}_{\varepsilon,B}$ obtained from the collocated first Bloch eigenmode; $f(x) = exp(x)$.

The system of Bloch shape functions obtained from the orthogonalization procedure by refining the collocation set geometrically towards $\eta = 0$ and selecting the first Bloch eigenfunction for each sampling point is presented in Figure 10. The collocation set is $\mathcal{N}_B = \{\eta_j = 10^{-j},\, j = 0, \ldots, \hat{\mu}\}$, with $\hat{\mu} = 10$, and the unit-cell problems discretization is based on the FE space $S^{\hat{p},1}_{\text{per}}(\widehat{Q}, \widehat{\mathcal{T}})$ of piecewise polynomials of degree $\hat{p} = 8$.

The performance of the generalized two-scale space $S^{p}_{0}(\Omega, \mathcal{T}; \mathcal{M}^{\mu}_{\varepsilon,B})$ for the model problem (57) with right hand side $f(x) = exp(x)$ has been investigated numerically. The macro mesh $\mathcal{T} = \mathcal{T}_b \cup \mathcal{T}_0$ consists of boundary elements $K \in \mathcal{T}_b$ of size $O(\varepsilon)$ and one interior element $K \in \mathcal{T}_0$ of size $O(1)$. The boundary mesh $\mathcal{T}_b$ covers 4 periods of length $2\pi\varepsilon$ at each boundary point. The micro degree $\mu_K = \mu$ is successively increased in the interior element $K \in \mathcal{T}_0$ whereas $\mu_k = 0$ in all boundary elements $K \in \mathcal{T}_b$. We choose a constant macro degree $p_K = p$ in all elements of $\mathcal{T}$ and we vary p from $p = 1$ up to $p = 9$.

The relative error in the energy versus the macro degree p is shown in Figure 11 for different micro degrees μ and for ε scales varying from 10^{-2} down to 10^{-5}. We observe that when increasing μ from $\mu = 0$ to $\mu = 3$ only a very slow convergence occurs (for $\varepsilon \cong 10^{-4}$ and $\varepsilon \cong 10^{-5}$, the error curves corresponding to $\mu = 0, 1$ and 2 are practically at the top at each other). Compared to the performance of the gFEM with $\mathcal{M}_\varepsilon^\mu$ based on oversampling in Figure 4, the micro scale shape functions derived from the first collocated Bloch eigenmode that correspond to the largest singular values are not able to represent correctly the micro structure of the solution. Exponential convergence is observed first when the micro degree $\mu \geq 3$ and for each μ a saturation occurs at p values depending on μ and ε. To achieve exponential convergence the macro polynomial degree has to be increased together with μ. Comparing the error plots for the different ε scales at fixed $\mu = 5$ we observe that the saturation occurs earlier for smaller ε. The convergence rate is therefore not robust.

The two-scale gFEM based on $\mathcal{M}_{\varepsilon,B}^\mu$ is quite sensitive with respect to the choice of the collocation set and selection of the eigenmodes. Our next experiment suggests that *if microspaces based on Bloch waves are used in the gFEM, it is sufficient to use the first Bloch eigenmode only.* In Figure 12 we plot the relative error in the energy versus p for different μ and for $\varepsilon \cong 10^{-2}$ and $\varepsilon \cong 10^{-3}$. The micro scale shape functions are derived from the first two Bloch eigenmodes $\phi_m(y, \eta)$, $m = 1, 2$, by collocation and orthogonalization (collocation set as before). We see that the convergence occurs even later and this procedure does not automatically improve the convergence. We conclude

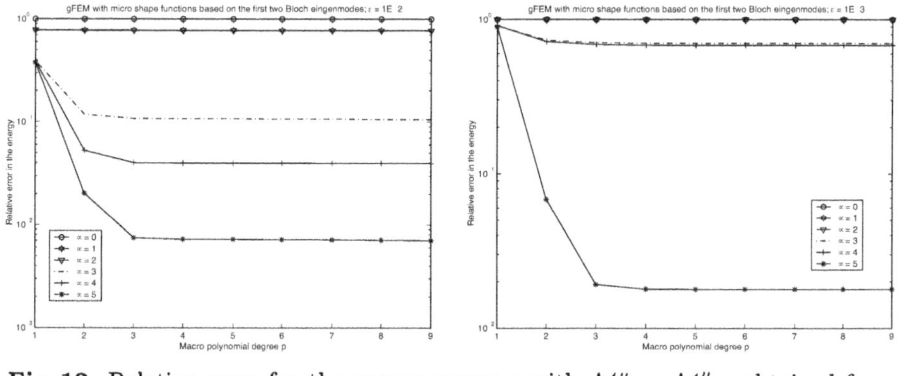

Fig. 12. Relative error for the energy versus p with $\mathcal{M}_\varepsilon^\mu = \mathcal{M}_{\varepsilon,B}^\mu$ obtained from the first two collocated Bloch eigenmodes $\phi_m(\cdot, \eta)$, $m = 1, 2$; $f(x) = exp(x)$.

that in the gFEM based on the micro-scale shape functions $\mathcal{M}_{\varepsilon,B}^\mu$ derived by collocating the Bloch eigenmodes essential information of the microstructure is missing. In addition, once discretizing the unit-cell problem, one has to solve parameter dependent eigenvalue problems which is more costly than solving linear systems.

References

1. G. Allaire, M. Amar, *Boundary Layer Tails in Periodic Homogenization*, European Series in Applied and Industrial Mathematics, 1999.
2. I. Babuška, B. Anderson, P. J. Smith, K. Levin, *Damage Analysis of Fiber Composites. Part I: Statistical Analysis on Fiber Scale*, Preprint No.**15**, The Aeronautical Research Institute of Sweden (1998).
3. I. Babuška, A. K. Aziz, *Survey lectures on the mathematical foundation of the finite element method*, in The Mathematical Foundations of the Finite Element Method with Applications to Partial Differential Equations, A. K. Aziz, ed., Academic Press, New York (1973) 5–359.
4. I. Babuška, G. Caloz, J. Osborn, *Special finite element methods for a class of second order elliptic problems with rough coefficients*, SIAM J. Numer. Anal., **31** No. **4**, (1994), 945-981.
5. I. Babuška, B. Q. Guo, *Regularity of the solution of elliptic problems with piecewise analytic data. Part 1: Boundary value problems for linear elliptic equation of second order*, SIAM J. Math. Anal., **19** (1988), pp. 172–203.
6. I. Babuška and J.M. Melenk, *The partition of unity Finite Element Method: Basic theory and applications* Comp. Meth. Appl. Mech. Engg. **40** (1997) 727–758.
7. N. Bakhvalov, G. Panasenko, *Homogenization: Averaging Process in Periodic Media*, Kluwer Publ., Dordrecht, 1989.
8. A. Bensoussan, J. L. Lions, G. Papanicolau, *Asymptotic Analysis for Periodic Structures*, North Holland, Amsterdam, 1978.
9. F. Bloch, *Über die Quantenmechanik der Electronen in Kristallgittern*, Z. Phys. **52** (1928), 555–600.
10. S.C. Brenner and L.R. Scott, *The Mathematical Theory of Finite Element Methods*, Texts in Applied Mathematics **15**, Springer Verlag, New York, 1994.
11. J. Céa, *Approximation variationelle des problèmes aux limites*, Ann. Inst. Fourier (Grenoble) 2, 1964, 610–620.
12. P.G. Ciarlet, *The Finite Element Method for Elliptic Problems*, North-Holland Publishing Company, 1976.
13. D. Cioranescu and Patrizia Donato, *An Introduction to homogenization*, Oxford University Press (1999).
14. D. Cioranescu and Jeanine Saint Jean Paulin, *Homogenization of Reticulated Structures*, Springer Applied Mathematical Sciences (1999).
15. C. Conca, M. Vanninathan, *Homogenization of periodic structures via Bloch decomposition*, SIAM J. Appl. Math. Vol. **57**, No. **6**, (1997) 1639–1659.
16. P. J. Davis, *Interpolation & Approximation*, Dover Publications, INC. New York, 1974.
17. I.S. Gradshteyn and I.M. Ryzhik, *Table of Integrals, Series, and Producs*, Corrected and enlarged edition. Wiley, New York, 1980.
18. W. Hackbusch, S. A. Sauter, *Composite Finite Elements for the Approximation of PDEs on Domains with complicated Micro-Structures*, Numer. Math. **75** (1997), 447–472.
19. W. Hackbusch and Stefan Sauter, *Composite finite elements for problems containing small geometric details Part II: Implementation and numerical results*, Computing and Visualization in Science 1 (1997), 15–25

20. T. Y. Hou, X. H. Wu, *A Multiscale Finite Element Method for Elliptic Problems in Composite Materials and Porous Media* J. Comput Phys., **134**, (1997) 169–189.
21. T. Hou, X.-H. Wu, Z. Cai, *Convergence of a multiscale finite element method for elliptic problems with rapidly oscillating coefficients*, to appear in Math. Comp.
22. Jonathan Aerospace materials corporation online brochure, Executive Summary, Wilmington, Ma USA (available via email from info@jamcorp.com).
23. G.E. Karniadakis, S.J. Sherwin, *Spectral hp Element Methods for CFD*, Oxford University Press, 1999.
24. C. Lage, *Concept Oriented Design of Numerical Software*, Technical Report 98-07, Seminar for Applied Mathematics, ETH Zürich http://www.sam.math.ethz.ch/Reports/prepr98.html/98-07.ps.Z.
25. P.D. Lax, A.N. Milgram, *Parabolic Equations*, volume 33 of Annals of Mathematics Studies, Princeton University Press, 1954, 167–190.
26. A.M. Matache, Spectral and p-Finite Elements for Problems with Microstructure, Doctoral Thesis, ETH Zürich 2000.
27. A.M. Matache and Ch. Schwab, *Homogenization via p-FEM for Problems with Microstructure*, Applied Numerical Mathematics, vol **33**, Issue 1–4, May 2000, available also as Report 99-04, Seminar for Applied Mathematics, ETH Zürich, http://www.sam.math.ethz.ch/Reports/ prepr99.html/99-09.ps.Z.
28. A.M. Matache, C. Schwab, *Two-Scale FEM for Homogenization Problems*, in preparation.
29. A.M. Matache, I. Babuška and Ch. Schwab, *Generalized p-FEM in Homogenization*, Numerische Mathematik, **86** Issue 2 (2000), 319–375.
30. J.M. Melenk, *On n-widths for elliptic problems*, Journal of Mathematical Analysis and Applications, Vol.**247**, No. 1, Jul 2000, pp. 272–289.
31. R. C. Morgan, I. Babuška, *An approach for constructing families of homogenized equations for periodic media. I: An integral representation and its consequences*, SIAM J. Math. Anal. Vol. 22, No. 1, pp.1–15, January 1991.
32. R. C. Morgan, I. Babuška, *An approach for constructing families of homogenized equations for periodic media. II: Properties of the kernel*, SIAM J. Math. Anal. Vol. 22, No. 1, pp.16–33, January 1991.
33. N. Neuss, W. Jäger, G. Wittum, *Homogenization and Multigrid*, Preprint 1998-04, SFB 359, University of Heidelberg.
34. M. Neuss-Radu, *Boundary Layers in the Homogenization of Elliptic Problems*, Doctoral Dissertation, Heidelberg University, 1999.
35. O. A. Oleinik, A.S. Shamaev, G. A. Yosifian, *Mathematical Problems in Elasticity and Homogenization*, North-Holland, 1992.
36. F.W.J. Olver, *Asymptotics and Special Functions*, Academic Press, 1974.
37. C. Schwab, *p- and hp- Finite Element Methods*, Oxford University Press, 1998.
38. C. Schwab, A.M. Matache, *High order generalized FEM for lattice materials*, Proceedings of the 3rd European Conference on Numerical Mathematics and Advanced Applications, Finland, 1999, ed. by P. Neittaanmäki, T. Tiihonen and P. Tarvainen, World Scientific, Singapore, 2000.
39. C. Schwab, M. Suri, *The p and hp versions of the Finite Element Method for problems with boundary layers*, Mathematics of Computation, Vol. **65**, No. **216**, (1996) 1403–1429.
40. F. Stenger, *Numerical Methods Based on Sinc and Analytic Functions*, Springer Verlag, 1993.

Nonlinear Multiscale Transforms

Jean-Luc Starck

[1] DAPNIA/SEI-SAP, CEA-Saclay, F-91191 Gif-sur-Yvette Cedex, France
[2] Statistics Department, Stanford University,
Sequoia Hall, Stanford, CA 94305 USA

Abstract. We present in this paper several approaches for performing nonlinear multiscale transforms to an image. Both redundant and non-redundant transformations are discussed. The median based multiscale transforms and some of their applications are detailed. Finally we show how a multiscale vision model can be used to decompose an image, and how several transformations can be combined in order to benefit of the advantages of each of them.

Keywords. Discrete Wavelet Transform, nonlinear transform, lifting scheme, combined transform, filtering.

1 Introduction

Multiscale methods have become very popular with the development of the wavelets in last ten years. Background texts on the wavelet transform include [13,18,48,32,6,45]. The most used wavelet transform algorithm is certainly the decimated bi-orthogonal wavelet transform (OWT). Using the OWT, a signal s can be decomposed by:

$$s(l) = \sum_k c_{J,k}\phi_{J,l}(k) + \sum_k \sum_{j=1}^{J} \psi_{j,l}(k) w_{j,k} \quad (1)$$

with $\phi_{j,l}(x) = 2^{-j}\phi(2^{-j}x - l)$ and $\psi_{j,l}(x) = 2^{-j}\psi(2^{-j}x - l)$, where ϕ and ψ are respectively the scaling function and the wavelet function. J is the number of resolutions used in the decomposition, w_j the wavelet (or details) coefficients at scale j, and c_J is a coarse or smooth version of the original signal s. Thus, the algorithm outputs $J+1$ subband arrays. The indexing is such that, here, $j = 1$ corresponds to the finest scale (high frequencies). Coefficients $c_{j,k}$ and $w_{j,k}$ are obtained by means of the filters h and g:

$$\begin{aligned} c_{j+1,l} &= \sum_k h(k-2l)c_{j,k} \\ w_{j+1,l} &= \sum_k g(k-2l)c_{j,k} \end{aligned} \quad (2)$$

where h and g verify:

$$
\begin{aligned}
\frac{1}{2}\phi(\frac{x}{2}) &= \sum_k h(k)\phi(x-k) \\
\frac{1}{2}\psi(\frac{x}{2}) &= \sum_k g(k)\phi(x-k)
\end{aligned}
\tag{3}
$$

and the reconstruction of the signal is performed with:

$$
c_{j,l} = 2\sum_k [\tilde{h}(k+2l)c_{j+1,k} + \tilde{g}(k+2l)w_{j+1,k}] \tag{4}
$$

where the filters $\tilde{h}$ and $\tilde{g}$ must verify the conditions of dealiasing and exact reconstruction:

$$
\begin{aligned}
\hat{h}(\nu+\frac{1}{2})\hat{\tilde{h}}(\nu) + \hat{g}(\nu+\frac{1}{2})\hat{\tilde{g}}(\nu) &= 0 \\
\hat{h}(\nu)\hat{\tilde{h}} + \hat{g}(\nu)\hat{\tilde{g}}(\nu) &= 1.
\end{aligned}
\tag{5}
$$

The application of the OWT to image compression, using the 7-9 filters [2] and the zerotree coding [41,38] has lead to impressive results, compared to previous methods like JPEG. However some problems related to the OWT may impact on their use in some applications, including the following:

1. Edges representation: If the OWT performs better than the FFT to represent edges in an image, it is still not optimal. There is only a fixed number of directional elements independent of scales, and there is no highly anisotropic elements [11]. For instance, the Haar 2D wavelet transform is optimal to find features with a ratio $length/width = 2$, and a horizontal,vertical, or diagonal orientation.
2. Isotropic features representation: the 2D OWT [33] leads to a wavelet transform with three wavelet functions (at each scale there are three wavelet coefficient sub-images) which does not simplify the analysis and the interpretation of the wavelet coefficients. An isotropic transform seems more appropriate for images containing features or objects with no favored orientation (astronomical images, etc).
3. Negative values: By definition, the wavelet coefficient mean is null. Every time we have a positive structure at a scale, we have negative values surrounding it. These negative values often create artifacts during the restoration process, or complicate the analysis.
4. Point artifacts: For example, cosmic ray hits in optical astronomy can "pollute" all the scales of the wavelet transform. The wavelet transform is non-robust relative to such real or detector faults.
5. Integer Values: The OWT produce floating values which are not easy to handle for lossless image compression.

These problems have lead to the development of other multiscale representations. Some of them are still linear transforms, like the ridgelet transform or the curvelet transform [11,10,43]. (both of them have been designed in order to better represent the edges in an image), while some others are nonlinear. We will describe in this article the second class.

Section 2 introduces the nonlinear multiscale transforms, in particular the lifting scheme [50] approach, which generalizes the standard filter bank decomposition. Using the lifting scheme, non linearity can be introduced in a straightforward way, allowing us to perform an integer wavelet transform, or a wavelet transform on a irregular sampled grid. Section 3 presents the median based multiscale representations [21,47] which discriminates well outliers in the data (non Gaussian noise, pixels with high intensity values, etc). Sections 4 and 5 show how a signal can be decomposed into several components, either using a Multiscale Vision Model, or by combining several transforms.

2 Multiscale Image Decomposition

2.1 The Lifting Scheme

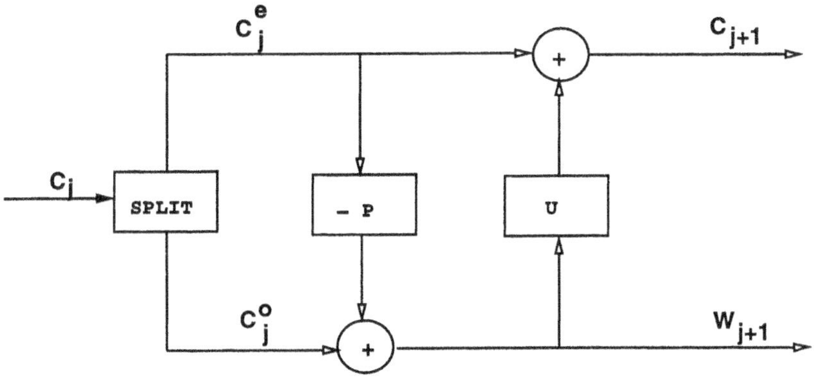

Fig. 1. The lifting scheme – forward direction.

The lifting scheme [50] is a flexible technique that has been used in several different settings, for easy construction and implementation of traditional wavelets [50], and of second generation wavelets [49] such as spherical wavelets [39].

Its principle is to compute the difference between a true coefficient and its prediction:

$$w_{j+1,l} = c_{j,2l+1} - \mathcal{P}(c_{j,2l-2L}, ..., c_{j,2l-2}, c_{j,2l}, c_{j,2l+2}, ..., c_{j,2l+2L}). \qquad (6)$$

A pixel at an odd location 2l+1 is then predicted using pixels at even locations.

The transformation is done in three steps:

1. **Split:** Split the signal into even and odd number samples:

$$\begin{aligned} c^e_{j,l} &= c_{j,2l} \\ c^o_{j,l} &= c_{j,2l+1}. \end{aligned} \tag{7}$$

2. **Predict:** Calculate the wavelet coefficient $w_{j+1,l}$ as the prediction error of $c^o_{j,l}$ from $c^e_{j,l}$ using the prediction operator $\mathcal{P}$:

$$w_{j+1,l} = c^o_{j,l} - \mathcal{P}(c^e_{j,l}). \tag{8}$$

3. **Update:** The coarse approximation c_{j+1} of the signal is obtained by using $c^e_{j,l}$ and $w_{j+1,l}$ and the update operator $\mathcal{U}$:

$$c_{j+1,l} = c^e_{j,l} + \mathcal{U}(w_{j+1,l}). \tag{9}$$

The lifting steps are easily inverted by:

$$\begin{aligned} c_{j,2l} &= c^e_{j,l} = c_{j+1,l} - \mathcal{U}(w_{j+1,l}) \\ c_{j,2l+1} &= c^o_{j,l} = w_{j+1,l} + \mathcal{P}(c^e_{j,l}). \end{aligned} \tag{10}$$

Some examples of wavelet transform via the lifting scheme are:

- Haar wavelet via lifting: the Haar transform can be performed via the lifting scheme by taking the predict operator equal to the identity, and an update operator which halves the difference. The transform becomes:

$$\begin{aligned} w_{j+1,l} &= c^o_{j,l} - c^e_{j,l} \\ c_{j+1,l} &= c^e_{j,l} + \frac{w_{j+1,l}}{2}. \end{aligned}$$

  All computation can be done in place.
- Linear wavelets via lifting: the identity predictor used before is correct when the signal is constant. In the same way, we can use a linear predictor which is correct when the signal is linear. The predictor and update operators are now:

$$\begin{aligned} \mathcal{P}(c^e_{j,l}) &= \frac{1}{2}(c^e_{j,l} + c^e_{j,l+1}) \\ \mathcal{U}(w_{j+1,l}) &= \frac{1}{4}(w_{j+1,l-1} + w_{j+1,l}). \end{aligned}$$

  It is easy to verify that:

$$c_{j+1,l} = -\frac{1}{8}c_{j,2l-2} + \frac{1}{4}c_{j,2l-1} + \frac{3}{4}c_{j,2l} + \frac{1}{4}c_{j,2l+1} - \frac{1}{8}c_{j,2l+2}$$

  which is the bi-orthogonal Cohen-Daubechies-Feauveau [16] wavelet transform.

- 9/7 filters: the lifting factorization of popular (9/7) filter pair leads to the following implementation [20]:

$$
\begin{aligned}
s_l^{(0)} &= c_{j,2l} \\
d_l^{(0)} &= c_{j,2l+1} \\
d_l^{(1)} &= d_l^{(0)} + \alpha(s_l^{(0)} + s_{l+1}^{(0)}) \\
s_l^{(1)} &= s_l^{(0)} + \beta(d_l^{(1)} + d_{l-1}^{(1)}) \\
d_l^{(2)} &= d_l^{(1)} + \gamma(s_l^{(1)} + s_{l+1}^{(1)}) \\
s_l^{(2)} &= s_l^{(1)} + \delta(d_l^{(2)} + d_{l-1}^{(2)}) \\
c_{j+1,l} &= u s_l^{(2)} \\
c_{j+1,l} &= u d_l^{(2)}
\end{aligned} \tag{11}
$$

with

$$
\begin{aligned}
\alpha &= -1.586134342 \\
\beta &= -0.05298011854 \\
\gamma &= 0.8829110762 \\
\delta &= 0.4435068522 \\
u &= 1.149604398.
\end{aligned} \tag{12}
$$

Every wavelet transform can be written via lifting.

Integer wavelet transform When the input data are integer values, the wavelet transform no longer consists of integers. For lossless coding, it is useful to have a wavelet transform which produces integer values. We can build an integer version of every wavelet transform [8]. For instance, denoting $\lfloor x \rfloor$ as the largest integer not exceeding x, the integer Haar transform (also called "S" transform) can be calculated by:

$$
\begin{aligned}
w_{j+1,l} &= c_{j,l}^o - c_{j,l}^e \\
c_{j+1,l} &= c_{j,l}^e + \lfloor \frac{w_{j+1,l}}{2} \rfloor
\end{aligned} \tag{13}
$$

while the reconstruction is

$$
\begin{aligned}
c_{j,2l} &= c_{j+1,l} - \lfloor \frac{w_{j+1,l}}{2} \rfloor \\
c_{j,2l+1} &= w_{j+1,l} + c_{j,2l}.
\end{aligned} \tag{14}
$$

More generally, the lifting operators for an integer version of the wavelet transform are:

$$
\mathcal{P}(c_{j,l}^e) = \lfloor \sum_k p_k c_{j,l-k}^e + \frac{1}{2} \rfloor
$$

$$\mathcal{U}(w_{j+1,l}) = \lfloor \sum_k u_k w_{j+1,l-k} + \frac{1}{2} \rfloor. \quad (15)$$

The linear integer wavelet transform is

$$\begin{aligned} w_{j+1,l} &= c^o_{j,l} - \lfloor \frac{1}{2}(c^e_{j,l} + c^e_{j,l+1}) + \frac{1}{2} \rfloor \\ c_{j+1,l} &= c^e_{j,l} + \lfloor \frac{1}{4}(w_{j+1,l-1} + w_{j+1,l}) + \frac{1}{2} \rfloor. \end{aligned} \quad (16)$$

Even if there is no filter that consistently performs better than all the other filters on all images, the linear integer wavelet transform performs generally better than others [8]: More filters can be found in [8].

Wavelet transform on irregular grid A wavelet transform on irregular samples can be performed by introducing weighted inner products [50,19], assuming that we are given weighted averages of some unknown function over interval:

$$c_{0,l} = \frac{1}{I_{0,l}} \int_{x_l}^{x_{l+1}} u(x) f(x) dx \quad (17)$$

where $[x_l, x_{l+1}]$ defines an interval, $u(x)$ is a positive weighting function, and $I_{0,l}$ is

$$I_{0,l} = \int_{x_l}^{x_{l+1}} u(x) dx. \quad (18)$$

In its simplest form, $u(x)$ can take the value 1 when a sample at position x is available, and 0 otherwise. If error measurements are also available, $u(x)$ can be derived from them.

In the case of the Haar wavelet transform, it leads to the unbalanced Haar transform which is obtained by [50]:

$$\begin{aligned} I_{j+1,l} &= I_{j,2l} + I_{j,2l+1} \\ w_{j+1,l} &= c^o_{j,l} - c^e_{j,l} \\ c_{j+1,l} &= c^e_{j,l} + \frac{I_{j,2l+1}}{I_{j+1,l}} w_{j+1,l}. \end{aligned} \quad (19)$$

Other kind of wavelets can be performed in a similar way. See [50] for more details.

Adaptive Wavelet Transform Adaptativity can be introduced in the wavelet transform by reversing the order of the predict and the update steps in the lifting scheme [14]. When the update operator is first applied, the prediction is based on the low-pass coefficients that are computed as in the

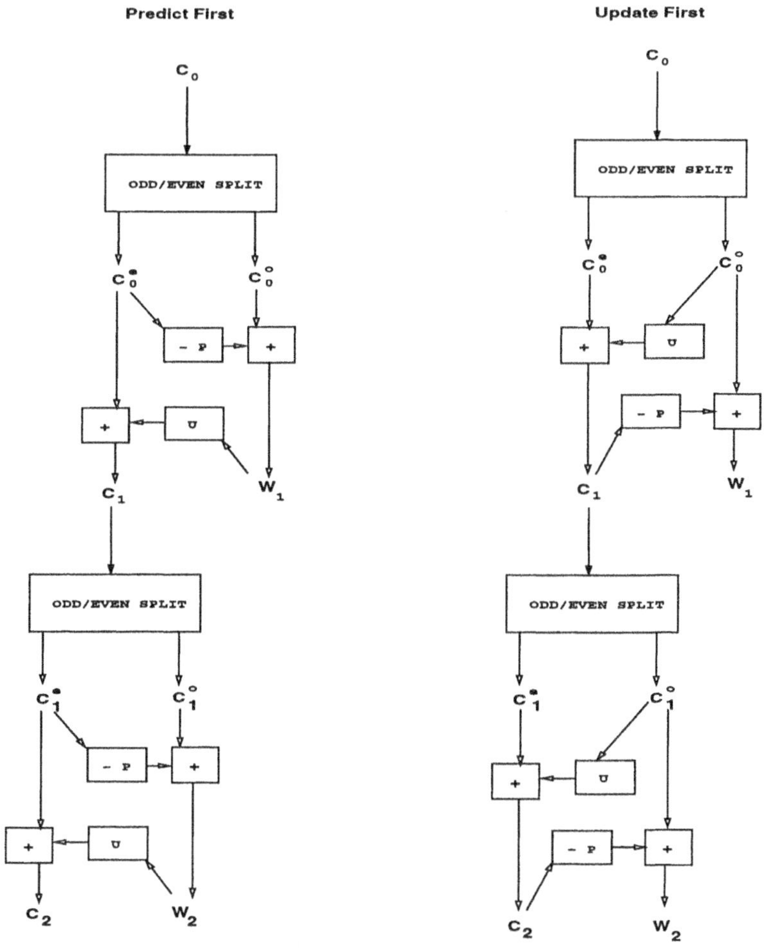

Fig. 2. The lifting scheme with two iterations: left, predict-first approach, and right, update-first approach.

standard wavelet transform. In the update-first approach, as illustrated in Figure 2, the detail coefficients are not in the loop for calculating the coefficients at a coarser scale. Hence, we can start the prediction process at the coarser scale, and working from coarse to fine scales. The idea is now to make the predictor data dependent. The prediction operator is chosen, based on the local properties of the data at a coarser scale. If an edge is detected, the order of the predictor is reduced, while if the data are smooth, a larger order is preferred.

2.2 Multiscale Morphological Transform

Mathematical Morphology Originally developed by Matheron [35,36] and Serra [40], the mathematical morphology is based on two operators: the *infimum* (denoted $\wedge$) and the *supremum* (denoted $\vee$). The infimum of a set of images is defined as the greatest lower bound while the *supremum* is defined as the least upper bound. The basic morphological transformations are erosion, dilation, opening and closing. For gray-level images, they can be defined in the following way:

- *erosion* consists of replacing each pixel of an image by the minimum of its neighbors:
$$\epsilon_B(f) = \bigwedge_{b \in B} f_{-b}$$
where f stands for the image, and B denote the structuring element, typically a small convex set such a square or disc.
- *dilation* consists of replacing each pixel of an image by the maximum of its neighbors:
$$\delta_B(f) = \bigvee_{b \in B} f_b.$$
- *opening* consists of doing an erosion followed by a dilations:
$$\alpha_B = \delta_B \epsilon_B \text{ and } \alpha_B(f) = f \circ B.$$
- *closing* consists of doing a dilation followed by an erosion:
$$\beta_B = \epsilon_B \delta_B \text{ and } \beta_B(f) = f \bullet B.$$

In a more general way, *opening* and *closing* referred to morphological filters which respect some specific properties [5].

Mathematical morphology have been until now considered as another way to analyze data, in competition with linear methods. But from a multiscale point of view [45,25,27], mathematical morphology or linear methods are just filters allowing us to go from a resolution to a coarser one, and the multiscale coefficients are then analyzed in the same way.

Undecimated Morphological Transform By choosing a set of structuring elements B_j having a size increasing with j, we can define an undecimated morphological multiscale transform by

$$\begin{aligned} c_{j+1,l} &= \mathcal{M}_j(c_j)(l) \\ w_{j+1,l} &= c_{j,l} - c_{j+1,l} \end{aligned} \tag{20}$$

where $\mathcal{M}_j$ is a morphological filter (erosion, opening, etc) using the structuring element B_j. An example of B_j is a box of size $(2^j + 1) \times (2^j + 1)$. As the detail signal w_{j+1} is obtained by calculating a simple difference between the c_j and c_{j+1}, the reconstruction is straightforward, and is identical to the reconstruction relative to the "à trous" wavelet transform [45]. An exact reconstruction of the image c_0 is obtained by:

$$c_{0,l} = c_{J,l} + \sum_{j=1}^{J} w_{j,l} \tag{21}$$

where J is the number of scales used in the decomposition. Each scale has the same number N of samples as the original data. The total number of pixels in the transformation is $(J+1)N$. Such a decomposition is obviously not dedicated to image compression. It is however interesting for the following reasons:

- The computational requirement is reasonable.
- The reconstruction algorithm is trivial.
- The transform is known at each pixel, allowing position detection without any error, and without interpolation.
- We can follow the evolution of the transform from one scale to the next.
- Invariance under translation is completely verified.
- The transform is isotropic (if the structuring element B is isotropic).

The last point is important if the image contain isotropic features, like cells in biologic images, or stars in astronomical images.

Pyramidal Transform A decimation can be introduced in order produce a pyramidal transform similar to the Burt and Adelson pyramidal [1], but by using morphological filters. We introduce two operators $\beta^{\downarrow}$, and $\beta^{\uparrow}$: $\beta^{\downarrow}$ (analysis operator) consists in filtering first the image and then to decimate it, and $\beta^{\uparrow}$ (synthesis operator) consists in interpolating the data.

$$\begin{aligned} c_{j+1} &= \beta^{\downarrow}(c_j) \\ w_{j+1} &= c_j - \beta^{\uparrow}(c_{j+1}). \end{aligned} \tag{22}$$

The reconstruction is obtained by:

$$c_j = w_{j+1} + \beta^{\uparrow}(c_{j+1}). \tag{23}$$

An example of operators is:

$$\begin{aligned} \beta^{\downarrow}(c_j)(l) &= c_{j,2l} \wedge c_{j,2l+1} \\ \beta^{\uparrow}(c_j)(2l) &= \beta^{\uparrow}(c_j)(2l+1) = c_{j+1,l}. \end{aligned} \tag{24}$$

In this scheme, $\beta^{\downarrow}$ and $\beta^{\uparrow}$ can be both linear or nonlinear. For example, choosing

$$\begin{aligned} \beta^{\downarrow}(c_j)(l) &= c_{j,2l} \wedge c_{j,2l+1} \\ \beta^{\uparrow}(c_j)(2l) &= c_{j+1,2l} \\ \beta^{\uparrow}(c_j)(2l+1) &= \frac{1}{2}(c_{j+1,l} + c_{j+1,l+1}) \end{aligned}$$

leads to a morphological analysis operator, and a linear synthesis operator. As for the Burt and Adelson pyramid, the redundancy factor, after transformation of the image, is $\frac{4}{3}N$ where N is the number of pixels.

Non Redundant Multiscale Transform A non-redundant multiscale morphological transform can easily be built via the lifting scheme, by introducing morphological operators. For example, the Haar morphological transform, also named "G" transform, is:

$$\begin{aligned} w_{j+1,l} &= c^o_{j,l} - \mathcal{P}(c^e_{j,l}) \\ c_{j+1,l} &= c^e_{j,l} + \mathcal{U}(w_{j+1,l}) \end{aligned} \tag{25}$$

where

$$\begin{aligned} \mathcal{P}(c^e_{j,l}) &= c_{j,2l} \\ \mathcal{U}(w_{j+1,l}) &= 0 \wedge w_{j+1,l}. \end{aligned} \tag{26}$$

It is easy to verify that:

$$\begin{aligned} w_{j+1,l} &= c_{j,2l+1} - c_{j,2l} \\ c_{j+1,l} &= c_{j,2l} + (0 \wedge (c_{j,2l+1} - c_{j,2l})) = c_{j,2l} \wedge c_{j,2l+1}. \end{aligned} \tag{27}$$

The Multiscale Operators Goutsias [25,27] has recently proposed a general multiscale decomposition scheme, by means of the *signal analysis operator* $\beta^{\downarrow}$ and the *detail analysis operator* $\gamma^{\downarrow}$:

$$\begin{aligned} w_{j+1,l} &= \gamma^{\downarrow}(c_j)(l) \\ c_{j+1,l} &= \beta^{\downarrow}(c_j)(l) \end{aligned} \tag{28}$$

and the reconstruction is done by means of the *synthesis operator* $\Psi^{\uparrow}$:

$$c_j = \Psi^{\uparrow}(c_{j+1}, w_{j+1}). \tag{29}$$

The perfect reconstruction condition is [27]:

$$c_{j,l} = \Psi^{\uparrow}(\beta^{\downarrow}(c_j), \gamma^{\downarrow}(c_j))(l). \tag{30}$$

We distinguish the case of the *uncoupled multiscale decomposition* when the operator $\Psi^{\uparrow}$ verifies the condition:

$$\Psi^{\uparrow}(c_{j+1}, w_{j+1}) = \beta^{\uparrow}(c_{j+1}) + \gamma^{\uparrow}(w_{j+1}) \tag{31}$$

where $\beta^{\uparrow}$ is the *signal synthesis operator*, and $\gamma^{\uparrow}$ is the *detail synthesis operator*.

Furthermore, the transformation is not redundant if the two following conditions are verified [27]:

$$\begin{cases} \beta_j^{\downarrow}(\Psi_j^{\uparrow}(c,w)) &= c \quad \text{if } c \in V_{j+1} \text{ and } w \in W_{j+1} \\ \gamma_j^{\downarrow}(\Psi_j^{\uparrow}(c,w)) &= w \quad \text{if } c \in V_{j+1} \text{ and } w \in W_{j+1} \end{cases}$$

where V_{j+1} and W_{j+1} are respectively the signal and the detail space at resolution level $j+1$.

For example, the undecimated morphological multiscale transform corresponds to the uncoupled multiscale decomposition, where

$$\begin{cases} \beta^{\downarrow} = \mathcal{M}_j \\ \gamma^{\downarrow} = Id - \beta^{\downarrow} \\ \beta^{\uparrow} = Id \\ \gamma^{\uparrow} = Id \end{cases} \tag{32}$$

and the standard bi-orthogonal decimated wavelet transform to:

$$\begin{cases} \beta^{\downarrow}(l) = \sum_k h(k-2l)c_{j,k} \\ \gamma^{\downarrow}(l) = \sum_k g(k-2l)c_{j,k} \\ \beta^{\uparrow}(l) = 2\sum_k \tilde{h}(k+2l)c_{j+1,k} \\ \gamma^{\uparrow}(l) = 2\sum_k \tilde{g}(k+2l)w_{j+1,k} \end{cases} . \tag{33}$$

Many morphological filters can be designed using the previously described scheme, but no solution has been proposed until now for choosing the nonlinear filters for a given application.

3 Median Based Multiscale Transforms

3.1 Multiscale Median Transform

The median transform is nonlinear, and offers advantages for robust smoothing (i.e. the effects of outliers pixel values are mitigated). Define the median transform of image f, with square kernel of dimensions $n \times n$, as $med_n(f)$. Let $n = 2s+1$; initially $s = 1$ or 2. The iteration counter will be denoted by j, and p is the user-specified number of resolution scales. the Multiscale Median Transform (MMT) algorithm is:

1. Let $c_j = f$ with $j = 0$.
2. Determine $c_{j+1} = med_{2s+1}(c_j)$.
3. The multiresolution coefficients w_{j+1} are defined as: $w_{j+1} = c_j - c_{j+1}$.
4. Let $j \longleftarrow j+1$; $s \longleftarrow 2s$. Return to Step 2 if $j < J$.

A straightforward expansion formula for the original image is given by:

$$f = c_J + \sum_{j=1}^{J} w_j \tag{34}$$

where c_J is the residual image.

In Step 3, the set of resolution levels associated with s lead to a dyadic analysis (i.e. the resolution is degraded in steps which are increasing integer exponents of 2). Other possibilities involving intermediate scales (e.g. $s \longleftarrow 1.5s$) can also be considered.

The multiresolution coefficient values, w_j, are evidently not necessarily of zero mean, and so the potential artifact-creation difficulties related to this aspect of wavelet transforms do not arise. Note of course that values of w can be negative.

The multiscale median transform corresponds to the uncoupled multiscale decomposition, where

$$\begin{cases} \beta^{\downarrow} = Med_{2j+1} \\ \gamma^{\downarrow} = Id - \beta^{\downarrow} \\ \beta^{\uparrow} = Id \\ \gamma^{\uparrow} = Id. \end{cases} \tag{35}$$

For integer image input values, this transform can be carried out in integer arithmetic only which may lead to computational savings.

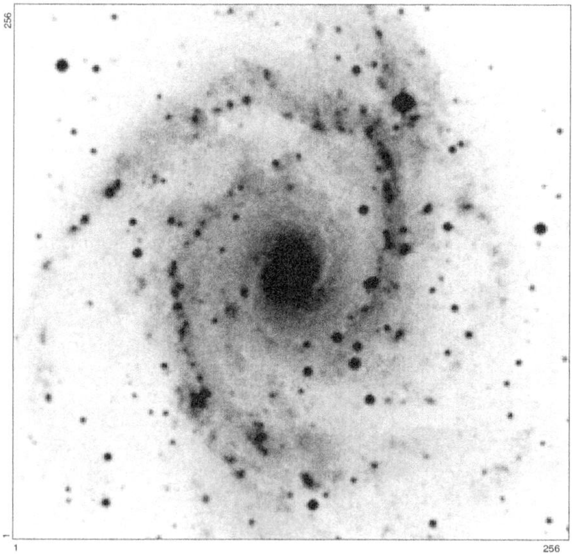

Fig. 3. Galaxy NGC2997.

Figure 3 shows the galaxy NGC2997, and figure 4 its multiscale decomposition by the multiscale median transform, with 6 scales (5 details images and the smoothed image). The coaddition of the six images reproduce exactly the

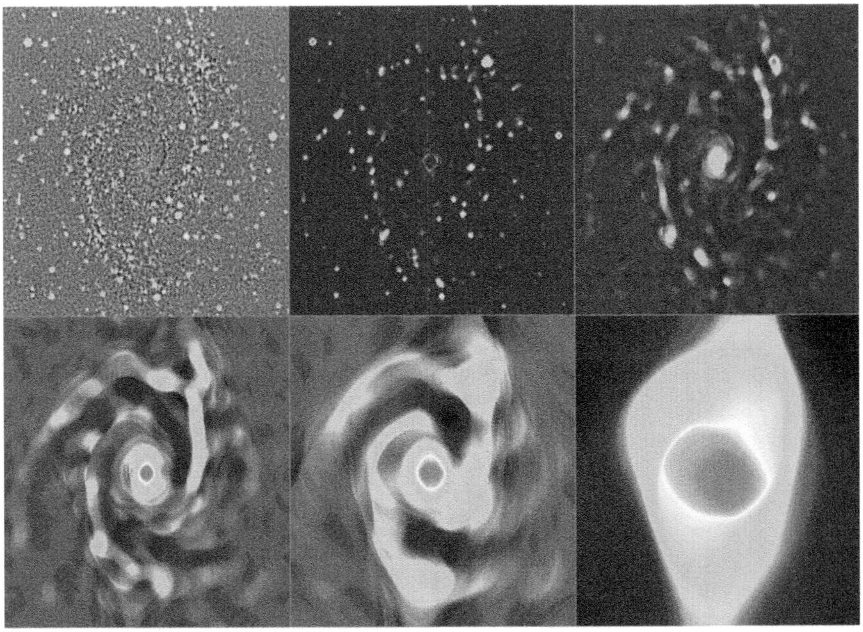

Fig. 4. Multiresolution median transform of NGC2997. See also Color Plate 4.

original image. We can see that even the brightest stars do not "pollute" the largest scales (from 4 to 6). This would not be true with a wavelet transform. Another point to notice is that there is no negative ring around the stars in the first scales. This point is of great importance, as discussed in the next example.

Figure 6 shows the comparison between the "à trous" wavelet transform (on the left) and the MMT (on the right) of the input signal displayed in Figure 5. In these diagrams, the scale is represented as a function of the time. The data were obtained with an infrared camera, on board of the ISO satellite (Infrared Space Observatory). A faint source is observed between pixels 120 to 150. Cosmic ray impacts appears around pixels 50 and 230. The second one has introduced a variation in the detector gain. Due to this kind of cosmic ray impacts (glitches), sources are not easy to identified. Figure 6 shows the decomposition of the signal with both the undecimated wavelet transform (UWT), and the MMT. The glitch and the source are clearly separated in the MMT and they merge in the UWT. Therefore, the MMT allows us a better separation between sources and glitches.

Computational requirements of the MMT are high, and these can be reduced by decimation: one pixel out of two is retained at each scale. Here the transform kernel does not change from one iteration to the next, but the

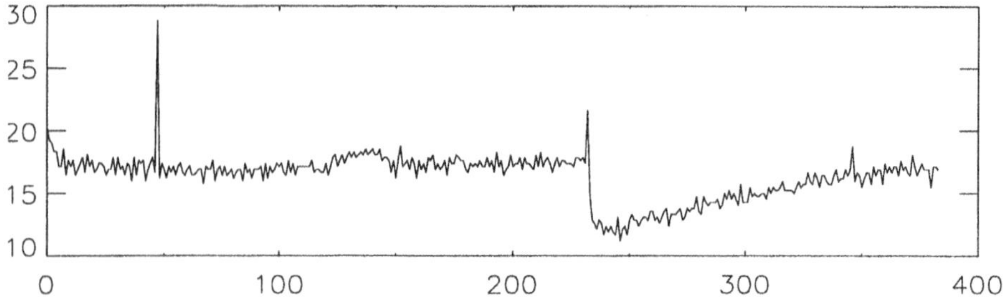

Fig. 5. Example of signal obtained with the ISOCAM infrared camera. A source is observed between pixels 120 to 150. Cosmic ray impacts appears around pixels 50 and 230. The second one has introduced a variation in the detector gain. Due to this kind of cosmic ray impacts (glitches), sources are not easy to identified.

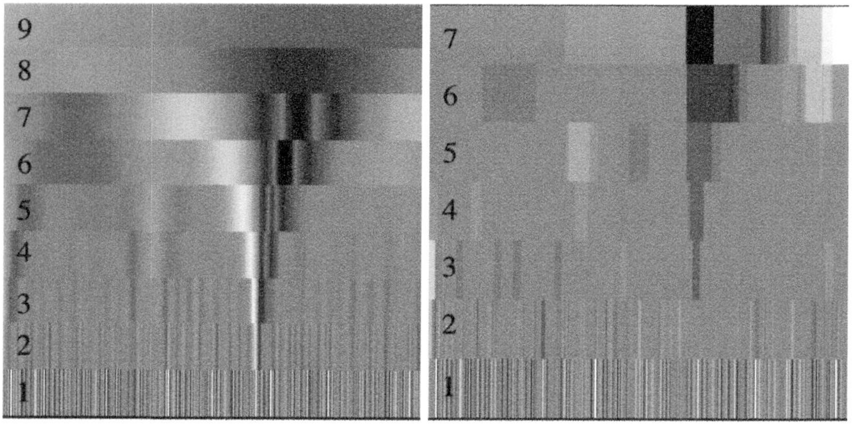

Fig. 6. Comparison between the "à trous" wavelet transform (left) and the multiresolution median transform (right) of above signal. Resolution scale is represented versus the time. We note that the separation between the source and the glitch is improved using the MMT. See also Color Plate 5.

image to which this transform is applied does. This pyramidal algorithm is looked at next.

3.2 Pyramidal Median Transform

The Pyramidal Median Transform (PMT) is obtained by the following algorithm:

1. Let $c_j = f$ with $j = 1$.
2. Determine $c^*_{j+1} = med_{2s+1}(c_j)$ with $s = 1$.

3. The pyramidal multiresolution coefficients w_{j+1} are defined as: $w_{j+1} = c_j - c^*_{j+1}$.
4. Let $c_{j+1} = dec(c^*_{j+1})$ where the decimation operation, dec, entails 1 pixel replacing each 2×2 sub-image.
5. Let $j \longleftarrow j + 1$. Return to Step 2 so long as $j < J$.

Here the kernel or mask of dimensions $(2s + 1) \times (2s + 1)$ remains the same during the iterations. The image itself, to which this kernel is applied, becomes smaller.

While this algorithm aids computationally, the reconstruction formula 34 is no longer valid. Instead we use the following algorithm based on B-spline interpolation:

1. Take the lowest scale image, c_j.
2. Interpolate c_j to determine the next resolution image (of twice the dimensionality in x and y). Call the interpolated image c'_j.
3. Calculate $c_{j-1} \longleftarrow c'_j + w_j$.
4. Set $j \longleftarrow j - 1$. Go to Step 2 if $j > 0$.

This reconstruction procedure takes account of the pyramidal sequence of images containing the multiresolution transform coefficients, w_j.

It presupposes, though, that a good reconstruction is possible. We ensure that by use of the following refined version of the Pyramidal Median Transform. Using iteration, the coefficients, $w_{j+1} = c_j - c_{j+1}$, are improved relative to their potential for reconstructing the input image.

3.3 Iterative Pyramidal Median Transform

An iterative scheme can be proposed for reconstructing an image, based on pyramidal multi-median transform coefficients. Alternatively, the PMT algorithm, itself, can be enhanced to allow for better estimates of coefficient values. The following is an iterative algorithm for this objective:

1. $i \longleftarrow 0$. Initialize f^i with the given image, f. Initialize the multiresolution coefficients at each scale j, w^f_j, to 0.
2. Using the Pyramidal Median Transform, determine the set of transform coefficients, $w^{f_i}_j$.
3. $w^f_j \longleftarrow w^f_j + w^{f_i}_j$.
4. Reconstruct image f_{i+1} from w^f_j (using the interpolation algorithm described in the previous section).
5. Determine the image component which is still not reconstructible from the wavelet coefficients: $f_{i+1} \longleftarrow f - f_{i+1}$.
6. Set $i \longleftarrow i + 1$, and return to Step 2.

The number of iterations is governed by when f_{i+1} in Step 5 approaches a null image. Normally 4 or 5 iterations suffice. Note that the additivity of the wavelet coefficients in Step 3 is justified by additivity of the image decomposition in Step 5 and the reconstruction formula used in Step 4, both of which are based on additive operations.

3.4 Non-Iterative Pyramidal Transform with Exact Reconstruction

A non-iterative version of the pyramidal median transform can be performed by decimating and interpolating the median images during the transform:

1. Let $c_j = f$ with $j = 1$.
2. Determine $c_{j+1} = dec[med_{2s+1}(c_j)]$.
3. Determine c^*_{j+1} = interpolation of c_{j+1} to size of c_j.
4. The pyramidal multiresolution coefficients w_{j+1} are defined as: $w_{j+1} = c_j - c^*_{j+1}$.
5. Let $j \longleftarrow j + 1$. Return to Step 2 so long as $j < J$.

This saves computation time in two ways. First, there is no need to iterate. Secondly, in step (2) one does not really calculate the median for all pixels and then decimate it; rather, one just calculates the median for the pixels to be left after decimation. Thus the medians are 4 times fewer. This algorithm is very close to the Laplacian pyramid developed by Burt and Adelson [7]. The reconstruction algorithm is the same as before, but the reconstructed image has no error. In the following, we will indicate this version by referring to PMT.

The PMT corresponds to the uncoupled multiscale decomposition:

$$\begin{cases} \beta^{\downarrow} = Med^{\downarrow}_{2j+1} \\ \gamma^{\downarrow} = Id - \beta^{\uparrow}\beta^{\downarrow} \\ \gamma^{\uparrow} = Id \\ \beta^{\uparrow} = \mathcal{I} \end{cases} \tag{36}$$

where $Med^{\downarrow}_{2j+1}$ is the median operator followed by a decimation, and $\mathcal{I}$ is an interpolation operator. A bilinear interpolation can be used [47,45], or an interpolation based on the morphological operator $\vee$ [25]:

$$\mathcal{I}(c)(2l) = c(l) \text{ and } \mathcal{I}(c)(2l+1) = c(l) \vee c(l+1).$$

We note in passing one particularly interesting use of the PMT: the final scale often provides a good estimate of the image background. This is especially so for images containing small structures, e.g. astronomical wide-field images.

3.5 Merging Wavelets and the PMT

One of the advantages of the pyramidal wavelet transform over the pyramidal median transform (PMT) is the ability to have robust noise estimation in the different scales, while the advantage of the PMT is a better separation of the structures in the scales. Using the PMT, a strong structure (like a bright star, cosmic rays, bad pixels, etc.) will not be spread over all scales as when using a wavelet transform. In fact, when there is no signal in a given region, a wavelet transform would be better, and if a strong signal appears, it is the PMT that we would like to use. So the idea arises to try to merge both transforms, and to adapt the analysis at each position and at each scale, depending on the amplitude of the coefficient we measure.

A possible algorithm to perform this on an image I is the following:

1. Let $c_j = I$, and $j = 0$.
2. Filter the image c_j by the median: we get m.
3. Set $c^* = c_j$ and $d = c_j - m$.
4. For all pixels k do
   if $\mid d(k) \mid > k\sigma_j$ then $c^*(k) = m(k)$.
5. Smooth c^* by a B_3-spline: we get b.
6. Set $w_{j+1} = c_j - b$.
7. $c_{j+1} = dec(b)$ (where the decimation operation, dec, entails 1 pixel replacing each 2 ×2 sub-image). Let $j = j+1$ and return to step 2 if $j < J$ (J being the number of scales).

In this algorithm, the linear filtering relative to the wavelet transform is not applied to the strong features contained in the image. Indeed, significant pixel values are detected at step 4, and are replaced by the median. Regions containing no bright object are treated as if the pyramidal wavelet transform is used. The parameter k used at step 4 must be large enough in order to be sure that noise is not filtered by the median ($k = 5$ seems high enough). As this transform merges the wavelet transform and the PMT, we call it the PMWT transform. PMWT takes more time than the PMT, but this algorithm has the advantages of the PMT without the drawbacks. The reconstruction is the same as for the PMT.

3.6 Non Redundant Multiscale Median Transform

A non-redundant multiscale median transform has been proposed independently by Egger and Li [23], and Florêncio and Schafer [24], which can easily be described using the lifting scheme by:

$$\begin{aligned} \mathcal{P}(c^e_{j,l}) &= med_N(c^e_j)(l) \\ \mathcal{U}(w_{j+1,l}) &= 0. \end{aligned} \tag{37}$$

It has also been suggested [26,15] that the update operator could also be defined using the median operator. This leads to the following equations:

$$\begin{aligned} \mathcal{P}(c^e_{j,l}) &= med_N(c^e)(l) \\ \mathcal{U}(w_{j+1,l}) &= med_{\tilde{N}}(w_{j+1})(l). \end{aligned} \tag{38}$$

$N = \tilde{N} = 4$ was used in [26], and $N = 5$ and $\tilde{N} = 3$ in [15].

3.7 Example 1 : filtering

Figure 7 shows the results of the MMT-multiscale entropy method on simulated data (2048 pixels). From top to bottom, the simulated data, the noisy data, the filtered data, and both noisy and filtered data overplotted.

The multiscale entropy method [46,44] consists of measuring the information h relative to wavelet coefficients, and of separating this into two parts h_s, and h_n. The expression h_s is called the signal information and represents the part of h which is certainly not contaminated by the noise. The expression h_n is called the noise information and represents the part of h which may be contaminated by the noise. We have $h = h_s + h_n$. Following this notation, the corrected coefficient $\tilde{w}$ should minimize:

$$J(\tilde{w}_j) = h_s(w_j - \tilde{w}_j) + \alpha h_n(\tilde{w}_j) \tag{39}$$

i.e. there is a minimum of information in the residual $(w - \tilde{w})$ which can be due to the significant signal, and a minimum of information which could be due to the noise in the solution $\tilde{w}_j$. In order to verify a number of properties, the following functions have been proposed for h_s and h_n in the case of Gaussian noise [46]:

$$\begin{aligned} h_s(w_j) &= \frac{1}{\sigma_j^2} \int_0^{|w_j|} u \operatorname{erf}\left(\frac{\mid w_j \mid - u}{\sqrt{2}\sigma_j} \right) du \\ h_n(w_j) &= \frac{1}{\sigma_j^2} \int_0^{|w_j|} u \operatorname{erfc}\left(\frac{\mid w_j \mid - u}{\sqrt{2}\sigma_j} \right) du. \end{aligned} \tag{40}$$

3.8 Example 2 : Astronomical image compression

The principle of the compression method is to select the information we want to keep, by using the PMT, and to code this information without any loss. Thus the first phase searches for the minimum set of quantized multiresolution coefficients which produce an image of "high quality". Criteria which can be retained for estimating the quality of a compression method in astronomy are

1. Visual aspect.
2. Signal to noise ratio.
3. Photometry (i.e. integrated pixel values in a given object).
4. Astrometry (i.e. object positions).
5. Detection of real and faint objects.
6. Objects morphology.

Lost information cannot be recovered, so if we do not accept any loss, we have to compress what we take as noise too, and the compression ratio will be low (3 or 4 only). The PMT compression method employed involves the following sequence of operations:

1. Determination of the significant coefficients.
2. Determination of the quantized multiresolution coefficients which give the filtered image.
3. Coding of each resolution level using the Huang-Bijaoui method [28]. This consists of quadtree-coding each image, followed by Huffman-coding the quadtree representation. There is no information lost during this phase.
4. Compression of the noise if this is wished.
5. Decompression consists of reconstituting the noise-filtered image (plus the compressed noise if this was specified).

The decompression is carried out scale by scale, starting from the low resolution, so it is not necessary to decompress the entire file if one is just interested in having a look at the image. Noise is decompressed and added at the end, if this is wanted. See the full description of the method in [47,37].

Various studies have been carried out in order to determine the best compression method for astronomical images, especially in the framework of the ALADIN project [4]. It has been shown, using a sample of nearly 2000 stars from an ESO Schmidt plate centered on the globular cluster M5, that the PMT can give compression ratios of up to 5 times the maximum ratio obtained from other methods [30], such JPEG or wavelet based methods.

Figure 8 top shows a subimage of a Schmidt photographic plate of the region of M5 (numbered ESO 7992v). This photographic plate was digitized using the MAMA ("Machine Automatique à mesurer pour l'Astronomie") machine of Paris Observatory (Paris, France).
Figures 8 bottom left and right allow the visual quality of both the JPEG (compression ratio 40:1) and PMT (compression ratio 260:1) to be compared.

3.9 Conclusion

The median based multiscale transforms present the advantage of being robust to strong singularities. These singularities can be due to non white noise, or from the observed data itself such as stars in astronomical images. In both cases, the median based multiscale transform is an adequate tool to separate these singularities from the rest of the signal. This is the reason of the success of the PMT for the compression of astronomical images.

Finally we would like to point out the work of Donoho [22] about a median based pyramidal transform, in which a decimation of 3 is used, instead of the standard dyadic scheme, in order to preserve the independence of the coefficients.

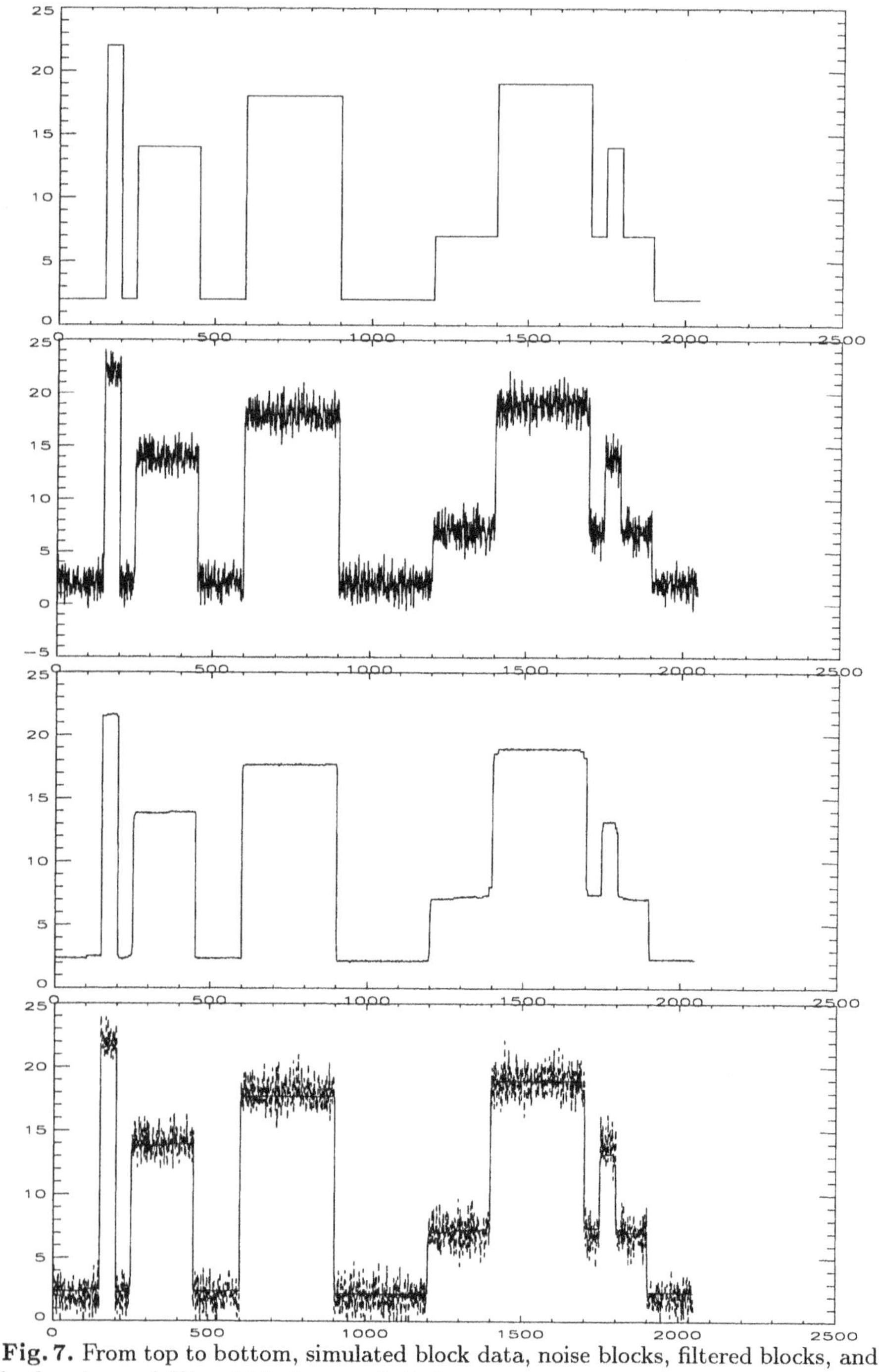

Fig. 7. From top to bottom, simulated block data, noise blocks, filtered blocks, and both noisy and filtered blocks overplotted.

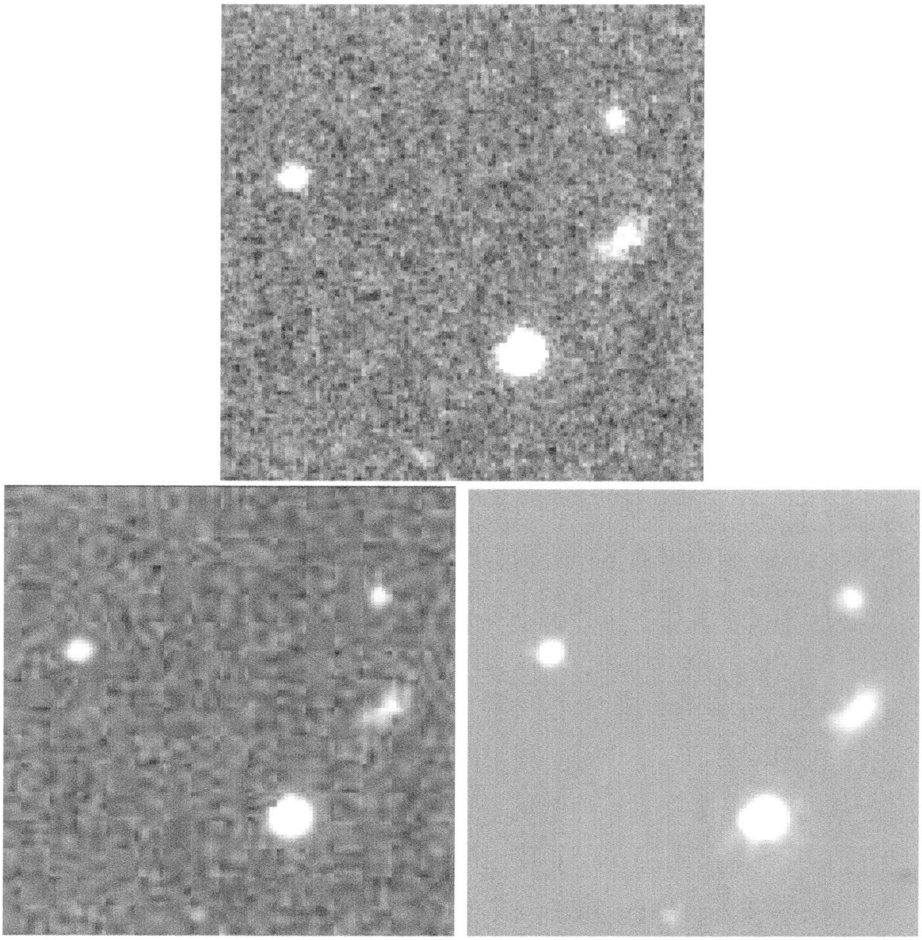

Fig. 8. Top, Original image: sub-image extracted from 1024x1024 patch, extracted in turn from the central region of ESO7992v. Bottom left, JPEG compressed image at 40:1 compression rate: small image patch extracted from central region of ESO7992v. Bottom right, PMT compressed image at 260:1 compression rate: small image patch extracted from central region of ESO7992v.

4 Decomposition of a signal into its main components

4.1 Introduction

The multiscale transform of an image by the à trous algorithm, or the Multiscale Median transform, produces, at each scale j, a set $\{w_j\}$. This has the same number of pixels as the image. The original image c_0 can be expressed as the sum of all the wavelet scales and the smoothed array c_J by the expression

$$c_{0,l} = c_{J,l} + \sum_{j=1}^{J} w_{j,l}. \tag{41}$$

Hence, we have a *multiscale pixel representation*, i.e. each pixel of the input image is associated to a set of pixels of the multiscale transform. A further step is to consider a *multiscale object representation*, which would associate to an object contained in the data, a volume in the multiscale transform. Such a representation obviously depends on the kind of image we need to analyze, and we present here a model which has been developed for astronomical data. It may however be used for other kind of data, if they present some similarities with astronomical data. We assume that an image I can be decomposed into a set of components:

$$c_0(k,l) = \sum_{i=1}^{N_o} O_i(k,l) + B(k,l) + N(k,l) \tag{42}$$

where N_o is the number of object, O_i are the objects contained in the data (stars galaxies, etc), B is the background image, and N is the noise.

To perform such a decomposition, we have to detect, to extract, to measure and to recognize the significant structures. This is done by first computing the multiresolution support of the image, and by applying a segmentation scale by scale. The wavelet space of a 2D direct space is a 3D one. An object has to be defined in this space. A general idea for object definition lies in the connectivity property. An object occupies a physical region, and in this region we can join any pixel to other ones. Connectivity in direct space has to be transported to wavelet transform space (WTS). In order to define the objects we have to identify the WTS pixels we can attribute to the objects. We describe in this section the different steps of this method.

4.2 Multiscale Vision Model Definition

The Multiscale Vision Model (MVM) [3,45] described an object as a hierarchical set of structures. It uses the following definitions:

- *significant wavelet coefficient*: a wavelet coefficient is significant when its absolute value is above a given detection limit. The detection limit depends on the noise model (Gaussian noise, Poisson noise, and so on). See Starck et al. [45] for a full description of the noise modeling.

- *structure*: a structure $\mathcal{S}_j$ is a set of significant connected wavelet coefficients at the same scale j.
- *object*: an object is a set of structures.
- *object scale*: the scale of an object is given by the scale of the maximum of its wavelet coefficients.
- *interscale relation*: the criterion allowing us to connect two structures into a single object is called the "interscale relation".
- *sub-object*: a sub-object is a part of an object. It appears when an object has a local wavelet maximum. Hence, an object can be composed of several sub-objects. Each sub-object can also be analyzed.

4.3 The multiresolution support and its segmentation

A multiresolution support of an image describes in a logical or Boolean way if an image I contains information at a given scale j and at a given position (x, y). If $M^{(I)}(j, x, y) = 1$ (or $=$ *true*), then I contains information at scale j and at the position (x, y). M depends on several parameters:

- The input image.
- The algorithm used for the multiresolution decomposition.
- The noise.
- All additional constraints we want the support to satisfy.

Such a support results from the data, the treatment (noise estimation, etc.), and from knowledge on our part of the objects contained in the data (size of objects, linearity, etc.). In the most general case, a priori information is not available to us.

The multiresolution support of an image is computed in several steps:

- Step one is to compute the wavelet transform of the image.
- Binarization of each scale leads to the multiresolution support (the binarization of an image consists of assigning to each pixel a value only equal to 0 or 1).
- A priori knowledge can be introduced by modifying the support.

This last step depends on the knowledge we have of our images. For instance, if we know there is no interesting object smaller or larger than a given size in our image, we can suppress, in the support, anything which is due to that kind of object. This can often be done conveniently by the use of mathematical morphology. In the most general setting, we naturally have no information to add to the multiresolution support.

The multiresolution support will be obtained by detecting at each scale the significant coefficients. The multiresolution support is defined by:

$$M(j, x, y) = \begin{cases} 1 & \text{if } w_j(x, y) \text{ is significant} \\ 0 & \text{if } w_j(x, y) \text{ is not significant} \end{cases} \tag{43}$$

In the case of Gaussian noise, it suffices to compare the wavelet coefficients $w_j(x, y)$ to a threshold level t_j. t_j is generally taken equal to $k\sigma_j$, where σ_j is the noise standard deviation at scale j, and k is chosen between 3 and 5. The value of 3 corresponds to a probability of false detection of 0.0027 for Gaussian statistics. If $w_j(x, y)$ is small, then it is not significant and could be due to noise. If $w_j(x, y)$ is large, it is significant:

$$\begin{array}{ll} \text{if } \mid w_j \mid \geq t_j & \text{then } w_j \text{ is significant} \\ \text{if } \mid w_j \mid < t_j & \text{then } w_j \text{ is not significant.} \end{array} \tag{44}$$

Many other kinds of noise can be considered in the wavelet space. See [45] for a review.

Multiresolution Support Segmentation The segmentation consists of labeling a boolean image (0 or 1). Each group of connected pixels having a "1" value gets a label value between 1 and L_{max}, L_{max} being the number of groups. This process is repeated at each scale of the multiresolution support. We define a "structure" S_j^i as the group of connected significant pixels which has the label i at a given scale j.

4.4 Interscale connectivity graph

An object is described as a hierarchical set of structures. The rule which allows us to connect two structures into a single object is called "interscale relation". Figure 9 shows how several structures at different scales are linked together, and form objects. We have now to define the interscale relation: let us consider two structures at two successive scales, S_j^k and S_{j+1}^l. Each structure is located in one of the individual images of the decomposition and corresponds to a region in this image where the signal is significant. Denoting p_m the pixel position of the maximum wavelet coefficient value of S_j^k, S_j^k is said to be connected to S_{j+1}^l if S_{j+1}^l contains the pixel position p_m (i.e. the maximum position of the structure S_j^k must also be contained in the structure S_{j+1}^l). Several structures appearing in successive wavelet coefficient images can be connected in such a way, which we call an object in the interscale connectivity graph.

4.5 Reconstruction

Hence, a set of structures defines an object $W = \{S_j^k, ...S_{j'}^{k'}\}$ which can be reconstructed separately from other objects. The coaddition of all reconstructed objects is a filtered version of the input data.

The reconstruction problem consists of searching for a signal O such that its wavelet coefficients are the same as those of the detected structures. If $\mathcal{T}$ describes the wavelet transform operator, and P_w the projection operator in

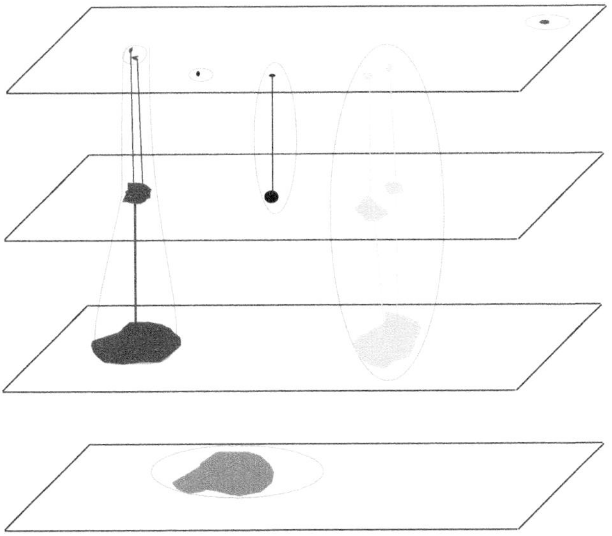

Fig. 9. Example of connectivity in wavelet space: contiguous significant wavelet coefficients form a structure, and following an interscale relation, a set of structures form an object. Two structures S_j, S_{j+1} at two successive scales belong to the same object if the position pixel of the maximum wavelet coefficient value of S_j is included in S_{j+1}.

the subspace of the detected coefficients (i.e. having set to zero all coefficients at scales and positions where nothing was detected), the solution is found by minimization of

$$J(O) = \| W - (P_w \circ \mathcal{T})O \| \tag{45}$$

where W represents the detected wavelet coefficients of the data. The object image size is generally relatively small, and this minimization does not present any computation problem. Furthermore, there is also no problem if instability, because only coefficients with high signal to noise ratio are used (generally larger than 3), and we do not need to add any regularization term.

Another interesting point to notice is the fact that the coaddition of all detected and reconstructed objects furnishes a filtered image of the input data. This restoration concept has been extended to the deconvolution problem [42]. Indeed, we can associate to an object its point spread function (PSF), and apply the following minimization:

$$J(O) = \| W - (P_w \circ \mathcal{T})(P * O) \| \tag{46}$$

where P is the point spread function. This combined detection-deconvolution method presents the following advantages:

1. Faint extended objects can be detected as well as point sources.
2. The analysis does not require background estimation.
3. Objects are easier to identify in the deconvolved map.
4. Morphological parameters (galaxy ellipticity and so on) are more accurate.
5. Since each object is deconvolved separately, a spatially variable point spread function can easily be taken into account.
6. Very large images can be deconvolved (each object has a small size).

Point 5 is of great interest. Deconvolution when the PSF is space variant is always problematic, while in this approach, we just have to associate to each object, the PSF at its center.

4.6 Example 1: Band Extraction

We have simulated a spectrum which contains an emission band at 3.50 μm and non stationary noise superimposed on a smooth continuum. The band is a Gaussian of width FWHM = 0.01 μm (FWHM = Full Width Half maximum), and normalized such that its maximum value equals ten times the local noise standard deviation.

Figure 10 (top) contains the simulated spectrum. The wavelet analysis results in the detection of an emission band at 3.50 μm above 3σ. Figure 10 (middle) shows the reconstruction of the detected band in the simulated spectrum. The real feature is over-plotted as a dashed line. Figure 10 (bottom) contains the original simulation with the reconstructed band subtracted. It can be seen that there are no strong residuals near the location of the band, which indicates that the band is well reconstructed. The center position of the band, its FWHM, and its maximum, can then be estimated via a Gaussian fit.

4.7 Example 2: Star subtraction in astronomical images

We applied the MVM to the galaxy NGC2997 (fig. 11 top left). Two images have been created by coadding objects detected from scale 1 to 2, and from scale 3 to 6. They are displayed respectively in figure 11 top right and bottom left. Figure 11 bottom right shows the difference between the input data and the image which contained the objects from scale 1 to 2. As we can see, all small objects have removed, and the galaxy can be better analyzed.

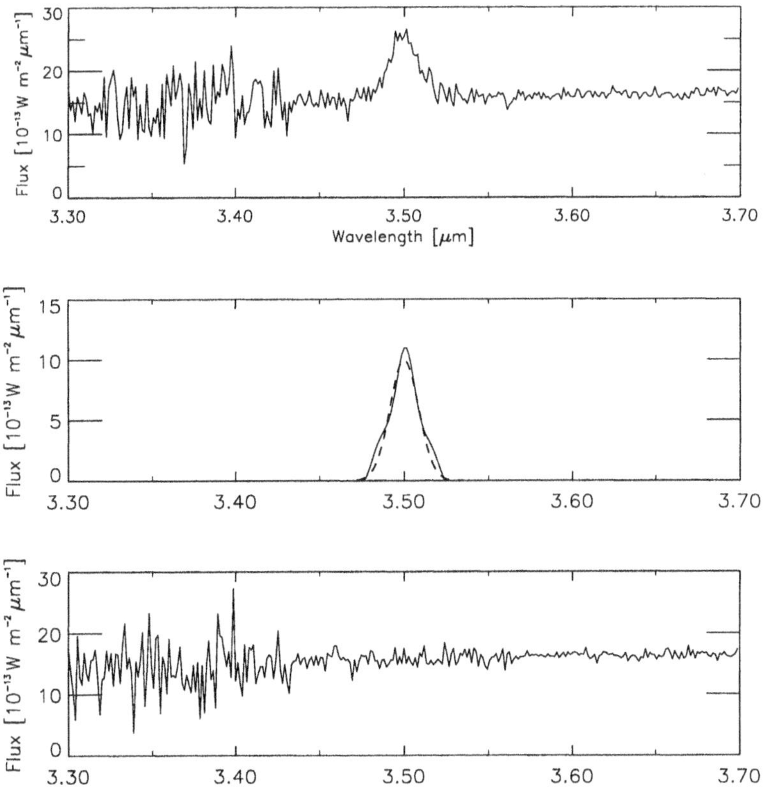

Fig. 10. Top: simulated spectrum. Middle: reconstructed simulated band (full line) and original band (dashed line). Bottom: simulated spectrum minus the reconstructed band.

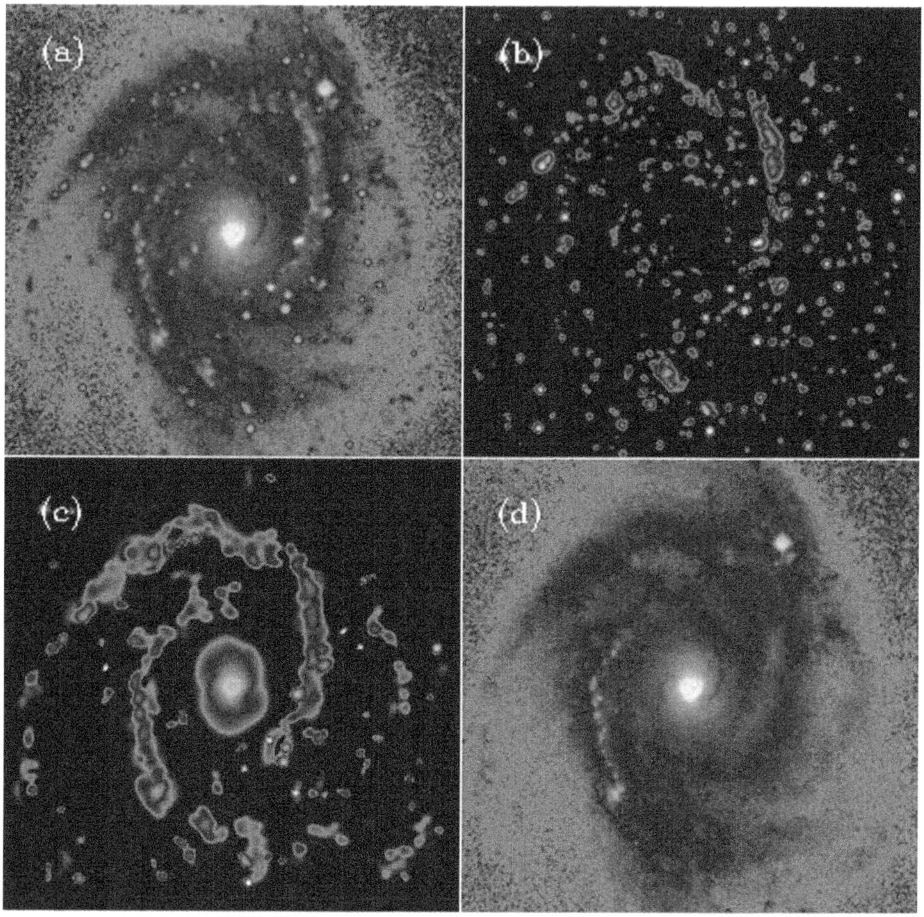

Fig. 11. (a) galaxy NGC2997, (b) objects detected from scale 1 to 2, (c) objects detected from scale 3 to 6, and (d) difference between (a) and (b). See also Color Plate 6.

5 Combined Transforms and Applications

5.1 Introduction

The content of an image is often complex, and there is not a single transform which is optimal to represent all the contained features. For example, the Fourier transform better represents some textures, while the wavelet transform better represents singularities. Even if we limit our class of transforms to the wavelet one, decision have to be taken between an isotropic wavelet transform which produce good results for isotropic objects (such stars and galaxies in astronomical images, cells in biological images, etc), or an orthogonal wavelet transform, which is better for images with edges. This has motivated the development of methods, such *matching pursuit* [34], or *basis pursuit* [12], which allow us to represent a signal with overcomplete dictionaries.

A dictionary $\mathcal{D}$ is defined as a collection of waveforms $(\phi_\gamma)_{\gamma \in \Gamma}$, and the goal is to obtain a representation of a signal s with a linear combination of a small number of basis such that:

$$s = \sum_\gamma a_\gamma \phi_\gamma \tag{47}$$

or an approximate decomposition

$$s = \sum_{i=1}^{m} a_{\gamma_i} \phi_{\gamma_i} + R^{(m)}. \tag{48}$$

Matching pursuit [34,31] method (MP) uses a greedy algorithm which adaptively refines the signal approximation with an iterative procedure:

- Set $s^0 = 0$ and $R^0 = 0$.
- Find the element $\alpha_k \phi_{\gamma_k}$ which best correlates with the residual.
- Update s and R:

$$\begin{aligned} s^{k+1} &= s^k + \alpha_k \phi_{\gamma_k} \\ R^{k+1} &= s - s^k. \end{aligned} \tag{49}$$

In case of non orthogonal dictionaries, it has been shown [12] that MP may spend most of the time correcting mistakes made in the first few terms, and therefore is suboptimal in term of sparsity.

Basis pursuit method [12] (BP) consists in minimizing the functional:

$$J(\alpha) = \| s - \Phi\alpha \|_2^2 + \lambda \| \alpha \|_1 . \tag{50}$$

Between all possible solutions, the chosen one has the minimum l^1 norm. This choice of l_1 norm is very important. A l_2 norm, as used in the method of frames [17], does not preserve the sparsity [12]. The minimization of equation 50, which involves linear programming, is however much more time consuming than MP, and remains prohibitive for large signals [31].

We present in the following an alternative approach, that we call *Combined Transforms Method* (CTM), which combines the different available transforms in order to benefit of the advantages of each of them.

5.2 Combined Transformation

Multiscale transforms which may be considered for a combined transformation are:

- Orthogonal wavelet transform. For filtering application, an undecimated wavelet transform must be preferred.
- A trous algorithm (isotropic objects)
- Ridgelet transform (lines or edges of a given size, and of any width)
- Curvelet transform (lines or edges of different sizes, and of any width)
- Fourier transform (oscillary pattern, etc)
- Multiresolution Median transform (strong singularities, such cosmic rays, very bright stars, saturated pixels, etc)
- Multiscale Morphological transform, wavelet packets, etc.

Depending of the content of the data, some of these transforms can be combined in order to get an optimal representation of all features contained in our data set. Compared to MP and BP, we are limited to the use of waveforms relative to the existing transforms, but in practice, these transforms furnish a very large panel of waveforms.

Two iterative methods, *soft-CTM* and *hard-CTM*, allowing us to realize such a combined transform, are describe in this section.

5.3 Soft-CTM

Noting $\mathcal{T}_1, ..., \mathcal{T}_{N_t}$ the N_t transform operators, a solution α is obtained by minimizing a functional of the form:

$$J(\alpha) = \parallel s - \sum_{k=0}^{N_t} \mathcal{T}_k^{-1} \alpha_k \parallel_2^2 + \lambda \sum_k \parallel \alpha_k \parallel_1 \tag{51}$$

where s is the original signal, and α_k are the coefficient obtained with the transform $\mathcal{T}_k$.

An simple algorithm to achieve such an solution is:

1. Set $\lambda = L_{\max}$.
2. While $\lambda >= 0$ do
3. For k = 1, .., N_t do
   - Calculate the residual $R = s - \sum_k \mathcal{T}_k^{-1} \alpha_k$.

- Calculate the transform $\mathcal{T}_k$ of the residual: $r_k = \mathcal{T}_k R$.
- Add the residual to α_k: $\alpha_k = \alpha_k + r_k$.
- Soft threshold the coefficient α_k with the λ threshold.

4. $\lambda = \lambda - \delta$, and goto 2.

Figure 12 illustrates the result in the case where the input image contains only lines and Gaussians. In this experiment, we have initialized $L_{\max}$ to 20, and δ to 2 (10 iterations). Two transform operators were used, the à trous wavelet transform and the ridgelet transform. The first is well adapted to the detection of Gaussian due to the isotropy of the wavelet function [45], while the second is optimal to represent lines [11]. Figure 12 top, bottom left, and bottom right represents respectively the original image, the reconstructed image from the à trous wavelet coefficient, and the reconstructed image from the ridgelet coefficient. The addition of both reconstructed images reproduces the original one.

In some specific cases where the data are sparse in all bases, it has been shown [29] that the solution is identical to the solution when using a $\| \cdot \|_0$ penalty term. This is however generally not the case. The problem we met in image restoration applications, when minimizing equation 51, is that both the signal and noise are split into the bases. The way the noise is distributed in the coefficients α_k is not known, and leads to the problem that we do not know at which level we should threshold the coefficients. Using the threshold we would have used with a single transform makes a strong over-filtering of the data. Using the l^1 optimization for data restoration implies to first study how the noise is distributed in the coefficients. The hard-CTM method does not present this drawback.

5.4 Hard-CTM

The following algorithm consists in hard thresholding the residual successively on the different bases.

1. Initialize $L_{\max}$ and the number of iterations Ni.
2. For noise filtering, estimate the noise standard deviation σ, and set $L_{\min} = k$. Otherwise, set $\sigma = 1$ and $L_{\min} = 0$.
3. Set $\delta_\lambda = \frac{L_{\max} - L_{\min}}{Ni}$.
4. Set $\lambda = L_{\max}$.
5. Set all coefficients α_k to 0.
6. While $\lambda >= L_{\min}$ do
7. for k = 1, .., N_t do
   - Calculate the residual $R = s - \sum_k \mathcal{T}_k^{-1} \alpha_k$.
   - Calculate the transform $\mathcal{T}_k$ of the residual: $r_k = \mathcal{T}_k R$.
   - For all coefficients $r_{k,i}$ do
     - Update the coefficients: if $\alpha_{k,i} \neq 0$ or $\mid r_{k,i} \mid > \lambda\sigma$ then $\alpha_{k,i} = \alpha_{k,i} + r_{k,i}$.

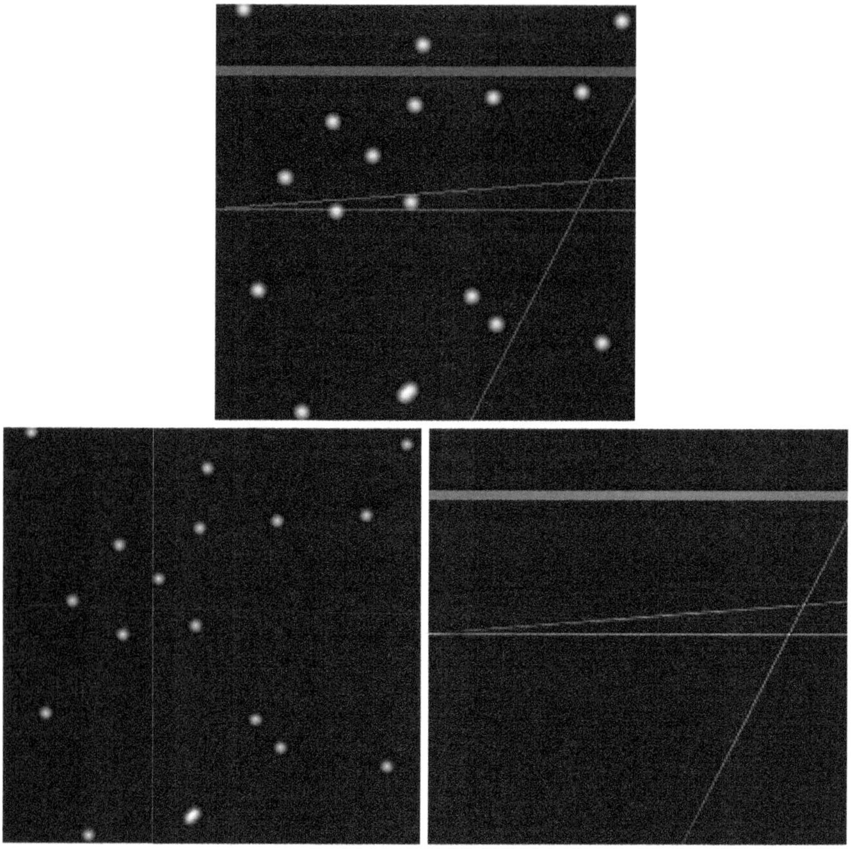

Fig. 12. Top, original image containing lines and gaussians. Botton left, reconstructed image for the à trous wavelet coefficient, bottom right, reconstructed image from the ridgelet coefficients.

8. $\lambda = \lambda - \delta_\lambda$, and goto 6.

For an exact representation of the data, k must be set to 0. Choosing $k > 0$ introduces a filtering. If a single transform is used, it corresponds to the standard $k\sigma$ hard thresholding.

It seems that starting with a high enough $L_{\max}$ and a high number of iterations would lead to the l^0 optimization solution, but this remains to be proved.

Figure 13 shows the decomposition of the classical `Lenna` image (512 by 512). Fig. 13 bottom left and right show respectively the reconstructed image from the wavelet and the ridgelet coefficients. The ridgelet transform [9] was used with a partitioning box of size 64. All the information relative to the long edges is in the ridgelet coefficients, while information relative to small edges or smooth regions is in the wavelet coefficients. The coaddition of both images reproduces the original one. This method does not present the drawback of the previous relative to the distribution of the noise in the different bases. As a hard thresholding is used, when a feature is detected in the residual by a given base, it is extracted. Hence, it cannot be diluted in several bases. This important for a restoration purpose. Indeed, it means that the detection level in a given base is unchanged by the fact that we use a set of bases. In our filtering simulations, we have set $L_{\max} = 40$, $Ni = 40$ and $L_{\min} = 4$.

Fig. 13. Combined transform using a ridgelet transform and an undecimated wavelet transform. Top, the classical Lenna image, bottom left, reconstructed image from the wavelet coefficients, and bottom right, reconstructed image from the ridgelet coefficients.

5.5 Applications

Elongated - point like object separation in astronomical images. Figure 14 shows the result of a decomposition of a spiral galaxy (NGC2997).

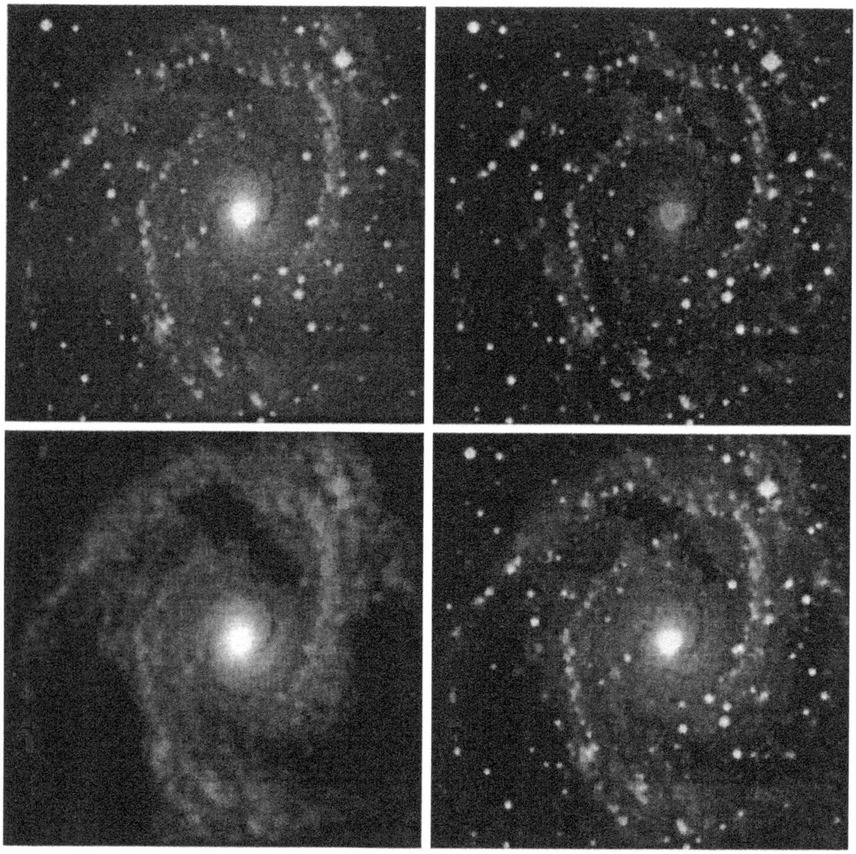

Fig. 14. Top left, galaxy NGC2997, top right reconstructed image from the à trous wavelet coefficients, bottom left, reconstruction from the ridgelet coefficients, and bottom right addition of both reconstructed images.

This image (figure 14 top left) contains many compact structures (stars and HII region), more or less isotropic, and large scale elongated features (NGC2997 spiral part). Compact objects are well represented by isotropic wavelets, and the elongated features are better represented by a ridgelet basis. In order to benefit of the optimal data representation of both transforms, the image has been decomposed on both the à trous wavelet transform and on the ridgelet transform by using the same method as described in section 5.4. When the functional is minimized, we get two images, and their coaddition is the filtered version of the original image. The reconstructions from the à trous coefficient, and from the ridgelet the ridgelet coefficient can be seen in figure 14 top right and bottom left. The addition of both images is presented in figure 14 bottom right.

Filtering. A simulation has been done with the classical `Lenna` 512×512 image. A white Gaussian noise with a standard deviation equal to 20 has been added to the original data. The noisy image has been filtered by the undecimated wavelet transform, the curvelet transform, and a combination of three methods (ridgelet, curvelet, and undecimated wavelet). Results are given in table 1

Fig. 15. Noisy image, and filtered images by the undecimated wavelet transform, the curvelet transform, and the combined method.

Figure 15 shows the noisy image (top left), the undecimated wavelet transform (UWT) filtering (top right), the curvelet filtering (bottom left), and the combined filtering (bottom right). Figure 16 shows a zoom of the same images. The residuals of the three methods are shown in figure 17.

For other examples, please check `http://www-stat.stanford.edu/~jstarck`.

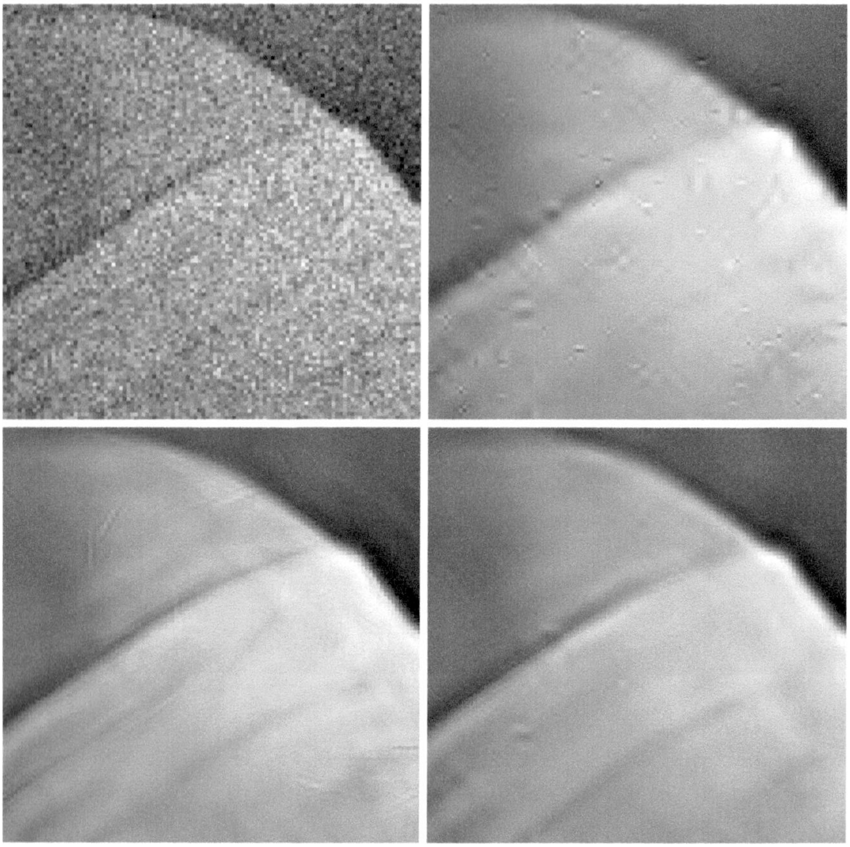

Fig. 16. Noisy image, and filtered images by the undecimated wavelet transform, the curvelet transform, and the combined method.

6 Conclusion

From these simulations, we can conclude that (i) the CTM-filtering leads to a real improvement for both the PSNR and the visual aspect, and (ii) only a combination of methods allows us to clean properly the residual.

The CTM approach costs obviously much more than a single transform, for both the computation time and the memory, but presents several advantages:

- It is easy to implement.
- Noise modeling can be addressed in the same way as when using a single transform.
- The reconstruction is straightforward.

We have seen in this paper that a separation point-like/extended objects can be done either by the use of a multiscale vision model, either by the

Fig. 17. Top, residual for the undecimated wavelet transform, and the curvelet transform, bottom, residual for the combined method.

combined transform method. But the first approach presents the drawback to need an explicit model of an object definition.

In our examples, we have built our combined transform from linear transforms (ridgelet, wavelets, curvelets), but any kind of nonlinear transform can be used in the same way.

Method	PSNR	Comments
Noisy image	22.13	
UWT7-9 + ksigma Hard thresh.	31.94	very few artifact
Curvelet (B=16)	31.95	no artifact
Combined filtering	32.17	no artifact

Table 1. PSNR after filtering the simulated image (Lena + Gaussian noise (sigma=20)). In the combined filtering, a ridgelet, a curvelet, and an undecimated wavelet transform have been used.

Acknowledgments. We wish to thank David Donoho and Emmanuel Candès for useful discussions and comments. This research was supported by National Science Foundation grants DMS 98–72890 (KDI) and DMS 95–05151; and by AFOSR MURI-95-P49620-96-1-0028.

References

1. E.H. Adelson, E. Simoncelli, and R. Hingorani. Optimal image addition using the wavelet transform. *SPIE Visual Communication and Image Processing II*, 845:50–58, 1987.
2. M. Antonini, M. Barlaud, P. Mathieu, and I. Daubechies. Image coding using wavelet transform. *IEEE Transactions on Image Processing*, 1(2):205–220, 1992.
3. A. Bijaoui and F. Rué. A multiscale vision model adapted to astronomical images. *Signal Processing*, 46:229–243, 1995.
4. F. Bonnarel, P. Fernique, F. Genova, J. G. Bartlett, O. Bienaymé, D. Egret, J. Florsch, H. Ziaeepour, and M. Louys. ALADIN: A reference tool for identification of astronomical sources. *Astronomical Data Analysis Software and Systems VIII, ASP Conference Series, Vol. 172. Ed. David M. Mehringer, Raymond L. Plante, and Douglas A. Roberts. ISBN: 1-886733-94-5 (1999), p. 229.*, 8:229–232, 1999.
5. E.J. Breen, R.J. Jones, and H. Talbot. Mathematical morphology: A useful set of tools for image analysis. *Statistics and Computing journal*, 10:105–120, 2000.
6. C.S. Burrus, R.A. Gopinath, and H. Guo. *Introduction to Wavelets and Wavelet Transforms*. Prentice Hall, 1998.
7. P.J. Burt and A.E. Adelson. The Laplacian pyramid as a compact image code. *IEEE Tranactions on Communications*, 31:532–540, 1983.
8. R. Calderbank, I. Daubechies, W. Sweldens, and B.-L. Yeo. Wavelet transforms that map integers to integers. *Appl. Comput. Harmon. Anal.*, 5(3):332–369, 1998.
9. E. J. Candès. Harmonic analysis of neural netwoks. *Applied and Computational Harmonic Analysis*, 6:197–218, 1999.
10. E.J. Candès and D. Donoho. Curvelets. Technical report, Statistics, Stanford University, 1999.

11. E.J. Candès and D. Donoho. Ridgelets: the key to high dimensional intermittency? *Phil. trans; R. Soc. Lond. A*, 357:2495–2509, 1999.
12. S.S. Chen, D.L. Donoho, and M.A. Saunder. Atomic decomposition by basis pursuit. *SIAM J. Math. Anal.*, 20(1):33–61, 1998.
13. C.H. Chui. *Wavelet Analysis and Its Applications.* Academic Press, 1992.
14. R. Claypoole, G.M. Davis, W. Sweldens, and R. Baraniuk. Nonlinear wavelet transforms for image coding via lifting. *IEEE Transactions on Image Processing*, 2000. submitted.
15. R.L. Claypoole, R.G. Baraniuk, and R.D. Nowak. Lifting construction of nonlinear wavelet transforms. In *IEEE International Conference IEEE-SP Time-Frequency and Time-Scale Analysis*, pages 49–52, 1998.
16. A. Cohen, I. Daubechies, and J.C. Feauveau. Biorthogonal bases of compactly supported wavelets. *Communications in Pure and Applied Mathematics*, 45:485–560, 1992.
17. I. Daubechies. Time-frequency localization operators: A geometric phase space approach. *IEEE Transactions on Information Theory*, 34(1):605–612, 1988.
18. I. Daubechies. *Ten Lectures on Wavelets.* Society for Industrial and Applied Mathematics (SIAM), Philadelphia, 1992.
19. I. Daubechies, I. Guskov, P. Schröder, and W. Sweldens. Wavelets on irregular point sets. *Phil. Trans. R. Soc. Lond. A*, To be published.
20. I. Daubechies and W. Sweldens. Factoring wavelet transforms into lifting steps. *J. Fourier Anal. Appl.*, 4(3):245–267, 1998.
21. D.L. Donoho. Nonlinear wavelet methods for recovery of signals, densities, and spectra from indirect and noisy data. *Proceedings of Symposia in Applied Mathematics*, 47, 1993.
22. D.L. Donoho. Nonlinear pyramid transforms based on median-interpolation. *SIAM J. Math. Anal.*, 60(4):1137–1156, 2000.
23. O. Egger and W. Li. Very low bit rate image coding using morphological operators and adaptive decompositions. In *IEEE International Conference ICIP-94 on Image Processing*, volume 2, pages 326–330, 1994.
24. D.A.F. Florêncio and R.W. Schafer. Perfect reconstructing nonlinear filter banks. In *IEEE International Conference ICASSP-96 on Acoustics, Speech, and Signal Processing*, volume 3, pages 1814–1817, 1996.
25. J. Goutsias and H. J. A. M. Heijmans. Nonlinear multiresolution signal decomposition schemes. part 1: Morphological pyramids. *IEEE Transactions on Image Processing*, 9(11):1862–1876, 2000.
26. F.J. Hampson and J.-C. Pesquet. A nonlinear subband decomposition with perfect reconstruction. In *IEEE International Conference ICASSP-96 on Acoustics, Speech, and Signal Processing*, volume 3, pages 1523–1526, 1996.
27. H.J.A.M. Heijmans and J. Goutsias. Multiresolution signal decomposition schemes. part 2: Morphological wavelets. *IEEE Transactions on Image Processing*, 9(11):1897–1913, 2000.
28. L. Huang and A. Bijaoui. Astronomical image data compression by morphological skeleton transformation. *Experimental Astronomy*, 1:311–327, 1991.
29. X. Huo. *Sparse Image Representation via Combined Transforms.* PhD thesis, Stanford Univesity, August 1999.
30. M. Louys, J.-L. Starck, S. Mei, F. Bonnarel, and F. Murtagh. Astronomical image compression. *Astronomy and Astrophysics, Suppl. Ser.*, 136:579–590, 1999.

31. S. Mallat. *A Wavelet Tour of Signal Processing.* Academic Press, 1998.
32. S. Mallat and F. Falzon. Analysis of low bit rate image transform coding. *IEEE Transactions on Signal Processing*, 46(4):1027–42, 1998.
33. S.G. Mallat. A theory for multiresolution signal decomposition: the wavelet representation. *IEEE Transactions on Pattern Analysis and Machine Intelligence*, 11(7):674–693, 1989.
34. S.G. Mallat and Z. Zhang. Atomic decomposition by basis pursuit. *IEEE Transactions on Signal Processing*, 41(12):3397–3415, 1993.
35. G. Matheron. *Elements pour une théorie des milieux poreux.* Masson, Paris, 1967.
36. G. Matheron. *Random Sets and Integral Geometry.* Willey, New York, 1975.
37. F. Murtagh, J.L. Starck, and M. Louys. Very high quality image compression based on noise modeling. *International Journal of Imaging Systems and Technology*, 9:38–45, 1998.
38. A. Said and W.A. Pearlman. A new, fast, and efficient image codec based on set partitioning in hierarchival trees. *IEEE Trans. on Circ. and Syst. for Video Tech.*, 6(3):243–250, 1996.
39. Peter Schröder and Wim Sweldens. Spherical wavelets: Efficiently representing functions on the sphere. *Computer Graphics Proceedings (SIGGRAPH 95)*, pages 161–172, 1995.
40. J. Serra. *Image Analysis and Mathematical Morphology.* Academic Press, London, 1982.
41. J.M. Shapiro. Embedded image coding using zerotrees of wavelet coefficients. *IEEE Transactions on Signal Processing*, 41(12):3445–3462, 1993.
42. J.L. Starck, A. Bijaoui, I. Vatchanov, and F. Murtagh. A combined approach for object detection and deconvolution. *Astronomy and Astrophysics, Suppl. Ser.*, 147:139–149, 2000.
43. J.L. Starck, E. Candès, and D.L. Donoho. The curvelet transform for image denoising. *IEEE Transactions on Image Processing*, 2001. submitted.
44. J.L. Starck and F. Murtagh. Multiscale entropy filtering. *Signal Processing*, 76(2):147–165, 1999.
45. J.L. Starck, F. Murtagh, and A. Bijaoui. *Image Processing and Data Analysis: The Multiscale Approach.* Cambridge University Press, Cambridge (GB), 1998.
46. J.L. Starck, F. Murtagh, and R. Gastaud. A new entropy measure based on the wavelet transform and noise modeling. *Special Issue on Multirate Systems, Filter Banks, Wavelets, and Applications of IEEE Transactions on CAS II*, 45(8), 1998.
47. J.L. Starck, F. Murtagh, B. Pirenne, and M. Albrecht. Astronomical image compression based on noise suppression. *Publications of the Astronomical Society of the Pacific*, 108:446–455, 1996.
48. G. Strang and T. Nguyen. *Wavelet and Filter Banks.* Wellesley-Cambridge Press, Box 812060, Wellesley MA 02181, fax 617-253-4358, 1996.
49. W. Sweldens. The lifting scheme: A construction of second generation wavelets. *SIAM J. Math. Anal.*, 29(2):511–546, 1997.
50. W. Sweldens and P. Schröder. Building your own wavelets at home. In *Wavelets in Computer Graphics*, pages 15–87. ACM SIGGRAPH Course notes, 1996.

Part II

Contributed Papers

Applications of Harten's Framework for Multiresolution: From Conservation Laws to Image Compression

F. Arandiga[1], G. Chiavassa[2], and R. Donat[1]

[1] Dept. Matematica Aplicada, University of Valencia, Spain.
[2] ESM2 and LATP, University of Marseilles II, France

Abstract. We briefly review Harten's framework for multiresolution decompositions and describe two situations in which two different instances of the general framework have been used with success.

In the numerical simulation of Hyperbolic Conservation Laws, the simple point-value setting with a linear centered interpolatory reconstruction is used to design a multilevel algorithm that effectively helps to reduce the computational expense associated to state of the art high resolution shock capturing schemes.

The possibility of using nonlinear, data dependent, reconstruction techniques is explored in image compression. For piecewise smooth images, ENO reconstruction techniques outperform classical algorithms based on biorthogonal wavelets.

1 Introduction: Harten's Multiresolution Framework

Multiscale decompositions aim at a 'rearrangement' of the information contents in a set of discrete data. To achieve such a 'rearrangement', Harten's general framework for multiresolution [16] relies on two operators that define the basic interscale relations: Decimation and Prediction. These operators act on finite dimensional linear vector spaces, V^j, that represent the different resolution levels (k increasing implies more resolution)

$$\text{(a) } D_j^{j-1} : V^j \to V^{j-1}, \qquad \text{(b) } P_{j-1}^j : V^{j-1} \to V^j, \tag{1}$$

and must satisfy two requirements of algebraic nature: D_j^{j-1} needs to be a *linear* operator and $D_j^{j-1} P_{j-1}^j = I_{V^{j-1}}$, i.e. the identity operator on the lower resolution level represented by V^{j-1}.

Using these two operators, a vector (i.e. a discrete sequence) $v^j \in V^j$ can be decomposed and reassembled as follows

$$\begin{array}{l} v^j \to v^{j-1} = D_j^{j-1} v^j \\ \quad \searrow \quad e^j = v^j - P_{j-1}^j v^{j-1} \end{array}, \qquad v^j = P_{j-1}^j v^{j-1} + e^j$$

The vector $D_j^{j-1} v^j = v^{j-1}$ represents the discrete information contents of v^j at the lower resolution level $j-1$, while $e^j = (I_{V^j} - P_{j-1}^j D_j^{j-1}) v^j := Q_j v^j$

represents de *prediction error*, that is the error committed in trying to predict v^j from the low-resolution vector v^{j-1} via P^j_{j-1} .

Notice that the consistency requirement $D^{j-1}_j P^j_{j-1} = I_{V^{j-1}}$ implies that $D^{j-1}_j e^j = 0$, therefore its representation as a vector of V^j is redundant. This redundancy can be eliminated by expressing e^j in terms of a basis of the null space $\mathcal{N}(D^{j-1}_j)$. The consistency requirement also implies that D^{j-1}_j has full rank, thus $\dim\mathcal{N}(D^{j-1}_j) = \dim V^j - \dim V^{j-1}$. Let us introduce the operator G_j that computes the coordinates of e^j in a basis of $\mathcal{N}(D^{j-1}_j)$ and call $d^j = G_j e^j$, then the sets v^j and $\{v^{j-1}, d^j\}$ have the same cardinality and are algebraically equivalent. The one to one correspondence between these two sets can be described using one more operator E_j, the canonical injection $\mathcal{N}(D^{j-1}_j) \hookrightarrow V^j$,

$$P^j_{j-1} v^{j-1} + E_j d^j = v^j \quad \longleftrightarrow \quad \{d^j, v^{j-1}\} \begin{cases} v^{j-1} = D^{j-1}_j v^j \\ d^j = G_j Q_j v^j. \end{cases} \tag{2}$$

This purely algebraic description can be recursively applied to 'rearrange' the information contents of a discrete sequence v^L, containing information on a very fine scale, as v^0, the information on a much coarser scale obtained by successive decimation of v^L, plus a sequence of (non-redundant) prediction errors at each resolution level

$$v^L \leftrightarrow M v^L = (v^0, d^1, \ldots, d^L), \qquad \begin{array}{ccccccc} v^L & \rightarrow & v^{L-1} & \rightarrow & v^{L-2} & \rightarrow & \ldots v^0 \\ & \searrow & d^L & \searrow & d^{L-1} & \searrow & \ldots d^1 \end{array}$$

In a multiresolution scheme *à la Harten*, D^{j-1}_j and P^j_{j-1} are constructed using two operators that relate discrete data to signals in a particular function space. These basic 'building blocks' are the *Discretization* and *Reconstruction* operators

$$\text{(a) } \mathcal{D}_j : \mathcal{F} \rightarrow V^j, \qquad \text{(b) } \mathcal{R}_j : V^j \rightarrow \mathcal{F} \tag{3}$$

The discretization operators $\mathcal{D}_j$ yield discrete information at the resolution level specified by V^j, and the reconstruction operators $\mathcal{R}_j$ relate a set of discrete data in V^j to functions in $\mathcal{F}$.

These operators also have to satisfy some basic requirements of algebraic nature: $\mathcal{D}_j$ must be a linear operator and $\mathcal{D}_j(\mathcal{F}) = V^j$. In addition, $\mathcal{D}_j \mathcal{R}_j = I_{V^j}$, which is again a consistency relation.

It is also necessary that the sequence of discretization operators $\{\mathcal{D}_j\}$ be *nested*, that is

$$\mathcal{D}_j f = 0 \Rightarrow \mathcal{D}_{j-1} f = 0, \quad \forall f \in \mathcal{F}. \tag{4}$$

Given a nested sequence of linear discretization operators $\{\mathcal{D}_j\}$ (3)-(a), and a sequence of consistent reconstruction operators, $\{\mathcal{R}_j\}$ (3)-(b), the interscale operators are defined as follows:

$$D^{j-1}_j = \mathcal{D}_{j-1} \mathcal{R}_j, \qquad P^j_{j-1} = \mathcal{D}_j \mathcal{R}_{j-1} \tag{5}$$

Note 1. The dependence of D_j^{j-1} on $\mathcal{R}_j$ is 'fictitious'. For a nested sequence of discretization, it is easy to prove that $\mathcal{D}_{j-1}\mathcal{R}_j \equiv \mathcal{D}_{j-1}\tilde{\mathcal{R}}_j$ for any two consistent sequences of reconstruction operators. Thus, D_j^{j-1} is completely specified by the sequence of discretization operators

2 The Role of $\mathcal{R}_j$: Linear versus Nonlinear

The description of the prediction operator as $P_{j-1}^j = \mathcal{D}_j\mathcal{R}_{j-1}$ opens up a tremendous number of possibilities in designing multiresolution schemes specifically adapted to particular applications. In principle, the general framework allows for nonlinear, data dependent, reconstruction operators, which in turn lead to nonlinear prediction operators and, thus, to nonlinear multiresolution transforms.

The reconstruction process becomes, then, a key step, while the discretization operator selects a particular setting for the multiresolution decomposition. Let us consider, for example the 1D signal displayed in Fig. 1 (solid line): a sinusoidal discontinuous function. There are several ways to assign a set of discrete data to this signal, i.e. we can choose different discretization operators to obtain a set of discrete data out of this function, and the choice of $\mathcal{D}_j$ determines the setting for the multiresolution representation. In what follows we shall briefly describe the two settings we shall use in the applications. In both applications we use 2D multiresolution transforms that are obtained as tensor products of 1D transforms, therefore, we simply review the 1D settings. For a full description of these settings we refer the reader to [14,16,6,23].

2.1 The Point-value setting: Interpolatory Multiresolution

Let us consider a set of nested dyadic grids $\{\mathcal{G}_l\}_{l=0}^L$ obtained by a dyadic coarsening of an initial fine grid $\mathcal{G}_L$ on a finite interval. The discretization operator that characterizes the setting is simply the point-wise evaluation at the points of each grid, i.e. $\mathcal{D}_j(f) = \{v_i^j\}$ where $v_i^j = f(x_i^j)$ and $\mathcal{G}_j = \{x_i^j\}$. Any *consistent* sequence of reconstruction operators must satisfy

$$\mathcal{D}_j\mathcal{R}_j v^j = v^j \quad \longrightarrow \quad (\mathcal{R}_j v^j)(x_k^j) = v_k^j = f(x_k^j),$$

that is, the reconstruction operators must interpolate the data on which they act.

In this framework, the one to one correspondence in (2) can be written as

$$\left\{ \begin{aligned} v_i^{l-1} &= v_{2i}^l \\ d_i^l &= v_{2i+1}^l - \mathcal{I}(x_{2i+1}^l, v^{l-1}) \end{aligned} \right\} \quad \longleftrightarrow \quad \left\{ \begin{aligned} v_{2i}^l &= v_i^{l-1} \\ v_{2i+1}^l &= \mathcal{I}(x_{2i+1}^l, v^{l-1}) + d_i^l \end{aligned} \right\}$$

where the notation $\mathcal{I}(x, v^j)$ has been used instead of $\mathcal{R}_j v^j(x)$. Notice that the prediction errors are simply interpolation errors and that they only need to be computed at odd points on the fine grid $\mathcal{G}_l$.

Data independent interpolatory techniques lead to linear multiresolution transforms, while data dependent techniques lead to nonlinear multiresolution transforms. In our first application we consider $\mathcal{I}(x, v^j)$ to be a piecewise polynomial function in which the polynomial piece corresponding to an interval, say $[x_{i-1}^j, x_i^j]$, on $\mathcal{G}_j$ has degree 3 and interpolates 4 values centered around the considered interval, i.e. $\{v_{i-2}^j, v_{i-1}^j, v_i^j, v_{i+1}^j\}$. This is an example of *centered* piecewise polynomial technique, a class of linear (data-independent) interpolatory techniques widely used in numerical analysis. Figure 1(a) shows $\mathcal{I}(x, v^l)$ where $v_i^l = f(x_i^l)$ and $\mathcal{G}_l = \{x_i^l\}$ is a uniformly spaced mesh with spacing $h_l = 1/16$ (the discrete values are shown by dots in the figures). The *quality* of the approximation should be compared with the one obtained in Fig. 1(b), where we used a nonlinear Essentially Non Oscillatory (ENO henceforth) interpolatory technique [12,6] of the same degree. Observe that the $\mathcal{I}^{ENO}(x, v^l)$ keeps a fully accurate approximation right up to the interval, on $\mathcal{G}_l$, where the discontinuity is located.

(a) (b)

Fig. 1. Solid line: A piecewise smooth signal (solid line) and its point-values on a uniform grid (dots on solid line). Dotted lines: (a) A linear Interpolatory reconstruction; (b) A nonlinear ENO interpolatory reconstruction.

2.2 The Cell-average Setting

As before, we consider a sequence of nested dyadic grids on a given interval, [0, 1] for simplicity. At each resolution level, the discrete values are defined as follows,

$$\mathcal{D}_j : L^1[0,1] \longrightarrow V^j, \qquad (\mathcal{D}_j f)_i = \frac{1}{h_j} \int_{x_{i-1}^j}^{x_i^j} f(x)dx,$$

where V^j is a space of sequences whose dimension is equal to the number of points on the jth grid and h_j is the uniform mesh spacing on $\mathcal{G}_j = \{x_i^j\}_i$.

The reconstruction operators (linear or nonlinear), have to satisfy the consistency requirement, $\mathcal{D}_j\mathcal{R}_j = I_{V^j}$, i.e.

$$v_i^j = \frac{1}{h_j}\int_{x_{i-1}^j}^{x_i^j} \mathcal{R}_j(x; v^j)dx. \qquad (6)$$

In one dimension, there is a simple way to obtain reconstruction operators satisfying (6). The basic observation is that if $\{v_i^j\}$ represent the cell averages of a function $f(x) \in L^1[0,1]$, we can define

$$F_i^j := h_j \sum_{s=1}^{i} v_s^j = \int_0^{x_i^j} f(x)dx = F(x_i^j)$$

where $F(x) := \int_0^x f(x)dx$ is a primitive of $f(x)$. Using the sequence $\{F_i^j\}_i$, we approximate $F(x)$ by interpolation, i.e. construct $\mathcal{I}(x, F^j)$, and obtain $\mathcal{R}_j$ satisfying (6) as follows [16,6]:

$$\mathcal{R}_j(x, v^j) = \frac{d}{dx}\mathcal{I}(x, F^j) \qquad (7)$$

In Fig. 2, the dots, which have been placed at the center of each subinterval, represent the cell-averages of $f(x)$ on the same grid as in Fig. 1. The dotted line on Fig. 2-(a) displays $\mathcal{R}(x, v^j)$ in (7) when $\mathcal{I}(x; F^j)$ is a centered piecewise polynomial function with polynomial pieces of degree 3. Again, the *quality* of the approximation should be compared with those in Figs. 2-(b) and (c) in which nonlinear interpolation techniques have been used in the reconstruction process. In Fig. 2(b), we used $\mathcal{I}^{ENO}(x; F^j)$ (same degree), and in Fig. 2(c) we used $\mathcal{I}^{ENO-SR}(x; F^j)$ a modification of the ENO interpolatory technique using Harten's *Subcell Resolution* technique [13]. The SR modification is based on the fact that the mean value of a function on an interval contains information on the *exact location* of a single discontinuity located in that interval. This information is then used to design a reconstruction which is fully accurate right up to the discontinuity (see [13,6] for specific details).

Figures 1 and 2 show that nonlinear techniques can be a powerful tool in approximating discontinuous functions. In particular, when the data are interpreted as cell-averages of a given piecewise smooth function, the combination of Harten's Subcell Resolution technique with the ENO interpolation (ENO-SR) allows for an almost perfect description of piecewise smooth functions with (sufficiently separated) jump discontinuities.

2.3 The question of Stability

Clearly, the purpose behind a multiresolution decomposition is not just to decompose and reconstruct, the goal is to do some processing (truncation or quantization) between the decomposition and reconstruction stages. Starting

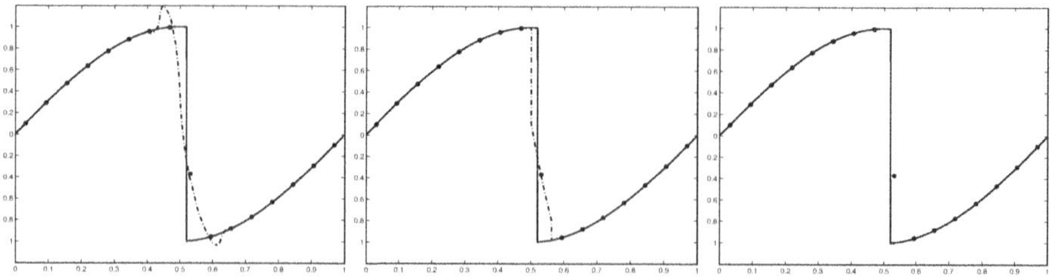

(a) (b) (c)

Fig. 2. Solid line: A piecewise smooth function and its cell-averages on a uniform grid (dots on solid line). Dotted lines: (a) Linear reconstruction technique; (b) Nonlinear ENO technique; (c) Nonlinear ENO-SR technique

with a sequence v^L, we compute its multiresolution decomposition $Mv^L = (v^0, d^1, \ldots, d^L)$ and *process* it to obtain a *compressed* representation that is to be stored, transmitted or used in some way. The processing step involves a sequence of pre-determined tolerance levels $\bar{\epsilon} = (\epsilon_0, \epsilon_1, \ldots, \epsilon_L)$ and it is designed in such a way that the compressed representation $M_{\bar{\epsilon}} v^L = (\hat{v}^0, \hat{d}^1, \ldots, \hat{d}^L)$ satisfies $||v^0 - \hat{v}^0|| \leq \epsilon_0$ and $||d^j - \hat{d}^j|| \leq \epsilon_j$.

The goal is to be able to control *also* the difference between the 'uncompressed' sequence $\hat{v}^L = M^{-1}(\hat{v}^0, \hat{d}^1, \ldots, \hat{d}^L)$ and the original one. Stability is a crucial issue in examining the effect of using 'perturbed values' $\hat{v}^0, \hat{d}^j$ instead of v^0, d^j in the input of M^{-1}. To ensure stability, we need to have

$$||v^L - M^{-1}(\hat{v}^0, \hat{d}^1, \ldots, \hat{d}^L)|| = \sigma(\bar{\epsilon}), \qquad \text{with } \lim_{\bar{\epsilon} \to 0} \sigma(\bar{\epsilon}) = 0$$

The stability of the repeated application of (2) is seen to be tightly connected to the properties of the reconstruction sequence [16]. When $\mathcal{R}_j$ are linear operators (i.e. data independent), the multiresolution transform can be seen as a change of basis functions in V^L, and the stability of M^{-1} can be studied with linear techniques as in wavelets or subdivision schemes [16,4]. However, when $\mathcal{R}_j$ are data-dependent, nonlinear operators, these techniques no longer apply and one needs a different approach to guarantee stability and control of the error.

In his very first paper on multiresolution [14], Harten proposed an interesting technique to achieve direct control on the difference $||v^L - M^{-1}\hat{v}^L||$. In Harten's algorithms for error-control the full sequence of decimated values $v^{L-1}, \ldots, v^1$ is computed first. Then, we start at the coarsest level and apply some perturbation process (usually thresholding or quantization) to v^0 to obtain $\hat{v}^0$ in such a way that $||v^0 - \hat{v}^0|| \leq \epsilon_0$, the user-set compression parameter for the coarsest level. From this point on, the idea consists in defining processed details $\hat{d}_j$ weaving them together with *processed values* $\hat{v}^j$, from coarse to fine scales in such a way that the accumulated compression error, $||v^j - \hat{v}^j||$, can be controlled at each step. For $j = 1, \ldots, L$ the processed details $\hat{d}^j$ represent a perturbation (again it is usually quantization or truncation) of the *real* prediction error $v^j - P^j_{j-1}\hat{v}^{j-1}$ using the processed data $\hat{v}^{j-1}$, the processed data $\hat{v}^j$ is then computed as $\hat{v}^j = P^j_{j-1}\hat{v}^{j-1} + E_j\hat{d}^j$.

It is important to notice that the prediction errors $v^j - P^j_{j-1}\hat{v}^{j-1}$ do not necessarily belong to the null space $\mathcal{N}(D^{j-1}_j)$ (in contrast to $e^j = v^j - P^j_{j-1}v^{j-1}$), and therefore one cannot simply define $\hat{d}^j$ as a perturbation of $G_j(v^j - P^j_{j-1}\hat{v}^{j-1})$. In the point-value and cell-average frameworks, there is a rather natural way to define the processed details $\hat{d}^j$ from $v^j - P^j_{j-1}\hat{v}^{j-1}$. In the point-value setting, the compressed details $\hat{d}^j_i$ are computed by processing (usually quantization or truncation) the differences $\tilde{d}^j_i = (v^j - P^j_{j-1}\hat{v}^{j-1})_{2i+1}$, but in the cell average setting $\hat{d}^j_i$ is obtained by processing instead $\tilde{d}^j_i = 1/2[(v^j - P^j_{j-1}\hat{v}^{j-1})_{2i} - (v^j - P^j_{j-1}\hat{v}^{j-1})_{2i+1}]$. In this last case, the coefficients $\tilde{d}^j_i$ coincide with the prediction errors e^j_{2i+1} only when $\epsilon_l = 0$, $l = 0, \ldots, L$, i.e. no compression takes place. If this is not the case, they represent a mean value between the prediction errors at odd and even points.

All in all, Harten's error-control strategy involves a modification of the analysis algorithm (the direct transform) that serves to ensure a prescribed accuracy after the application of M^{-1}. We refer the reader to [14], and more specifically to [6] for full details on the 1-D algorithms and to [3] for a full description of the 2-D error-control algorithms as well as the error bounds on the compression error $||v^L - M^{-1}(\hat{v}^0, \hat{d}^1, \ldots, \hat{d}^L)||$ in various norms.

3 The Point-Value Setting: A Multilevel Strategy for the Numerical Simulation of Hyperbolic Conservation Laws

3.1 Introduction and Motivation

The solutions of Hyperbolic systems of Conservation Laws (HCL henceforth) are known to develop discontinuities (shock waves and contact discontinuities) in finite time. The discontinuities may develop spontaneously, and this is a major difficulty in designing numerical schemes that can accurately approximate these solutions.

Numerical schemes that attempt to obtain high order numerical approximations by traditional techniques, i.e. based on Taylor expansions, lead to oscillations around the discontinuities in the solution [18]. This oscillatory behavior is not only highly inaccurate, and very unpleasant, but also extremely dangerous (see [17]).

State of the art numerical schemes for HCL combine high order approximation in smooth regions with sharp transition profiles at discontinuities, and are known as High Resolution Shock Capturing (HRSC henceforth) methods. There are nowadays a number of HRSC schemes that provide reliable numerical approximations in many situations, however, the *perfect scheme* has yet to be found (and may very well not even exist [19]), and the search continues for reliable schemes that can cope with (at least some of) the flaws displayed by well known HRSC schemes actually in use.

To keep our description simple, we shall consider a system of conservation laws in 1D

$$u_t + f(u)_x = 0 \tag{8}$$

where u is the vector of conserved variables and $f(u)$ the flux vector. The basic structure of a HRSC scheme for (8) on a uniform grid of mesh size δx is quite simple. In a semi-discrete formulation, it looks like

$$\frac{d}{dt}U_i + (\mathcal{D}U)_i = 0 \tag{9}$$

with the *numerical divergence* $(\mathcal{D}U)_i$ in *conservation form*

$$(\mathcal{D}U)_i = \frac{F_{i+1/2} - F_{i-1/2}}{\delta x}. \tag{10}$$

Here $F_{i+1/2}$ is the *numerical flux function*: the trademark of the scheme and where most of the computational work goes. Some HRSC schemes demand the computation of one spectral decomposition (or even two) of the Jacobian matrix $A(u) = \partial f / \partial u$ at each cell interface. In addition, some sort of reconstruction technique, possibly nonlinear, is also involved in the computation of $F_{i+1/2}$ if one wants to obtain high resolution in the presence of discontinuities.

The high resolution, un-oscillatory behavior that characterizes a HRSC scheme is a direct consequence of the sophisticated, highly complex and often very expensive numerical flux function used. However, every user of HRSC codes is painfully aware that these expensive flux evaluations are only necessary in a neighborhood of an existing discontinuity or in a region where compression, leading to shock formation, is taking place. In smooth regions, sufficiently far from discontinuities (existing or ready to form), a more traditional approach would lead to equally good results.

Any tool that can determine the regions of non-smooth behavior from a careful examination of the discrete data at a given time step could, in principle, be used to reduce the computational expense associated to a HRSC scheme. It is absolutely imperative to use a sophisticated flux formula in critical regions if one wants the full benefits of the scheme, but in smoothness regions one can do something cheaper, as long as it is equally accurate.

The link with multiresolution decompositions comes out naturally, and it was first explored by Harten [15] and pursued by other authors [1,7,10,22]. In [8], we propose an attractive alternative to the algorithms developed in these works.

3.2 The Basic Strategy: A Multilevel Evaluation of the Numerical Divergence

Let us consider again the simplest fully discrete realization of (9),

$$U_i^{n+1} = U_i^n - \delta t(\mathcal{D}U)_i^n \tag{11}$$

on a grid $\mathcal{G}_L$, where the HRSC scheme we have chosen gives us a numerical solution which we consider completely satisfactory. Our goal is to compute this numerical solution (or a 'high resolution' sufficiently close analog) at a much lower cost. In order to do this, we design a strategy that will allow us to obtain the values $\mathcal{D}(U^n)_i$, necessary to compute U^{n+1}, the numerical solution at the next time step, using the numerical flux computation prescribed by the HRSC *only when strictly necessary.*

The main ingredients in our multilevel scheme are the following:

• The multiresolution analysis of U^n, the numerical data available at the beginning of the time step, i.e. MU^n.

The multiresolution framework used to analyze the smoothness of the data is the simplest of all multiresolution frameworks: the interpolatory multiresolution framework. In this case, the *scale coefficients* (prediction errors) are simply interpolation errors, thus, its relation with the smoothness of the underlying function is well understood. Since we are using uniform grids, with h_j being the mesh spacing in grid $\mathcal{G}_j$, and a centered interpolatory technique of degree 3 (see section 2.1), we know that

$$d_i^j = O(h_{j-1}^4) \qquad \text{in smooth regions}$$
$$d_i^j = O(1) \qquad \text{around jump discontinuities}$$

• The thresholding algorithm which associates to each scale coefficient (therefore to a particular location in space) a boolean flag b_i^l, whose value (0 or 1) will determine the choice of procedure in the evaluation of the numerical divergence at that location. The thresholding algorithm uses the smoothness information contained in d_i^l to mark out the critical regions (discontinuities and compression regions) of *both* U^n *and* U^{n+1} (unknown at this stage of the computation). The details of the thresholding algorithm can be found in [8], but it is worth mentioning here that the propagation of information is limited by the CFL condition number *on the finest grid*, which bounds the location of moving singularities from one time step to the next.

It should be mentioned that since the thresholding algorithm measures the size of interpolation errors, the thresholding parameters should be solution-dependent (and independent of the geometry of the computational domain).

• The multilevel evaluation of the numerical divergence To compute the values $(\mathcal{D}U)_i$ on the finest grid $\mathcal{G}_L$, we start by computing them on the lowest resolution grid $\mathcal{G}_0$ using our chosen HRSC scheme . Once the numerical divergence is known on $\mathcal{G}_{l-1}$, i.e. $(\mathcal{D}U)^{l-1}$ has been computed, we only need to determine $(\mathcal{D}U)^l$ at points in $\mathcal{G}_l \setminus \mathcal{G}_{l-1}$. To do this we examine the flag vector:

- if $b_i^l = 1$ compute $(\mathcal{D}U)_i^l$ directly with the HRSC scheme
- if $b_i^l = 0$, $(\mathcal{D}U)_i^l = \mathcal{I}(x_i^l, (\mathcal{D}U)^{l-1})$

letting l go from 1 to L gives the values of $\mathcal{D}(U^n)$ on $\mathcal{G}_L$.

3.3 Numerical Tests: Quality and Efficiency

We remark that the purpose of the multilevel strategy is to be as close as possible to the *reference simulation*, i.e. to the solution that would be obtained with the underlying HRSC scheme on the same fine grid. Hence, the numerical results of the multilevel scheme are to be evaluated in terms of *Quality*, i.e. the difference between the outcome of the multilevel algorithm and the reference simulation, and *Efficiency*, i.e. net savings of the multilevel computation with respect to the reference simulation.

In [8], we use (a 2D tensor product version of) the multilevel scheme to compute numerical approximations to the solution of several benchmark tests for the Euler Equations of Gas Dynamics in 2D. The numerical results reported in [8] indicate that the quality of the multilevel approximation is directly controlled by the tolerance parameter used in the thresholding algorithm. The thresholding tolerance is a user-dependent parameter which is set at the beginning of the computation according to the order of the interpolation technique used in the multiresolution transform, the mesh spacing of the finest grid, and the solution itself.

The efficiency of the multilevel scheme is, of course, problem dependent. In a typical simulation involving a system of HCL, the non-smooth structures of the solution occupy a small percentage of the total computational domain, and in this situation the multilevel strategy is very efficient. We refer again to [8] for a specific evaluation of the efficiency of the multilevel scheme in several situations.

Figures 3 and 4, display high resolution numerical approximations to the solution of 2D-Riemann problems for the Euler equations of gas dynamics in 2D. The initial data is given on the unit square and involves 4 different constant states, one in each quadrant (see [21] for the initial data and a description of the exact the solution).

In Fig. 3, the initial configuration gives rise to 4 strong shocks plus a number of interior structures that are very hard to compute accurately unless a high-order, high-resolution scheme is used on a fine mesh. The numerical results shown have been obtained with our multilevel strategy and Marquina's third order scheme [11] as the underlying HRSC scheme. Figure 3-(b) shows a zoom of the region at the left corner of the domain, where Kelvin-Helmholtz type instabilities start to develop, due to the low numerical viscosity in the simulation. It is worth mentioning that the percentage of mesh points on the finest grid where the numerical divergence has been computed with the HRSC scheme grows from 3% to 22% for this test case. In practice, this leads

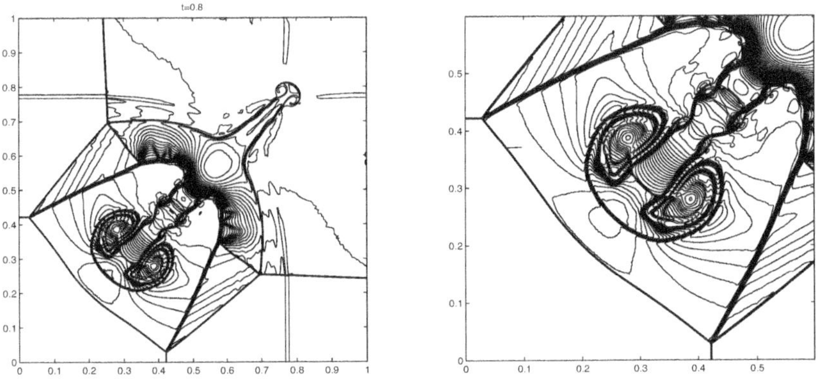

(a) (b)

Fig. 3. 4-shock configuration: High Resolution numerical approximation to the density obtained with the Multilevel-Marquina scheme on a 1024 × 1024 grid

to a decrease in cpu-time by a factor of 4.2, with respect to the reference simulation (i.e. without multiresolution).

Decreasing the numerical viscosity in the simulation can have a dramatic effect on the computational results. The initial configuration for Fig. 4 leads to the development of 4 contact discontinuities separating the initial states. Refining the grid, and thus reducing the numerical viscosity in the simulation, accentuates the Kelvin-Helmholtz instabilities at contact discontinuities. The 'roll-up' typical of this type of instability is more visible in finer grids. In this case, the percentage of HRSC divergence evaluation goes from 4% to 28 % for the 1024 × 1024 grid, and an associated decrease in cpu time by a factor of 3.6. In practice, this means that the simulation (on a PC at 350Mhz) can be run in a few days instead of a few weeks.

One of the nicest features of our multilevel algorithm is its simplicity: it can be introduced into an existing HRSC code without modifying its essential structure. This feature makes it into a useful tool to explore the possibilities (and the flaws) of state of the art HRSC schemes. It gives the user the possibility to use very fine (uniform) grids to run test problems at low cost: the cost of the user's own numerical technique on a much coarser mesh.

4 The Cell-Average Setting: Nonlinear Multiresolution Transforms for Image Compression

Images can be understood as discrete data corresponding to piecewise smooth functions. When designing a compression scheme, the approximation properties of the reconstruction sequence play a key role. Highly accurate reconstruction operators lead to small prediction errors, which can be truncated or quantized with very little loss in *real* information contents.

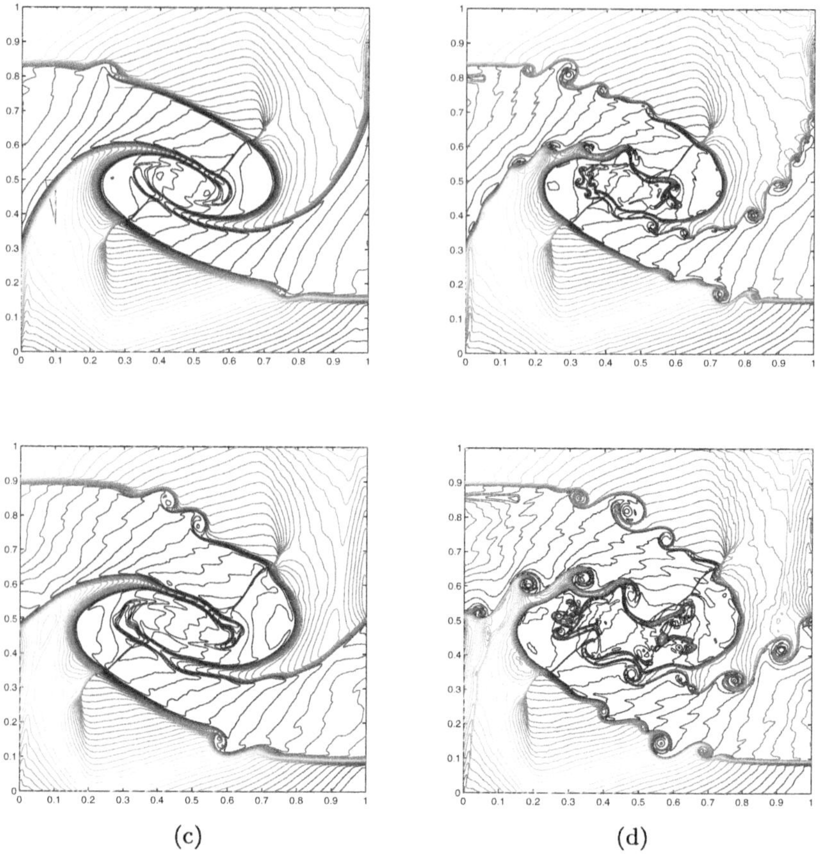

Fig. 4. 4-contact configuration: High Resolution numerical approximation to the density obtained with the Multilevel-Marquina scheme. (a) and (b) t=.8, (c) and (d) t=1. (a) and (c) 512×512 mesh-points, (b) and (d) 1024×1024 mesh-points. See also Color Plate 7.

In section 2, we have seen that nonlinear, data dependent reconstruction techniques can be used to obtain fully accurate approximations almost up to discontinuities. In particular, ENO interpolation plus Harten's Subcell Resolution technique in the cell-average setting maximize the region in which the approximation is fully accurate, and will in turn lead to multiresolution schemes with good compression capabilities.

Let us examine first the approximation capabilities of the various reconstruction techniques considered in this paper. For this, let us consider a purely geometric image, the one displayed in Fig. 5-(a) (512×512 pixels). By repeated decimation (in the cell-average setting) we obtain a coarse version of the image with only 32×32 pixels (not shown); in the notation of this paper

v^L is the original image, v^0 is the coarse representation and $L = 4$ in this case.

To show the advantages of nonlinear reconstruction processes, we compute $M^{-1}(v^0, 0, \ldots, 0)$, where the prediction operator, $P^j_{j-1} = \mathcal{D}_j \mathcal{R}_{j-1}$, is constructed as specified in section 2.2. In particular, in constructing $\mathcal{R}_j$, we shall consider piecewise polynomial techniques of degree 3 for the primitive function and

- A centered piecewise polynomial technique. We recall [16,4] that the resulting multiresolution transform is that of the *BOW* scheme of [9] $(N = 1, \tilde{N} = 3)$.
- A nonlinear ENO piecewise polynomial technique.
- A nonlinear ENO-SR technique.

The Gibbs-like oscillations typical of linear schemes in the presence of discontinuities lead to the blurring of the edges observed in Fig. 5-(c). The edges are much sharper in the non-linear case and, in this special situation in which the discontinuities are aligned with the grid, the ENO-SR prediction scheme gives back the exact original image.

The limitations of the nonlinear techniques can be observed by repeating this same experiment but with an image with geometric features not aligned with the tensor-product grids, as in Fig. 6. Since $d^j \equiv 0, j = 1, \ldots, L$, the data shown in Figs. 5 and 6 allow us to see those regions where the reconstruction process leads to large prediction errors. The blurred region is larger in the linear scheme, more localized in the plain ENO alternative and even more so in the ENO-SR case. It is clear that the number of coefficients kept in compressing the original image would be smaller for the nonlinear schemes.

We refer the reader to [2,3] for a thorough comparison of the capabilities of ENO-type reconstruction operators for image compression with respect to traditional wavelet-based alternatives.

As pointed out in section 2.3, stability is not granted when using nonlinear multiresolution transforms. To ensure stability for ENO-based compression

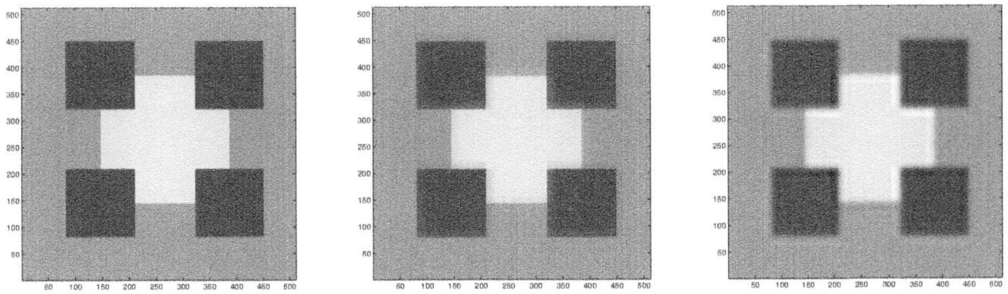

(a) (b) (c)

Fig. 5. A purely geometric image. (a) original and reconstruction by ENO-SR multiresolution scheme; (b) reconstruction by ENO multiresolution scheme; (c) reconstruction from linear (*bow*) scheme.

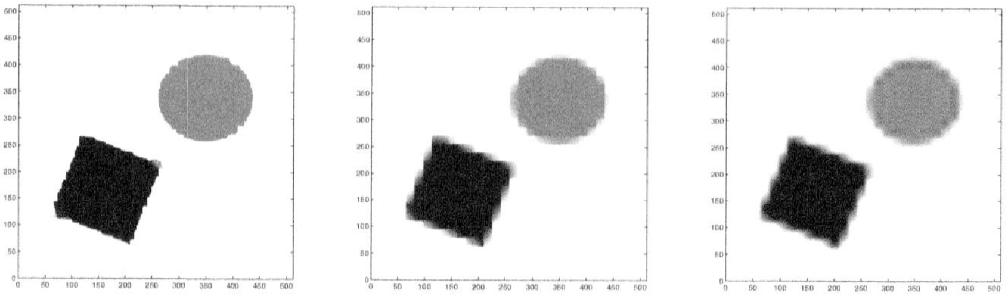

(a) (b) (c)

Fig. 6. A purely geometric image. (a) reconstruction by ENO-SR multiresolution scheme; (b) reconstruction by ENO multiresolution scheme; (c) reconstruction from linear (*bow*) scheme.

schemes, we implement the error-control algorithms described in [3] (it is not known at the moment whether or not the *direct* ENO-based nonlinear multiresolution transforms are stable). We end this section with an example that shows the differences in the outcome of the 2D error-control algorithm [3] versus the direct algorithm.

As usual when dealing with images, we consider the cell-average setting to compress the familiar image of Lena (512 × 512 pixels) using the 3-cell ENO technique of the previous sections for the prediction step. We use the direct algorithm with $L = 4$ to compute Mv^L and compress by truncating the scale coefficients i.e. if $|d_i^j| < \epsilon_j$ then $\hat{d}_i^j = 0$, otherwise $\hat{d}_i^j = d_i^j$. In this case we have used $\epsilon_L = 12$ and $\epsilon_j = \epsilon_{j-1}/2$. Figure 7-(b) shows $M^{-1}(v^0, \hat{d}^1, \ldots, \hat{d}^L)$. The number of non-zero elements in the compressed image $M_{\bar{\epsilon}}v^L = (v^0, \hat{d}^1, \ldots, \hat{d}^L)$ is 16852, which corresponds to a compression rate of 9.34:1 (.85 *bpp*), and $||v^L - M^{-1}M_{\bar{\epsilon}}v^L||_2 = 8.4$. On the other hand, Figs. 7-(a) shows $M^{-1}(v^0, \tilde{d}^1, \ldots, \tilde{d}^L)$, where $M_{\bar{\epsilon}}^m v^L = (v^0, \tilde{d}^1, \ldots, \tilde{d}^L)$, is the outcome of the *modified encoding* that characterizes the error-control algorithms (see [3] for details). Now, the number of non-zero elements in $M_{\bar{\epsilon}}^m v^L$ is 17314, i.e. a compression rate of 9.12:1 (.88 *bpp*) and $||v^L - M^{-1}M_{\bar{\epsilon}}^m v^L||_2 = 5.0$. We remark that the error-control algorithm leads to comparable compression rates, while keeping at the same time a noticeable better quality in the reconstructed image.

The stability of the direct nonlinear multiresolution transform is still being investigated, but it should be mentioned that the error-control algorithms lead to compression schemes in which it is possible to guarantee a certain accuracy *a priori*, and thus could be useful even when using linear multiresolution transforms. We refer the interested reader to [3,2].

Acknowledgments The authors would like to thank A. Marquina for many enlightening conversations on numerical solutions of conservation laws and specifically on the simulations presented in this paper. We also like to

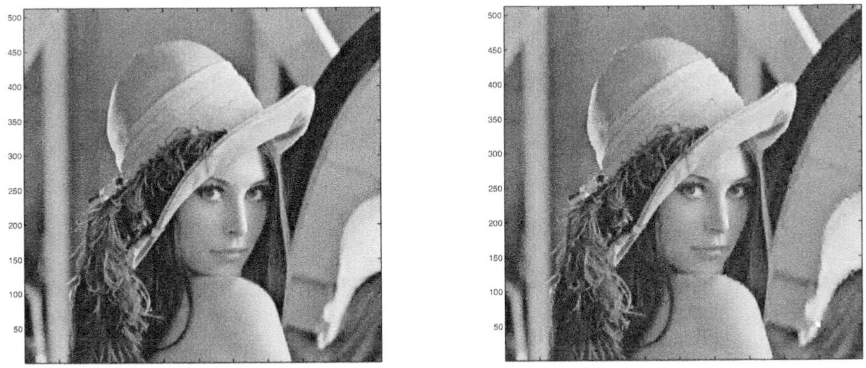

(a) (b)

Fig. 7. Images of Lena reconstructed from a compressed nonlinear multiresolution decomposition. (a) with an error-control algorithm; (b) with a direct algorithm

acknowledge support from the TMR project ERBFMRX-CT98-0184 and the first and last authors also partial support from y DGCYT PB97-1402.

References

1. Abgrall, R. Multiresolution in unstructured meshes: Application to CFD. NUmerical Methods for Fluid Dynamics, Oxford University Press, **5** (1996).
2. Amat, S., Aràndiga, F., Cohen, A., Donat, R., García G., Von Oehsen, M.: Data compression with ENO schemes. Submitted to ACHA (2000).
3. Amat, S., Aràndiga, F., Cohen, A., Donat, R.: Tensor product multiresolution analysis with error control for compact image representation. Submitted to Signal Processing (2000).
4. Aràndiga, F., Donat, R., Harten, A.: Multiresolution based on weighted averages of the hat function I: linear reconstruction techniques. SIAM J. Numer. Anal. **36** (1999) 160-203.
5. Aràndiga, F., Donat, R., Harten, A.: Multiresolution based on weighted averages of the hat function II: non-linear reconstruction techniques. SIAM J. Sci. Comput. **20** (1999) 1053-1093.
6. Aràndiga, F., Donat, R.: Nonlinear multi-scale decompostitions: The aproach of A. Harten, Numer. Algorith. **23** (2000) 175-216.
7. Bihari, B., Harten, A.: Multiresolution schemes for the numerical solution of 2-D conservation laws. I. SIAM J. Sci. Comput. **18** (1997) 315-354.
8. Chiavassa, G., Donat, R.: Point-value multi-scale algorithm for 2D compressible flows. To appear in SIAM J. Sci. Comput. (2000).
9. Cohen, A., Daubechies, I., Feauveau, J.C.: Biorthogonal bases of compactly supported wavelets. Comm. Pure Applied Math. **45** (1992) 485-560.
10. Dahmen, W., Gottschlich-Müller, B., Müller S.: Multiresolution schemes for Conservation Laws, Technical Report 159, IGMP, RWTH Aachen (1998). To appear in Numerische Mathematik.

11. Donat, R., Marquina, A.: Capturing shock reflections: An improved flux formula. J. Comp. Phys. **125** (1996) 42-58.
12. Harten, A., Engquist B., Osher S., Chakraverthy S.: Uniformly high order accurate essentially non-oscillatory schemes III. J. Comp. Phys. **71** (1987) 231-303.
13. Harten, A.: ENO Schemes with Subcell Resolution. J. Comp. Phys. **83** (1989) 148-184.
14. Harten, A.: Discrete Multiresolution Analysis and Generalized Wavelets. J. Appl. Numer. Math. **12** (1993) 153-192.
15. Harten, A.: Multiresolution algorithms for the numerical solution of hyperbolic conservation laws. Comm. Pure Appl. Math. **48** (1995) 1305-1342.
16. Harten, A.: Multiresolution representation of data: a general framework. SIAM J. Numer. Anal. **33** (1996) 1205-1256.
17. Harten, A., Hyman, J. M., Lax, P. D.: On finite-difference approximations and entropy conditions for shocks. Comm. Pure Appl. Math. **29** (1976) 297-322.
18. LeVeque, R.: Numerical Methods for Conservation Laws. Birkhäuser Verlag, 1992.
19. Quirk, J.: A contribution to the great Riemann solver debate. Int'l J. Num. Meth. Fluids **18** 555-574.
20. Shu, C.W., Osher, S. J.: Efficient implementation of essentially non-oscillatory shock-capturing schemes II. J. Comp. Phys. **83** (1989) 32-78.
21. Schulz-Rinne, C. W., Collins J. P., Glaz, H. M.: Numerical solution of the Riemann problem for two-dimensional gas dynamics. SIAM J. Sci. Comput. **14** (1993) 1394-1414.
22. Sjögrenn, B.: Numerical experiments with the multiresolution scheme for the compressible euler equations. J. Comput. Phys. **117** (1995) 251-261.
23. Sweldens, W.: The lifting scheme: A construction of second generation wavelets. SIAM J. Math. Anal. **29** (1997) 511-546.

Two Level Finite Element Technique for Pressure Recovery from Stream Function Formulation of the Navier-Stokes Equations

Faisal Fairag

Mathematics Department, King Fahd University of Petroleum & Minerals, Dhahran, 31261 Saudi Arabia (ffairag@kfupm.edu.sa).

Abstract. We consider two-level finite element discretization methods for the stream function formulation of the Navier-Stokes equations. The two-level method consists of solving a small nonlinear system on the coarse mesh, then solving a linear system on the fine mesh. It is shown in [8] that the errors between the coarse and fine meshes are related superlinearly. This paper presents an algorithm for pressure recovery and a general analysis of convergence for the algorithm. The numerical example for the 2D driven cavity fluid is considered. Streamfunction contours are displayed showing the main features of the flow.

1 Introduction

Two level finite element method is one of the promising approaches for solving nonlinear problems that arise in areas such as fluid mechanics. The computational attractions of the methods are that they require the solution of only a small system of nonlinear equations on coarse mesh and one linear system of equations on fine mesh. Apparently, the two-level method was proposed first in [16,15,14] and used for semilinear elliptic problems. The method was implemented for the velocity-pressure formulation of the Navier-Stokes equations in [11–13] and for the streamfunction formulation of the Navier-Stokes equations in [8,17].

The Navier-Stokes equations may be solved using either the primitive variable or stream function formulation. Here, we use the stream function formulation. The attractions of the stream function formulation are that the incompressibility constraint is automatically satisfied, the pressure is not present in the weak form, and there is only one scalar unknown to solve for. The standard weak formulation of the stream function version first appeared in 1979 in [10]. In this direction, Cayco and Nicolaides [6,5,4] studied a general analysis of convergence for this standard weak formulation. Once the approximated streamfunction is obtained, the momentum equations can be used to approximate the pressure.

The goal of this paper is to demonstrate an algorithm that uses the two-level finite element method to approximate the streamfunction and pressure recovery

The organization of the paper is as follows. Section 2 presents the model equations and the weak forms. Section 3 presents a pressure recovery algorithm and general analysis of convergence for this algorithm are presented in Section 4. Finally, in Section 4 we apply our method to one test problem.

2 Governing Equations

Consider the Navier-Stokes equations describing the flow of an incompressible fluid

$$-\mathrm{Re}^{-1}\Delta \boldsymbol{u} + (\boldsymbol{u}\cdot\Delta)\boldsymbol{u} + \nabla p = \boldsymbol{f} \quad \text{in } \Omega, \tag{1}$$
$$\nabla\cdot\boldsymbol{u} = 0 \quad \text{in } \Omega, \tag{2}$$
$$\boldsymbol{u} = 0 \quad \text{in } \partial\Omega, \tag{3}$$

where $\boldsymbol{u} = (u_1, u_2)$ and p denote, respectively, the unknown velocity and pressure field in a bounded, simply-connected polygonal domain $\Omega \subseteq R^2$; $\boldsymbol{f}$ is a given body force, and Re is the Reynolds number.

The introduction of a stream function $\psi(x, y)$ defined by

$$u_1 = -\frac{\partial\psi}{\partial y}, \quad u_2 = \frac{\partial\psi}{\partial x}$$

means that the continuity condition (2) is satisfied identically. The pressure may then be eliminated from (1) to give

$$\mathrm{Re}^{-1}\Delta^2\psi_y\Delta\psi_x + \psi_x\Delta\psi_y = \mathrm{curl}\ \boldsymbol{f} \text{ in } \Omega, \tag{4}$$
$$\psi = 0 \text{ on } \partial\Omega, \tag{5}$$
$$\frac{\partial\psi}{\partial\hat{n}} = 0 \text{ on } \partial\Omega, \tag{6}$$

where $\hat{n}$ represents the outward unit normal to Ω. In order to write (4) – (6) in a variational form, we define the Sobolev spaces

$$H^1(\Omega) = \{v : v \in L^2(\Omega), Dv \in L^2(\Omega)\}, \tag{7}$$
$$H_0^1(\Omega) = \{v : v \in H^1(\Omega), v = 0 \text{ on } \partial\Omega\}, \tag{8}$$
$$H^2(\Omega) = \{v : v \in L^2(\Omega), Dv \in L^2(\Omega), D^2v \in L^2(\Omega)\}, \tag{9}$$
$$H_0^2(\Omega) = \{v : v \in H^2(\Omega) : v = \frac{\partial v}{\partial n} = 0 \text{ on } \partial\Omega)\}, \tag{10}$$
$$[H_0^1(\Omega)]^2 = H_0^1(\Omega) \times H_0^1(\Omega), \tag{11}$$
$$H^{-m}(\Omega) = \text{ the dual of } H_0^m(\Omega) \tag{12}$$

where $L^2(\Omega)$ is the space of square integrable functions on Ω and D represents differentiations with respect to x or y. For each $\phi \in H^1(\Omega)$, define curl $\phi = \begin{pmatrix}\phi_y \\ -\phi_x\end{pmatrix}$.

The standard weak form of equations (1) – (3) is

Find $\boldsymbol{u} \in [H_0^1]^2$, $p \in L_0^2(\Omega)$, such that $\forall\ \boldsymbol{w} \in [H_0^1(\Omega)]^2$, $q \in L_0^2(\Omega)$

$$\left.\begin{aligned} \mathrm{Re}^{-1}\boldsymbol{a}(\boldsymbol{u},\boldsymbol{w}) + \tilde{b}(\boldsymbol{u};\boldsymbol{u},\boldsymbol{w}) + \tilde{c}(\boldsymbol{w},p) &= \langle \boldsymbol{f},\boldsymbol{w}\rangle \\ \tilde{c}(\boldsymbol{u},q) &= 0, \end{aligned}\right\} \tag{13}$$

where

$$\left.\begin{aligned} \tilde{a}(\boldsymbol{u},\boldsymbol{w}) &= \mathrm{Re}^{-1}\int_\Omega \nabla\boldsymbol{u} : \nabla\boldsymbol{w}, \\ \tilde{b}(\boldsymbol{u};\boldsymbol{v},\boldsymbol{w}) &= \int_\Omega ((\boldsymbol{u}\cdot\nabla)\boldsymbol{v})\cdot\boldsymbol{w}, \\ \tilde{c}(\boldsymbol{w},q) &= \int_\Omega q \ \mathrm{div}\ \boldsymbol{w}, \end{aligned}\right\} \tag{14}$$

and $\langle\cdot,\cdot\rangle$ denotes the duality pairing in $L^2(\Omega)$.

The standard weak form of equations (4) – (6) is

Find $\psi \in H_0^2(\Omega)$ such that for all $\phi \in H_0^2(\Omega)$,

$$a(\psi\phi) + b(\psi;\psi,\phi) = \ell(\phi), \tag{15}$$

where

$$\left.\begin{aligned} a(\psi,\phi) &= \mathrm{Re}^{-1}\textstyle\int_\Omega \Delta\psi\cdot\Delta\phi d\Omega, \\ b(\xi;\psi,\phi) &= \int_\Omega \Delta\xi(\psi_y\phi_x - \psi_x\phi_y)d\Omega, \\ \ell(\phi) &= (\boldsymbol{f},\ \mathrm{curl}\ \phi) = \int_\Omega \boldsymbol{f}\cdot\mathrm{curl}\ \phi d\Omega. \end{aligned}\right\} \tag{16}$$

3 Pressure Recovery Algorithm

We consider the approximate solution of (4) by a two-level, finite-element procedure. Let $X^h, X^H \subset H_0^2(\Omega)$ denote two conforming finite-element meshes with $H \gg h$. The method we consider computes an approximate solution ψ^h in the finite-element space X^h by solving one linear system for the degrees of freedom on X^h. This particular linear problem requires the construction of a finite-element space X^H upon a very coarse mesh of width '$H \gg h$', and then the solution of a much smaller system of nonlinear equations for an approximation in X^H.

Once the approximated ψ^h is obtained, the momentum equations can be used to approximate the pressure. The basic idea is to solve an equation of the form

$$\tilde{c}(\boldsymbol{v},p) = g(\boldsymbol{v}), \tag{17}$$

where

$$\left.\begin{aligned} g(\boldsymbol{v}) &= g(\psi;\boldsymbol{f},\ \mathrm{Re})(\boldsymbol{v}) \\ &= \langle \boldsymbol{f},\boldsymbol{v}\rangle - \mathrm{Re}^{-1}\tilde{a}(\mathrm{curl}\ \psi,\boldsymbol{v}) - \tilde{b}(\mathrm{curl}\ \psi;\ \mathrm{curl}\ \psi,\boldsymbol{v}). \end{aligned}\right\} \tag{18}$$

It is clear that g depends on the solution ψ of (13), the data $\boldsymbol{f}$ and Re. For any $g \in [H_0^1(\Omega)]'$, a unique $p \in L_0^2(\Omega)$ exists such that (17) holds for each $\boldsymbol{v} \in [H_0^1(\Omega)]^2$ (see [2]). It follows directly from the continuity of $\tilde{a}$ and $\tilde{b}$ (see [10]) that

Lemma 1. For $\boldsymbol{f} \in \boldsymbol{H}^{-1}(\Omega), \psi \in H_0^2(\Omega)$, g as defined above is a bounded linear functional on $H_0^1(\Omega)$. Moreover, for all $\psi, \varphi \in H_0^2(\Omega), \boldsymbol{v} \in [H_0^1(\Omega)]^2$

$$\left.\begin{aligned} &|g(\psi; \boldsymbol{f}, \mathrm{Re})(\boldsymbol{v}) - g(\varphi; \boldsymbol{f}, \mathrm{Re})(\boldsymbol{v})| \leq \\ &\quad (\mathrm{Re}^{-1} + \Gamma_1|\psi|_2 + \Gamma_1|\varphi|_2)|\psi - \varphi|_2|\boldsymbol{v}|_1 . \end{aligned}\right\} \tag{19}$$

Then we have

Theorem 1. Given $\psi \in H_0^2(\Omega), \boldsymbol{f} \in H^{-1}(\Omega)$, there exists a unique $p \in L_0^2(\Omega)$ such that

$$\tilde{c}(\boldsymbol{v}, p) = -g(\psi; \boldsymbol{f}, \mathrm{Re})(\boldsymbol{v}) \text{ for all } \boldsymbol{v} \in [H_0^1(\Omega)]^2. \tag{20}$$

Note that $\tilde{c}$ is continuous and coercive on $[H_0^1(\Omega)]^2 \times L_0^2(\Omega)$ ([10], Theorem 3.7, p. 35). The coercivity condition does not necessarily hold for arbitrary subspace $V^h \subset [H_0^1(\Omega)]^2$ and $S^h \subset L_0^2(\Omega)$. Also, in discretizing (17), not only will P and $\boldsymbol{v}$ be discretized, but so will g. We mean that $g(\psi; \boldsymbol{f}, \mathrm{Re})(\boldsymbol{v})$ has to be replaced by $g(\psi^h; \boldsymbol{f}, \mathrm{Re})(\boldsymbol{v}^h)$. Since the null space of $g(\psi^h; f, \mathrm{Re})$ does not necessarily coincide with the discretely div-free functions in V^h, the discretized analog of (17) can only hold in some subspace of V^h. Thus subspace is generally quite difficult to find. Because of this, Cayco and Nicolaides [5] introduced the following equivalent problem:

Find $\boldsymbol{w} \in [H_0^1(\Omega)]^2, p \in L_0^2(\Omega)$ such that for all $\boldsymbol{v} \in [H_0^1(\Omega)]^2, q \in L_0^2(\Omega)$

$$\left.\begin{aligned} \mathrm{Re}^{-1}\tilde{a}(\boldsymbol{w}, \boldsymbol{v}) + \tilde{c}(\boldsymbol{v}, p) &= g(\psi; \boldsymbol{f}, \mathrm{Re})(\boldsymbol{v}) \\ \tilde{c}(\boldsymbol{w}, q) &= 0 \end{aligned}\right\} \tag{21}$$

(20) is uniquely solvable (See [10]), and by Theorem 3.2, $w = 0$. We discretize (20) as follows:

Find $\boldsymbol{w}^h \in V^h, p^h \in S^h$ such that for all $\boldsymbol{v}^h \in V^h, q^h \in S^h$

$$\left.\begin{aligned} \mathrm{Re}^{-1}\tilde{a}(\boldsymbol{w}^{\cdot}\boldsymbol{v}^h) + \tilde{c}(\boldsymbol{v}^h, p^h) &= g(\psi^h; \boldsymbol{f}, \mathrm{Re})(\boldsymbol{v}^h) \\ \tilde{c}(\boldsymbol{w}^h, q^h) &= 0 \end{aligned}\right\} \tag{22}$$

The solution procedure is then given as follows:

Algorithm

<u>Step 1</u>: Solve the nonlinear system on coarse mesh for $\psi^H \in X^H$

$$a(\psi^H, \varphi^H) + b(\psi^H, \psi^H, \varphi^H) = (\boldsymbol{f}, \mathrm{curl}\ \varphi^H) \text{ for all } \varphi^H \in X^H. \tag{23}$$

Step 2: Solve the linear system on fine mesh for $\psi^h \in X^h$

$$a(\psi^h, \varphi^h) + b(\psi^H, \psi^h, \varphi^h) = (\boldsymbol{f}, \operatorname{curl} \varphi^h) \text{ for all } \varphi^h \in X^h. \tag{24}$$

Step 3: Solve the linear system for $\boldsymbol{w}^h \in V^h, p^h \in S^h$ such that for all $\boldsymbol{v}^h \in V^h, q^h \in S^h$

$$\left.\begin{aligned} \mathrm{Re}^{-1}\tilde{a}(\boldsymbol{w}^h, \boldsymbol{v}^h) + \tilde{c}(\boldsymbol{v}^h, p^h) &= g(\psi^h; \boldsymbol{f}, \mathrm{Re})(\boldsymbol{v}^h) \\ \tilde{c}(\boldsymbol{w}^h, q^h) &= 0 \end{aligned}\right\} \tag{25}$$

Existence, uniqueness and error bound of Steps 1 and 2 are given in the following theorems. The proofs of these theorems are given in [8].

Theorem 2 (Step 1). (a) If the global uniqueness condition $\mathrm{Re}^2 N|f|_* < 1$ holds, ψ and ψ^H both exist uniquely, then the error $|\psi - \psi^H|_2$ satisfies

$$|\psi - \psi^H|_2 \le C(\mathrm{Re}) \inf_{w^H \in X^h} |\psi - w^H|_2, \tag{26}$$

where

$$\begin{aligned} N &:= \sup_{\xi,\psi,\varphi \in H_0^2(\Omega)} \frac{|b(\xi, \psi, \varphi)|}{|\psi|_2 \cdot |\varphi|_2} \\ |f|_* &:= \sup_{\varphi \in H_0^2(\Omega)} \frac{(f,\ \operatorname{curl} \varphi)}{|\varphi|_2} \\ C(\mathrm{Re}) &= (1 + 2N|f|_* \cdot \mathrm{Re}^2)(1 - N|f|_* \mathrm{Re}^2)^{-1} \le C(\sqrt{N|f|_*}). \end{aligned}$$

(b) If the uniqueness condition fails, and ψ is a non-singular solution of (4), then there is an $H_0 = H_0(\psi, f, \mathrm{Re})$ and $c = c(\psi, f, \mathrm{Re}, N)$ such that for $H \le H_0$,

$$|\psi - \psi^H|_2 \le c(\psi, f, \mathrm{Re}, N) \inf_{w^H \in X^h} |\psi - w^H|_2, \tag{27}$$

where $c(\psi, f, \mathrm{Re}, N) = \gamma^{-1}(\mathrm{Re}^{-1} + N \cdot \mathrm{Re}|f|_*) + 1$.

Theorem 3 (Step 2). Let $X^{h,H} \subset H_0^2(\Omega)$ be two finite element spaces, ψ a solution to (4) and ψ^H a solution to (23). Then the solution to (24) exists uniquely and satisfies

$$|\psi - \psi^h|_2 \le c_1 \inf_{\varphi^h \in X^h} |\psi^h - \varphi^h|_2 + c_2\sqrt{|\ln h|}\, |\psi - \psi^H|_1, \tag{28}$$

where $c_1 = 2 + N|f|_* \mathrm{Re}^2$ and $c_2 = 2N \cdot \mathrm{Re}^2 |f|_* c$.

Our aim in this paper is to give an error analysis for Step 3. This analysis will be explained in the next section.

4 Main Result

In this section, we establish an error estimate for Step 3 in the above algorithm. The following theorem is the main theorem in this paper which gives an error estimate.

Theorem 4. Let ψ be the solution to (4–6) and ψ^h the solution to (24). Let the space $V^h \subset [H_0^1(\Omega)]^2$ and the space $S^h \subset L_0^2(\Omega)$ be chosen such that

$$\tilde{C}(v^h, q^h) \geq \beta ||q^h||_0 \cdot |v^h|. \tag{29}$$

Then (25) is uniquely solvable and

$$||p - p^h||_0 \leq K_0 |\psi - \psi^h|_2 + K_1 \inf_{q^h \in S^h} ||p - q^h||_0,$$

where K_0, K_1 are positive constants independent of p, p^h, w^h, ψ and ψ^h.

Proof. . It follows from the primitive variable theory that (22) is uniquely solvable. Subtracting (21) from (22), we get

$$\tilde{a}(w - w^h, v^h) = -\tilde{c}(v^h, p - p^h) + g(\psi; f, \mathrm{Re})(v^h) - g(\psi^h, f, \mathrm{Re})(v^h). \tag{30}$$

Also, we have

$$\tilde{c}(w^h, p^h) = \tilde{c}(w^h, q^h) = 0 \text{ for all } q^h \in S^h.$$

Setting $v^h = w^h$ in (30), we get

$$|w^h|_1 \leq \alpha_1 |\psi - \psi^h|_2 + ||p - q^h||_0, \tag{31}$$

where $\alpha_1 = \mathrm{Re}^{-1} + 2||f||\mathrm{Re}^2$.

Now using (29) for any $q^h \in S^h$,

$$\begin{aligned} \beta ||q^h - p^h|| &\leq \sup_{v^h \in V^h} \frac{|\tilde{c}(v^h, q^h - p^h)|}{|v^h|_1} \\ &= \sup_{v^h \in V^h} \frac{|\tilde{c}(v^h, q^h - p) + \tilde{c}(v^h, p - p^h)|}{|v^h|_1}. \end{aligned}$$

Using (19), we get

$$\begin{aligned} &\beta ||q^h - p^h||_0 \leq \\ &\leq \sup_{v^h \in V^h} \frac{|\tilde{c}(v^h, q^h - p) - \tilde{a}(w - w^h, v^h) + g(\psi, f, \mathrm{Re})(v^h) - g(\psi^h, f, \mathrm{Re})(v^h)|}{|v^h|_1} \\ &\leq \gamma_1 ||q^h - p||_0 + \gamma_2 |w - w^h|_1 + \gamma_3 |\psi - \psi^h|_2, \end{aligned}$$

which implies

$$\|q^h - p^h\|_0 \le \frac{\gamma_1}{\beta}\|q^h - p\|_0 + \frac{\gamma_2}{\beta}|w^h|_1 + \frac{\gamma_3}{\beta}|\psi - \psi^h|_2. \tag{32}$$

Triangle inequalities and (31) give

$$\begin{aligned}\|p - p^h\|_0 &\le \|p - q^h\|_0 + \|p^q - p^h\|_0 \\ &\le \|p - q^h\|_0 + \frac{\gamma_1}{\beta}\|q^h - p\|_0 + \frac{\gamma_2}{\beta}|w^g|_1 + \frac{\gamma_3}{\beta}|\psi - \psi^h|_2 \\ &\le \|p - q^h\|_0 + \frac{\gamma_1}{\beta}\|q^h - p\|_0 + \frac{\gamma_2}{\beta}\alpha_1|\psi - \psi^h|_2 \\ &\quad + \frac{\gamma_2}{\beta}\|q^h - p\|_0 + \frac{\gamma_3}{\beta}|\psi - \psi^h|_2 \\ &\le K_0|\psi - \psi^h|_2 + K_1\|q^h - p\|_0,\end{aligned}$$

where $K_0 = \frac{\gamma_2}{\beta}\alpha_1 + \frac{\gamma_3}{\beta}$ and $K_1 + 1 + \frac{\gamma_1}{\beta} + \frac{\gamma_2}{\beta}$.
Hence $\|p - p^h\|_0 \le K_0|\psi - \psi^h|_2 + K_1 \inf_{q^h \in S^h} \|q^h - p\|_0$.

5 Numerical Examples

In this section we describe some numerical results obtained by implementing the two-level algorithm for the well-known driven cavity problem. We chose this problem because there are numerous results in the literature with which to compare. Cavity flows have been a subject of study for some time. These flows have been widely used as test cases for validating incompressible fluid dynamics algorithm. Corner singularities for two-dimensional fluid flows are very important since most examples of physical interest have corners. For example, singularities of most elliptic problems develop when the boundary contour is not smooth. In this example, we consider the driven flow in a rectangular cavity when the top surface moves with a constant velocity along its length. The upper corners where the moving surface meets the stationary walls are singular points of the flow at which the horizontal velocity is multi-valued. The lower corners are also weakly singular points.

For this problem, the region Ω is the unit square $\{0 < x < 1,\ 0 < y < 1\}$ with no-slip boundary conditions, i.e., $u = v = 0$ in all boundaries except $y = 1$, where $u = 1$. This problem has been studied and addressed by many researchers including Ghia, Ghia, Shin [9], and J.E. Akin [1]. The numerical computation of this example was obtained using a Sun Ultra 2 with 2 200 MHz ultrasparc processor running Solaris 2.5.1. and for the finite element discretization we use the Bogner-Fox-Schmit elements. Bogner-Fox-Schmit elements are bicubic polynomials within each rectangle. The degrees of freedom are chosen to be the function value, the first derivatives, and the mixed second derivative at the vertices. We set the function and the normal derivative values equal to zero at all vertices on the boundary. Bogner-Fox-Schmit elements are used with 17×17 grid points on the coarse mesh and 33×33 grid points on the fine mesh. The streamlines for Re = 1, 10, 50, 100 are plotted in Fig.1. All nonlinear problems were solved by Newton's method until the norm of the difference in successive iterates and the norm of residual were within a fixed tolerance. In each Newton's iteration, we need to solve a linear system. The resulting linear system is non-symmetric whose symmetric part is positive definite. Moreover, the resulting matrix is sparse matrix. We choose the Bi-Conjugate Gradient Stabilized method (BICGSTAB) which requires two matrix-vector products and four inner products in each iteration. BICGSTAB is given and discussed in [3]. When solving the linearized problem with a mesh spacing h we need the solution u^H generated on a mesh with spacing H. To do this we interpolate the solution u^H onto the grid with spacing h.

Acknowledgment

The auther is greatful for the financial and facilities support provided by KFUPM.

References

1. J. Akin. *finite element for analysis and design.* Academic Press, San Diego, 1994.
2. I. Babuska and A. K. Aziz. *The Mathematical Foundations of the Finite Element Method with Applications to Partial Differential Equations.* Academic Press, New York, 1972.
3. R. Barrett, M. Berry T. Chan J. Demmel J. Donator J. Doncarra V. Eijkhout R. Pozo C. Romine and H. Van dev Vorst, *Templates: for the solution of linear systems: building blocks for iterative methods.* e-mail : templates@cs.utk.edu.
4. M. Cayco. *Finite Element Methods for the Stream Function Formulation of the Navier-Stokes Equations.* PhD thesis, CMU, Pittsburgh, PA., 1985.
5. M. Cayco and R.A. Nicolaides. Analysis of nonconforming stream function and pressure finite element spaces of the Navier-Stokes equations. *Comp. and Math. Appl.*, (8): 745–760, 1989.
6. M. Cayco and R.A. Nicolaides. Finite element technique for optimal pressure recovery from stream function formulation of viscous flows. *Math. Comp.*, (56): 371–377, 1986.
7. Ph. G. Ciarlet, *the finite element method for elliptic problems*, North Holland, Amsterdam, 1978
8. F. Fairag. Two-level finite element method for the stream function formulation of the Navier-Stokes equations. *Computers Math. Applic.*, 36(2): 117–127, 1998.
9. K.N. Ghia U. Ghia and C.T. Shin. High Re solutions for incompressible flow using the Navier-Stokes equations and a multigrid method. *J. Comput. Phys.*, 48: 387–411, 1982.

10. V. Girault and P. A. Raviart. *Finite Element Approximation of the Navier-Stokes Equations*, volume 749. Springer, Berlin, 1979.
11. W. Layton. A two-level discretization method for the Navier-Stokes equations. *Computers Math. Applic.*, 26:33–38, 1993.
12. W. Layton and W. Lenferink. Two-level Picard-defect corrections for the Navier-Stokes equations at high Reynolds number. *Applied Math. and Computing*, 1995.
13. W. Layton and W. Lenferink, A multilevel mesh independence principle for the Navier-Stokes equations, *SIAM J. N. A.* (1996)
14. J. Xu. A novel two-grid method for semilinear elliptic equations. *SIAM J. Sci. Comput.* 15(1): 231–237, 1994.
15. J. Xu, *Some Two-Grid Finite Element Methods*, chapter 157, pages 79–87. Number 157 in In Domain Decomposition Methods in Science and Engineering. Amer. Math. Soc., Providence, RI, 1994.
16. J. Xu. Some two-grid finite element methods. Technical report, P. S.U., 1992.
17. X. Ye. Two-level discretizations of the stream function form of the Navier-Stokes equations. University of Pittsburgh.

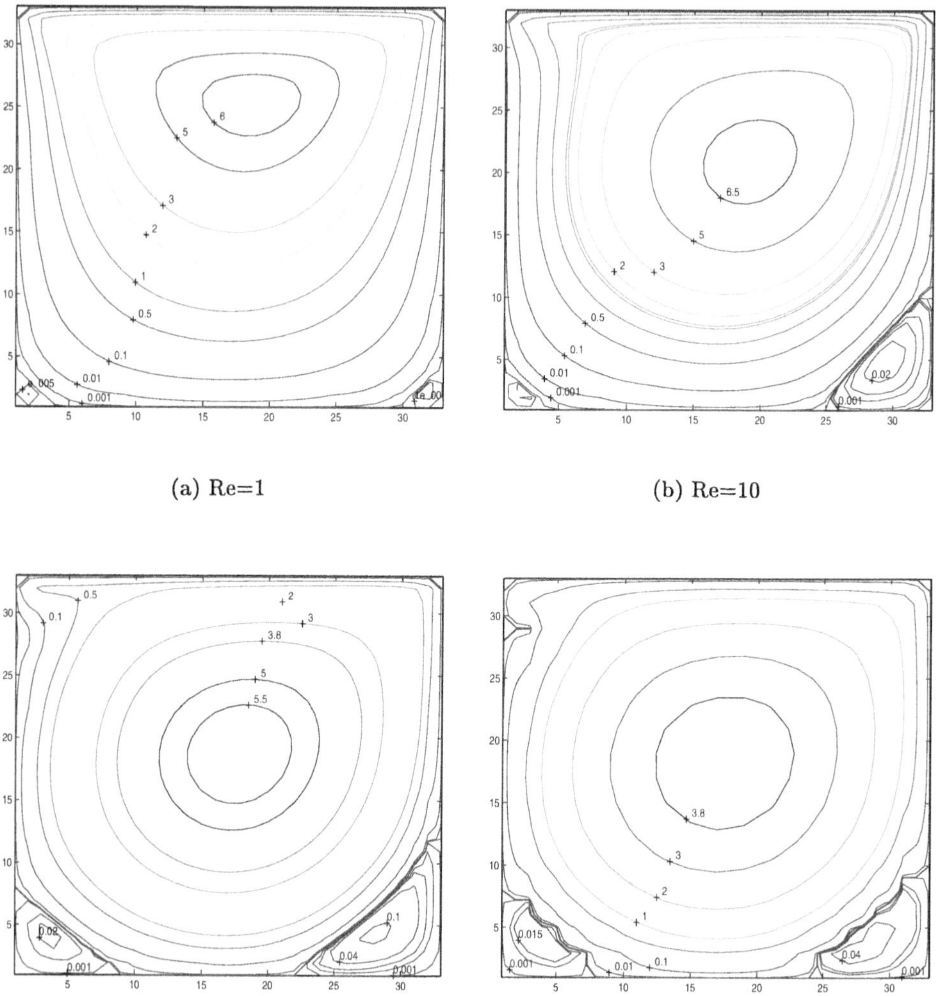

(a) Re=1 (b) Re=10

(c) Re=50 (d) Re=100

Fig. 1. Cavity Problem : Streamlines for $H = \frac{1}{16}, h = \frac{1}{32}$ with different values of Re numbers using two level method

The Role of Multiresolution in Mining Massive Image Datasets

Imola K. Fodor and Chandrika Kamath

Center for Applied Scientific Computing
Lawrence Livermore National Laboratory
P.O. Box 808, Livermore, CA 94551

Abstract. Scientists are collecting data from observations and simulations at an ever increasing pace. In order to extract useful information from these massive datasets, they are turning to data mining techniques as an attractive solution approach. Data mining is an iterative and interactive process that consists of data pre-processing and pattern recognition. Pre-processing the raw data in order to transform it into a form suitable for pattern recognition is an important and time-consuming first step. In this paper, we discuss the crucial role multiresolution techniques can play in the pre-processing of massive datasets. Using both simulated and real images, we describe our work in de-noising image data using wavelet-based multiresolution techniques. Our initial experiences show that a judicious choice of wavelet transforms, threshold selection methods, and threshold application schemes can effectively reduce the noise in the data without a significant loss of the signal.

1 Introduction

Our ability to generate data far outstrips our ability to explore, analyze, and understand it. Data that was measured in gigabytes until recently, is now being measured in terabytes and petabytes. The sheer size of the data, as well as its complexity, have made manual data analysis impractical. As a result, scientists are increasingly turning to semi-automated techniques, such as data mining, to extract useful information from these massive datasets.

Data mining is the non-trivial extraction of implicit, previously unknown, and potentially useful information from data. As the size of the datasets has increased, multiresolution techniques are playing an ever important role in their analysis. These techniques can be invaluable in compression and transmission of data, in reduction of the size of the data being analyzed, in de-noising data, and in extracting relevant features representing the data at different resolutions.

In this paper, we describe our use of multiresolution techniques within the Sapphire project in large-scale data mining [23]. As part of this project, we are implementing several components of data mining in a parallel, object-oriented, software toolkit. This includes image processing functions that specifically target the exploration of large image datasets. For the reasons stated above, multiresolution techniques form an integral part of our research and software.

In order to better understand the role that multiresolution techniques, in particular, wavelets, can play in data mining, we have initially focused on the task of de-noising image data. In this context, we have implemented various wavelet decomposition transforms, threshold calculation methods, and wavelet coefficient thresholding schemes. In this paper, we present the results of our experiments using both images with simulated noise and noisy images from a real problem in astronomy, namely, the Faint Images of the Radio Sky at Twenty-cm (FIRST) survey.

In Section 2, we briefly describe the process of data mining and discuss, in greater detail, the role that multiresolution techniques can play in mining image datasets. Section 3 describes various options in de-noising data through the thresholding of wavelet coefficients. We summarize our experimental results in Section 4, and conclude with a summary in Section 5.

2 Data Mining and Multiresolution Methods

Data mining is a process concerned with uncovering patterns, associations, anomalies and statistically significant structures in data [9,10]. As illustrated in Figure 1, it consists of two main phases: data pre-processing, during which relevant features are extracted from the data, and pattern recognition, in which these features are used to identify patterns in the data. It is an interactive and iterative process. Input from domain scientists is an integral part of this process, and frequently results in the refinement of one or more steps.

The data pre-processing phase consists of several steps. If the raw data is very large, we may use sampling to work with fewer instances, or multiresolution techniques to work with data at a coarser resolution. This first step may also include data fusion, if required. Next, noise is removed from the data and relevant features are extracted. This results in a feature vector for each data instance or item. Depending on the problem and the data, we may need to reduce the number of features using dimension reduction techniques such as principal component analysis. After pre-processing, patterns in the data are identified using techniques such as classification and clustering. The patterns are then presented to the user for appropriate validation.

Multiresolution methods are gaining importance for analyzing data in several signal and image processing applications in diverse fields such as astronomy, biology, and medicine [28,24]. These methods view the data at different resolutions, enabling us to exploit the fact that signals and images often contain relevant features at different scales or resolutions. There are different, but closely related, ways in which one can view multiresolution techniques. For example, it is well known that the pyramid scheme of Burt and Adelson [2] is closely related to wavelet theory and multiresolution analysis, and that filter banks can be used for the computation of wavelet decompositions [5,19,25].

In the field of data mining, multiresolution techniques can be exploited in several different ways:

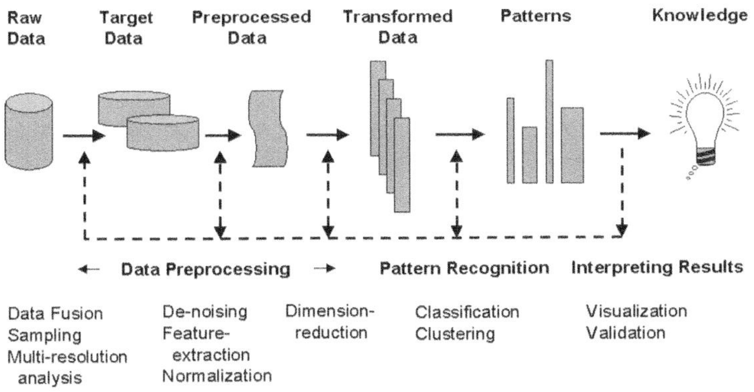

Fig. 1. Data mining: an iterative and interactive process.

- **Data compression and transmission:** Wavelet-based methods have been extensively used for data compression and transmission [11,14,21]. The successive approximation property of these techniques has also been used in progressive transmission of large images.
- **Image registration for data fusion:** Data fusion is the process of combining data from several different sensors in order to make more accurate decisions based on the complementary information. A key task in image data fusion is image registration, which determines the best match of two or more images acquired at the same time or at different times by different or identical sensors. When the images are at different resolutions, multiresolution techniques can be used for image registration [15,16].
- **Feature Extraction:** Multiresolution techniques can be very useful in extracting features when there are several patterns of interest at different scales [1,17,20]. A wavelet decomposition can be applied either to the image itself, or to some derived feature from the image. An example of the latter case is a two-dimensional character which can be represented as a one-dimensional curvature function. Once a decomposition has been obtained, either some, or all of the wavelet coefficients at a single level, or at multiple levels, can be used as representative features. It is also possible to use features that are derived from the wavelet coefficients such as the energy measure for the whole image or at each level.
- **De-noising:** The remainder of the paper discusses the role of the statistics-based wavelet thresholding approach for de-noising data.

3 A Statistical Approach to Wavelet De-noising

De-noising of images is an active research area and several diverse techniques have been proposed to address the problem. Included in these are image processing methods using spatial filters [30,27], partial differential equations-based techniques using level set [18] and total variation methods [4], computational fluid dynamics approaches using essentially non-oscillatory (ENO) schemes [3], and statistical procedures based on wavelet thresholding. In this section, we discuss this last technique in more detail.

Wavelet de-noising by thresholding of wavelet coefficients consists of three main steps:

1. Apply a discrete wavelet transform to the observed noisy data.
2. Threshold the detail wavelet coefficients.
3. Inverse transform the thresholded coefficients to obtain the de-noised data.

Calculating and applying the threshold in step 2. involves selecting the shrinkage function (Section 3.2), estimating the noise in some cases (Section 3.3), and selecting the shrinkage rule (Section 3.4). The arithmetic complexity of the overall de-noising process is linear in the number of data points and on modern workstations, moderate-sized images can be processed interactively.

There is extensive literature on de-noising methods for one-dimensional data [22,29]. These techniques can be easily extended to higher dimensions as well. However, most standard signal processing software products lack extensive two-dimensional implementations. In order to better understand the behavior of these methods, we are undertaking a comprehensive study to compare the performances of different two-dimensional de-noising techniques. A more detailed study is forthcoming [13].

Most one-dimensional rules in the literature propose to calculate thresholds in either a global or in a level-dependent manner. However, in practice, the rules can be implemented in both ways. We have extended the level-dependent idea to include the following options for two-dimensional data, when using decimated wavelet transforms: (a) provide one global threshold, (b) provide J different thresholds corresponding to the J multiresolution levels, and (c) provide $3J$ different thresholds corresponding to the J levels and to the three directions (vertical, horizontal, diagonal).

For the remainder of this section, we assume that the observed data $Y_{i,j}$ is a noisy version of the signal $X_{i,j}$:

$$Y_{i,j} = X_{i,j} + \epsilon_{i,j}, \quad i, j, = 1, \cdots, N, \tag{1}$$

where $\{\epsilon_{i,j}\}$ is independent and identically distributed (iid) Gaussian $N(0, \sigma^2)$, independent of the signal. Our goal is to estimate $X_{i,j}$ with minimal mean square error (MSE).

In the next sections, we describe some of the options available in the three steps mentioned above. This list in not intended to be exhaustive; rather, it represents the current state of our multiresolution work in the Sapphire toolkit. Our software is flexible enough to let the user experiment with many different possibilities. Future releases of the software will include enhancements, as outlined in Section 5.

Figure 3 presents an example FIRST image and three wavelet de-noised versions of it, corresponding to methods in Figure 2.

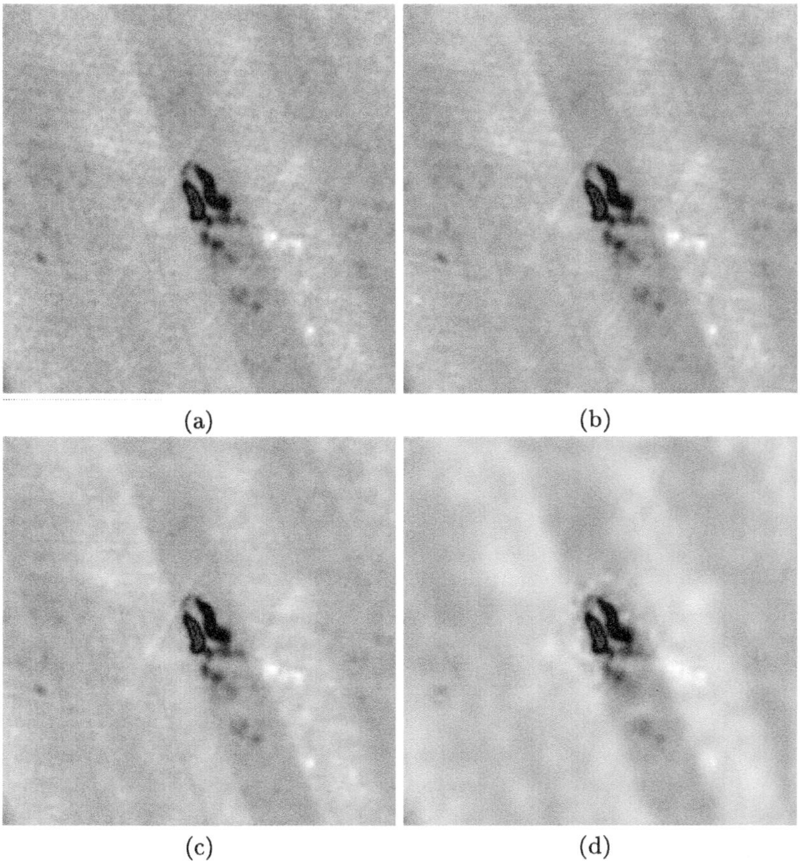

Fig. 3. Example of a FIRST image. (a) Observed. (b) Global min. false discovery with soft. (c) Global universal with soft. (d) Level-dependent top with soft.

The two wavy lobes of the bent-double galaxy are connected by a fainter bridge. This bridge is an important feature that must be preserved in de-noising. As the bridge lies on one of the noise streaks, we have to be careful in the use of de-noising. This task is made more challenging by the fact that the number of radio galaxies precludes individual examination of the effects of de-noising on each image. We are currently investigating various de-noising options on test images from FIRST to characterize which method works best. Upon completion of this work, we will choose an optimal combination for the FIRST dataset.

5 Summary and Future Directions

In this paper, we described the roles multiresolution analysis can play in mining massive datasets. We discussed the work we are doing in wavelet de-noising and presented some of our early results using data with real and simulated noise.

Our immediate goal for the future is to complete the comparison of the methods already implemented in our software, using different test images, different noise models, and different SNRs. We will also investigate, in more detail, various level- and direction-dependent de-noising methods, including the application of different methods on different levels and directions. Next, we plan to include additional wavelets (e.g. biorthogonal), additional de-noising methods (e.g. Bayes), and additional multiresolution methods (e.g. wavelet packets, matching pursuit algorithms, curvelets, steerable pyramids).

Our long term plans include automating the de-noising by tailoring the methods to the signal and noise distribution of the observed data. Finally, we plan to study the merits and pitfalls of parallelizing the de-noising methods for massive datasets.

Acknowledgments

We would like to thank the members of the Sapphire team for their contributions to this work. We also thank the reviewers for their helpful and constructive comments. UCRL-JC-139713. This work was performed under the auspices of the U.S. Department of Energy by University of California Lawrence Livermore National Laboratory under contract W-7405-Eng-48.

References

1. Anant, K., Dowla, F., Rodrigue, G.: Detection of the electrocardiogram P-wave using wavelet analysis. Proc. of the SPIE **2242** (1994) 744–749.
2. Burt, P., Adelson, E.: The Laplacian pyramid as a compact image code. IEEE Trans. Commun. **31** (1983) 532–540.
3. Chan, T., Zhou, H.: Adaptive ENO-wavelet transforms for discontinuous functions. UCLA Technical Report 99-21 (1999) June.
4. Chan, T., Zhou, H.: Optimal construction of wavelet coefficients using total variation regularization in image compression. UCLA Technical Report CAM 00-27 (2000) July.
5. Daubechies, I.: Orthonormal bases of compactly supported wavelets. Commun. on Pure and Applied Mathematics **41** (1988) 909–996.
6. Daubechies, I.: Ten Lectures on Wavelets. SIAM (1992).
7. Donoho, D. L., Johnstone, I. M.: Ideal spatial adaptation via wavelet shrinkage. Biometrika **81** (1994) 425–455.
8. Donoho, D. L., Johnstone, I. M.: Adapting to unknown smoothness via wavelet shrinkage. JASA **90** (1995) 1200–1224.

9. Fayyad, U. M., Piatetsky-Shapiro, G., Smyth, P.: The KDD process for extracting useful knowledge from volumes of data. Communications of the ACM Special Issue on Data Mining **39** (1996) 27–34.
10. Fayyad, U. M., Piatetsky-Shapiro, G., Smyth, P., Uthurusamy, R.: Advances in Knowledge Discovery and Data Mining. MIT Press, Cambridge, Mass. (1996).
11. FBI Fingerprint Image Compression Standard Web Page: http://www.c3.lanl.gov/~brislawn/FBI/FBI.html.
12. Fodor, I. K., Cantú-Paz, E., Kamath, C., Tang, N. A.: Finding bent-double radio galaxies: a case study in data mining. Computing Science and Statistics **23** (2000).
13. Fodor, I.K., Kamath, C.: On de-noising images using wavelet-based statistical techniques. Manuscript in preparation (2001).
14. JPEG 2000 Standard Web Page: http://www.jpeg.org.
15. LeMoigne, J.: Parallel registration of multi-sensor remotely sensed imagery using wavelet coefficients. Proc. SPIE Wavelet Applications Conference (1994) 432–443.
16. Li, H., Manjunath, B. S., Mitra, S. K.: Multisensor image fusion using the wavelet transform. Proc. First International Conference on Image Processing (1994) 51–55.
17. Ma, W.Y., Manjunath, B.S.: A comparison of wavelet transform features for texture image annotation. Proc. Second International Conference on Image Processing (1995) 256–259.
18. Malladi, R., Sethian, J.: A unified approach to noise removal, image enhancement, and shape recovery. IEEE Trans. Image Processing **5** (1996) 1154–1168.
19. Mallat, S.: Multiresolution approximation and wavelet orthonormal bases of L2. Trans. American Mathematical Society **315** (1989) 69–87.
20. Meng, Q., Thompson, W., Flachs, G., Jordan, J.: Wavelet transform application in human face recognition. Proc. of the SPIE **3068** (1997) 124–135.
21. MPEG-4 Standard Web Page: http://www.cselt.it/mpeg/.
22. Ogden, R.T.: Essential Wavelets for Statistical Applications and Data Analysis. Birkhäuser (1997).
23. Sapphire Web Page: Sapphire: Large-scale Data Mining and Pattern Recognition. http://www.llnl.gov/casc/sapphire.
24. Starck, J.-L., Murtagh, F., Bijaoui, A.: Image and Data Analysis: The Multiscale Approach. Cambridge University Press (1998).
25. Strang, G., Nguyen, T.: Wavelets and Filter Banks. Wellesley-Cambridge Press (1996).
26. Uhl, A.: Wavelets: adaptive and parallel methods in image coding and signal processing. PhD thesis, University of Salzburg (1996).
27. Umbaugh, S.: Computer Vision and Image Processing: A Practical Approach using CVIPtools. Prentice Hall (1998).
28. Unser M., Aldroubi,A.: A review of wavelets in biomedical applications. Proc. of the IEEE **84** (1996) 626–638.
29. Vidakovic, B.: Statistical Modeling by Wavelets. Wiley Series in Probability and Statistics. John Wiley & Sons (1999).
30. Weeks, A.: Fundamentals of Electronic Image Processing. SPIE Optical Engineering Press and IEEE Press (1996).

Dynamic Subgrid Modeling for Scalar Convection-Diffusion-Reaction Equations with Fractal Coefficients

Johan Hoffman

Mathematical Department, Chalmers University of Technology
SE-412 96 Göteborg, Sweden

Abstract. In this paper we propose and study a subgrid model for linear convection-diffusion-reaction problems with fractal rough coefficients. The subgrid model is based on extrapolation of a modeling residual from coarser scales using a computed solution without subgrid model on a finest scale as reference. We present a priori and a posteriori error estimates, and we show in experiments that a solution with subgrid model on a scale h corresponds to a solution without subgrid model on a scale less than $h/4$.

1 Introduction

In recent years, new approaches to the subgrid modeling problem for turbulent flow problems have been taken based on *dynamic subgrid modeling*, an idea first introduced by Germano et al. [5]. The basic assumption here is that a particular subgrid model applies on different scales with the same value on the model parameters. Using this assumption, one seeks to find a subgrid model by computing approximations of the subgrid model on coarser scales using a fine scale computed solution without subgrid model as reference, and then finally extrapolating the so obtained model to the finest computational scale, with the hope of being able to extrapolate from the finest resolvable scales to unresolvable scales. In order for such a dynamic subgrid modeling process based on extrapolation to work, it is necessary that the underlying problem has some "scale regularity", so that the experience gained by fitting the model on a coarse scale with a fine scale solution as reference, may be extrapolated to the finer scale. It is conceivable that many problems involving a range of scales from large to small, such as fluid flow at larger Reynolds numbers, in fact does have such a regularity, once the larger scales related to the geometry of the particular problem have been resolved.

The purpose of this note is to study the feasibility of the indicated dynamic subgrid modeling procedure in the context of some simple model problems related to linear convection-diffusion-reaction with irregular coefficients with features on many scales. The scale regularity in this case appears to be close to assuming that the coefficients have a "fractal nature" and that the solution inherits this structure to some degree. Thinking of the model problem, for example, as a reacting scalar in a turbulent velocity field, this assumption can be motivated by the abundant empirical observations [4],[11],[12],[13],[14] about fractality in hydrodynamic turbulence.

2 Formulation of the Problem

As a model problem we consider the scalar convection-diffusion-reaction problem of the form

$$Lu(x) = -\nabla \cdot (\epsilon(x)\nabla u(x)) + \beta(x) \cdot \nabla u(x) + \alpha(x)u(x) = f(x), \qquad (1)$$

for $x \in \Omega \subset \mathbb{R}^n$, together with boundary conditions for the inflow and outflow boundaries $\partial\Omega_- = \{x \in \partial\Omega : \ \beta(x) \cdot n(x) < 0\}$ and $\partial\Omega_+ = \{x \in \partial\Omega : \beta(x) \cdot n(x) > 0\}$, where n is an unit outward normal from Ω. $\epsilon(x)$, $\beta(x)$, and $\alpha(x)$ are given coefficients depending on x, $f(x)$ is a given force, and $u(x)$ is the solution. We assume that the coefficients ϵ, β, and α are piecewise continuous, and we seek a weak solution u that satisfies the corresponding variational formulation of (1), and the boundary conditions in the sense of traces.

We assume that the coefficients and the force vary on a range of scales from very fine to coarse scales, and we expect the exact solution u in general to vary on a related range of scales. We denote by h the finest possible scale we allow us to use, which may be the finest possible scale in a computation of a solution, and we denote the corresponding approximate solution u_h. We assume for now that u_h is the weak solution of the following simplified problem

$$L_h u_h = -\nabla \cdot ([\epsilon]^h \nabla u_h) + [\beta]^h \cdot \nabla u_h + [\alpha]^h u_h = [f]^h, \quad x \in \Omega, \qquad (2)$$

together with boundary conditions, where $[\epsilon]^h$, $[\beta]^h$, $[\alpha]^h$, and $[f]^h$ are approximations of the corresponding functions on the scale h, with the finer scales left out. We may think of u_h as an approximation of the exact solution u obtained by simplifying the model by simplifying the coefficients in the model removing scales finer than h. Typically, the coefficient $[\beta]^h$ is some local average of β on the scale h, etc. The difference $u - u_h$ thus represents a modeling error resulting from averaging the coefficients on the scale h.

We now consider a situation where u_h is not a sufficiently good approximation of u, and we would like to improve the quality of u_h without computing using finer scales than h. The equation $Lu = f$ satisfied by the exact solution can be written in the form

$$L_h u = [f]^h + F_h(u), \qquad (3)$$

where

$$F_h(u) = f - [f]^h - (L - L_h)u \qquad (4)$$

acts as a *modeling residual*. The subgrid modeling problem is to model $F_h(u)$ on the scale h. There are a variety of possibilities to approach this problem.

We may use $F_h(u)$ as a correction on the force and replace the model $L_h u_h = [f]^h$ by the model

$$L_h \tilde{u}_h = [f]^h + \tilde{F}_h, \tag{5}$$

with solution $\tilde{u}_h$, where $\tilde{F}_h$ is supposed to approximate $F_h(u)$. Alternatively, we may seek to model $L - L_h$ as a correction $\tilde{L}_h$ of the operator L_h and solve a modified problem of the form

$$(L_h + \tilde{L}_h)\tilde{u}_h = f, \tag{6}$$

where thus the correction $\tilde{L}_h$ acts as a model of $L - L_h$. In the first approach the subgrid model takes the form of a corrective force $\tilde{F}_h$ independent of $\tilde{u}_h$, and in the second approach the subgrid model also contains a correction $\tilde{L}_h \tilde{u}_h$ depending on $\tilde{u}_h$.

3 Extrapolation of $F_h(u)$ Using MRA

The notion of Multiresolution Analysis (MRA) was introduced in the early 90's by Meyer [10] and Mallat [9] as a general framework for construction of wavelet bases. For simplicity, we will only consider the one dimensional case in this section. The multi dimensional case is presented in [7], and for a more detailed presentation of the MRA concept we refer to [1] or [8].

An orthonormal MRA of $L_2([0,1])$ is a decomposition of $L_2([0,1])$ into a chain of closed subspaces

$$V_0 \subset V_1 \subset ... \subset V_j \subset ... \qquad \text{such that} \qquad \overline{\bigcup_{j \geq 0} V_j} = L_2([0,1]).$$

Each V_j is spanned by the dilates and integer translates of one *scale function* $\varphi \in V_0$, that is, $V_j = \text{span}\{\varphi_{j,k}(x) = 2^{j/2}\varphi(2^j x - k)\}$. The functions $\varphi_{j,k}$ form an L_2-orthonormal basis in V_j, and we denote the orthogonal complement of V_j in V_{j+1} by W_j, which is generated by another orthonormal basis (the *wavelets*) $\psi_{j,k}(x) = 2^{j/2}\psi(2^j x - k)$, where $\psi \in W_0$ is called the *mother wavelet.* The space $L_2([0,1])$ can now be represented as a direct sum

$$L_2([0,1]) = V_0 \oplus W_0 \oplus ... \oplus W_j \oplus ...$$

and each $f \in L_2([0,1])$ has a unique decomposition

$$f = f_\varphi \varphi + \sum_{i,k} f_{i,k}\psi_{i,k} = f_\varphi + \sum_i f_i, \tag{7}$$

where the f_i represent the contributions on the different scales 2^{-i} corresponding to subdivisions S_i of $[0,1]$ with mesh points $x_{i,k} = k2^{-i}$, $k = 0, 1, ..., 2^i$, and subintervals $I_{i,k} = (k2^{-i}, (k+1)2^{-i})$. The coefficients $f_{i,k}$ are given as the L_2-inner product of the function f and the corresponding wavelet: $f_{i,k} = \int_0^1 f(x)\psi_{i,k}(x)\,dx$, and $f_\varphi = \int_0^1 f(x)\varphi(x)\,dx$. In the case of the one dimensional Haar MRA in $L_2([0,1])$, the scale function is defined by $\varphi(x) = 1$ for $x \in [0,1]$ and 0 else, and the mother wavelet is defined by $\psi(x) = 1$ for $x \in (0,1/2)$, -1 for $x \in (1/2,1)$ and 0 else.

Definition 1. For $f \in L_2([0,1])$ we define

$$[f]^h = f_\varphi + \sum_{j<i} f_j, \tag{8}$$

where we let $h = 2^{-i}$ in the rest of this paper.

Remark 2. For the one dimensional Haar MRA we have that V_i is the space of piecewise constant functions on S_i, and the linear mapping $L_2 \ni f \to [f]^h \in V_i$ can then be identified with the L_2-projection of f onto V_i.

Assuming that ϵ is constant, so that $[\epsilon]^h = \epsilon$, the modeling residual $F_h(u)$ is given by

$$F_h(u) = f - [f]^h - (L - L_h)u = f - [f]^h - (\beta Du + \alpha u - [\beta]^h Du - [\alpha]^h u).$$

In the rest of this section we will restrict ourselves to the case of the Haar MRA, where we have that $[[f]^H g]^h = [f]^H [g]^h$ for $H \geq h$ ($H = 2^{-j}$, $h = 2^{-i}$ with $j < i$). This gives that

$$[F_h(u)]^h = [L_h u]^h - [Lu]^h = [\beta]^h [Du]^h - [\beta Du]^h + [\alpha]^h [u]^h - [\alpha u]^h,$$

and we denote the projection $[F_h(u)]^h$ of $F_h(u)$ onto V_i by $\bar{F}_h(u)$. We shall now seek to extrapolate $\bar{F}_h(u)$, and we are thus led to study in particular quantities of the form

$$E_h(v,w) = [vw]^h - [v]^h [w]^h \tag{9}$$

for given functions v and w, which has the form of a covariance. Using the Haar MRA, the covariance $E_h(v,w)$ takes a simple form (proof in [6]):

Lemma 3. $v, w \in L_2([0,1]) \Rightarrow$ *For a given* $x \in I_{i,k}$

$$E_h(v,w)(x) = 2^i \sum_{j \geq i} v_{j,l} w_{j,l}, \quad \forall l : I_{j,l} \subset I_{i,k}. \tag{10}$$

Lemma 3 asserts that $E_h(v,w)(x)$ equals the sum of the mean of the wavelet coefficients over $I_{i,k}$, on each scale finer than and equal to h. The asymptotic behaviour of the wavelet coefficients determine the micro scale structure of a function. An interesting situation is when both v and w are "scale regular" in the sense that $v_{j,k} = \alpha\ 2^{-j(1/2+\delta)}$ and $w_{j,k} = \beta\ 2^{-j(1/2+\gamma)}$, where $\alpha, \beta, \delta, \gamma$ are functions of x, which corresponds to v and w having a simple fractal structure ($v_{j+1}(x) = 2^{-(1/2+\delta(x))} v_j(x)$ and $w_{j+1}(x) = 2^{-(1/2+\gamma(x))} w_j(x)$). In that particular situation we find that $E_h(v,w)$ has a certain form (proof in [6]):

Corollary 4. *If* $v, w \in L_2(I)$ *with* $v_{j,k} = \alpha\, 2^{-j(1/2+\delta)}$ *and* $w_{j,k} = \beta\, 2^{-j(1/2+\gamma)}$, *where* $\alpha, \beta, \delta, \gamma \in V_i$, *then for* $x \in [0,1]$

$$E_h(v,w)(x) = C(x)h^{\mu(x)}, \tag{11}$$

where $C = \frac{\alpha\beta}{1-2^{-(\delta+\gamma)}}$ *and* $\mu = \delta + \gamma$.

Assuming that the coefficients in (1) are "scale regular" in the sense of Corollary 4, and that the solution inherits this local fractal structure to some degree, we can formulate the following simple Ansatz: For a given $x \in [0,1]$, $E_h(v,w)(x) = C(x)h^{\mu(x)}$, where $C(x)$ and $\mu(x)$ are functions independent of the cut-off h. If $E_h(v,w)$ has this form, extrapolation of $E_h(v,w)$ will be possible from knowledge of $E_H(v,w)$ and $E_{\hat{H}}(v,w)$ with $h < H < \hat{H}$, from which the coefficients $C(x)$ and $\mu(x)$ may be determined.

Remark 5. The usefulness of higher order wavelet bases to extrapolate (9) is not investigated in this paper but might be of interest, for example, using the same basis for the numerical method as for the extrapolation would certainly seem like the natural choice. In this work we use a special property of the Haar basis, but it may be possible to also use higher order wavelets although the analogy of Corollary 4 is not obvious. On the other hand, the accuracy of the Haar basis seems sufficient in this context and the simplicity is appealing in itself since, for example, the extension to unstructured meshes is straightforward (using hierarchical meshes).

4 Error Analysis

In the first theorem of this section we will for simplicity assume that $[\beta]^h$ is differentiable and satisfies $\nabla\cdot[\beta]^h = 0$. For the case when $[\beta]^h$ is the truncated Haar expansion of β corresponding to Definition 1 in one dimension, we can think of $[\beta]^h$ as being the piecewise linear function we get by linearly connecting the midpoint values of the subdomains on the scale h. We let $||w||$ denote the L_2-norm of w, and we let $||w||_\rho$ denote the weighted L_2-norm, defined by $||w||_\rho = (\int_\Omega \rho(x)w^2(x)\,dx)^{1/2}$, for $v, w \in L_2(\Omega)$, where the weight $\rho : \Omega \to \mathbb{R}_+$ is locally integrable on Ω. The proofs of Theorem 6-7 are given in [7].

Theorem 6. *Assume that* ϵ *is constant,* $[\beta]^h$ *is differentiable with* $\nabla\cdot[\beta]^h = 0$*, and* $0 < c \leq [\alpha]^h$*. If* u*,* u_h *and* $\tilde{u}_h$ *are the solutions to the corresponding variational formulations of (1), (2) and (5) respectively, and* $\tilde{F}_h$ *is an approximations of* $F_h(u)$ *then*

$$||\nabla(u-u_h)||_\epsilon^2 + ||u-u_h||^2 \leq C||F_h(u)||^2,$$
$$||\nabla(u-\tilde{u}_h)||_\epsilon^2 + ||u-\tilde{u}_h||^2 \leq C||F_h(u)-\tilde{F}_h||^2,$$

where $C = \max(1, \frac{1}{c})$ *and* $F_h(w) = f - [f]^h - (\beta\cdot\nabla w + \alpha w - [\beta]^h\cdot\nabla w - [\alpha]^h w)$.

Theorem 6 relates the modeling error in u_h and $\tilde{u}_h$ to the modeling residual $F_h(u)$, and indicates, for example, that if $F_h(u)$ is large, then the modeling error $u - u_h$ is large and subgrid modeling is needed. On the other hand, if $F_h(u)$ is small, then the modeling error $u - u_h$ is also small and subgrid modeling is not needed. The quality of the corrected solution $\tilde{u}_h$ depends on how well we can approximate $F_h(u)$ by $\tilde{F}_h$.

To make the extrapolation of the modeling residual possible at all, the numerical errors in the computations underlying the extrapolation have to be small enough. Thus, it will be of central importance to accurately balance the errors from numerical computation and subgrid modeling. Denoting a computed solution to the simplified problem (2) by U_h, we have

$$e = u - U_h = (u - u_h) + (u_h - U_h) = e_h + E_h, \tag{12}$$

where e_h represents a modeling error and E_h a numerical error. From (12) we have that

$$[e]^h = [e_h]^h + [E_h]^h. \tag{13}$$

Similarly, denoting the computed approximation of the corrected problem (5) by $\tilde{U}_h$, we have

$$[\tilde{e}]^h = [u - \tilde{U}_h]^h = [u - \tilde{u}_h]^h + [\tilde{u}_h - \tilde{U}_h]^h = [\tilde{e}_h]^h + [\tilde{E}_h]^h. \tag{14}$$

If $||[e_h]^h|| > ||[E_h]^h||$, then subgrid modeling may be attempted with the goal of computing a corrected solution $\tilde{U}_h$ with $||[\tilde{e}_h]^h|| << ||[e_h]^h||$. Theorem 6 is not useful to estimate $||[e_h]^h||$ and $||[E_h]^h||$ since the bounds are in terms of the unknown modeling residual $F_h(u)$.

We now present a posteriori error representations allowing us to estimate $||[e]^h||$ and $||[E_h]^h||$, and thereby $||[e_h]^h||$, and similarly $||[\tilde{e}]^h||$ and $||[\tilde{E}_h]^h||$, and hence $||[\tilde{e}_h]^h||$. The error representations are expressed in terms of residuals and the solutions to certain dual problems related to (1) of the form:

$$L^*\varphi = -\nabla \cdot (\epsilon \nabla \varphi) - \nabla \cdot (\beta \varphi) + \alpha \varphi = \Psi. \tag{15}$$

A posteriori analysis based on duality is carefully described in [3], and the basic calculation is

$$(e, \Psi) = (e, L^*\varphi) = (Le, \varphi) = (R, \varphi), \tag{16}$$

where $(\cdot,\cdot)$ is the usual L_2-inner product, e is some error, and R is the corresponding residual. To estimate the error $||[e]^h||$ we set $\Psi = [e]^h$ and since $[\cdot]^h$ is a projection we get that $(e, \Psi) = (e, [e]^h) = ([e]^h, [e]^h) = ||[e]^h||^2$. This leads to the following error representations:

Theorem 7. *We have the following error representations*

$$||[e]^h||^2 = (R(U_h), \varphi), \tag{17}$$
$$||[\tilde{e}]^h||^2 = (R(\tilde{U}_h), \varphi), \tag{18}$$
$$||[E_h]^h||^2 = (R_h(U_h), \varphi), \tag{19}$$
$$||[\tilde{E}_h]^h||^2 = (R_h(\tilde{U}_h), \varphi), \tag{20}$$

where $R(w) = f - \beta \cdot \nabla w - \alpha w$, $R_h(w) = [f]^h - [\beta]^h \cdot \nabla w - [\alpha]^h w$, *and* φ *is the solution to the problem (15) with* $\Psi = [e]^h$, $\Psi = [\tilde{e}_h]^h$, $\Psi = [E_h]^h$, *and* $\Psi = [\tilde{E}_h]^h$ *respectively.*

Remark 8. We note that we can decompose the residual $R(w)$ in Theorem 7 as $R(w) = R_h(w) + F_h(w)$, where $R_h(w)$ is a *numerical residual* and $F_h(w)$ is the modeling residual from Theorem 6.

Remark 9. Using Theorem 7 to estimate $[e]^h$ and $[E_h]^h$ involves the solution of the dual problem (15), for which we refer to [3]. However, estimating the inner product in (17) and (19) by Cauchy-Schwarz inequality suggests that the relative size of $||R(U_h)||$ and $||R_h(U_h)||$ could possibly serve as a crude indicator of whether the modeling error in U_h is significant or not.

5 A Numerical Example

We now study the performance of the indicated subgrid model, based on extrapolation of $F_h(u)$, in a numerical experiment (for more numerical tests see [6],[7]). We consider the problem (1) when $\epsilon << \beta$, for which the problem has a hyperbolic character corresponding to a flow problem with a high Reynolds number. Motivated by empirical observations about fractality in hydrodynamic turbulence [4],[11],[12],[14], and in particular [13], we use coefficients that are tensor products of the one dimensional fractal functions of the form

$$\eta_{\gamma,\delta}(x) = 1 + \sum_{\substack{j \geq 0 \\ l \,:\, x \in I_{j,l}}} \gamma \; 2^{-j(1/2+\delta)} \psi_{j,l}(x), \quad x \in [0,1], \tag{21}$$

which are scale regular in the sense of Corollary 4 (see Fig.1-2). We use a Streamline diffusion method [2] to solve the problem

$$-\epsilon \Delta u + \beta \cdot \nabla u + \alpha u = f, \tag{22}$$

with boundary conditions $u|_{\partial\Omega_-} = \frac{\partial u}{\partial n}|_{\partial\Omega_+} = 0$. $\epsilon = 10^{-3}$ is constant, and we have the corresponding simplified problem

$$-\epsilon \Delta u_h + [\beta]^h \cdot \nabla u_h + [\alpha]^h u_h = [f]^h, \tag{23}$$

with the same boundary conditions as for (22). $[\cdot]^h$ is the L_2-projection onto piecewise constant functions on the scale h, allowing us to use the results for the Haar MRA in Sect.3 and the references cited therein. The exact solution u to the problem (22) satisfies

$$-\epsilon \Delta u + [\beta]^h \cdot \nabla u + [\alpha]^h u = [f]^h + F_h(u), \tag{24}$$

where the modeling residual $F_h(u)$ projected onto V_i is given by

$$\begin{aligned} \bar{F}_h(u) &= [\beta]^h \cdot [\nabla u]^h - [\beta \cdot \nabla u]^h + [\alpha]^h [u]^h - [\alpha u]^h \\ &= [\beta_1]^h [\frac{\partial u}{\partial x}]^h - [\beta_1 \frac{\partial u}{\partial x}]^h + [\beta_2]^h [\frac{\partial u}{\partial y}]^h - [\beta_2 \frac{\partial u}{\partial y}]^h + [\alpha]^h [u]^h - [\alpha u]^h \\ &= \bar{F}_h^1(u) + \bar{F}_h^2(u) + \bar{F}_h^3(u). \end{aligned} \tag{25}$$

Now we let $\beta = (\beta_1, \beta_2)$, where $\beta_1(x,y) = \beta_2(x,y) = \eta_{\gamma_\beta,\delta_\beta}(x)\eta_{\gamma_\beta,\delta_\beta}(y)$, and $\alpha(x,y) = \eta_{\gamma_\alpha,\delta_\alpha}(x)\eta_{\gamma_\alpha,\delta_\alpha}(y)$. We set $(\gamma_\beta, \delta_\beta) = (0.2, 0.25)$, and $(\gamma_\alpha, \delta_\alpha) = (0.3, 0.08)$. Following the discussion in Sect.3, the simple fractal structure of the coefficients α and β, together with the assumption that the solution inherits this structure to some degree, motivates the Ansatz: For given $(x,y) \in \Omega$,

$$\bar{F}_h^k(u)(x,y) \approx C_k(x,y)h^{\mu_k(x,y)}, \quad k = 1,2,3, \tag{26}$$

where $C_k, \mu_k : \Omega \to \mathbb{R}$ are independent of the cut-off parameter h. By computing approximations $\bar{F}_H^k(u) \approx \bar{F}_H^k(u_h)$ on coarser scales H and $\hat{H}$ we can independently determine (C_k, μ_k) and since we are using the Haar MRA $\bar{F}_H^k(u_h)$ will be piecewise constant, and thus will also (C_k, μ_k) be piecewise constant. We denote the extrapolated approximations $\tilde{F}_h^k$, and we denote the sum $\tilde{F}_h$. We define the corrected solution $\tilde{u}_h$ to be the solution to the problem

$$-\epsilon \Delta \tilde{u}_h + [\beta]^h \cdot \nabla \tilde{u}_h + [\alpha]^h \tilde{u}_h = [f]^h + \tilde{F}_h. \tag{27}$$

The domain of the problem is the unit square, and first we will compute on a very fine mesh (uniform quadratic with 512×512 elements), and we will then assume the numerical errors to be neglible compared to the modeling errors. We will denote by u the solution to the problem with the finest possible resolved coefficients on this mesh, and we will use u as an approximation of the exact solution to the problem in the computation of the error. To measure the relative error in U_h and $\tilde{U}_h$, projected on the scale h, we define a "gain-factor" GF. We also relate the reduction of the modeling error we get from using the extrapolated modeling residual to the reduction of the error we get by refining the mesh, by introducing a "mesh-factor" MF_p.

$$GF = \frac{\|[u - U_h]^h\|}{\|[u - \tilde{U}_h]^h\|} \qquad MF_p = \frac{\|[u - U_{h/p}]^h\|}{\|[u - \tilde{U}_h]^h\|}$$

We use the cut-off $h = 2^{-5} \approx \sqrt{\epsilon}$, and when extrapolating from $(H, \hat{H}) = (2h, 4h)$ we get $GF = 1.9$ and $MF_2 = 1.7$. This means that the modeling error in the corrected solution $\tilde{U}_h$ is less than the modeling error in a non corrected solution on a uniformly refined mesh, so $\tilde{u}_h$ corresponds to an "extrapolation of u_h beyond the unresolved scale $h/2$". In Fig.3 we have plotted $\bar{F}_{\hat{H}}^1(u_h)$, $\bar{F}_H^1(u_h)$, $\tilde{F}_h^1$, and in Fig.4 we have plotted the approximation of the total modeling residual $\tilde{F}_h$.

In [6] it was suggested that the extrapolations could be improved by extrapolating from coarser scales than $H = 2h$ and $\hat{H} = 4h$. The extrapolations are based on computed functions v_h and w_h on the scale h, which means that all wavelet coefficients corresponding to finer scales than h will be zero. Thus by Lemma 3 the approximation of $E_H(v_h, w_h)$ improves when the difference $H - h$ increases (in case of the Haar MRA $E_h(v_h, w_h) = 0$). On the other hand, extrapolating from coarser scales involves coarser averaging which leads to a poor ability to represent spatial variations in the extrapolated approximation

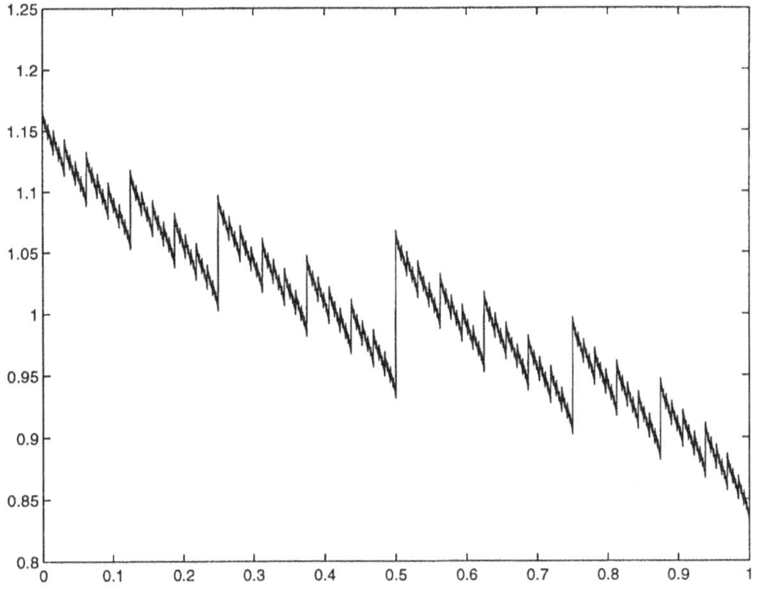

Fig. 1. $\eta_{\gamma,\delta}(x)$ with $\gamma = 0.05$ and $\delta = 0.5$.

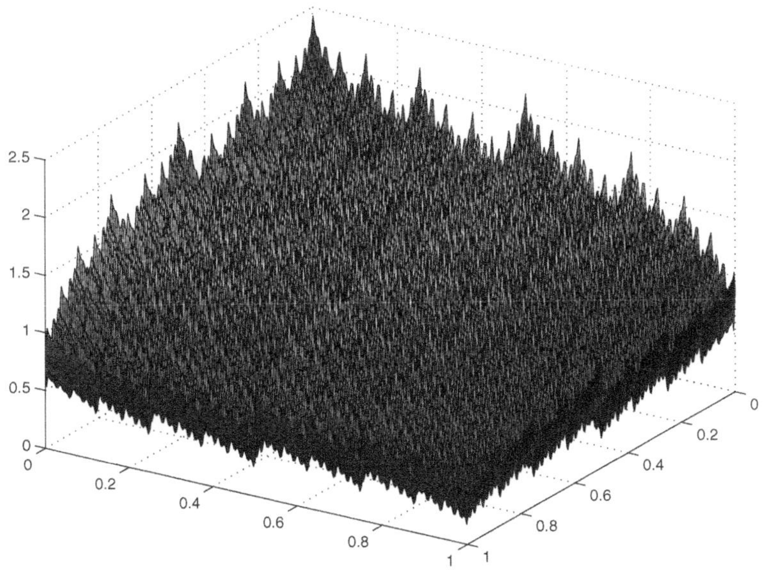

Fig. 2. Two dimensional tensor product $\eta_{\gamma,\delta}(x)\eta_{\gamma,\delta}(y)$ with $\gamma = 0.05$ and $\delta = 0.5$. See also Color Plate 8.

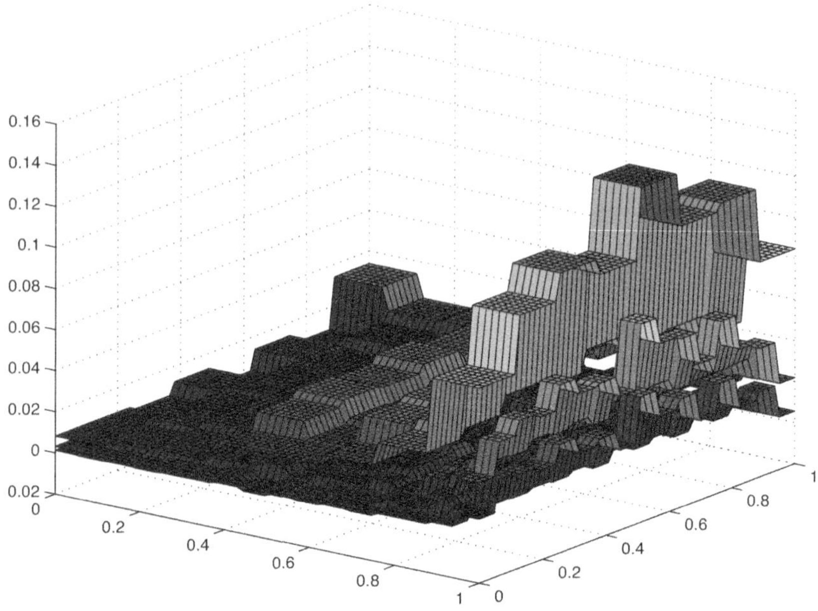

Fig. 3. $\bar{F}^1_{\hat{H}}(u_h)$, $\bar{F}^1_H(u_h)$ and $\tilde{F}^1_h$. See also Color Plate 9.

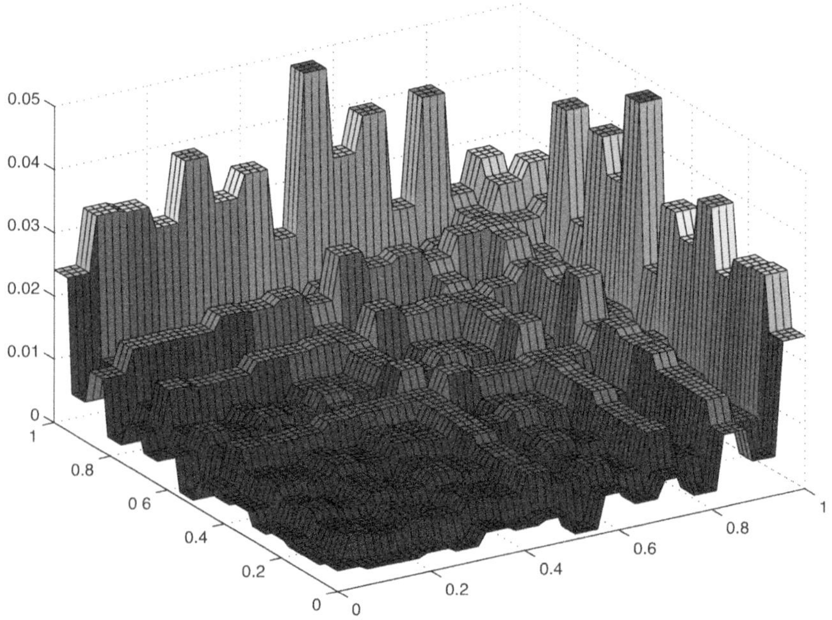

Fig. 4. Extrapolated modeling residual $\tilde{F}_h$. See also Color Plate 10.

of $E_h(v,w)$. We also have to be sure that the scale regularity assumption is still valid for the coarser scales (here this is clear by the construction of $\eta_{\gamma,\delta}$). By extrapolating from $H = 4h$ and $\hat{H} = 8h$ we get $GF = 7.5$ and $MF_4 = 2.3$, so now $\tilde{u}_h$ corresponds to an "extrapolation of u_h beyond the unresolved scale $h/4$".

We then introduce numerical errors by computing on a uniform quadratic mesh with 32×32 elements, corresponding to the cut-off scale $h = 2^{-5}$. Following the discussion in Remark 9 we compute $||R(U_h)||/||R_h(U_h)|| = 1.4$, indicating that in this case the modeling errors can be significant since $||R(U_h)|| > ||R_h(U_h)||$. By extrapolating from $H = 2h$ and $\hat{H} = 4h$ we get $GF = 1.6$ and $MF_2 = 0.91$, so now the total error (modeling error + numerical error) in the corrected solution approximately corresponds the total error in a non corrected solution on the scale $h/2$. We find that extrapolation from greater H and $\hat{H}$, up to a certain level, further reduces the error in $\tilde{u}_h$ (this is summarised in Tab.1).

Table 1. extrapolation level dependence

error	$(H,\hat{H})$	GF	MF_2	MF_4
modeling	$(2h,4h)$	2.1	1.2	0.65
	$(2h,8h)$	2.4	1.4	0.75
	$(4h,8h)$	7.5	4.5	2.3
	$(4h,16h)$	7.2	4.3	2.2
	$(8h,16h)$	1.1	0.63	0.33
modeling + numerical	$(2h,4h)$	1.6	0.91	0.47
	$(2h,8h)$	1.8	1.0	0.52
	$(4h,8h)$	3.9	2.2	1.1
	$(4h,16h)$	5.5	3.1	1.6
	$(8h,16h)$	1.4	0.79	0.40

6 Conclusions

In this paper we have proposed a subgrid model based on extrapolation of a modeling residual, in the case of a linear convection-diffusion-reaction problem with features on a range of scales. We assumed a certain "fractal scale regularity" in the coefficients, and for this case we motivated an Ansatz on the modeling residual using a Haar MRA. We then showed that it was possible to approximate the modeling residual on coarser scales with a fine scale computed solution without subgrid model as a reference, and then extrapolate the modeling residual to the computational scale. We showed in numerical experiments for a hyperbolic model problem that by extrapolating from $2h$ and $4h$, where h is the computational scale, the sum of the numerical and modeling errors in the corrected solution typically corresponded to a non corrected solution on the scale $h/2$, and by extrapolating from $4h$ and $8h$ the corrected solution corresponded to a non corrected solution on the scale $h/4$.

References

1. I.Daubechies. *Ten Lectures on Wavelets.* Society for Industrial and Applied Mathematics, Philadelphia, Pennsylvania, USA, 1992.
2. K.Eriksson, D.Estep, P.Hansbo, C.Johnson. *Computational Differential Equations.* Studentlitteratur, Lund, Sweden, 1996.
3. K.Eriksson, D.Estep, P.Hansbo, C.Johnson. Introduction to Adaptive Methods for Differential Equations. *Acta Numerica, 105-158*, 1995.
4. U.Frisch. *Turbulence, the Legacy of A.N.Kolmogorov.* Cambridge University Press, Cambridge, Great Britain, 1995.
5. M.Germano, U.Poimelli, P.Moin and W.Cabot. A dynamic subgrid scale eddy-viscosity model. *Phys. Fluids A* **3**, *1760*, 1991.
6. J.Hoffman, C.Johnson, S.Bertoluzza. Dynamic Subgrid Modeling I. *Preprint. Chalmers Finite Element Center*, 1999.
7. J.Hoffman. Dynamic Subgrid Modeling II. *Preprint. Chalmers Finite Element Center*, 2000.
8. A.K.Louis, P.Maaß, A.Rieder. *Wavelets, Theory and application.* John Wiley & Sons Ltd, New York, USA, 1997.
9. S.Mallat. Multiresolution approximation and wavelets. *Technical report, GRASP Lab, Dept. of Computer and Information Science, University of Pennsylvania.*
10. Y.Meyer. Ondelettes sur l'intervalle. *Revista Matematica Iberoamericana* **7(2)**, *115-133*, 1991.
11. A.A.Praskovsky, W.F.Dabberdt, E.A.Praskovskaya, W.G.Hoydysh, O.Holynskyj. Fractal geometry of isoconcentration surfaces in a smoke plume. *J. Atmos Sci.* **53**, *5*, 1996.
12. I.Procaccia, A.Brandenburg, M.H.Jensen, A.Vincent. The fractal dimension of iso-vorticity structures in 3-dimensional turbulence. *Europhys. Lett.* **19**, *183*, 1992.
13. A.Scotti, C.Meneveau. Fractal dimension of velocity signal in high-Reynolds-number hydrodynamic turbulence. *Phys. Review E* **51**, *5594*, 1995.
14. K.R.Sreenivasan, C.Meneveau. The fractal facets of turbulence. *J. Fluid Mech.* **173**, *357*, 1986.

Multilevel Methods for Inverse Bioelectric Field Problems

C.R. Johnson[1], M. Mohr[2], U. Rüde[2], A. Samsonov[1], and K. Zyp[1]

[1] Scientific Computing and Imaging Institute, University of Utah
[2] Lehrstuhl für Informatik 10 (Systemsimulation), Friedrich–Alexander–Universität Erlangen–Nürnberg

Abstract. The reconstruction of bioelectric fields from non-invasive measurements can be used as a powerful new diagnostic tool in cardiology and neurology. Mathematically, the reconstruction of a bioelectric field can be modeled as an inverse problem for a potential equation. This problem is ill-posed and requires special treatment, in particular either regularization or an otherwise suitable restriction of the solution space.

The differential equation itself can be discretized by finite differences or finite elements and thus gives rise to a large sparse linear systems for which multigrid is one of the most efficient solvers, but regularization, adaptive mesh refinement, and efficient solution techniques must be combined to solve the inverse bioelectric field problem efficiently. While multigrid algorithms can reduce the compute times substantially, new local regularization techniques can be used to improve the quality of the reconstruction. Local mesh refinement can be used to increase the resolution in domains of increased activity, but must be used with care because refined meshes worsen the ill-conditioning of the inverse problem.

1 Introduction

Hundreds of times each second, the brain sends electrical impulses racing through the body's web of nerve cells to the motor neurons, where they initiate the chemical reactions that cause muscles to contract. Several decades ago, scientists recognized that these excitation currents produce an electrical field that can be detected as small voltages in the skin or scalp. By measuring changes in the pattern of the body's electrical activity, they could detect some forms of heart diseases and neurological disorders. Electrocardiograms (ECGs), for the heart, or electroencephalograms (EEGs), for the brain, measure these voltages. However, because they are relatively coarse and imprecise techniques, they provide physicians with mere snapshots of heart or brain activity. These glimpses help doctors spot disorders but are often insufficient for accurate diagnosis. For the latter, doctors turn to other techniques — even, in rare cases, to exploratory surgery. For example, to determine whether a patient who is not responding to medication has an operable form of the disorder known as temporal lobe epilepsy, neurosurgeons open the cranium and attach electrodes directly to the brain to identify whether the abnormal electrical activity is highly localized (thus operable) or diffused over the entire brain. Similarly, in clinical cardiology there are forms of abnormal heart rhythm (arrhythmias) which do not respond to drug treatment or to artificial stimulation and must be treated surgically. Determining just where to operate, however, is seldom possible with standard ECGs. Instead, such a determination requires either further investigation via a catheter or an open chest examination with a roving, hand-held electrode or a flexible grid of surface electrodes. The localization process is often lengthy and may be unsuccessful [10].

Electrocardiographic imaging, uses an inverse solution applied to electric potentials recorded on the body surface. The goal of the computation is then to reconstruct the complete distribution of electric potentials throughout the torso and the cardiac current source characteristics that produce those distributions. This gives a more complete and informative description of the source of abnormal cardiac activity. To produce such images it is necessary to have accurate individual geometric patient models, to compute the solution of field equations for the potential distributions. Regularization is needed to deal with the extremely ill-posed nature of the problem.

The detection of focal electrical activity in the brain can be characterized as *an inverse source localization* problem. Given a subset of electrostatic potentials measured on the scalp and the geometry and electrical conductivity properties within the head, the inverse problem is to calculate the current vectors and potential fields within the brain. Mathematically this can be stated as *inversely* solving Poisson's equation for the internal current sources. Clinical applications of inverse source localization and imaging from the EEG are already part of practice in advanced neurological and radiological departments, but there remain significant limitations to the technique. Present estimates of the best-case accuracy of a localized dipole in the brain lie in the range of 7–10 mm, a relatively coarse error given the size scale of the brain.

In this paper we will describe the models used in bioelectric field computations and will report on research applying multilevel methods to solve large-scale ill-posed bioelectric field problems using realistic geometries. In particular, multigrid methods can be used as very efficient solvers for the associated forward problem and we will present our experience with algebraic multigrid methods (AMG) which are well suited to deal with complex geometries, as are characteristic of this application. In Sect. 5 we will extend standard regularization to become a local regularization technique and in Sect. 4 we will discuss adaptive mesh refinement methods and their use in inverse problems. In Sect. 7 we will present our numerical results.

2 Mathematical Models

In this article we discuss two types of inverse problems. The first inverse problem we consider is *electrocardiographic imaging*, which can be formulated

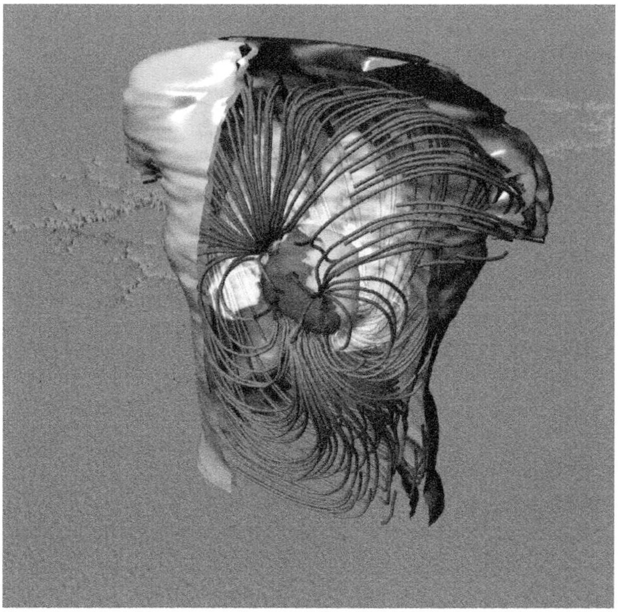

Fig. 1. A visualization of the geometry and electrical current flow in a model of the human thorax for a single time step in the cardiac cycle. See also Color Plate 11.

as the problem of finding the potentials (voltages) on the epicardial surface as a function of those on the surface of the thorax, together with knowledge of the geometric and resistive features of the intervening volume conductor.

The second inverse problem we consider is *source localization* and arises in neurology and neurosurgery planning and intervention. Given a subset of electrostatic potentials measured on the scalp and the geometry and electrical conductivity properties within the head, the inverse problem is to calculate internal current sources that are mathematically modeled as dipoles with unknown magnitude, position, and orientation.

In both cases the physical and mathematical model is based on a general formulation in terms of the primary current sources within the torso or brain described by a Poisson-like equation for electrical conduction:

$$\nabla \cdot \sigma \nabla \Phi = -I_v \quad \text{in } \Omega \tag{1}$$

with the boundary condition:

$$\sigma \nabla \Phi \cdot \mathbf{n} = 0 \quad \text{on } \Gamma_T = \partial \Omega \tag{2}$$

where Φ is the electrostatic potential, σ is the conductivity tensor, I_v are the current sources per unit volume, and Γ_T and Ω represent the surface and the volume of the thorax or head, respectively.

Typically, we can assume Cauchy boundary conditions:

$$\Phi = \Phi_0 \quad \text{on } \Sigma \subseteq \Gamma_T \quad \text{and} \quad \sigma\nabla\Phi \cdot \mathbf{n} = 0 \quad \text{on } \Gamma_T \tag{3}$$

where the Neumann condition is describing that no current flows out of the body, and the Dirichlet condition is a mathematical abstraction of potential measurements on Σ, which are in reality obtained from only a finite number of electrodes.

The simplest source localization problem uses the assumption that $I_v = Q(m, x, y, z, \phi, \theta)$ is a single dipole, characterized by six parameters, corresponding to magnitude m, position (x, y, z), and orientation (ϕ, θ). These parameters are now adjusted such that the resulting electrostatic potential best matches the given measured data. More general models allow for several dipoles and consequently increased number of unknown parameters, or — even more generally — the reconstruction of I_v without assuming anything specific about its shape. Unfortunately this latter problem is generally not uniquely solvable. Solutions can only be found with suitable regularization or a restriction of the solution space by assuming I_v of a special form, where the dipole model is the most common one.

Electrocardiographic imaging describes cardiac sources in terms of the epicardial surface potentials such that instead of Poisson's equation we solve a generalized Laplace's equation:

$$\nabla \cdot \sigma\nabla\Phi = 0 \tag{4}$$

again with boundary conditions (3), but now Γ_T is only a part of $\partial\Omega$ and recovering the unknown (Dirichlet) boundary values on the epicardial surface Γ_E is the primary task.

The solution to the inverse ECG problem in the above form is unique. However, all these problems share a characteristic of all electrocardiographic inverse problems in that they are ill-posed in the Hadamard sense; the solution does not depend continuously on the data. Thus small errors in the measurement of the torso potentials or thorax geometry can yield unbounded errors in the solution. The origins of this ill-posedness are biophysical and lie in the attenuation of potentials as we move away from the source and the fact that the potential at any point on the torso surface is a weighted superposition of all the individual sources within the heart. Hence, the ECG or EEG represents an integration of many sources, the influence of each of which decreases sharply with distance. To find the inverse solution, we must perform the complementary operations of amplification and differentiation on the data, but also on the inevitable noise that accompanies it. The result is exquisite sensitivity to any fluctuations in the input signals or geometric models [13].

3 Discretization and Geometric Models

In order to solve the boundary value problem in (1) or (4) in terms of the bioelectric potentials we need to pose the problem in a computationally tractable way. With current technology it has become feasible to employ the real geometry of an individual patient's head or torso in the setup of Ω. The necessary information is obtained from magnetic resonance imaging (MRI scans), which in some cases is supplemented by computer tomography (CT). The data are used to form a 3D model (e.g. a mesh of tetrahedra, or a rectangular box of voxels).

MRI scans return a collection of 2D slices containing pixel information. Thus a simple and obvious approach to get to a volume representation is to extend pixels from each 2D slice to 3D voxels. In the segmentation process each such voxel is then assigned a tissue information. Usually one restricts oneself to a limited number of tissue types and assumes isotropic conductivities in each compartment, see Fig. 2. For our purposes at least air, scalp, skull and brain have to be distinguished, where the conductivity of air is assumed to be zero. For a more sophisticated segmentation, see e.g. [24].

The next step is to find an approximate solution Φ_h of (1). Using a derivation similar to finite volume theory, we assume that Φ_h is a constant Φ_l within each voxel V_l for $l = 1, \ldots, N$. Using the technique of boxed integration we approximate (1) for every voxel V_l with a difference equation

$$\sum_{V_k \in \mathcal{N}(V_l)} c_{l,k}\Phi_k + c_l\Phi_l = F_l \quad \text{for } l = 1, \ldots, N \tag{5}$$

Here $\mathcal{N}(V_l)$ denotes the *orthogonal* neighbors of V_l, i.e voxels which share a common face with V_l.

The linear system (5) has the form $A\xi = b$ where the system matrix A is sparse, symmetric, and positive semidefinite. Since we are solving a pure Neumann problem, the matrix is singular with a one-dimensional null space. However, the problem can still be solved by standard iterative methods. Due to the sparsity each relaxation step is cheap, but the convergence rate of relaxation methods deteriorates quickly with the problem size.

For the work on the inverse ECG problem a finite element discretization has been used. The torso was modeled using a tetrahedral mesh the details of which can be found in Sect. 7. A 2D variant is shown in Fig. 2. The finite element approximation of the original boundary value problem (1) can be stated as:

$$\text{Find } \Phi_h \in V_h \text{ such that } (\sigma\nabla\Phi_h, \nabla\overline{\Phi}) = -(I_v, \overline{\Phi}) \ . \tag{6}$$

In our case $(.,.)$ denotes the L_2 inner product and V_h is a finite dimensional space of functions living on the tetrahedral mesh. The problem can equivalently be expressed as a system of N equations with N unknowns, $\xi_1, \ldots, \xi_N$ (the electrostatic potentials at the nodes in the domain). In matrix form, the above system is again expressed as $A\xi = b$, where here $A = (a_{ij})$ is the global stiffness matrix and has elements $(a_{ij} = (\sigma_{ij} \nabla \Psi_i, \nabla \Psi_j))$, while $b_i = -(I_v, \Psi_i)$ is the vector of source contributions. The major advantage of finite elements compared to finite difference is a greater flexibility in handling complicated geometries and an extensive theory for error estimation.

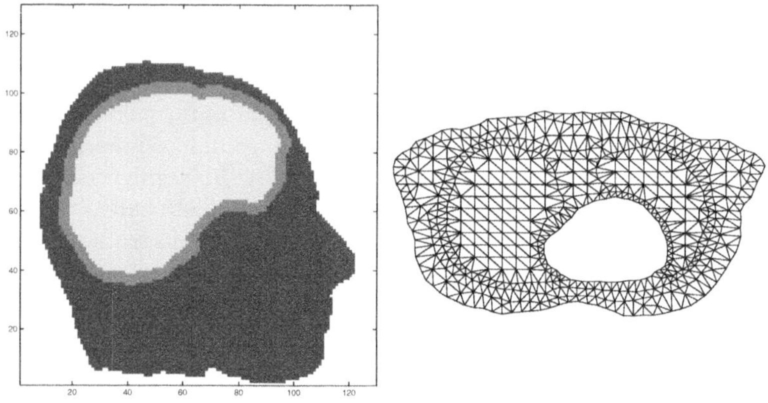

Fig. 2. 3-compartment voxel model of a human head (left), triangular mesh of a torso cross-section (right).

4 Adaptive Mesh Refinement

The most straightforward technique to define a computational grid is to set it up once and then use it throughout the computation. Typically all voxels or elements are then roughly of the same size and they are held constant over time, independent of any changes in the bioelectric potentials.

More advanced techniques provide for an *adaption* of the meshes. Finer meshes are to be placed in regions of increased activity, where otherwise discretization errors could sharply increase. This is especially true in source localization problems where naturally much of the potential activity is focused around the location of the dipoles. But also in cardiac imaging there is more activity close to the heart and therefore seemingly a good reason to use finer meshes there.

Generally, automatic mesh refinement techniques are based on *a posteriori* error *indicators* or error *estimators* [18,22]. A wide variety of such error estimators has been suggested in the literature. In contrast to the standard a-priori estimates for discretization methods, such techniques take the actual shape of the solution into account and can be used to identify regions with large error, where consequently a finer mesh will lead to an improved accuracy in the overall problem.

Additionally, the remeshing has to be done carefully to avoid a too high computational cost on the one side and also avoid a degeneration of the meshes. For the benefit of multilevel solvers the refinement process should generate a sequence of nested meshes. This implies that the finite element spaces are also nested, and this in turn can be used to define an multilevel iterative solver. However, finite element theory also requires that the interior angles of a mesh not become to large. This puts additional constraints on any automatic mesh refinement algorithm [20].

It has always been assumed that improving the accuracy of the forward solution also improves the subsequent inverse solution. This seems to be true up to a point since further refinement increases the number of degrees of freedom and correspondingly increases the condition number of the transfer matrix. Finding the "optimal" point at which the added degrees of freedom stop increasing the accuracy of the inverse solution is an open research problem.

5 Inverse Problems and Regularization

Dealing with the ill-posed nature of the inverse problem has become the primary goal of a great deal of recent research as it remains the single largest obstacle to medical implementation. The most common approach is to apply constraints to the inverse solution in order to reduce its dependence on boundary conditions. The points of difficulty lie in the choice of constraints and the corresponding weight each receives in determining the best — in whatever sense we wish to define "best" — solution (see [12] for a recent review).

Traditional schemes for solving the inverse problem, (4), have involved reformulating the linear equation, $A\xi = b$, into $\xi_T = K\xi_E$, where ξ_T and ξ_E are the voltages on the body surface (torso) and heart's surface (epicardium) respectively, and K is the $T \times E$ transfer (lead field) matrix of coefficients relating the measured torso voltages to the voltages measured on the heart. From this statement of the forward problem, the inverse problem can then be expressed as $\xi_E = K^{-1}\xi_T$.

Unfortunately, K is ill-conditioned, and regularization techniques are necessary to restore continuity of the solution back onto the data. Most regularization schemes treat all parts of K uniformly and apply Tikhonov, singular value decomposition (SVD), or Twomey algorithms with a single regularization parameter. Briefly, one wishes to find an approximate vector, $\xi_{E\varepsilon}$, that represents a balance between the residual error $\|\xi_T - K\xi_{E\varepsilon}\|$, which is corrupted by error and the ill-posed nature of the problem, and whatever side constraints we place on the solution based on *a priori* knowledge. Rather than applying the same level of regularization to the entire matrix K, we have found that it is possible, and advantageous, to first decompose K into submatrices and then apply regularization differently to each component of the resulting expression.

The rationale for such a local approach is that the discontinuities in the inverse solution appear irregularly distributed throughout the solution domain. Tikhonov and other regularization schemes are, in effect, filters, which restore continuity by attenuating the high (spatial) frequency components of the solution. Since regularization is usually applied globally, the results can leave some regions overly damped or smoothed while others remain poorly constrained.

We have two ideas regarding multilevel methods applied to ill-posed problems. The first idea is to apply regularization only to regions (represented by the sub-matrices of K or some other partition) that require it and to apply different amounts of regularization to different sub-matrices/partition, in effect a hierarchical, more local regularization scheme. Our second idea is to use the natural filtering of high frequencies in multigrid and construct a multigrid method that can effectively solve ill-conditioned systems.

Since the goal of the bioelectric inverse problems is to accurately and noninvasively (i.e., non-surgically) estimate accurate solutions, we need a method to choose an *a priori* optimal regularization parameter. Traditional schemes have been based on discrepancy principles [16] and quasi-optimality criterion [16]. Recently, Hansen [6] has extended the observations of Lawson and Hanson [11] and proposed a new method for choosing the regularization parameter based on an algorithm that locates the "corner" of a plot of the norm (or semi-norm) of the side constraint, $\|Lx_\alpha\|$, versus the norm of the corresponding residual vector, $\|Ax_\alpha - b\|$. In this way, one can evaluate the compromise between the residual error and the effect of the side constraint on the solution. We have utilized this *L-curve* algorithm along with our local regularization scheme to improve the accuracy of solutions to the inverse problem. We have also utilized an AMG approach (described in more detail in Section 7) to help in regularizing the ill-conditioned inverse problems.

6 Multilevel Methods

Many approaches to solve inverse bioelectric field problems involve the repeated numerical solution of related forward problems, usually for different boundary conditions and/or right hand sides. Prominent examples are the setup of the lead field matrix [21,23] or the preconditioning of Krylov subspace methods for the inverse problem. In both our applications, the forward problem is of elliptic type. Multigrid algorithms belong to the fastest solution algorithms for such problems and thus can be used to speed up the solution process. Nevertheless, multigrid algorithms have not yet found widespread use for this application. While theoretically well suited for the task, the implementation is complicated by the problem geometry and the jumping and/or anisotropic conductivities. These problems have been individually solved with multigrid in other application fields [2,3], but to get full multigrid efficiency, a substantial development effort in each individual case is required.

A possibility to simplify the development of multigrid is the use of so-called algebraic multigrid methods. These algorithms solely work on the system matrix and do not need further geometric information. AMG was originally developed in the early 1980's [19], but has only recently become a mainstream topic of multigrid research. AMG deals with many of the difficulties that arise in employing standard multigrid in an automatic way and is thus much easier to apply in practical situations. In particular, AMG automatically finds coarser grids: there is no need for the user to provide a mesh hierarchy as is necessary for conventional multigrid algorithms. AMG is also robust against jumps in the coefficients, the type of boundary conditions, or geometric features of the solution domain. Therefore this paper will focus on exploring the suitability of AMG methods for the bioelectric field problems. Compared to standard multigrid, the AMG storage requirements are usually significantly higher than for a comparable geometric multigrid. General sparse matrix structures are also less storage efficient than grid-oriented data structures, which could otherwise be used in our finite difference models. Furthermore, the operation count for AMG is higher and in most cases the convergence rates are not quite as good as for geometric multigrid. The further development of AMG is currently a very active field of research.

The discretization of elliptic PDEs leads in most cases to linear systems that can be handled with Conjugate Gradient methods [5]. The convergence of this method crucially depends on the spectrum (ratio and distribution) of eigenvalues of the problem matrix. Thus preconditioning is employed to improve convergence. Geometric and algebraic multigrid methods can be used as preconditioners, and arguably, this combination of CG and multilevel techniques is currently the most advanced technique to solve sparse linear systems. For more details, see Section 7. Techniques for applying multilevel methods directly to the inverse problem are described in [15].

7 Results

Finite Difference Head Model. Here we will report on results for different iterative solvers for the EEG forward problem discretized by finite differences, as described in Sect. 3. The original task is to locate a dipole inside the brain from EEG data. The algorithm proposed for this task in [21] consists of two stages. In a preparatory stage the potential distribution for $(N-1)$ forward problems has to be computed, where N is the number of electrodes / measurement points. In the second stage the dipole localization can be performed without further forward solves using the reciprocity theorem of Helmholtz.

We used a three-compartment (scalp, skull, brain) geometric head model with constant, isotropic conductivities in each compartment, cf. Sect. 3. Our experiments use two different data sets, a small one (A) with a resolution of 65^3 cells and a larger one (B) with 129^3 cells. Note that in both cases the head occupies only about 25% of the surrounding cube of voxels. We compare the following linear solvers:

- SOR: Successive over-relaxation on the pure Neumann problem with a relaxation parameter $\omega = 1.94$.
- CG: The conjugate gradient method applied to the pure Neumann problem (1), or to the case, where the value at one cell is fixed at zero (2).
- AMG: Algebraic multigrid method applied to the definite problem.
- PCG(AMG): AMG preconditioned CG with one AMG-V-cycle per CG step with one pre- and post-smoothing Gauss-Seidel step.

The SOR method is implemented in a highly optimized code described in [25] and specifically developed for the given problem, while the AMG/PCG solver is a general purpose code and is taken from the Pebbles package [1]. The performance comparison is summarized in Tab. 1. In all cases, the iteration is stopped when the L_2-norm of the residual drops below 10^{-8}. The iteration counts are given for a single forward solve, while the user times are taken for the whole first part of the algorithm. This includes setup of system matrix/stencil coefficients, the solution of 26 forward problems corresponding to 27 electrodes, and, in case of AMG and PCG(AMG), additionally the construction of the matrix hierarchy. Creation of the latter takes about 6s for data set (A) and 79s for (B). The figures in Tab. 1 show that at least for smaller problems a simple, but well implemented SOR algorithm is still reasonably competitive, at least in terms of user time. It outperforms the CG method without preconditioning even in the number of iterations. Clearly, however, the number of iterations for SOR is much higher than for the multilevel algorithms. On the other hand, a single SOR iteration is much cheaper than an AMG cycle. Additionally the Pebbles AMG implementation has larger memory requirements than the SOR code. Overall, at least for the larger data set, AMG in combination with CG acceleration is more efficient

User time in seconds				
Method	Data set			
	A		B	
	Setup	Total	Setup	Total
SOR	0.2	139	1.2	5857
CG(1)	0.2	400	1.2	5358
AMG	6.2	218	79.0	8156
PCG(AMG)	6.2	136	79.0	1924

Number of Iterations		
Method	Data set	
	A	B
SOR	314	1644
CG(1)	408	741
CG(2)	533	987
AMG	18	98
PCG(AMG)	11	20

Table 1. User times (left) for an AMD Athlon K7, 700MHz, 768MB, running linux, iteration counts (right) for the two data sets

than either SOR or plain CG. Note, however, that the Pebbles code does not yet perform with asymptotic optimality, as multigrid theory predicts for an optimally designed multigrid method. The number of iterations is not independent of the mesh size. For plain AMG, almost 100 iterations are required for the large data set. CG acceleration cuts that down to 20, but this is still twice as much as for the small data set and thus clearly not yet fully scalable.

Thus these experiments show that multilevel algorithms have a good potential for accelerating the forward solution process. However, we believe that the full potential of multigrid for this application is not yet exploited. We therefore plan to implement a geometric multigrid method for the forward problem, using data structures similar to those used in the SOR implementation. This should lead to faster run times for each iteration and reduce the memory requirements. Using techniques described e.g. in [2] for treating discontinuous coefficients will also lead to improved convergence rates.

Finite Element Heart Model. To study forward and inverse problems in electrocardiography, we developed a series of two- and three-dimensional boundary element and finite element models based upon magnetic resonance images from a human subject. Each of 116 MRI scans were segmented into contours defining torso, fat, muscle, lung, and heart regions. Additional node points were added to each layer and pairs of layers tessellated into tetrahedra using a Delaunay triangulation algorithm [8]. The resulting model of the human thorax contained approximately 675,000 volume elements, each with a corresponding conductivity tensor [4]. A visualization of a forward simulation from epicardial potentials is shown in Fig. 1.

We created a two-dimensional model from the full 3D model shown in Fig. 1, located approximately 4 cm above the apex of the heart, Fig. 2. This provided a more tractable geometry for the initial studies on the effects of adaptive control of errors and local regularization.

Local Regularization. To test the local regularization technique, we applied epicardial potentials recorded during open chest surgery from a cardiac arrhythmia patient as the Dirichlet boundary conditions of the two-dimensional model described above. The tissue conductivities were assigned as follows: fat = 0.045 S/m, epicardial fat-pad = 0.045 S/m, lungs = 0.096 S/m, skeletal muscle (in the fiber direction) = 0.3 S/m, skeletal muscle (across the fiber direction) = 0.1 S/m, and an average thorax value = .24 S/m. Forward solutions calculated using the adapted mesh served as torso boundary conditions, where $\Phi = \Phi_0$ on $\Sigma \subseteq \Gamma_T$ for the inverse solution both with and without 10% added Gaussian noise.

The application of the local regularization technique using a generalized singular value decomposition (GSVD) recovered the voltages to within 6.1% relative root-mean-squared (RMS) error of the true solution. This compares with a previous best from our work of 12.1% relative error [7]. Previous studies have reported the recovery of epicardial potentials with errors in the range of

20–100%, [14,17], although in these cases investigators used three-dimensional geometric models and in some cases measured potentials on both epicardial and torso surfaces. Fig. 3 shows the inverse solution calculated using the local regularization technique compared with the recorded heart voltages as a function of position on the epicardium. The global solution tended to be smoother, not able to follow the extrema as well as the local solution could. The local regularized solution also showed a few areas of local error which suggest that a different partitioning of the sub-matrices might provide even better accuracy.

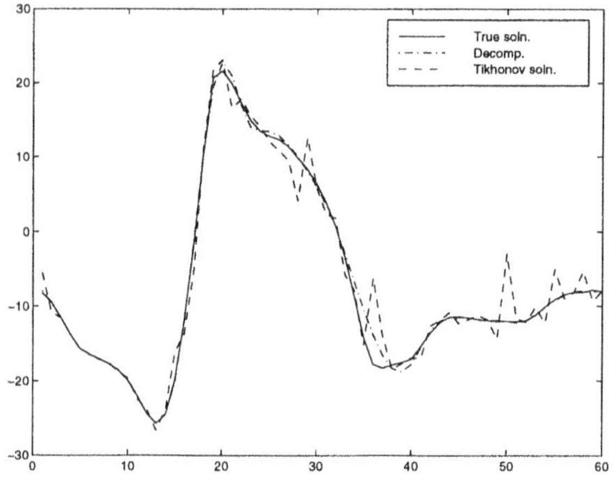

Fig. 3. Local GSVD-based regularization technique. The figure shows the GSVD local regularization solution (marked Decomp), the best Tikhonov solution, and true solution.

AMG Solutions for the Inverse ECG Problem. In this section we report on preliminary work in using AMG for solving the inverse ECG problem. The algorithm used for coarsening in the experiment was based on the graph of the finite element stiffness matrix [9]. The coarsening results in fast computation of the coarse grid matrices, which is a benefit for large-scale problems if the stiffness matrix is known.

For our test problem we used the same 2D slice of the patient torso as used with the hierarchical finite element with local regularization application. Several approaches to coarsening were tested. One of the most valuable was based on the following algorithm:

1. First of all, matrix graph G was created: $G = (v, E)$ where $v = \{i\}$ and $E = \{(i, j) : A[i, j] \neq 0\}$, where $i, j = 1, \ldots, n$

2. Then, graph G was subdivided into trees by the following method: Starting with the first vertex k, not marked as visited, put the vertex into the interpolation graph I, make it a root of the new tree and mark it as visited. Then, for each vertex m in G connected to k and not yet visited, put (k, m) into I and mark m as visited. Repeat these steps until all the nodes are marked as visited.
3. Given interpolation graph I, build a form of a Galerkin interpolation operator: For every sub-tree in I, make its root a coarse grid point, and then build a Galerkin interpolation operator:

$$I_g(i,j) = \begin{cases} 0 & \text{node is not in sub-tree of j-node} \\ 1 & \text{for a coarse grid point} \\ 1/n & \text{otherwise, where n = the number of} \\ & \quad \text{subtrees this node belongs to and} \\ & \text{where } j \text{ belongs to the coarse grid point set.} \end{cases}$$

After construction of the interpolation operator, the restriction operator is defined by $R = I_g^T$, and the final system on the coarse grid will be $A_H x_H = b_H$ where $A_H = RAI_g, b_H = Rb$ and H denotes the original (heart) linear system.

Having constructed the interpolation and restriction operations and coarse grid matrices, we can apply our AMG scheme to the 2D inverse ECG problem. After only four iterations of the AMG method, the reconstruction yields results similar to an exact finite element solution which is obtained at much higher cost, see the comparison in in Fig. 4. This seems to verify an observation made in [15] that multigrid posseses an inherent regularization property. It seems to provide fast convergence for the well-conditioned modes, but is converging poorly for the ill-conditioned modes.

8 Conclusions

In this paper we have described research applying multilevel methods to solve large-scale ill-posed bioelectric field problems using realistic geometries. 3D models based upon patient MRI scans result in meshes that may contain millions of elements, such that even a single forward solution is very compute intensive. Classical multigrid methods can be used as very efficient solvers for the associated forward problem and can then be used as either a preconditioner or for constructing the lead field matrix for the inverse problem.

Finite elements and in particular finite elements on unstructured grids on complicated geometries lead to less sparse, unstructured linear systems, for which a natural grid hierarchy, as needed for classical multigrid, does not exist or would be difficult to construct. Multigrid will in principle be able to solve the forward problem with less than 100 operations per unknown, provided all details within the multigrid algorithm are handled in exactly the right way, but this will not be easy to achieve. Therefore we have used AMG as a black box sparse system solver. As the results indicate, there is still room

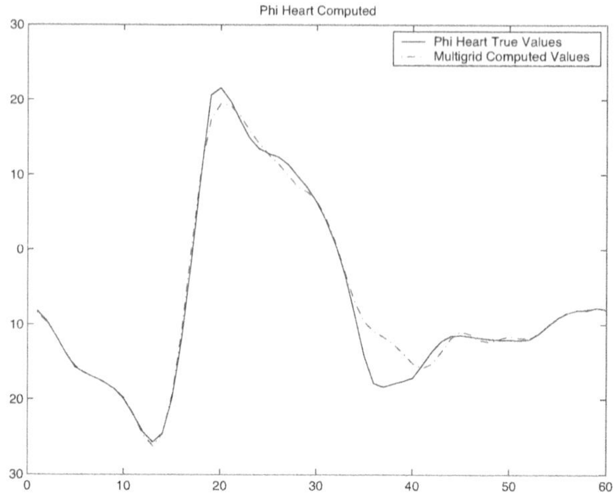

Fig. 4. Results from applying AMG to the ill-posed inverse ECG problem. After one iteration the RMS error is 40.1% and after four iterations the RMS error was reduced to 12.7%.

for improvement, so that we plan to implement geometric multigrid methods to further improve the method.

Even more intriguing is the use of multilevel methods to apply multiple levels to regularize the ill-posed problem, thus resulting in a more *local* regularization technique that provides more control and tuning of the regularization process. Current methods for solving inverse bioelectric field problems are primarily aimed at global Tikhonov regularization methods. Compared to these currently used methods, the application of multilevel methods can provide significant improvement that can make many presently intractable problems solvable.

Acknowledgments

This work was supported in part by awards from the NIH NCRR and the NSF. The authors are also in debt to Bart Vanrumste and the MEDISIP group at the University of Ghent for providing the head data sets and Stefan Reitzinger for the Pebbles AMG implementation.

References

1. *Pebbles – User's Guide.* www.sfb013.uni-linz.ac.at/~reitz/pebbles.html.
2. R. E. Alcouffe, A. Brandt, J. E. Dendy, and J. W. Painter. The multi–grid methods for the diffusion equation with strongly discontinuous coefficients. *SIAM J. Sci. Stat. Comput.*, 2:430–454, 1981.

3. A. Brandt. *Multigrid techniques: 1984 guide with applications to fluid dynamics.* GMD–Studien Nr. 85. Gesellschaft für Mathematik und Datenverarbeitung, St. Augustin, 1984.
4. K.R. Foster and H.P. Schwan. Dielectric properties of tissues and biological materials: A critical review. *Critical Reviews in Biomed. Eng.*, 17:25–104, 1989.
5. A. Greenbaum. *Iterative Methods for Solving Linear Systems.* Frontiers in Applied Mathematics. SIAM, 1997.
6. P.C. Hansen. Analysis of discrete ill-posed problems by means of the L-curve. *SIAM Review*, 34(4):561–580, 1992.
7. C.R. Johnson and R.S. MacLeod. Inverse solutions for electric and potential field imaging. In R.L. Barbour and M.J. Carvlin, editors, *Physiological Imaging, Spectroscopy, and Early Detection Diagnostic Methods*, volume 1887, pages 130–139. SPIE, 1993.
8. C.R. Johnson, R.S. MacLeod, and M.A. Matheson. Computer simulations reveal complexity of electrical activity in the human thorax. *Comp. in Physics*, 6(3):230–237, May/June 1992.
9. F. Kickinger. Algebraic multigrid for discrete elliptic second order problems. Technical report, MGNet, November 1997.
10. H. Korab. Mapping the body's electrical fields. *Access: High-performance computing magazine*, pages 14–15, 1995.
11. C.L. Lawson and R.J. Hanson. *Solving Least Squares Problems.* Prentice-Hall, Englewood Cliffs, NJ, 1974.
12. R.S. MacLeod and D.H. Brooks. Recent progress in inverse problems in electrocardiology. *IEEE Eng. in Med. & Biol. Soc. Magazine*, 17(1):73–83, January 1998.
13. R.S. MacLeod and D.H. Brooks. Validation approaches for electrocardiographic inverse problems. In *Advances in Computational Biomedicine.* WIT Press, 2000.
14. B.J. Messinger-Rapport and Y. Rudy. Regularization of the inverse problem in electrocardiography: A model study. *Math. Biosci.*, 89:79–118, 1988.
15. M. Mohr and U. Rüde. Multilevel Techniques for the Solution of the Inverse Problem of Electrocardiography. In E. Dick, K. Riemslagh, and J. Vierendeels, editors, *Multigrid Methods VI, Proceedings of the Sixth European Multigrid Conference*, volume 14 of *Lecture Note in Computational Science and Engineering*, pages 186–192. Springer, 1999.
16. V.A. Morozov. *Methods for Solving Incorrectly Posed Problems.* Springer-Verlag, New York, 1984.
17. H.S. Oster and Y. Rudy. Regional regularization of the electrocardiographic inverse problem: A model study using spherical geometry. *IEEE Trans Biomed. Eng.*, 44(2):188–199, 1997.
18. U. Rüde. *Mathematical and Computational Techniques for Multilevel Adaptive Methods*, volume 13 of *Frontiers in Applied Mathematics.* SIAM, Philadelphia, 1993.
19. J. W. Ruge and K. Stüben. Algebraic multigrid (AMG). In S. F. McCormick, editor, *Multigrid Methods*, volume 3 of *Frontiers in Applied Mathematics*, pages 73–130. SIAM, Philadelphia, PA, 1987.
20. J.A. Schmidt, C.R. Johnson, J.C. Eason, and R.S. MacLeod. Applications of automatic mesh generation and adaptive methods in computational medicine. In I. Babuska, J.E. Flaherty, W.D. Henshaw, J.E. Hopcroft, J.E. Oliger, and T. Tezduyar, editors, *Modeling, Mesh Generation, and Adaptive Methods for Partial Differential Equations*, pages 367–390. Springer-Verlag, 1995.

21. B. Vanrumste, G. Van Hoey, P. Boon, M. D'Havé, and I. Lemahieu. Inverse calculations in EEG source analysis applying the finite difference method, reciprocity and lead fields. In *Proceedings of 20th Annual International Conference of the IEEE Engineering in Medicine and Biology Society*, volume 20, part 4/6, pages 2112–2115, Hong Kong, 1998.
22. R. Verfürth. *A Review of A Posteriori Error Estimation and Adaptive Mesh–Refinement Techniques.* Wiley Teubner, New York, Stuttgart, 1996. ISBN 0-471-96795-5.
23. D.M. Weinstein, L. Zhukov, and C.R. Johnson. Lead-field bases for EEG source imaging. *Annals of Biomedical Engineering*, 28:1–7, 2000.
24. C. Wolters, S. Reitzinger, A. Basermann, S. Burkhardt, U. Hartmann, F. Kruggel, and A. Anwander. Improved tissue modelling and fast solver methods for high-resolution FE-modelling in EEG/MEG-source localization. In J. Nenonen, R.J. Ilmoniemi, and T. Katila, editors, *Biomag 2000, Proc. of the 12th Internat. Conf. on Biomagnetism.*
25. M. Zetlmeisl. Performance Optimization of Numerically Intensive Codes – A Case Study from Biomedical Engineering. Studienarbeit, Lehrstuhl Informatik 10, FAU Erlangen–Nürnberg, 2000.

Multiscale Eigenbasis Calculations: N Eigenfunctions in $O(N \log N)$

Oren E. Livne[1,2] and Achi Brandt[1,3]

[1] Department of Computer Science and Applied Mathematics, Weizmann Institute of Science, Rehovot 76100, Israel
[2] Email address: livneo@wisdom.weizmann.ac.il
[3] Email address: achi@wisdom.weizmann.ac.il

Abstract. Motivated by quantum chemical calculations, we explore a novel multiscale approach for computing, storing, and expanding in many eigenfunctions of differential operators. This approach leads to efficient multiscale eigenbasis algorithms, which typically scale as $O(N \log N)$, where N is the number of eigenfunctions. In particular, they provide a vast generalization of the Fast Fourier Transform (FFT) algorithm, which expands in Fourier series, to expansions in terms of eigenfunctions of a general 1D linear differential operator with general boundary conditions. Generalizations to higher dimensional eigenproblems are discussed. A complete and detailed discussion of the methods can be found in [14,15].

1 Motivation: Electronic Structure Computations

The basic equations of condensed matter, e.g., the Kohn-Sham equations in ab-initio quantum chemistry [22] have been known for several decades. In principle, one could utilize them for *computerizing chemistry*: controlling chemical interactions and designing materials with prescribed properties (from drugs to missile coating), instead of conducting expensive empirical experiments.

Despite their dazzling speed, even modern supercomputers cannot surmount the highly complicated eigenbasis computations involved. Current numerical eigenbasis algorithms are slow, thus limited to computing moderately complex electronic structures. Linear-complexity algorithms ("$O(N)$ methods") [3,11] have been constructed only under localization assumptions on the electronic orbitals.

Our research goal was to design linear-complexity multiscale algorithms for computing, storing, and manipulating many eigenfunctions of the periodic Schrödinger operator and other related differential operators, *without localization assumptions on the eigenfunctions.*

We have focused mainly on the periodic Schrödinger eigenproblem

$$(-\Delta + V(\boldsymbol{x}))\psi(\boldsymbol{x}) = \lambda\psi(\boldsymbol{x}), \quad \boldsymbol{x} \in \Omega \subset \mathbf{R}^d , \tag{1}$$

where $V(x)$ is an Ω-periodic potential function containing $O(N)$ general "features", such as wells representing screened atomic nuclei [20].

Our work has four computational aims:

1. *Computing* and *storing* N distinct eigenfunctions $\psi_j(x)$ and eigenvalues of (1) in $O(N \log N)$ operations;
2. *Integrating (or summing)* N eigenfunctions of (1) into the electronic density function $\varrho(x) = \sum_j |\psi_j(x)|^2$ in $O(N \log N)$ operations;
3. *Expanding* a given function in N eigenfunctions of (1) in $O(N \log N)$ operations;
4. *Expanding* a given function in N eigenfunctions of *a general 1D linear differential operator*, in $O(N \log N)$ operations.

The last computational task stems from a natural generalization of the 1D Schrödinger operator to Sturm-Liouville operators and general parameter-dependent ODEs.

1.1 Example: The 1D Linear Case

In actual ab-initio electronic structure computations we encounter the nonlinear (self-consistent) 3D Schrödinger eigenproblem, where V depends on non-local integrals involving the electronic density ϱ [18]. Nevertheless, we chose to start with the 1D linear case ((1) for $d = 1$), because of the following reasons:

- *Isolation of the main obstacle: the multiplicity of eigenfunctions.* Other aspects of the general equations, such as nonlinearity, systems of PDEs, and singularities, have already been treated by multiscale methods in other contexts [4,7].
- *Feasibility.* One cannot hope to construct collective representations of eigenfunctions in 2D and 3D if it cannot be accomplished in 1D. It is true that the 1D implementation includes devices that do not seem to manifest direct higher-dimensional generalizations, most important of which is the concept of monodromy. But indirect generalizations do exist; the monodromy concept can be replaced by coarse-level difference equations (see Sect. 6).
- *Applications to other important 1D problems.* We have generalized our fast expansion and integration algorithms to expansions in eigenfunctions of general linear differential operators and, in particular, to Sturm-Liouville transforms. Although efficient transform procedures already exist for very special cases (such as FFT and FLT – Fast Legendre Transform [15]), our fast transform algorithm is robust and generic, and hence should have broad applications. The 1D Schrödinger eigenproblem itself has important physical applications to 1D conductors and soliton theory [2,16].

2 The Main Complexity Factors

There are three main complexity factors in computing, storing, and manipulating N eigenfunctions of the Schrödinger operator:

1. *Many features in* $V(x)$. The number of spatial points needed for a good resolution of $V(x)$ is $O(N)$; hence each eigenfunction needs at least an $O(N)$-point resolution in x.
2. *Many distinct eigenfunctions.* A large number N of distinct eigenfunctions and eigenvalues need to be computed, representing N electrons.
3. *Many orthogonalization steps.* In order to orthonormalize the eigenbasis, $O(N)$ orthogonalization steps per eigenfunction are needed, each requiring $O(N)$ operations.

Whereas the first two factors result in $O(N^2)$ storage, which is needed to represent $O(N)$ eigenfunctions in the current eigenbasis algorithms (from early algebraic solvers [17] to modern multigrid eigensolvers [9,10]), the third factor raises their complexity to $O(N^3)$.

Our novel approach and methodology is based on the observation that *"neighboring" eigenfunctions (with close "momenta") are distinguishable from each other only at large scales.* Using *collective representations* of eigenfunctions that share the common description of their details at finer scales, and progressively separate them out only on increasingly coarse grids, we have developed algorithms that require only $O(N \log N)$ computer operations and storage, *without localization assumptions on the electronic orbitals* (the theoretical complexity is $O(N(\log N)^2)$ at worst, but it reduces to $O(N \log N)$ in most problems). Even in localized systems, our approach requires only $O(N(\log N_{\text{local}})^2)$ operations in the worst case, versus $O(N N_{\text{local}}^2)$ needed by current localization methods [11], where N_{local} is the number of orbitals in the localized cell.

3 The Main Principle

The 1D collective-eigenfunction approach involves the concept of *monodromy transformation*, which represents the propagation of solutions of

$$\partial_x \boldsymbol{U}(x) = \begin{pmatrix} 0 & 1 \\ V(x) - \lambda & 0 \end{pmatrix} \boldsymbol{U}(x) , \qquad \boldsymbol{U}(x) := \begin{pmatrix} \psi(x) \\ \partial_x \psi(x) \end{pmatrix} , \tag{2}$$

where $\partial_x = d/dx$. Equation (2) is equivalent to (1) in one dimension.

3.1 The Monodromy Transformation

Equation (2) can be regarded as a parameter-dependent ODE. Hence, every solution $\boldsymbol{U}(x)$ is uniquely determined by its *initial value*, say, at $x = a$.

In other words, the value $U(b)$ depends uniquely on $U(a)$ for any interval $[a,b]$. Since (2) is linear, the transformation $U(a) \longmapsto U(b)$ is a 2×2 matrix $M([a,b];\lambda)$, called the *monodromy matrix* [19]. Some important properties of the monodromy transformation are:

1. *Locality:* the monodromy depends on λ and the values of $V(x)$ inside $[a,b]$ only.
2. *Non-degeneracy:* $\det(M([a,b];\lambda)) = 1$; in particular, $M([a,a];\lambda) = I$.
3. *Multiplicativity:* for every $a < b < c$,

$$M([a,c];\lambda) = M([b,c];\lambda) \cdot M([a,b];\lambda) . \tag{3}$$

4. *Relation to boundary value problems:* the function $M(\lambda) := M(\Omega;\lambda)$ and the specific boundary conditions determine the spectrum of (2) [19]. Two examples:
   - Dirichlet B.C.: $\psi(0) = \psi(L) = 0$. Equivalently, the eigenvalues are the roots of the equation

$$M_{12}(\lambda) = 0 . \tag{4}$$

   - Quasiperiodic B.C.: given a periodic potential $V(x+L) = V(x)$, $-\infty < x < \infty$, Bloch-type solutions [19]

$$\psi(x+L) = e^{i\beta L}\psi(x), \qquad \beta \in \mathbf{R}, \forall x \in \mathbf{R}$$

   form a *band spectrum* [19]. They represent all the bounded eigenfunctions on the entire space, as required in electronic structure calculations. This spectrum is characterized by the inequality

$$|D(\lambda)| := |M_{11}(\lambda) + M_{22}(\lambda)| \le 2 .$$

3.2 Complementary Smoothness of the Monodromy

The monodromy function $M(I;\lambda)$ of a fixed interval I of length x is an analytic function of λ [19]. Furthermore, its scale of smoothness is inversely proportional to the size of I. For example, in regions where $\lambda - V(x) > 0$, each of its entries is an oscillating function of λ, with a frequency *directly proportional to the length of I.*

This can be easily observed in the special case of $V \equiv 0$, for which

$$M([0,x];\lambda) = \begin{pmatrix} \cos(\omega x) & \omega^{-1}\sin(\omega x) \\ -\omega \sin(\omega x) & \cos(\omega x) \end{pmatrix}, \qquad \omega := \sqrt{\lambda} . \tag{5}$$

The frequency of the entries as functions of ω is x. This holds for all nonnegative λs and for the monodromy $M(I;\lambda)$ of *any* interval I of length x, since for a constant potential the monodromy does not depend on the starting point of the spatial interval, but on its length only.

This property of monodromies is called the *complementary smoothness principle*, resembling the *Heisenberg Principle* of quantum mechanics, where x and λ are the complementary dimensions: the *product* of the spatial and eigenvalue resolutions depends *only* on the desired accuracy in the monodromy transformation.

This principle is proved in [14] for monodromies of *any 1D linear differential eigenproblem*. It does not depend on the smoothness of the coefficient functions (such as the potential function $V(x)$). We apply this principle to the construction of efficient multiscale eigenbasis algorithms.

4 The Multiscale Eigenbasis Structure

Our basic tool in fast eigenbasis computations is the Multiscale Eigenbasis (MEB) structure, which implicitly represents $O(N)$ eigenfunctions of a differential operator to accuracy ε in $O(N \log N \log(1/\varepsilon))$ computer operations and storage. In Sect. 5 we present some applications of MEB to the efficient execution of some typical computational tasks involving the Schrödinger eigenproblem, as well as more general classes of eigenproblems.

In the spirit of electronic structure computations [11,18], we consider (1) with a potential $V(x)$ containing N "features" (e.g., wells), over a spatial periodicity cell $\Omega = [0, L], L = O(N)$, and a fixed energy domain $[\lambda_{\min}, \lambda_{\max}]$ of interest.

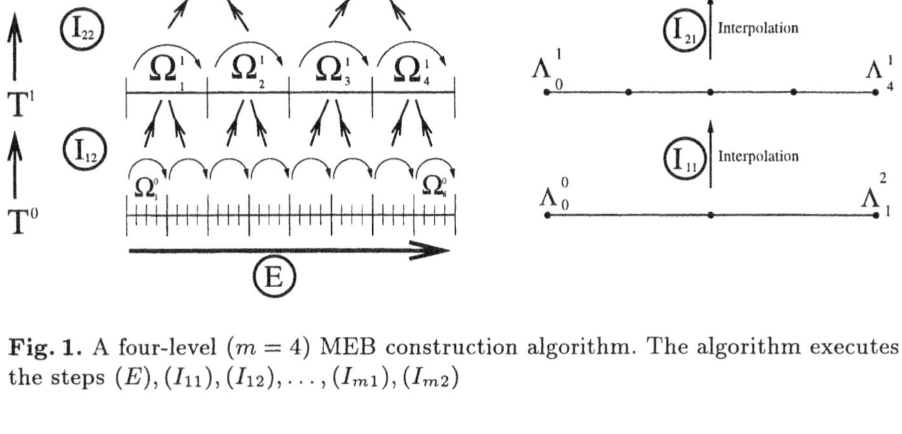

Fig. 1. A four-level ($m = 4$) MEB construction algorithm. The algorithm executes the steps $(E), (I_{11}), (I_{12}), \ldots, (I_{m1}), (I_{m2})$

The MEB consists of $m+1=O(\log N)$ levels $\{T^l\}_{l=0}^m$; each T^l embodies $n_l\times(k_l+1)$ monodromy functions $\{\{M(\Omega_j^l,\Lambda_t^l)\}_{j=1}^{n_l}\}_{t=0}^{k_l}$, where $\cup_{j=1}^{n_l}\Omega_j^l=\Omega$, $\Lambda_t^l=\lambda_{\min}+t(\lambda_{\max}-\lambda_{\min})/k_l$, $n_l=2^{-l}n_0$ and $k_l=2^lk_0$, so that $n_l\cdot k_l$ is constant throughout stages $l=0,\ldots,m$.

The MEB construction consists of the following stages (the precise algorithm can be found in [14,15]):

(E) *Evaluation:* for each Λ_t^0, we discretize (2) in each interval Ω_j^0 and use a marching scheme to approximate the monodromy matrix $M(\Omega_j^0,\Lambda_t^0)$. By the complementary smoothness principle, a small k_0 suffices to resolve well the dependence on λ, since Ω_j^0 are of short length (see Fig. 1). Consequently, the computed monodromies can be interpolated to any other $\lambda\in[\lambda_{\min},\lambda_{\max}]$ with any desirably small interpolation error.

(I_{11}) *Interpolation:* in particular, $\{M(\Omega_j^0;\Lambda_t^0)\}_{j=1,t=0}^{n_0,k_0}$ can be interpolated to a λ-lattice twice as fine, obtaining $\{M(\Omega_j^0;\Lambda_t^1)\}_{j=1,t=0}^{n_0,k_1}$.

(I_{12}) *Merging:* since $\Omega_j^1:=\Omega_{2j-1}^0\cup\Omega_{2j}^0$, we can calculate

$$M(\Omega_j^1;\Lambda_t^1):=M(\Omega_{2j}^0;\Lambda_t^1)\cdot M(\Omega_{2j-1}^0;\Lambda_t^1) \tag{6}$$

for $j=1,\ldots,n_1$, $t=1,\ldots,k_1$. The new monodromies, of spatial intervals twice as large, are smooth on the finer λ lattice.

(I_{21}),... *Recursion:* since the monodromies of level 1 are smooth on their lattice, they can be interpolated to a finer lattice and then merged to monodromies of intervals twice as large, yielding $\{M(\Omega_j^2;\Lambda_t^2)\}_{j,t}$. The stages of interpolating and merging are repeated until the monodromy of Ω is reached on the appropriate λ resolution.

A precise analysis of errors [14, Sect. 4.2.1] indicates that a minimal amount of work per fixed error ε is attained for $p\le O(\log(N/\varepsilon)), q\le O(\log(N/\varepsilon))$, where p is the discretization order at step (E), and q is the interpolation order of monodromies (steps (I_{l1})). The complexity of the MEB construction is bounded by $O(N(\log N)^2\log(1/\varepsilon))$. In actual cases (where errors do not always reinforce each other), $p=O(q)=O(\log(1/\varepsilon))$ prove to be satisfactory. Consequently, the MEB construction requires in practice only $O(N\log N\log(1/\varepsilon))$ operations.

5 MEB Applications

The MEB structure is the basis for the efficient execution of general eigenbasis computations. Here we briefly list some of its applications. For a complete description, see [14, Chaps. 4–5].

– *Fast Electronic Density Integration (FEDI).* The Kohn-Sham equations of Density Functional Theory [3,18] involve the computation of the *electronic density* $\varrho(x)$ of the periodic Schrödinger operator, whose integrals are used to determine the self-consistent potential function $V(x)$ in the nonlinear version of the Schrödinger equation. It is possible to compute ϱ in only $O(N \log N \log(1/\varepsilon))$ operations, using the MEB structure and the fast multilevel summation algorithm [5].

- *Discrete eigenvalue evaluations.* A common computational task in eigenproblems is evaluating the smallest N eigenvalues of a general 1D differential operator with general boundary conditions. The eigenvalues are the roots of an equation involving the entries of the monodromy transformation on the full spatial domain (e.g., (4)). Once the MEB structure has been constructed for the operator, we can evaluate any eigenvalue to an accuracy ε in $O(\log(1/\varepsilon))$ computer operations, regardless of the specific boundary conditions [14, Sect. 4.2.1].
- *Fast expansion in discrete eigenfunctions.* The FEDI algorithm can be applied for computing N coefficients

$$\hat{f}_n := (f, u_n) = \frac{1}{L} \int_0^L f(x) \overline{u_n(x)} dx, \quad n = 1, \ldots, N$$

of the expansion

$$f(x) = \sum_{n=1}^{N} \hat{f}_n u_n = \sum_{n=1}^{N} (f, u_n) u_n$$

to an accuracy ε in $O(N \log N \log(1/\varepsilon))$ operations, where $\{u_n\}_{n=1}^N$ are the eigenfunctions corresponding to the smallest N eigenvalues $\{\lambda_n\}_{n=1}^N$ of a general 1D linear differential operator. Similarly, inverse transforms can be computed to an ε-accuracy in $O(N \log N \log(1/\varepsilon))$ operations [14, Sect. 5.3]. In particular, it is possible to expand fast in eigenfunctions of general self-adjoint Sturm-Liouville eigenproblems

$$(P(x)u'(x))' + Q(x)u(x) = \lambda R(x)u(x) \,, \tag{7}$$

which are of particular interest in mathematical physics. These fast transforms also generalize the FFT algorithm (the case of $Q \equiv 0, P = R \equiv 1$) [14, Sect. 4.2.1].
- *Fast computation of moments.* The FEDI algorithm can be extended to the computation of moments, namely,

$$\mu_n := \frac{1}{L} \int_0^L |u(x; \lambda_n)|^\kappa dx, \quad n = 1, \ldots, N \,,$$

which are important in applications to photonic and phononic crystals [13] (typical values for κ are 2, 4). The computational complexity of calculating $\{\mu_n\}_{n=1}^N$ to accuracy ε is $O(\kappa^2 N \log N \log(1/\varepsilon))$ [14, Sect. 5.4].

6 Summary and Generalizations

In this paper we presented a class of fast eigenbasis algorithms, based on the multiscale eigenbasis structure for the collective representation of eigenfunctions.

These algorithms are *not limited to localized electronic structures* of quantum chemistry, and can be shown to be *highly parallelizable.* Although we are mainly motivated by ab-initio chemical problems, these algorithms address a much broader variety of 1D expansions in orthogonal systems of general linear differential operators, with general boundary conditions, on general (non-uniform) grids.

Our main future research direction is concerned with the generalization of our 1D algorithms to higher dimensional problems, which are of great importance in condensed matter physics. Although not trivial, indirect higher-dimensional generalizations of our 1D devices do exist.

- In a higher dimension, the concept of a monodromy on a given interval should be replaced by that of a *finite-difference equation* on a comparable meshsize. Analogous to the construction of monodromies of larger intervals from those of smaller intervals (by multiplication), a coarse grid difference equation can be constructed from the next-finer-grid difference equations, using only a certain local set of the latter. The error in the constructed coarse equations decreases *exponentially* with the size of that local set. General algebraic approaches and examples for such constructions are described in [6].
- Finite-difference equations could of course be also used in the *one-dimensional* problem. In the present study the main reason that monodromies were used instead is a certain complication associated with the construction of coarse equations for *highly indefinite* equations. Indeed, for a large meshsize H and for the main range of λ values (those for which $(\lambda - V(x))H^2$ is not small), (1) is highly indefinite on scale H. Specifically, for each such value of λ, solutions of (1) are too oscillatory to be directly represented, even locally, on a grid with such a meshsize. Instead, MEB algorithms should employ a construction similar to the method developed in [8]. The yet unknown solution $u^H(x)$ is expanded as

$$u^H(x) = \sum_{m=1}^{M} a_m^H(x) w_m^H(x) . \tag{8}$$

The $w_m^H(x)$ are *known oscillatory* "basis functions", found by relaxing at the fine levels the "root equation"

$$\partial_{\boldsymbol{\xi}} w(x) + i(\lambda - V(x))^{1/2} w(x) = 0 , \tag{9}$$

where $\boldsymbol{\xi}$ is the "propagation direction" associated with the particular basis function w; each basis function of a coarse level is constructed by *locally* recombining *neighboring* basis functions of the next finer level ("locally" meaning in space $\boldsymbol{x}$, whereas "neighboring" denotes close eigenvalue and propagation direction). The functions $a_m^H(\boldsymbol{x})$ are *smoothly varying, yet unknown* "amplitudes". These *amplitudes are to be calculated on grid* H; for each of them we can construct an equation on grid H, using a local set of the next-finer-grid equations.

- The expansion of the form (8) for each eigenfunction should include only a number (M) of terms which is independent of H, even for $d > 1$. This can be achieved since the known functions $w_m^H(\boldsymbol{x})$ provide a suitable basis of local approximate eigenfunctions on each scale. The local accuracy obtained by relaxation is enough, since the basis functions are subsequently modified by the coarse-level smooth amplitudes. However, the number of different basis functions (or different finite-difference equations) for each value of λ scales linearly with H^{d-1}. At the same time, the number of different λs represented on grid H will be $O(H)$, as with 1D. Thus, the total number of basis functions represented on grid H will be $O(H^d)$.

This outline implies that before tackling higher dimensional extensions of the MEB algorithms, special studies should be first conducted on the following two types of tasks:

1. 1D MEB algorithms based on finite-difference equations instead of monodromies.
2. Extensions of the wave-equation algorithms of [8] to non-constant coefficients, and possibly to more general boundary conditions. The cases of discontinuous coefficients and/or boundary singularities are particularly challenging. However, in applications to electronic-structure problems we do not usually encounter such extreme conditions, since the effective potential is very smooth and boundaries are practically absent.

Having developed MEB structures in higher dimensions, an intriguing question will be whether $O(N \log N)$ computational work is indeed really necessary. It is reasonable to expect that $O(N)$, and sometimes even less, would be enough. First because for many low values of λ, the eigenfunctions are essentially local, lying only in areas where the potential $V(\boldsymbol{x})$ is sufficiently low. On the other hand, for high values of λ, the eigenfunctions approach an asymptotic behavior that can be used to describe them all. At that range of high λs, the main difference between eigenfunctions belonging to different eigenvalues is only in the frequency associated with the base functions, a quantity that can be singled out of the equations by choosing an appropriate form for those functions. Thus, it is expected that the amount of work will not really depend on $\lambda_{\min}$ and $\lambda_{\max}$, but instead will be only proportional to the number of discrete variables needed to resolve the potential $V(\boldsymbol{x})$; that is, *the work is expected to be proportional to the number of atoms.*

For very large structures, even that amount of work may not be needed, since such structures are usually highly *repetitive*. Once coarse-level *equa-*

tions have been constructed, they can be used in other areas where the potential repeats itself (unlike coarse-level *solutions*, which cannot generally be re-used). This aspect is emphasized in other papers on multiscale methods [1,12,21]. In [1], for example, coarse-level Hamiltonians for polymer chains are constructed using local simulations with fine-level Hamiltonians, which are assumed to be known. In a similar manner, the fine-level Hamiltonians themselves (the force field) could presumably be constructured, once and for all, by using local simulations with electronic-structure dynamics.

7 Acknowledgments

This research was supported by grant no. 696/97 from the Israel Science Foundation, by the US Air Force (Contract F33615-97-D5405 and EOARD Contract F61775-00-WE067), and by the Carl F. Gauss Minerva Center for Scientific Computation at the Weizmann Institute of Science.

References

1. Bai, D. and Brandt, A.: Multiscale computation of polymer models, in: Proceedings of the NATO Advanced Research Workshop on Multiscale Methods in Chemistry and Biology, Eilat, April 5–11 (in press)
2. Bernasconi J., and Schneider, T. (Eds.) Physics in One Dimension, Proceedings of an International Conference at Fribourg, Switzerland, August 25–29, 1980. Springer, Berlin
3. Bernholc, J. and Fattebert, J. L. : Towards grid-based $O(N)$ DFT methods: optimized non-orthogonal orbitals and multigrid acceleration. Physical Review B **62(3)** (2000) 1713–1722
4. Brandt, A.: Multigrid Techniques: 1984 Guide, with Applications to Fluid Dynamics, Available as GMD Studien Nr. 85, from GMD-AIW, postfach 1240, D-5205, St. Augustin 1, West Germany, 1984
5. Brandt, A.: Multilevel computations of integral transforms and particle interactions with oscillatory kernels. Comp. Phys. Comm. **65** (1991) 24–38
6. Brandt, A.: General highly accurate algebraic coarsening. Elec. Trans. Num. Anal. **10** (2000) 1–10
7. Brandt, A.: Multiscale Scientific Computation 2000. These Proceedings
8. Brandt, A. and Livshits, I.: Wave-ray multigrid method for standing wave equations. Elec. Trans. Num. Anal. **6** (1997) 162–181
9. Brandt, A., McCormick, S. and Ruge, J.: Multigrid methods for differential eigenproblems, SIAM J. Sci. Stat. Comput. **4** (1983) 244–260
10. Costiner, S. and Ta'asan, S.: Adaptive multigrid techniques for large-scale eigenvalue problems: Solutions of the Schrödinger problem in two and three dimensions. Physical Review E **51(4)** (1995) 3704–3717
11. Gödecker, S.: Linear scaling electronic structure methods. Reviews of Modern Physics **71(4)** (1999) 1085–1123
12. Ilyin, V. and Brandt, A.: Multilevel approach in statistical physics of liquids, in: Proceedings of the NATO Advanced Research Workshop on Multiscale Methods in Chemistry and Biology, Eilat, April 5–11 (in press)

13. Joannopolous, J. D., Meade, R. D. and Winn, J. N.: Photonic Crystals, Princeton University Press, Princeton, New Jersey, 1995
14. Livne, O. E.: Multiscale Eigenbasis Algorithms. Ph. D. Thesis, Weizmann Institute of Science, Rehovot, 2000
15. Livne, O. E. and Brandt, A.: $O(N \log N)$ multilevel calculation of N eigenfunctions, in: Proceedings of the NATO Advanced Research Workshop on Multiscale Methods in Chemistry and Biology, Eilat, April 5–11 (in press)
16. Manakov, S. V., Novikov, S .P., Pitaevskij, L. P. and Zakharov, V.E. Theory of Solitons: The Inverse Problem Method, Nauka, Moscow. English transl.: Consultants Bureau, New York, 1984
17. Parlett, B. N.: The Symmetric Eigenvalue Problem, Prentice-Hall, Englewood Cliffs, New Jersey, 1980
18. Parr, R.G. and Yang, W.: Density-Functional Theory of Atoms and Molecules, Oxford University Press, New York, 1989
19. Shubin, M.A. (Ed.): Partial Differential Equations VII: Spectral Theory of Differential Operators, Springer, Berlin, 1994
20. Slater, J.: Quantum Theory of Molecules and Solids, McGraw-Hill, New York, 1963
21. Ron, D. and Brandt, A.: Renormalization multigrid (RMG): Coarse-to-fine Monte Carlo acceleration and optimal derivation of macroscopic actions, in: Proceedings of the NATO Advanced Research Workshop on Multiscale Methods in Chemistry and Biology, Eilat, April 5–11 (in press)
22. Ziman, J.M.: Principles of the Theory of Solids, Cambridge University Press, Cambridge, 1972

Wavelet Galerkin BEM on Unstructured Meshes by Aggregation*

Gregor Schmidlin and Christoph Schwab

Seminar für Angewandte Mathematik, ETH Zentrum, 8092 Zürich, Switzerland

Abstract. We investigate the numerical solution of strongly elliptic boundary integral equations on unstructured surface meshes Γ in $\mathbb{R}^3$ by Wavelet-Galerkin boundary element methods (BEM). They allow complexity-reduction for matrix setup and solution from quadratic to polylogarithmic (i.e. from $O(N^2)$ to $O(N(\log N)^a)$ for some small $a \geq 0$, see, e.g. [2,3,9,10] and the references there). We introduce an agglomeration algorithm to coarsen arbitrary surface triangulations on boundaries Γ with possibly complicated topology and to construct stable wavelet bases on the coarsened triangulations in linear complexity. We describe an algorithm to generate the BEM stiffness matrix in standard form in polylogarithmic complexity. The compression achieved by the agglomerated wavelet basis appears robust with respect to the complexity of Γ. We present here only the main results and ideas - full details will be reported elsewhere.

1.1 Introduction

1.1.1 Problem formulation

We consider the numerical solution of the boundary integral equation

$$\mathcal{A}u = f \quad \text{on } \Gamma. \tag{1.1}$$

Here Γ is the boundary of a bounded Lipschitz domain $\Omega \in \mathbb{R}^d$ $(d = 2, 3)$, $f \in L^2(\Gamma)$ and $\mathcal{A}$ a boundary integral operator of order 0. We assume that $\mathcal{A}$ is injective and strongly elliptic in $L^2(\Gamma)$ in the sense that it satisfies a Gårding inequality. Then, under these conditions, Galerkin discretizations of (1.1) based on a dense sequence of subspaces of $L^2(\Gamma)$ converge quasioptimally.

Galerkin discretizations of (1.1) lead to fully populated stiffness matrices due to the nonlocal operator $\mathcal{A}$ entailing $O(N^2)$ complexity. It was shown in [9,10] and the references there that this complexity can be reduced to polylogarithmic order by the use of suitable wavelet bases on Γ and by compression of the stiffness matrix, i.e. by dropping most of the $O(N^2)$ entries in the wavelet stiffness matrix. In [9,10], certain compression strategies were shown to preserve the optimal asymptotic rate of convergence. The analysis of [9,10] indicated, however, a strong dependence of the constants in the compression

* This work was supported in part under the TMR network "Multiscale Methods in Numerical Analysis" of the EC by the Swiss government under grant number BBW 97.0404

estimates on the geometry of the surface and required a nested sequence of meshes on Γ in an essential way. Numerical experiments confirmed that this was not an artifact of the analysis, but that indeed the performance of these wavelet algorithms deteriorates on complex surfaces. In practice, the assumption of a nested sequence of triangulations on Γ is unrealistic, since for complex surfaces the coarsest possible mesh already contains a large number of degrees of freedom. The present paper presents algorithms which coarsen arbitrary triangulations on surfaces Γ with possibly complicated topology and then construct stable wavelet bases on the coarsened triangulations in linear complexity. We sketch an algorithm which directly generates the compressed stiffness matrix in standard form in polylogarithmic complexity.

1.1.2 Boundary Element Method

We illustrate our approach for the simplest boundary elements: we triangulate Γ into a quasiuniform mesh $\mathcal{T}_N(\Gamma)$ of N panels $\{\pi_i\}_{i=1}^N$ with

$$\Gamma = \overline{\bigcup_{1\le i\le N} \pi_i} \quad \text{and} \quad \pi_i \cap \pi_j = \emptyset, \quad i \neq j. \tag{1.2}$$

To each panel π we assign the radius r_π and the center c_π

$$(r_\pi, c_\pi) = \inf\{(r,c) : \pi \subset B_r(c), \;\; r,c \in \mathbb{R}^d\} \tag{1.3}$$

where $B_r(c)$ is the ball with radius r and center c. With

$$h := \max\{r_\pi : \pi \in \mathcal{T}_N(\Gamma)\} \tag{1.4}$$

the uniformity condition of the panels reads

$$r_\pi > \frac{h}{C_u} \quad \forall \pi \in \mathcal{T}_N \tag{1.5}$$

with a constant C_u. We further assume that there is a constant C_μ such that the part of $\mathcal{T}_N$ in $B_r(c)$ satisfies

$$|B_r(c) \cap \mathcal{T}_N| \le C_\mu r^{d-1} \quad \forall r \ge 0, c \in \mathbb{R}^d. \tag{1.6}$$

Throughout, $|S|$ denotes the surface measure of the panels in the subset $S \subset \mathcal{T}_N$.

We consider the Galerkin Boundary Elements based on piecewise constant boundary elements with basis functions given by

$$\phi_i(\underline{x}) = \begin{cases} 1, & \underline{x} \in \pi_i \\ 0, & \text{else} \end{cases} \qquad i = 1, \ldots, N \tag{1.7}$$

to discretize the integral equation (1.1). This leads to the linear system

$$\underline{\underline{A}}^N \underline{u}^N = \underline{b}^N \tag{1.8}$$

for the coefficients of the approximate solution $u^N(\underline{x}) = \sum_{i=1}^N u_i^N \phi_i(\underline{x}) \in V_N$ where $V_N = \text{span}\{\phi_i, i = 1, \ldots, N\} \subset L^2(\mathcal{T}_N)$. Here

$$\underline{\underline{A}}_{ij}^N = \int_\Gamma \phi_i(\underline{x}) c(x) \phi_j(\underline{x}) ds_{\underline{x}} + \int_\Gamma \phi_i(\underline{x}) \int_\Gamma k(\underline{x}, \underline{y}) \phi_j(\underline{y}) ds_{\underline{y}} ds_{\underline{x}} = (\phi_i, \mathcal{A}\phi_j)$$

$$\underline{u}^N = (u_1^N, u_2^N, \ldots, u_N^N)^\top, \quad \underline{b}_i^N = \int_{\pi_i} \phi_i(\underline{x}) f(\underline{x}) ds_{\underline{x}}$$

and the integrals in the definition of $(\phi_i, \mathcal{A}\phi_j)$ must be understood, in general, as Cauchy principal values. We assume that the kernel $k(\underline{x}, \underline{y})$ satisfies the estimates

$$\left| D_{\underline{x}}^\alpha D_{\underline{y}}^\beta k(\underline{x}, \underline{y}) \right| \leq \frac{C}{\left| \underline{x} - \underline{y} \right|^{2+|\alpha|+|\beta|}} \qquad |\alpha|, |\beta| \leq 1 \tag{1.9}$$

where $D_{\underline{x}}, D_{\underline{y}}$ denote partial derivatives with respect to the Cartesian coordinates $x, y \in \mathbb{R}^3$ in a fixed open neighborhood of the boundary Γ. We generalize the wavelet algorithm and the compression analysis of [7,9,10] to general, nonnested triangulations on Γ.

1.2 Construction of the Wavelet basis

On the possibly unstructured mesh $\mathcal{T}_N$, we construct an agglomerated wavelet basis $\{\psi_{\tau,j}\}$ with vanishing moments of order 1, i.e. the mean value of the wavelets with respect to the surface measure vanishes. It is well known that this implies the smallness of certain entries of the stiffness matrix [7,9,10]. We emphasize, however, that in the compression estimates for the wavelets constructed here the parametric boundary representation and its derivatives do not enter. Only the regularity of the kernel function $k(x, y)$ in the ambient space matters.

To construct the agglomerated wavelet basis, we group the panels $\{\pi\}$ to clusters thereby introducing a hierarchical structure among them. As in [6] we define

Definition 1.2.1 *1. A cluster τ is defined as a non empty set of panels*

$$\tau = \bigcup_k \pi_{i_k}, \qquad \pi_{i_k} \in \mathcal{T}_N.$$

Each cluster is assigned its radius r_τ and its center c_τ

$$(r_\tau, c_\tau) = \inf\{(r, c) : \tau \subset B_r(c), \ \ r, c \in \mathbb{R}^d\}. \tag{1.10}$$

2. A cluster pool $P(\mathcal{T}_N) = P$ defines an arbitrary but fixed selected subset of all clusters of $\mathcal{T}_N$ with $\mathcal{T}_N \subset P$ and $\pi_i \subset P$, $\forall \pi_i \in \mathcal{T}_N$.

3. A cluster tree $\mathcal{P}(\mathcal{T}_N) = \mathcal{P}$ is a cluster pool P with a hierarchical structure

$$\forall \tau, \tilde{\tau} \in \mathcal{P} : \quad \tau \cap \tilde{\tau} \in \{\emptyset, \tau, \tilde{\tau}\}.$$

4. *For $\tau \in \mathcal{P}$, with a cluster tree $\mathcal{P}$, we define* child(τ) *by the set of all children of the cluster $\tau \in \mathcal{P}$*

$$\text{child}(\tau) := \{\tau' \in \mathcal{P} : \tau' \subset \tau \wedge (\tau' \subset \tau'' \Rightarrow \tau \subset \tau'', \forall \tau'' \in \mathcal{P})\}.$$

5. *The level $L_{\mathcal{P}} \leq 0$ of a cluster tree $\mathcal{P}$ is defined by*

$$L_{\mathcal{P}} := \\ -\max_{\tau \in \mathcal{P}} \{l \in \mathbb{N}_0 : \tau \subsetneq \tau_{i_1} \subsetneq \ldots \subsetneq \tau_{i_{l-1}} \subsetneq \tau_{i_l} = \mathcal{T}_N, \tau_{i_1}, \ldots, \tau_{i_l} \in \mathcal{P}\}$$

6. *The level $l_\tau \leq 0$ of a cluster τ in the cluster tree $\mathcal{P}$ is defined by*

$$l_\tau := L_{\mathcal{P}} \\ + \max\{l \in \mathbb{N}_0 : \tau \subsetneq \tau_{i_1} \subsetneq \ldots \subsetneq \tau_{i_{l-1}} \subsetneq \tau_{i_l} = \mathcal{T}_N, \tau_{i_1}, \ldots, \tau_{i_l} \in \mathcal{P}\}.$$

7. *The root of the cluster tree $\mathcal{P}$ is the cluster* root$(\mathcal{P})$ *which fulfills*

$$l_{\text{root}(\mathcal{P})} = L_{\mathcal{P}}.$$

The example in Fig. 1.1 and 1.2 illustrates these notions. In these figures, $\mathcal{T}_{18} = \{\pi\}_{i=1}^{18}$. The cluster tree $\mathcal{P}$ of Fig. 1.2 consists of the clusters $\{\tau\}_{i=1}^{35}$ shown in the lower part of Fig. 1.1. In this example we have

$$\begin{aligned} L_{\mathcal{P}} = l_{\tau_1} &= -6 \quad (\tau_1 \text{ is the root of } \mathcal{P}) \\ l_{\tau_2} = l_{\tau_3} &= -5 \\ l_{\tau_4} = l_{\tau_5} = l_{\tau_6} = l_{\tau_7} &= -4 \\ l_{\tau_{32}} = l_{\tau_{33}} = l_{\tau_{34}} = l_{\tau_{35}} &= 0 \\ \text{child}(\tau_3) &= \{\tau_6, \tau_7\}. \end{aligned}$$

With the hierarchical structure of the mesh given by the cluster tree $\mathcal{P}$ we can construct a wavelet basis of V_N. For this construction we need the following conditions. The function $\hat{\mu}$ on $\mathcal{P}$ is defined recursively by

$$\hat{\mu}(\mathcal{T}_N) := 1, \quad \hat{\mu}(\tau') := \frac{\hat{\mu}(\tau)}{|\text{child}(\tau)|} \qquad \text{for } \tau' \in \text{child}(\tau).$$

This means that the division of $\tau \in \mathcal{P}$ into $|\text{child}(\tau)|$ sons should correspond to a division of the area into $|\tau'| \approx |\tau|/|\text{child}(\tau)|$. The precise condition that there are constants $0 < k_l$, k and $C_{\mathcal{P}}$ such that

$$|\text{child}(\tau)| = \begin{cases} 0, & \text{or} \\ k_{l_\tau} \geq 2 & \end{cases} \tag{1.11}$$

$$k_{l_\tau} \leq k \tag{1.12}$$

$$|\mathcal{T}_N| \hat{\mu}(\tau) \leq C_{\mathcal{P}} \tag{1.13}$$

$$r_\tau \leq C_{\mathcal{P}} r_{\text{root}(\mathcal{P})} (\hat{\mu}(\tau))^{1/(d-1)} \tag{1.14}$$

for all $\tau \in \mathcal{P}$.

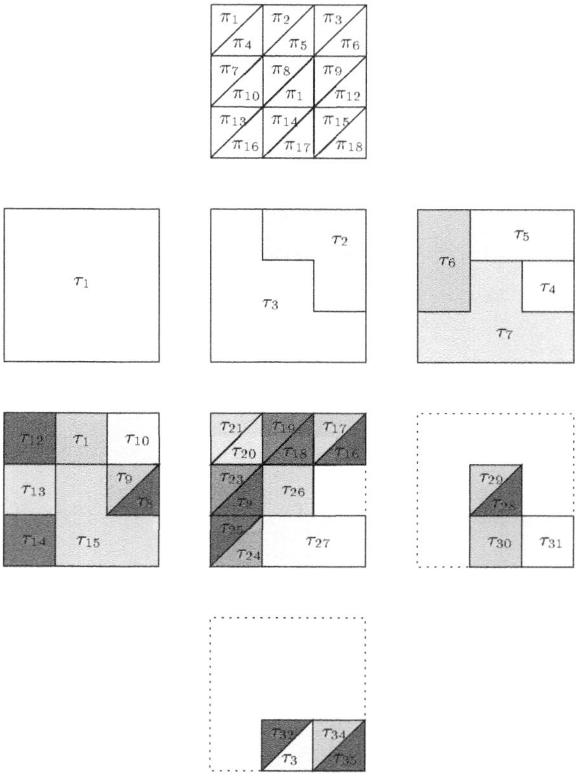

Fig. 1.1. Partition $\mathcal{T}_N = \{\pi_i\}_{i=1}^{18}$ with a cluster tree $\mathcal{P} = \{\tau_i\}_{i=1}^{35}$.

Remark 1.2.2 Note that the conditions (1.2), (1.6) and (1.12)-(1.14) are analogous to those in [5] for the multipole method. They are not really a restriction for reasonable surfaces Γ. Condition (1.11) is not a restriction on the surface but depends on the type of tree used in the implementation.

The wavelet basis $\{\psi_{\tau,j}\}$ is constructed from the scaling functions $\{\phi_i\}_{i=1}^N$ in (1.7) by the following agglomeration procedure. In a first step, we compute for each $\tau \in \mathcal{P}$ with child$(\tau) = \emptyset$ the piecewise constant orthonormal basis functions ϕ_τ $((\phi_\tau, \phi_{\tau'}) = \delta_{\tau\tau'})$

$$\phi_\tau(\underline{x}) = \begin{cases} 1/\sqrt{|\tau|}, & \underline{x} \in \tau \\ 0, & \text{else} \end{cases} \tag{1.15}$$

with $|\tau| = \int_\tau ds_{\underline{x}}$. Afterwards for each cluster τ with child$(\tau) \neq \emptyset$ the wavelets $\psi_\tau = \{\psi_{\tau,j}\}_{j=1}^{|\text{child}(\tau)|-1}$ and the coarsened scaling function ϕ_τ are computed by the local wavelet transformation $\underline{\underline{M}}_\tau$. The constant normed function ϕ_τ with $\text{supp}(\phi_\tau) = \tau$ is needed to get the wavelets on the next lower level. At the

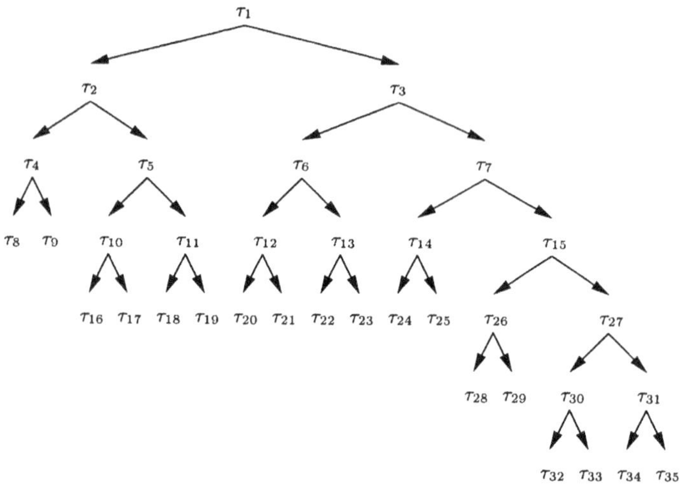

Fig. 1.2. Cluster tree $\mathcal{P}$ corresponding to Fig. 1.1.

end of this recursive bottom up, the wavelet basis consists of the functions

$$\phi_{\mathrm{root}(\mathcal{P})} \text{ and } \{\psi_{\tau,j}\}_{(\tau,j)\in\mathcal{J}} \tag{1.16}$$

where $\mathcal{J} = \{(\tau, j) : \tau \in \mathcal{P} \wedge \mathrm{child}(\tau) \neq \emptyset,\ 1 \leq j < |\mathrm{child}(\tau)|\}$. The construction of the agglomerated wavelet basis is realized by Algorithm 1.2.3.

Algorithm 1.2.3

for $l = 0$ *down to* $L_{\mathcal{P}}$ {
  for all $\tau \in \mathcal{P}$ *with* $l_\tau = l$ {
    if $\mathrm{child}(\tau) = \emptyset$ {
      compute ϕ_τ;
    } ***else*** {

$$\text{compute} \begin{pmatrix} \phi_\tau \\ \psi_\tau \end{pmatrix} = \underline{\underline{M}}_\tau^\top \begin{pmatrix} \phi_{\tau_{i_1}} \\ \vdots \\ \phi_{\tau_{i_{|\mathrm{child}(\tau)|}}} \end{pmatrix}, \tau_{i_\lambda} \in \mathrm{child}(\tau)$$

    }
  }
}

1.3 Local transformation $\underline{\underline{M}}_\tau$

For each tree $\mathcal{P}(\mathcal{T})$ which satisfies the conditions 1.11 and 1.12 we can construct the agglomerated wavelet basis of the space V_N of piecewise constants by applying Algorithm 1.2.3 and the local transformations $\underline{\underline{M}}_\tau$.

The local transformation $\underline{\underline{M}}_\tau$ converts the scaling functions $\phi_{\tau_{i_\lambda}}$ with $\tau_{i_\lambda} \in \text{child}(\tau)$ into the local wavelet basis $\psi_\tau = \{\psi_{\tau,j}\}_{j=1}^{|\text{child}(\tau)|-1}$ and the coarsened scaling function ϕ_τ:

$$\begin{pmatrix} \phi_\tau \\ \psi_\tau \end{pmatrix} = \underline{\underline{M}}_\tau^\top \begin{pmatrix} \phi_{\tau_{i_1}} \\ \vdots \\ \phi_{\tau_{i_{|\text{child}(\tau)|}}} \end{pmatrix}. \tag{1.17}$$

To get an orthonormal wavelet basis of V_N the local transformation matrix $\underline{\underline{M}}_\tau = (\underline{m}_1^\tau, \ldots, \underline{m}_{\text{child}(\tau)}^\tau)$ has to satisfy

$$\underline{\underline{M}}_\tau^\top \underline{\underline{M}}_\tau = I. \tag{1.18}$$

For each cluster τ in the tree $\mathcal{P}$ we compute the local transformation matrix $\underline{\underline{M}}_\tau$ by defining $\underline{\underline{\tilde{M}}}_\tau = (\underline{m}_1^\tau, \underline{\tilde{m}}_2^\tau, \ldots, \underline{\tilde{m}}_{|\text{child}(\tau)|}^\tau)$ with

$$(\underline{m}_1^\tau)^\top = \frac{1}{\sqrt{|\tau|}}(\sqrt{|\tau_{i_1}|}, \ldots, \sqrt{|\tau_{i_{|\text{child}(\tau)|}}|}), \quad \tau_{i_\lambda} \in \text{child}(\tau)$$

and $\underline{\tilde{m}}_2^\tau, \ldots, \underline{\tilde{m}}_{|\text{child}(\tau)|}^\tau$ such that

$$\det(\underline{\underline{\tilde{M}}}_\tau) \neq 0, \quad (\underline{m}_1^\tau)^\top \underline{\tilde{m}}_i^\tau = 0, \quad 2 \leq i \leq |\text{child}(\tau)|.$$

Applying the singular value decomposition (SVD) to the vectors $\underline{\tilde{m}}_2^\tau$,..., $\underline{\tilde{m}}_{|\text{child}(\tau)|}^\tau$, we get the matrix $\underline{\underline{M}}_\tau$

$$(\underline{\underline{U}}_\tau, \underline{\underline{S}}_\tau, \underline{\underline{V}}_\tau) = \text{SVD}((\underline{\tilde{m}}_2^\tau, \ldots, \underline{\tilde{m}}_{|\text{child}(\tau)|}^\tau)) \tag{1.19}$$

$$\underline{\underline{M}}_\tau = (\underline{m}_1^\tau, \underline{\underline{U}}_\tau) \tag{1.20}$$

which satisfies (1.18).

Proposition 1.3.1 *With the local transformation matrix $\underline{\underline{M}}_\tau$ defined above, we get a local basis ϕ_τ, ψ_τ which is orthonormal with respect to the surface measure on Γ, i.e.*

$$\|\phi_\tau\|_{L^2(\Gamma)} = 1 \tag{1.21}$$

$$(\phi_\tau, \psi_{\tau,j}) = 0 \qquad 1 \leq j < |\text{child}(\tau)| \tag{1.22}$$

$$(\psi_{\tau,j}, \psi_{\tau,j'}) = \delta_{jj'} \qquad 1 \leq j, j' < |\text{child}(\tau)|. \tag{1.23}$$

In our implementation, we use the particular choice of $\underline{\underline{\tilde{M}}}_\tau$ given in Assertion 1.3.2.

Assertion 1.3.2 *The matrix*

$$\underline{\underline{\tilde{M}}}_\tau = \begin{bmatrix} \tilde{m}_1^\tau & -\tilde{m}_2^\tau & 0 & 0 & \cdots & 0 \\ \tilde{m}_2^\tau & \tilde{m}_1^\tau & -\tilde{m}_3^\tau & 0 & \cdots & 0 \\ \tilde{m}_3^\tau & 0 & \tilde{m}_2^\tau & -\tilde{m}_4^\tau & & \vdots \\ \vdots & \vdots & & \ddots & \ddots & 0 \\ \tilde{m}_{n-1}^\tau & 0 & & & \tilde{m}_{n-2}^\tau & -\tilde{m}_n^\tau \\ \tilde{m}_n^\tau & 0 & & \cdots & 0 & \tilde{m}_{n-1}^\tau \end{bmatrix} \tag{1.24}$$

with $\tau \in \mathcal{P}$, $\tau_{i_\lambda} \in \text{child}(\tau)$, $\tilde{m}_\lambda^\tau = \sqrt{|\tau_{i_\lambda}|/|\tau|}$ *and* $n = |\text{child}(\tau)|$ *has full rank.*

Remark 1.3.3 The local transformation can be seen as an application of the lifting scheme presented in [11,12] or the stable completion in [1,2]. We start with the scaling function ϕ_τ and the wavelets

$$\tilde{\psi}_{\tau,j} = (\underline{\tilde{m}}_{j+1}^\tau)^\top \begin{pmatrix} \phi_{\tau_{i_1}} \\ \vdots \\ \phi_{\tau_{i_{|\text{child}(\tau)|}}} \end{pmatrix} \qquad 1 \leq j < |\text{child}(\tau)|$$

defined by the matrix $\underline{\underline{\tilde{M}}}_\tau$ and $\{\phi_{\tau_\lambda}\}$. These wavelets are lifted to an orthonormal basis on Γ with vanishing mean value with respect to the surface measure ds on Γ. This is done using a singular value decomposition.

Remark 1.3.4 In case of a binary tree, the local transformation matrix $\underline{\underline{M}}_\tau$ is

$$\underline{\underline{M}}_\tau = \begin{bmatrix} \sqrt{|\tau_1|/|\tau|} & -\sqrt{|\tau_2|/|\tau|} \\ \sqrt{|\tau_2|/|\tau|} & \sqrt{|\tau_1|/|\tau|} \end{bmatrix}$$

for $\tau_1, \tau_2 \in \text{child}(\tau)$. Then the singular value decomposition can be omitted.

Remark 1.3.5 Note that the support of the wavelets is not necessarily connected. Therefore, we can collect the surface patches in a certain region of $\mathbb{R}^d$ in a cluster independent of the boundary Γ. This leads to a compression rate of the stiffness matrix which is not spoiled by the geometry.

1.4 Properties of the agglomerated wavelet basis

For any triangulation $\mathcal{T}_N$ on Γ, Algorithm 1.2.3 generates an agglomerated wavelet basis. The forward and backward transformation of the pyramid scheme for vectors $\underline{f}^\phi$ and $\underline{f}^\psi$ are standard and omitted, since they are almost identical to Algorithm 1.2.3. The agglomerated wavelet basis has the desirable properties:

Proposition 1.4.1 *The wavelet basis obtained from Algorithm 1.2.3 is an* $L^2(\Gamma, ds)$ *orthonormal basis of* V_N.

a) The wavelets fulfill the vanishing moment property

$$\int_\Gamma \psi_{\tau,j}(\underline{x}) ds_{\underline{x}} = 0 \qquad \text{for } (\tau, j) \in \mathcal{J} \tag{1.25}$$

with $\mathcal{J}$ *defined as in (1.16).*

b) The agglomerated wavelet basis satisfies Paseval's equation in $L^2(\Gamma, ds)$.

$$\|u\|^2_{L^2(\Gamma)} = (u, \phi_{\mathrm{root}(\mathcal{P})})^2 + \sum_{(\tau,j)\in\mathcal{J}} (u, \psi_{\tau,j})^2 \tag{1.26}$$

for all $u \in V_N$.

c) With $W_l = \mathrm{span}\{\psi_{\tau,j} : l_\tau = l-1, 1 \le j < |\mathrm{child}(\tau)|\}$, $l > L_{\mathcal{P}}$, *it holds*

$$V_N = W_{L_{\mathcal{P}}} \oplus W_{L_{\mathcal{P}}+1} \oplus \ldots \oplus W_0 \tag{1.27}$$

where we have set $W_{L_{\mathcal{P}}} = \mathrm{span}\{\phi_{\mathrm{root}(\mathcal{P})}\}$

Equation (1.26) is a consequence of the clusters on the same level being disjoint and of the orthogonality of the wavelets between different levels. The orthogonality between different levels follows from (1.22).

Remark 1.4.2 If we refine $\mathcal{T}_N$ beyond level $l = 0$ by regular subdivision of each $\pi_i \in \mathcal{T}_N$, the corresponding spaces of piecewise constants are dense in $L^2(\Gamma, ds)$ and the finite sum in Parseval's equality (1.26) can be extended to an infinite one as for example in [7,10].

1.5 Examples

We give examples of the agglomerated wavelets and exhibit some special properties of them.

In the first example, we computed some wavelet coefficients of the function $f(\underline{x}) = (1 - x_1)(1 - x_2)$ on the unit square. As shown in Fig. 1.3, we chose a uniform triangulation with $N = 200$ triangles. We effect the agglomeration of the panels by a binary tree (other types of trees are also possible). We construct the tree recursively by computing the Čebyšev center and radius of a set of the panels and a dyadic subdivision of this panel set. An algorithm to compute the Čebyšev center and radius efficiently can be found in [6]. The recursion starts with the whole geometry and stops when only two panels belong to the panel set of the node. To each node in the binary tree we can assign a node of the cluster tree $\mathcal{P}$. In Fig. 1.4 the approximation f_N of f is plotted. Figure 1.5 shows the different levels of the wavelets. In the left picture the approximation of f is restricted to the wavelets on level $L_{\mathcal{P}} + 4$ on the right hand side the level $L_{\mathcal{P}} + 5$ is plotted.

The second example is the surface shown in Fig. 1.6 left. The agglomeration is done as in the previous example. The clusters are assigned to Čebyšev balls of the binary tree instead of Čebyšev circles of the 2d example. A single wavelet is shown in Fig. 1.7 – one clearly sees that the support consists of two connected parts, since with the binary tree the panels are aggregated in space and not on the surface anymore. The agglomerated wavelets are not restricted to hierarchical meshes anymore. In addition, our scheme generates wavelets with supports of "small" diameter. This pays in the matrix compression.

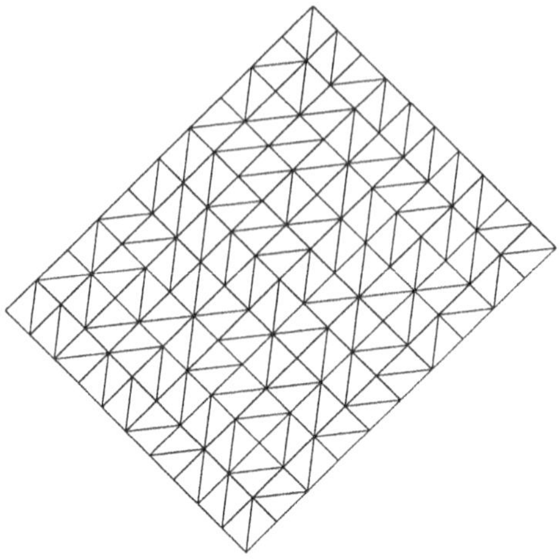

Fig. 1.3. Unit square with uniform, nonhierarchical mesh – 200 triangles.

Fig. 1.4. Function $f_N(\underline{x}) = \sum_{i=1}^{N} f_i \phi_i(\underline{x})$ with $N = 200$ on the unit square

1.6 Complexity of the transformation

In this section the complexity of Algorithm 1.2.3 is estimated. In a first step of the algorithm we generate the cluster tree $\mathcal{P}$. With the assumptions in section 1.1 and 1.2 on the boundary Γ, we see that we can compute a cluster tree $\mathcal{P}$ with $L_{\mathcal{P}} = O(\log N)$ (see [5]). To insert the leafs π_i into the cluster tree we need for each leaf at most $L_{\mathcal{P}} = O(\log N)$ operations. Therefore, we

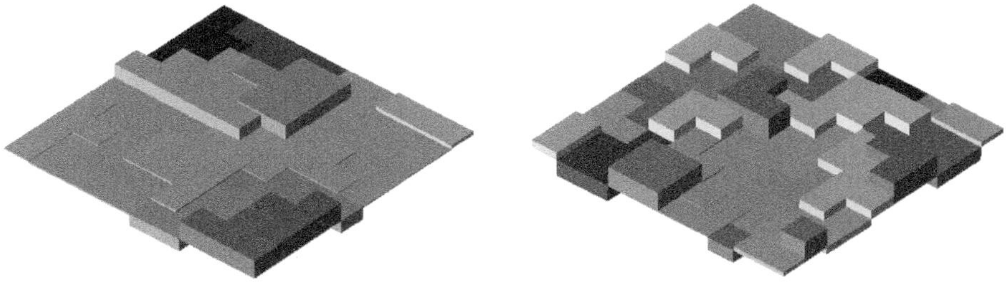

Fig. 1.5. Contribution to the wavelet representation of the function f by the wavelets on level l. Left: $l = L_{\mathcal{P}} + 4$ and Right: $l = L_{\mathcal{P}} + 5$.

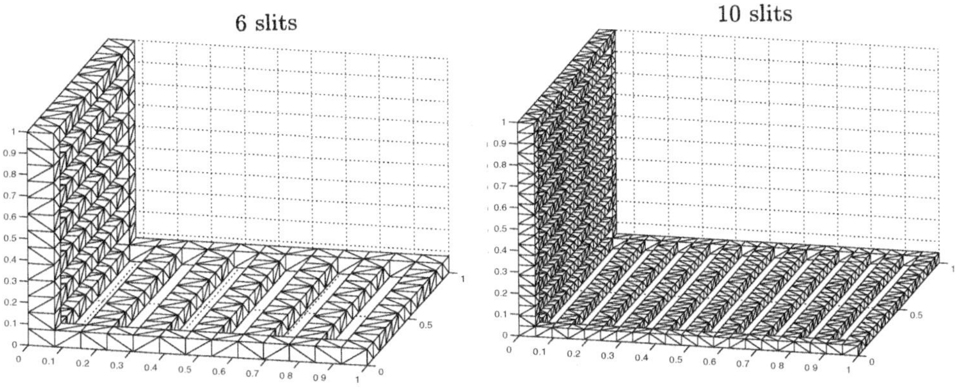

Fig. 1.6. L-shaped domains with slits.

can compute the cluster tree $\mathcal{P}$ for N leafs in $O(N \log N)$ operations. In [6] a binary tree is presented which fulfills the conditions (1.11) and (1.12). With this cluster tree we apply Algorithm 1.2.3.

Proposition 1.6.1 *Algorithm 1.2.3 takes* $W_{tot} = O(N)$ *operations.*

Proof Each leaf in the cluster tree $\mathcal{P}$ corresponds to a degree of freedom in the one scale basis $\{\phi\}_{i=1}^{N}$. Therefore we have

$$|\{\tau \in \mathcal{P} : \mathrm{child}(\tau) = \emptyset\}| = N. \tag{1.28}$$

Using (1.11), we find that each cluster τ with $\mathrm{child}(\tau) \neq \emptyset$ corresponds to at least one degree of freedom in the wavelet basis. This leads to

$$|\{\tau \in \mathcal{P} : \mathrm{child}(\tau) \neq \emptyset\}| \leq N. \tag{1.29}$$

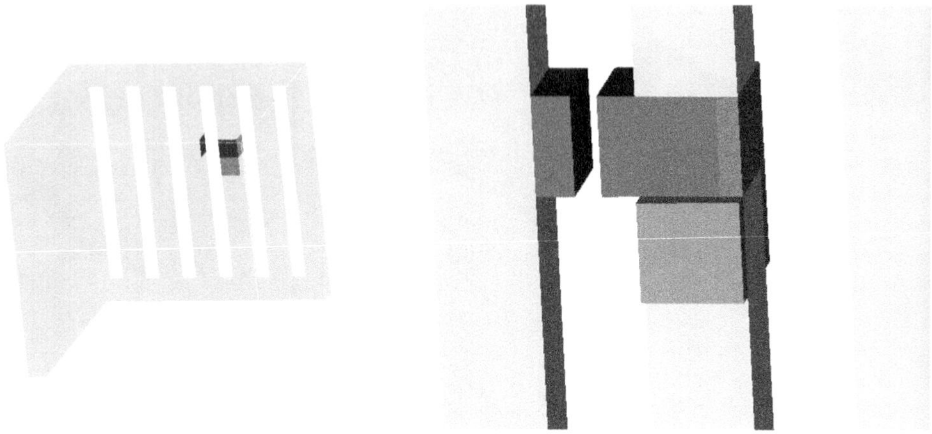

Fig. 1.7. Wavelet with non connected support.

With

$$|\{\tau \in \mathcal{P}\}| = |\{\tau \in \mathcal{P} : \text{child}(\tau) = \emptyset\}| + |\{\tau \in \mathcal{P} : \text{child}(\tau) \neq \emptyset\}|$$

and the equations (1.28), (1.29) we find

$$|\{\tau \in \mathcal{P}\}| \leq 2N. \tag{1.30}$$

The work for the local transformations in each cluster is bounded by Ck^3 where C is independent of k and N. The term k^3 stems from the singular value decomposition of the $|\text{child}(\tau)| \times |\text{child}(\tau)|$ local transformation matrix $\underline{\underline{\tilde{M}}}_\tau$ (see [4]) where $|\text{child}(\tau)|$ is estimated by k in (1.12). With equation (1.30) we get

$$W_{\text{loc}} \leq Ck^3 \Rightarrow W_{\text{tot}} \leq Ck^3 \cdot |\{\tau \in \mathcal{P}\}| \;\leq\; Ck^3 2N.$$

□

We illustrate in Fig. 1.8 the result of Proposition 1.6.1. The forward and backward transformation is applied to a vector in the one scale basis. The diagram shows the cpu-time for the pyramid scheme for the unit-square problem and for the L-shaped domain with slits. The thickness of the L-shaped domain and the slits is always one triangle. Therefore an increase of the degrees of freedom in the L-shaped domain problem induces also more slits and changes consequently the geometry. This leads to a series of domains where in Fig. 1.6 the domains with 6 ($N = 1500$) and 10 slits ($N = 3772$) are plotted. The increase in the number of degrees of freedom in the square is due to the subdivision of the domain into smaller triangles. We see that the cpu-time of the algorithm for the unit-square problem and for the L-shaped domain with slits is almost equal and behaves like $O(N)$ as stated in Proposition 1.6.1.

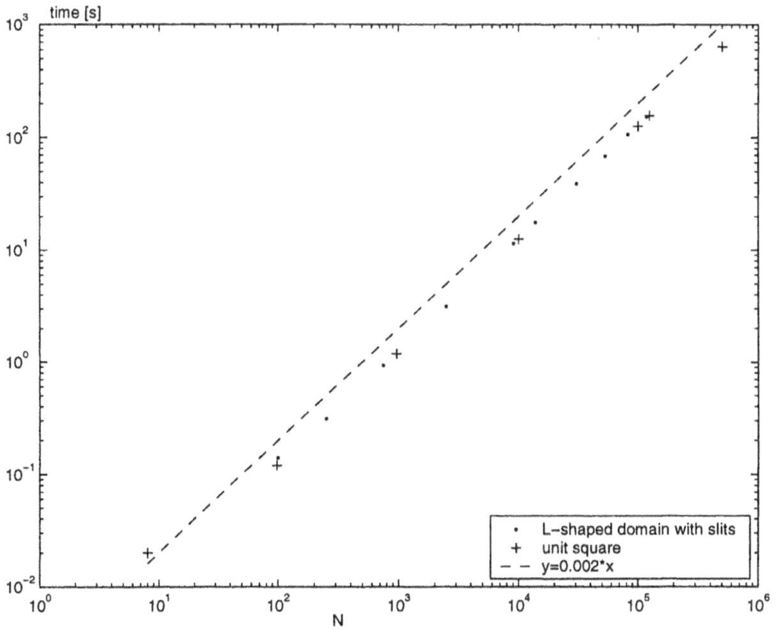

Fig. 1.8. cpu-time to compute the cluster tree and to apply the pyramid scheme to a vector versus the number of degrees of freedom. In the diagram we compare the L-shaped domains with slits (Fig. 1.6) and the unit square problem (Fig. 1.3).

1.7 Evaluation of the compressed stiffness matrix

In the preceding sections we discussed an algorithm to construct the agglomerated wavelet basis. In this section an algorithm is derived to compress the stiffness matrix using the new wavelets. The algorithm to locate the nonzero entries is essentially the same as presented in [7] but the computation of the entries is different. Contrary to [7], we do not require a hierarchical mesh and the computation of entries in the compressed stiffness matrix is more difficult, since through the aggregation process the shape of the clusters is a priori unknown and no standard integration formula can be applied. A naive splitting of the cluster into panels leads to a $O(N^2)$ algorithm. We overcome this difficulty by approximating the compressed stiffness matrix entries by a multipole-type algorithm. We directly obtain the standard form of the compressed stiffness matrix in contrast to [13] where the non-standard form is computed.

1.7.1 Localization of entries in the compressed stiffness matrix

To describe our localization algorithm for the compressed stiffness matrix, we begin by presenting some needed definitions. Analogously to (1.8) and (1.9), we write the linear equation system in the wavelet basis as

$$\underline{\underline{A}}^L \underline{u}^L = \underline{b}^L \tag{1.31}$$

with

$$\begin{aligned} \underline{\underline{A}}^L_{(\tau,j)(\tau',j')} &= \int_\tau \psi_{\tau,j}(\underline{x})(A\psi_{\tau',j'})(\underline{x})ds_{\underline{x}} \\ \underline{u}^L &= (u_{\text{root}(\mathcal{P})}, u^L_{\tau_1,1}, \ldots, u^L_{\tau_t,|\text{child}(\tau_t)-1|}) \\ \underline{b}^N_{\tau,j} &= \int_\tau \psi_{\tau,j}(\underline{x})f(\underline{x})ds_{\underline{x}} \end{aligned} \tag{1.32}$$

where $t = |\{\tau \in \mathcal{P} : \text{child}(\tau) \neq \emptyset\}|$. The matrix $\underline{\underline{A}}^L$ can be subdivided into the subblocks $\underline{\underline{A}}^L_{ll'}$ which represents all entries $\underline{\underline{A}}^L_{(\tau,j)(\tau',j')}$, $1 \leq j < |\text{child}(\tau)|$, $1 \leq j' < |\text{child}(\tau')|$ with $l_\tau = l$ and $l_{\tau'} = l'$. For each blockmatix $\underline{\underline{A}}^L_{ll'}$ there exists a truncation parameter $\delta_{ll'}$ at our disposal. Setting

$$\delta_{ll'} \geq \max\{a \prod_{\mu=L_\mathcal{P}}^{0} k_\mu^{-\frac{1}{2}} \prod_{\mu=l}^{0} k_\mu^{\frac{\alpha}{2}} \prod_{\mu=l'}^{0} k_\mu^{\frac{\tilde{\alpha}}{2}}, \prod_{\mu=L_\mathcal{P}}^{l-1} k_\mu^{-\frac{1}{2}}, \prod_{\mu=L_\mathcal{P}}^{l'-1} k_\mu^{-\frac{1}{2}}\} \tag{1.33}$$

for $a > 1$, $s, \tilde{s} \in [0,1]$ and $\alpha \geq \frac{s+1}{2}$, $\tilde{\alpha} \geq \frac{\tilde{s}+1}{2}$ we approximate the matrix $\underline{\underline{A}}^L$ by the sparse one

$$\tilde{\underline{\underline{A}}}^L_{ll',(\tau,j)(\tau',j')} := \begin{cases} \underline{\underline{A}}^L_{ll',(\tau,j)(\tau',j')} & \text{if } \text{dist}(\tau,\tau') \leq \delta_{ll'} \\ 0 & \text{otherwise} \end{cases} . \tag{1.34}$$

Note that $\delta_{ll'}$ corresponds to the truncation parameter in [10].

Proposition 1.7.1 *The compressed wavelet Galerkin matrix* $\tilde{\underline{\underline{A}}}^L$ *has*

$$\mathcal{N} = \begin{cases} O(N\log^2(N)) & \text{if } \alpha = \tilde{\alpha} = 1 \\ O(N\log(N)) & \text{otherwise} \end{cases}$$

nonzero entries (see also [10]).

To compute the entries efficiently, we need an algorithm which locates these entries in $O(\mathcal{N})$ operations.

Remark 1.7.2 Later on we will show that once we have located the entries that we will be able to compute them in $O(\log^3(N))$ operations.

The observation

$$\delta_{ll'} \geq \delta_{\tilde{l}\tilde{l}'}, \quad \text{for } l \leq \tilde{l}, l' \leq \tilde{l}'.$$

leads to a fast algorithm because with this we can skip the whole subtree $\tilde{\tau}' \subset \tau'$ if $\text{dist}(\tau,\tau') > \delta_{l_\tau l_{\tau'}}$. In Algorithm 1.7.4 we also use the symmetry of the matrix $(\delta_{ll'})$ which leads to a symmetric sparsity pattern of $\tilde{\underline{\underline{A}}}^L$.

To define the location algorithm we need a lexicographical order of the clusters. Because of the hierarchical structure of the cluster tree $\mathcal{P}$ we can introduce a breadth-first ordering.

Definition 1.7.3 *We denote by (τ, τ') the cluster in $\mathcal{P}$ on the highest possible level such that it contains τ and τ'.*

$$\tau, \tau' \subset (\tau, \tau') \wedge (\tau, \tau' \subset \hat{\tau} \in \mathcal{P} \Rightarrow (\tau, \tau') \subset \hat{\tau}). \tag{1.35}$$

A lexicographical order of the clusters $\tau, \tau' \in \mathcal{P}$ can now be defined by

$$\tau < \tau' \Leftrightarrow l_\tau < l_{\tau'} \text{ or } (l_\tau = l_{\tau'} \text{ and } \lambda < \lambda') \tag{1.36}$$

for $\tau \subset \hat{\tau}_{i_\lambda}, \tau' \subset \hat{\tau}_{i_{\lambda'}}$ and $\hat{\tau}_{i_\lambda}, \hat{\tau}_{i_{\lambda'}} \in \mathrm{child}((\tau, \tau'))$, $\lambda, \lambda' \in \{1, \dots, |\mathrm{child}((\tau, \tau'))|\}$.

With this definition we get for example for the cluster tree in Fig. 1.2 the ordering

$$\begin{gathered}
\tau_1 < \tau_2 < \tau_3 < \tau_4 < \tau_5 < \tau_6 < \tau_7 < \tau_8 < \tau_9 < \tau_{10} < \tau_{11} < \tau_{12} < \tau_{13} \\
\tau_{13} < \tau_{14} < \tau_{15} < \tau_{16} < \tau_{17} < \tau_{18} < \tau_{19} < \tau_{20} < \tau_{21} < \tau_{22} < \tau_{23} < \tau_{24} \\
\tau_{24} < \tau_{25} < \tau_{26} < \tau_{27} < \tau_{28} < \tau_{29} < \tau_{30} < \tau_{31} < \tau_{32} < \tau_{33} < \tau_{34} < \tau_{35}
\end{gathered}$$

With the lexicographical order of the clusters the algorithm to locate the nonzero entries is [7]:

Algorithm 1.7.4

assemble(τ, τ') {
  if $(\mathrm{dist}(\tau, \tau') \leq \delta_{l_\tau l_{\tau'}})$ *and* $(\tau' < \tau$ *or* $\tau = \tau')$ {
    evaluate $E^\psi_{\tau\tau'}$, $E^\psi_{\tau'\tau}$ *and update* $\tilde{\underline{\underline{A}}}^L$

    for all $\tilde{\tau}' \in \mathrm{child}(\tau')$ *with* $\mathrm{child}(\tilde{\tau}') \neq \emptyset$
      assemble$(\tau, \tilde{\tau}')$
  }
}

with the calling sequence

for all $\tau \in \{\tau' \in \mathcal{P} : \mathrm{child}(\tau') \neq \emptyset\}$
  assemble$(\tau, \mathrm{root}(\mathcal{P}))$

The matrices $E^\psi_{\tau\tau'}$ and $E^\psi_{\tau'\tau}$ denote the element stiffness matrices in the wavelet basis.

Because of the symmetric sparsity pattern of $\tilde{\underline{\underline{A}}}^L$ we have to call *assemble* at most $\mathcal{N} + N < 2\mathcal{N}$ times (at most $\mathcal{N}$ recursive calls and N calls from the calling sequence). Without computing the matrices $E^\psi_{\tau\tau'}$ and $E^\psi_{\tau'\tau}$ each call of *assemble* takes at most $O(k)$ operations (condition (1.12)). This leads to

Assertion 1.7.5 *Algorithm 1.7.4 locates the nonzero entries of $\tilde{\underline{\underline{A}}}^L$ in $O(\mathcal{N})$ operations.*

1.7.2 Computation of the entries in the compressed stiffness matrix

To get polylogarithmic complexity $O(N\log^5(N)) = O(\mathcal{N})O(\log^3(N))$ for the matrix setup, we have to compute any entry in the stiffness matrix $\underline{\underline{\tilde{A}}}^L$ in $O(\log^3(N))$ work. This is done with the panel clustering method with $\eta \in (0,1)$. The nearfield and farfield of a cluster $\tau \in \mathcal{P}$ are defined by

$$\mathcal{N}_\eta(\tau) = \{\tau' \in \mathcal{P} : r_\tau + r_{\tau'} \geq \eta \|c_\tau - c_{\tau'}\|\} \tag{1.37}$$

$$\mathcal{F}_\eta(\tau) = \{\tau' \in \mathcal{P} : r_\tau + r_{\tau'} < \eta \|c_\tau - c_{\tau'}\|\}. \tag{1.38}$$

We assume that all integrations can be executed exactly. That means that there is no error in the evaluation of the multipole coefficients and in the interaction between two panels (this assumption is generally not valid in practice but with an appropriate numerical integration scheme as presented in [10] the error becomes sufficiently small while preserving the polylogarithmic work-estimates).

The algorithm to compute the compressed stiffness matrix can be subdivided into four steps:

Step 1 The multipole coefficients Φ_p^τ, $0 \leq p \leq m$, of order m in c_τ are computed for each leaf of the cluster tree $\mathcal{P}$ introduced in section 1.2.

Algorithm 1.7.6
  for all $\{\tau \in \mathcal{P} : \mathrm{child}(\tau) = \emptyset\}$
    compute Φ_p^τ, $0 \leq p \leq m$

Step 2 For each cluster τ which is not a leaf compute the multipole coefficients Φ_p^τ, $0 \leq p \leq m$, of order m in c_τ by shifting the coefficients of its children to the cluster center and multiplication with the local transformation matrix.

Algorithm 1.7.7
  for $l = 0$ *down to* $L_\mathcal{P}$
    for all $\{\tau \in \mathcal{P} : \mathrm{child}(\tau) \neq \emptyset$ *and* $l_\tau = l\}$
      for all $\tau_{\lambda_j} \in \mathrm{child}(\tau)$, $j = 1, \ldots, |\mathrm{child}(\tau)|$
        shift the multipole coefficients
        from c_{τ_j} *to* c_τ $(\Phi_p^{\tau_j} \to (\underline{\hat{\Phi}}_p^\tau)_j)$
      $\Phi_p^\tau = (\underline{m}_1^\tau)^\top \underline{\hat{\Phi}}_p^\tau$

Step 3 Compute the nearfield entries of the one scale matrix on each level bottom up. To simplify notation we use for $\tau, \tau' \in \mathcal{P}$, $\tau_{i_\lambda} \in \mathrm{child}(\tau)$, $1 \leq$

$\lambda \le |\text{child}(\tau)|$ and $\tau'_{j_{\lambda'}} \in \text{child}(\tau')$, $1 \le \lambda' \le |\text{child}(\tau')|$ the abbreviations

$$
\begin{aligned}
\underline{\underline{a}}_{\phi_\tau,\phi_{\tau'}} &= \left(\int_{\tau_{i_\lambda}} \phi_{\tau_{i_\lambda}} (\mathcal{A}\phi_{\tau'_{j_{\lambda'}}})ds\right)_{\tau_{i_\lambda}\tau'_{j_{\lambda'}}} \\
\underline{\underline{a}}^{\mathcal{N}}_{\phi_\tau,\phi_{\tau'}} &= \left(\int_{\tau_{i_\lambda}} \phi_{\tau_{i_\lambda}} (\mathcal{A}\phi_{\tau'_{j_{\lambda'}}})ds\right)_{\tau_{i_\lambda}\tau'_{j_{\lambda'}}} \qquad \tau'_{j_{\lambda'}} \in \mathcal{N}_\eta(\tau_{i_\lambda}) \\
\underline{\underline{a}}^{\mathcal{F}}_{\phi_\tau,\phi_{\tau'}} &= \left(\int_{\tau_{i_\lambda}} \phi_{\tau_{i_\lambda}} (\mathcal{A}\phi_{\tau'_{j_{\lambda'}}})ds\right)_{\tau_{i_\lambda}\tau'_{j_{\lambda'}}} \qquad \tau'_{j_{\lambda'}} \in \mathcal{F}_\eta(\tau_{i_\lambda}) \\
a_{\phi_\tau,\phi_{\tau'}} &= \int_\tau \phi_\tau(\mathcal{A}\phi_{\tau'})ds \\
\underline{\underline{MPE}}^{\mathcal{F}}_{\tau,\tau'} &= \left(\sum_{p,p'=0}^{m} \Phi_p^{\tau_{i_\lambda}} F_{pp'}^{\tau_{i_\lambda}\tau'_{j_{\lambda'}}} \Phi_{p'}^{\tau'_{j_{\lambda'}}})\right)_{\tau_{i_\lambda}\tau'_{j_{\lambda'}}} \qquad \tau'_{j_{\lambda'}} \in \mathcal{F}_\eta(\tau_{i_\lambda})
\end{aligned}
$$

where the matrices are expanded with zeros up to a dimension of $|\text{child}(\tau)| \times |\text{child}(\tau')|$. $F_{pp'}^{\tau_{i_\lambda}\tau'_{j_{\lambda'}}}$ denotes the (p,p') derivative of the kernel function evaluated at the centers $c_{\tau_{i_\lambda}}$ and $c_{\tau'_{j_{\lambda'}}}$ and $\Phi_p^{\tau_{i_\lambda}}$ is the p-th multipole moment with respect to the cluster center $c_{\tau_{i_\lambda}}$.
With $l_\tau = l_{\tau'}$ we can write the entries of the one scale matrix on each level by the following formula

$$
\begin{aligned}
a_{\phi_\tau,\phi_{\tau'}} &= (\underline{m}_1^\tau)^\top \underline{\underline{a}}_{\phi_\tau,\phi_{\tau'}} \underline{m}_1^{\tau'} \\
&= (\underline{m}_1^\tau)^\top \underline{\underline{a}}^{\mathcal{N}}_{\phi_\tau,\phi_{\tau'}} \underline{m}_1^{\tau'} + (\underline{m}_1^\tau)^\top \underline{\underline{a}}^{\mathcal{F}}_{\phi_\tau,\phi_{\tau'}} \underline{m}_1^{\tau'} \\
&\approx (\underline{m}_1^\tau)^\top \underline{\underline{a}}^{\mathcal{N}}_{\phi_\tau,\phi_{\tau'}} \underline{m}_1^{\tau'} + (\underline{m}_1^\tau)^\top \underline{\underline{MPE}}^{\mathcal{F}}_{\tau,\tau'} \underline{m}_1^{\tau'}
\end{aligned}
$$

The algorithm for step three looks like

Algorithm 1.7.8
  for $l = 0$ *down to* $L_{\mathcal{P}}$
    for all $\{\tau \in \mathcal{P} : l_\tau = l\}$
      for all $\tau' \in \mathcal{N}_\eta(\tau)$ *and* $l_{\tau'} = l_\tau$ {
        $a_{\phi_\tau,\phi_{\tau'}} = (\underline{m}_1^\tau)^\top \underline{\underline{a}}^{\mathcal{N}}_{\phi_\tau,\phi_{\tau'}} \underline{m}_1^{\tau'} + (\underline{m}_1^\tau)^\top \underline{\underline{MPE}}^{\mathcal{F}}_{\tau,\tau'} \underline{m}_1^{\tau'}$

Step 4 For $\tau, \tau' \in \mathcal{P}$ and $l_\tau \ge l_{\tau'}$ compute the approximate element matrix $\underline{\underline{E}}^{\psi}_{\tau\tau'}$. To describe the algorithm we need in addition to the abbreviations of step 3 the notation

$$
\begin{aligned}
\underline{\underline{a}}_{\psi_\tau,\psi_{\tau'}} &= \left(\int_{\tau_i} \psi_{\tau,i}(\mathcal{A}\psi_{\tau',j})ds\right)_{i,j} \\
\underline{\underline{\hat{M}}}_\tau &= (\underline{m}_2^\tau, \ldots, \underline{m}_{|\text{child}(\tau)|}^\tau)
\end{aligned}
$$

with $1 \le i < |\text{child}(\tau)|$ and $1 \le j < |\text{child}(\tau')|$. The element matrices can be approximated by the formula

$$\begin{aligned} \underline{\underline{E}}^{\psi}_{\tau\tau'} &= \underline{\underline{a}}_{\psi_\tau,\psi_{\tau'}} = \underline{\underline{\hat{M}}}_{\tau}^{\top} \underline{\underline{a}}_{\phi_\tau,\phi_{\tau'}} \underline{\underline{\hat{M}}}_{\tau'} \\ &= \underline{\underline{\hat{M}}}_{\tau}^{\top} \underline{\underline{a}}^{\mathcal{N}}_{\phi_\tau,\phi_{\tau'}} \underline{\underline{\hat{M}}}_{\tau'} + \underline{\underline{\hat{M}}}_{\tau}^{\top} \underline{\underline{a}}^{\mathcal{F}}_{\phi_\tau,\phi_{\tau'}} \underline{\underline{\hat{M}}}_{\tau'} \\ &\approx \underline{\underline{\hat{M}}}_{\tau}^{\top} \underline{\underline{a}}^{\mathcal{N}}_{\phi_\tau,\phi_{\tau'}} \underline{\underline{\hat{M}}}_{\tau'} + \underline{\underline{\hat{M}}}_{\tau}^{\top} \underline{\underline{MPE}}^{\mathcal{F}}_{\tau,\tau'} \underline{\underline{\hat{M}}}_{\tau'} \end{aligned} \tag{1.39}$$

The term $\underline{\underline{\hat{M}}}_{\tau}^{\top} \underline{\underline{a}}^{\mathcal{N}}_{\phi_\tau,\phi_{\tau'}} \underline{\underline{\hat{M}}}_{\tau'}$ is split recursively in a sum over the nearfield and the farfield of the children cluster of τ'. This splitting is done up to level l_τ. On level l_τ the nearfield entries were computed in step three and the recursion ends.

The algorithm of step four consists of two parts. The first part computes the element matrices $\underline{\underline{E}}^{\psi}_{\tau\tau'}$ and $\underline{\underline{E}}^{\psi}_{\tau'\tau}$ needed by Algorithm 1.7.4. The second part is the recursive function *elementMatrix* which performs equation (1.39).

Algorithm 1.7.9

$\underline{\underline{E}}^{\psi}_{\tau\tau'} = \underline{\underline{\hat{M}}}_{\tau}^{\top}$ *elementMatrix*$(\tau, \tau')\underline{\underline{\hat{M}}}_{\tau'}$

$\underline{\underline{E}}^{\psi}_{\tau'\tau} = \underline{\underline{\hat{M}}}_{\tau'}^{\top}$ *elementMatrix*$(\tau', \tau)\underline{\underline{\hat{M}}}_{\tau}$

Algorithm 1.7.10

```
E ← elementMatrix(τ, τ') {
  E = MPE^F_{τ,τ'}
  if l_τ > l_τ'
    for all {τ'_{i_λ} ∈ child(τ') : τ'_{i_λ} ∈ N_η(τ̂), τ̂ ∈ child(τ),
               1 ≤ λ ≤ |child(τ')|}
      E = E + elementMatrix(τ, τ'_{i_λ}) m_1^{τ'_{i_λ}} e_λ^T
  else if l_τ < l_τ'
    for all {τ_{i_λ} ∈ child(τ) : τ_{i_λ} ∈ N_η(τ̂'), τ̂' ∈ child(τ'),
               1 ≤ λ ≤ |child(τ)|}
      E = E + e_λ (m_1^{τ_{i_λ}})^T elementMatrix(τ_{i_λ}, τ')
  else
      E = E + a^N_{φ_τ,φ_τ'}
}
```

$\underline{e}_\lambda$ denotes the λ'th unit vector $(0,\dots,0,1,0,\dots,0)^\top$.

1.8 Numerical Experiments

In this section we present a numerical experiment obtained with the described implementation of the multiscale scheme (see Section 1.7). In a series of L-shaped domains with slits (see Section 1.6 and Fig. 1.6) we consider the interior Dirichlet Problem

$$\begin{aligned} \Delta U &= 0 \qquad \text{on } \Omega \\ U &= f \qquad \text{on } \Gamma \end{aligned} \tag{1.40}$$

where f is the function

$$f(\underline{x}) = \frac{1}{\sqrt{(x_1-2)^2+(x_2-2)^2+(x_3-2)^2}} .$$

This problem was solved with the double layer ansatz $U(\underline{x}) = \int_\Gamma k(\underline{x},\underline{y})u(\underline{y})ds_{\underline{y}}$ with

$$k(\underline{x},\underline{y}) = \frac{1}{4\pi}\frac{\underline{n}(\underline{y})\cdot(\underline{y}-\underline{x})}{|\underline{y}-\underline{x}|^3} . \tag{1.41}$$

where k satisfies equation (1.9). We measured the cpu-time to compute the compressed stiffness matrix. The result was obtained on a single processor (*SUN UltraSPARC-II*, 336 MHz) with 1 GB RAM using MATLAB and the *g++ 2.95* Compiler for some C routines. In Fig. 1.9 we plotted the cpu-time versus the number of unknowns for the parameters $\eta = 0.5$, $m = 6$, $a = 1$ and $\alpha = \tilde{\alpha} = 0.5$ described in (1.33), (1.37), (1.38) and on page 374. This time includes the construction of the cluster tree, the computation of the multipole expansion and of the matrix entries. The cpu-time grows with $O(N\log(N))$ (lower dashed line) which confirms Proposition 1.7.1. The upper dashed line corresponds to $O(N\log^2(N))$. This leads to the supposition that our estimations in Section 1.7.2 were too pessimistic and that the cpu-time increases linearly with the number of non-zero entries.

1.9 Conclusions

In this report we presented an agglomerated Haar wavelet basis based on piecewise constant functions. These wavelets have one vanishing moment and satisfy Parseval's equation on the boundary. With these ingredients we are able to compress the wavelet stiffness matrix. The compression rate with the new wavelets is geometry insensitive because the wavelets are orthogonal to constant functions in $\mathbb{R}^3$ with respect to the surface measure.

In the second part we introduced an algorithm to compute directly the compressed wavelet stiffness matrix in the standard form for complex geometries. The algorithm is a combination of the panel clustering and the standard wavelet algorithm.

References

1. J.M. Carnicer, W. Dahmen, J.M. Peña. *Local decomposition of refinable spaces and wavelets*, Applied and Computational Harmonic Analysis, 3, pp. 127-153, 1996.

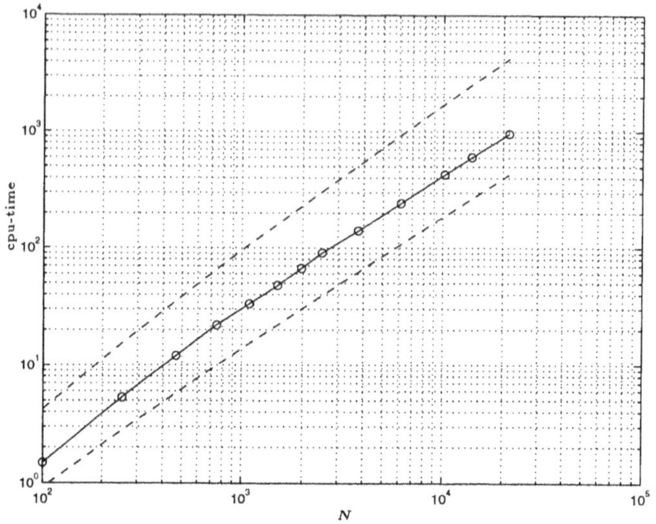

Fig. 1.9. Generating the compressed stiffness matrix in standard form for L-shaped domains with slits: cpu-time versus number of unknowns.

2. W. Dahmen. *Wavelet and Multiscale Methods for Operator Equations* Acta Numerica, pp. 55-228, 1997
3. W. Dahmen, S. Prössdorf, R. Scheider. *Wavelet approximation methods for pseudodifferential equations II: Matrix compression and fast solution* Adv. Comput. Math. 1, pp. 259-335, 1993
4. G.H. Golub, C.F. van Loan. *Matrix Computations*, The Johns Hopkins University Press, 1989
5. W. Hackbusch, Z. P. Nowak. *On the Fast Matrix Multiplication in the Boundary Element Method by Panel Clustering*, Numer. Math. 54, pp. 463-491, 1989.
6. Ch. Lage. *Softwareentwicklung zur Randelementmethode: Analyse und Entwurf effizienter Techniken*, Dissertation at Christian-Albrechts-Universität Kiel, 1995.
7. Ch. Lage, Ch. Schwab. *Wavelet Galerkin Algorithms for Boundary Integral Equations*, SIAM J. Sci. Comput., Vol 20, No. 6, pp. 2195-2222, 1999.
8. P. Oswald. *Multilevel Finite Element Approximation: Theory and Applications* Teubner Skripte zur Numerik, B. G. Teubner Stuttgart, 1994.
9. R. Schneider. *Multiskalen- und Wavelet-Matrixkompression* Advances in Numerical Mathematics, B.G. Teubner Stuttgart, 1998.
10. T. von Petersdorff, Ch. Schwab. *Fully Discrete Multiscale Galerkin BEM*, Wavelet Analysis and its Application, Volume 6 (287-346), Academic Press 1997.
11. W. Sweldens. *The Lifting Scheme: A Custom-Design Construction of Biorthogonal Wavelets*, Appl. Comput. Harmon. Anal., Vol. 3, 2, pp. 186-200, 1996.
12. W. Sweldens. *The Lifting Scheme: A Construction of Second Generation Wavelets*, SIAM J. Math. Anal., Vol. 29, 2, pp. 511-546, 1998.
13. J. Tausch, J White *Multiscale bases for the sparse representation of boundary integral operators on complex geometry*, SMU Math Report 2000-01

Appendix

Collected Color Plates

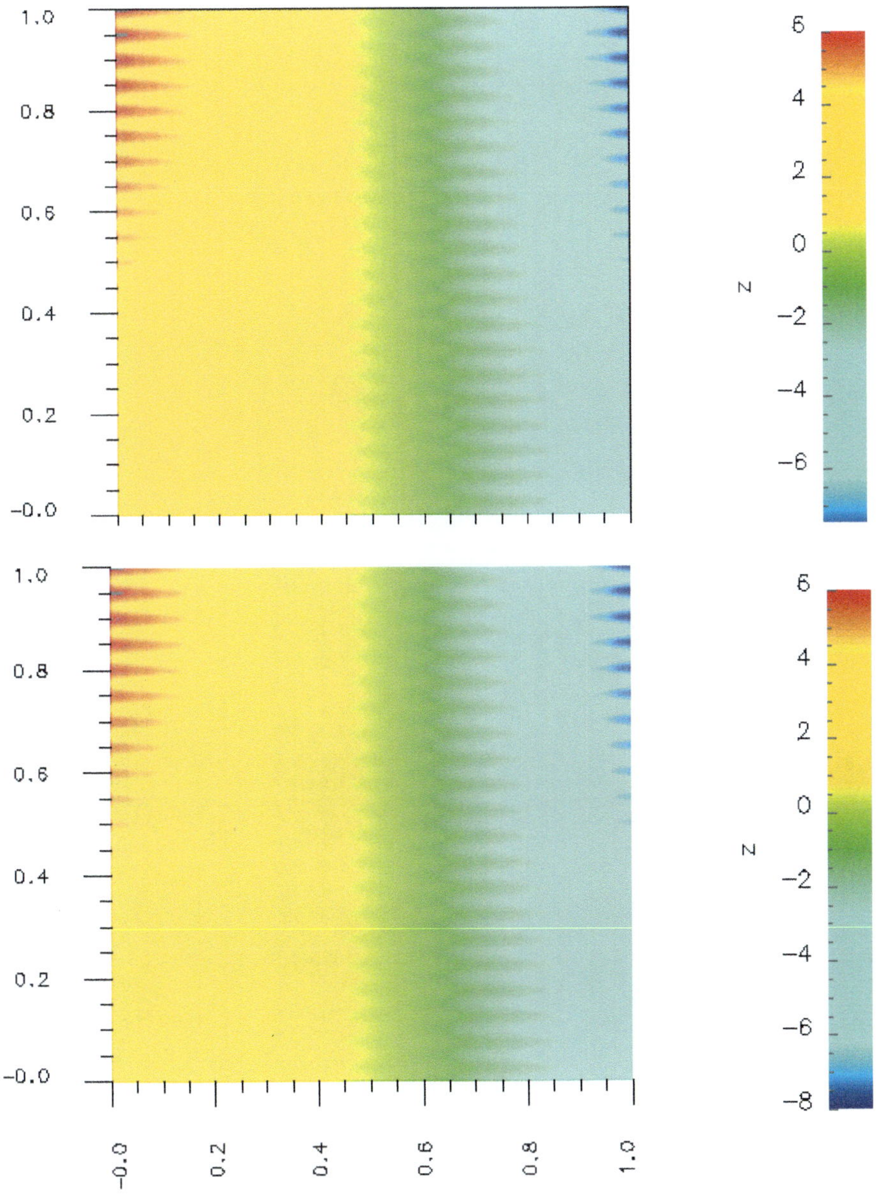

Color Plate 1. The first component of the flux $a(x/\varepsilon)\nabla u^{\varepsilon}$ obtained by the generalized FEM based on oversampling (top) and by scale resolution (bottom); $\widehat{Q} = (0, 2\pi)^2$, $\varepsilon = 1/(40\pi)$. (See Fig. 7 on page 227.)

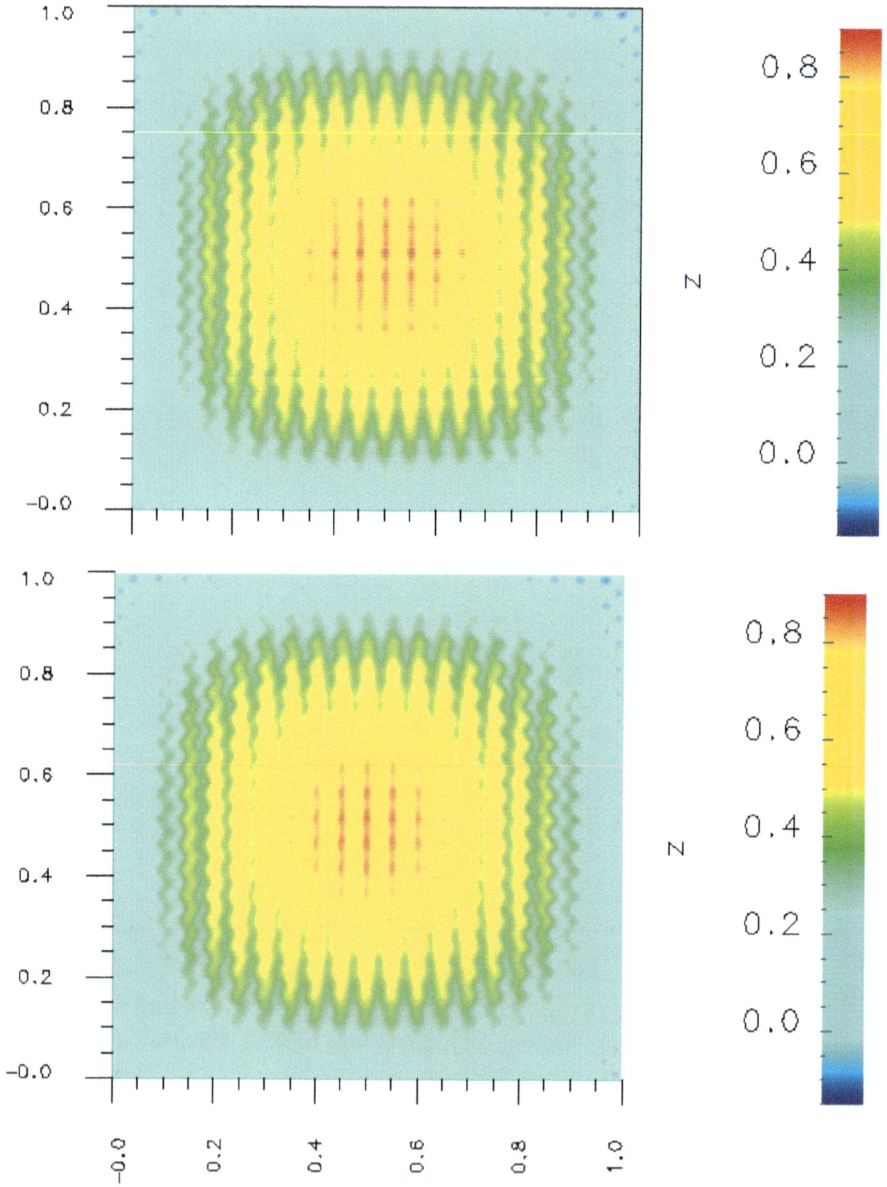

Color Plate 2. The second component of the flux $a(x/\varepsilon)\nabla u^\varepsilon$ obtained by the generalized FEM based on oversampling (top) and by direct FEM computation (bottom); $\widehat{Q} = (0, 2\pi)^2$, $\varepsilon = 1/(40\pi)$. (See Fig. 8 on page 228.)

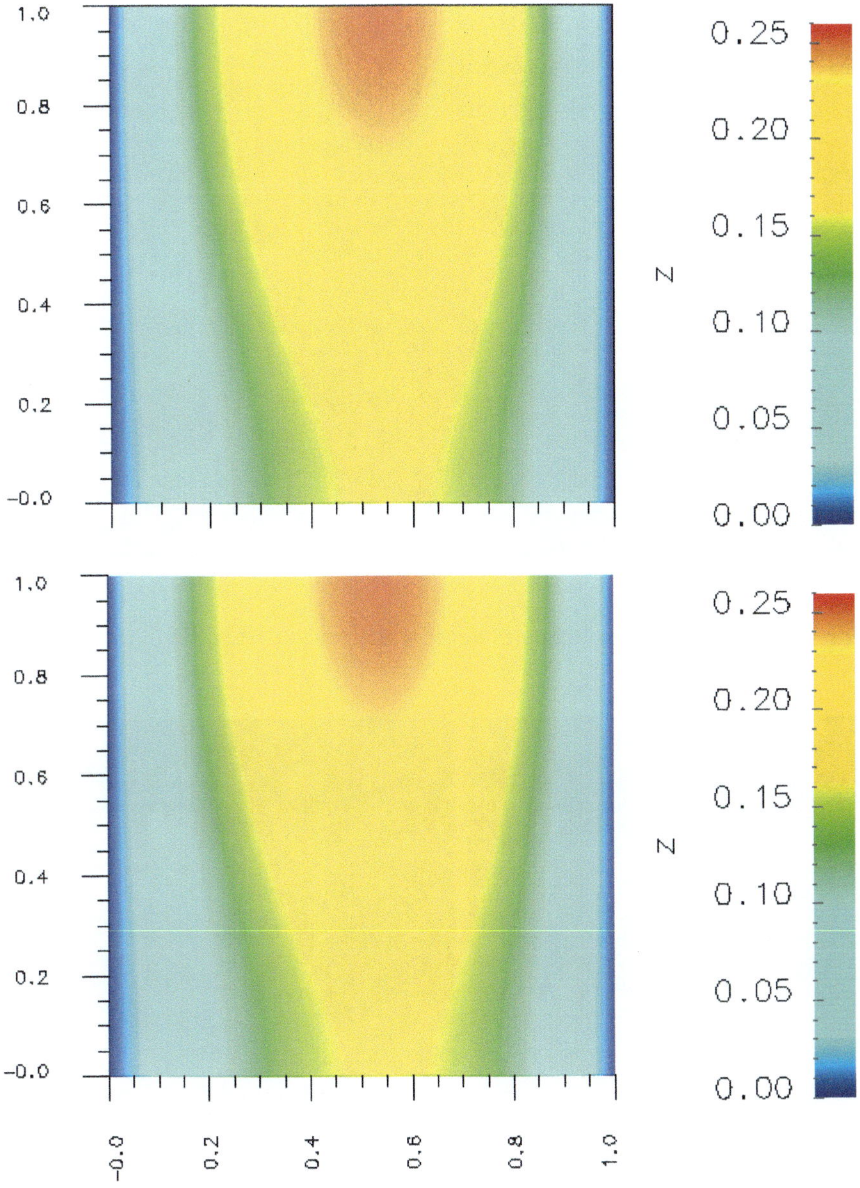

Color Plate 3. The numerical solutions obtained by the generalized FEM based on oversampling (top) and by direct FEM computation (bottom); $\widehat{Q} = (0, 2\pi)^2$, $\varepsilon = 1/(40\pi)$. (See Fig. 9 on page 229.)

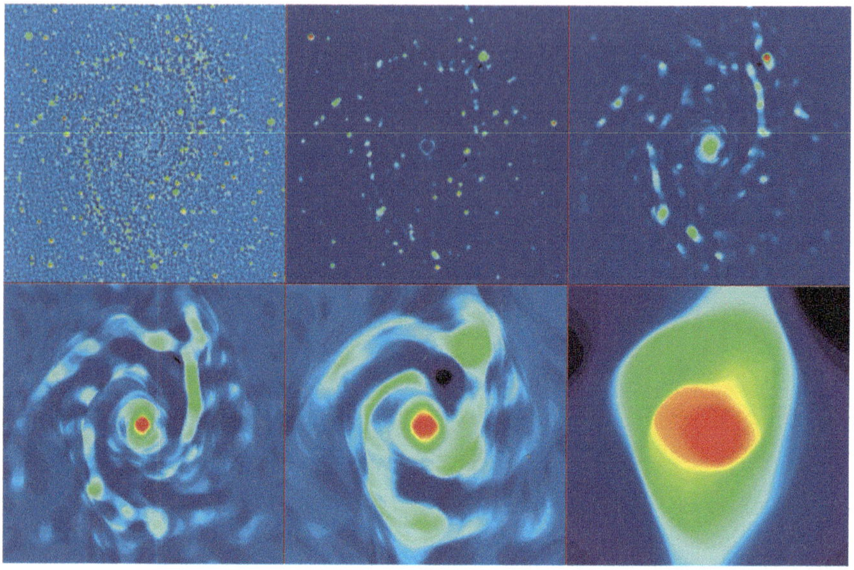

Color Plate 4. Multiresolution median transform of NGC2997. (See Fig. 4 on page 251.)

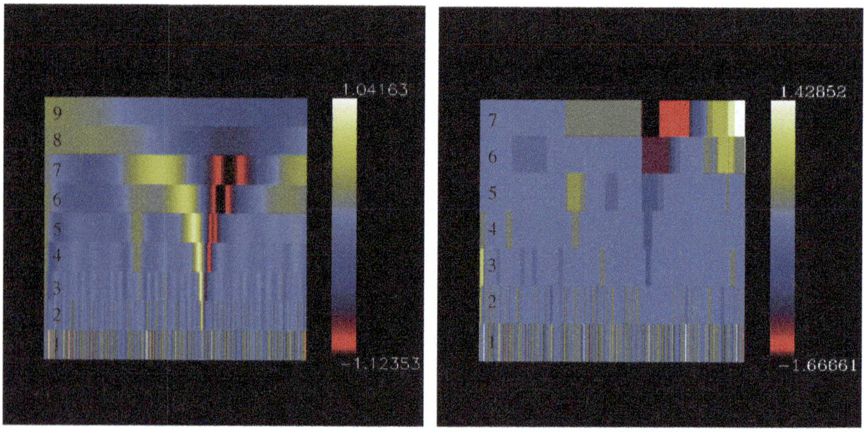

Color Plate 5. Comparison between the "á trous" wavelet transform (left) and the multiresolution transform (right) of the above signal. Resolution scale is represented versus time. We note the separation between the source and the glitch is improved using the MMT. (See Fig. 6 on page 252.)

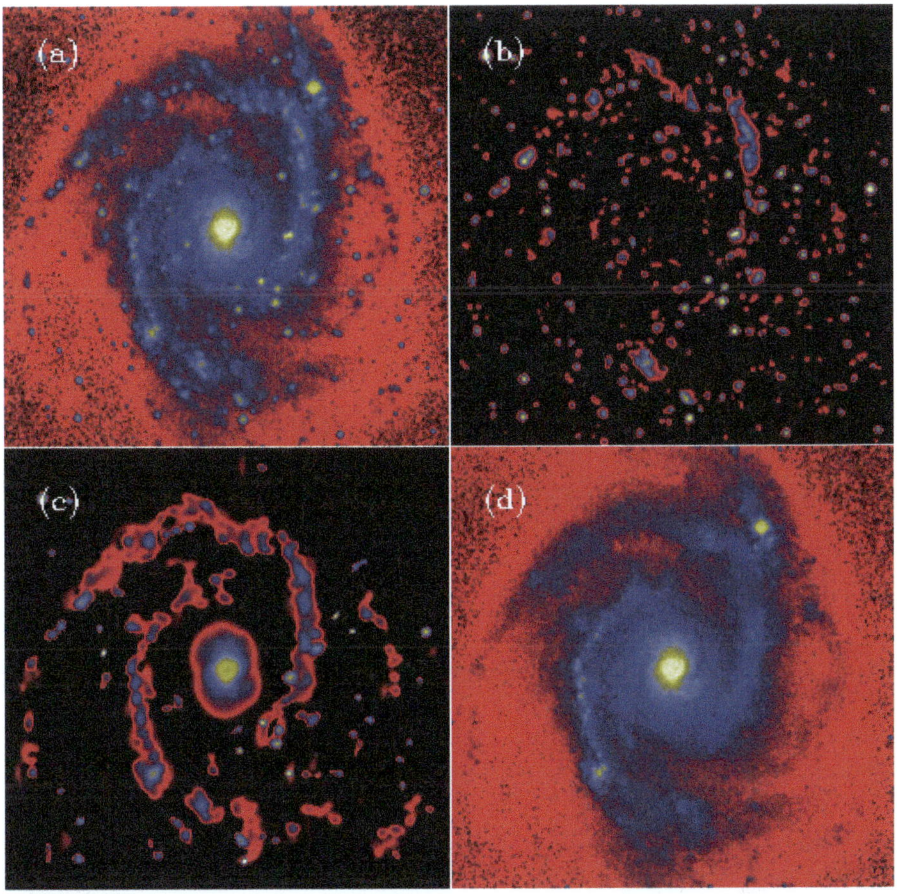

Color Plate 6. (a) galaxy NGC2997, (b) objects detected from scale 1 to 2, (c) objects detected from scale 3 to 6, and (d) difference between (a) and (b). (See Fig. 11 on page 266.)

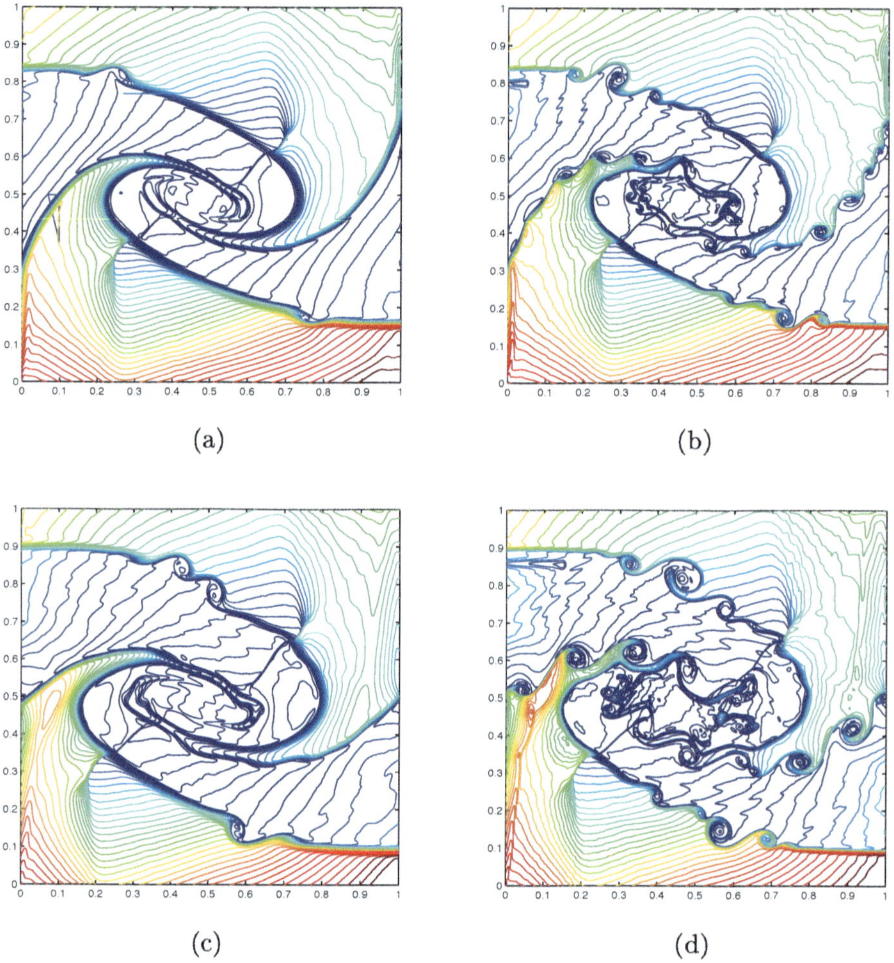

Color Plate 7. 4-contact configuration: High Resolution numerical approximation to the density obtained with the Multilevel-Marquina scheme. (a) and (b) $t = .8$, (c) and (d) $t = 1$. (a) and (c) 512×512 mesh-points, (b) and (c) 1024×1024 mesh-points. (See Fig. 4 on page 292.)

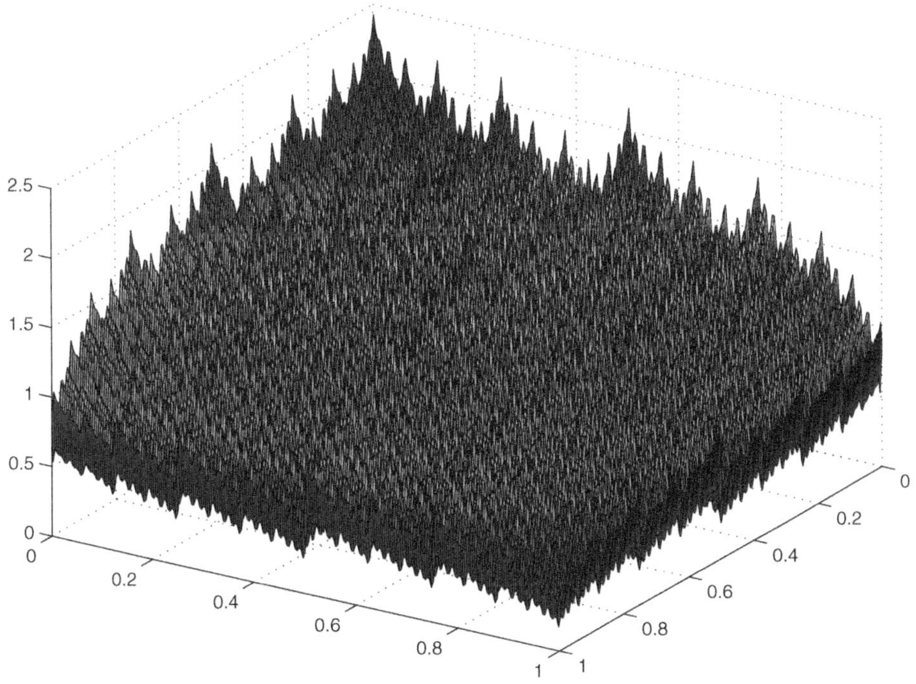

Color Plate 8. Two dimensional tensor product $\eta_{\gamma,\delta}(x)\eta_{\gamma,\delta}(y)$ with $\gamma = 0.05$ and $\delta = 0.5$. (See Fig. 2 on page 327.)

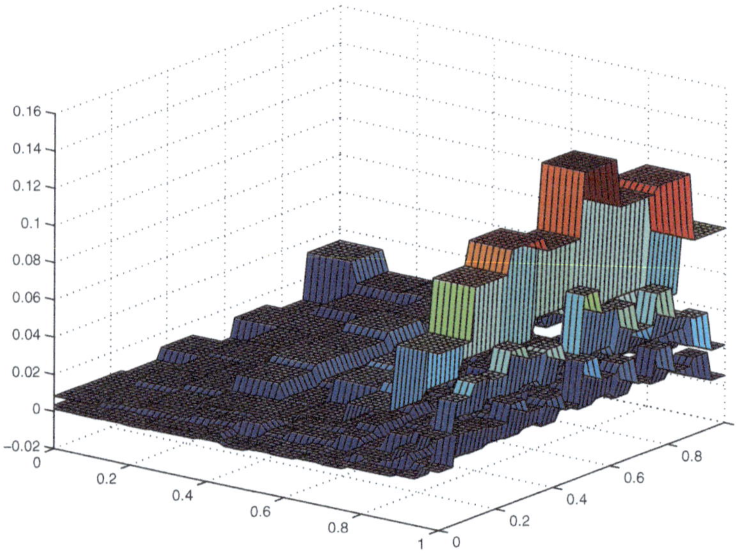

Color Plate 9. $\overline{F}^1_{\widehat{H}}(u_h)$, $\overline{F}^1_H(u_h)$ and $\widetilde{Fq}^1_h$. (See Fig. 3 on page 328.)

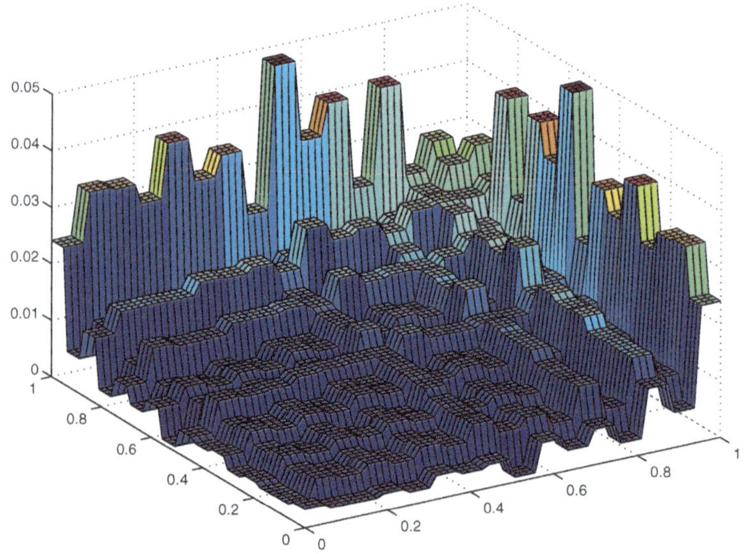

Color Plate 10. Extrapolated modeling residual $\widetilde{F}_h$. (See Fig. 4 on page 328.)

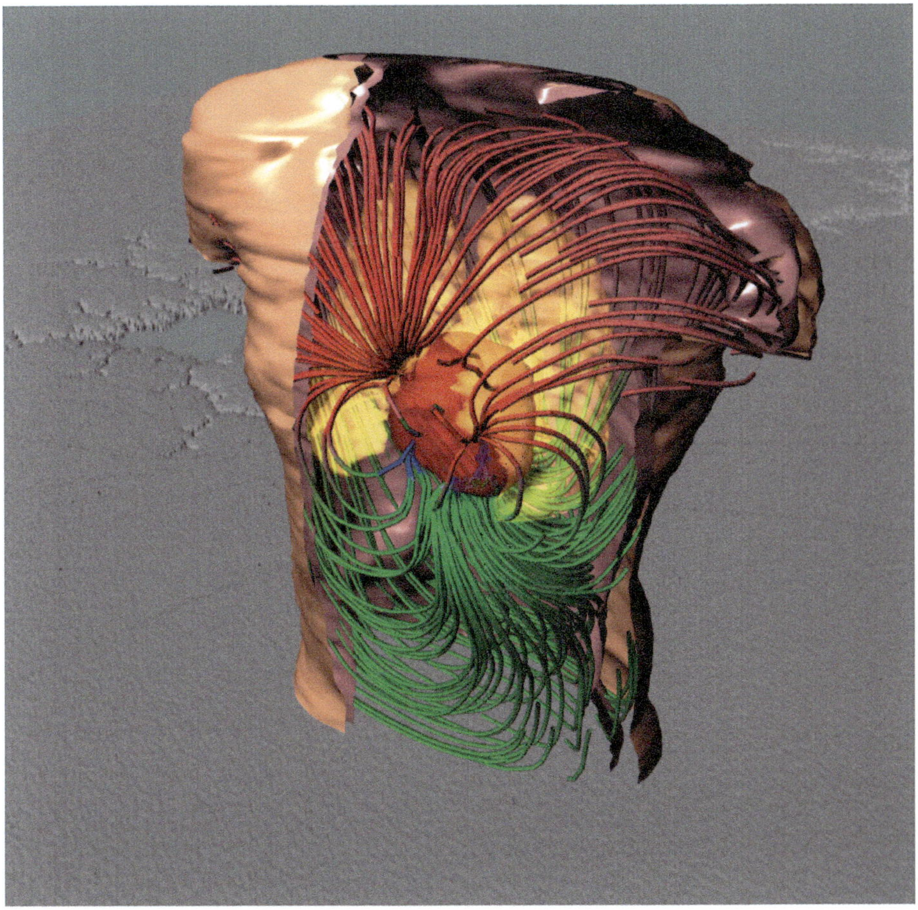

Color Plate 11. A visualization of the geometry and electrical current flow in a model of the human thorax for a single time step in the cardiac cycle. (See Fig. 1 on page 333.)

Editorial Policy

§1. Volumes in the following four categories will be published in LNCSE:

i) Research monographs
ii) Lecture and seminar notes
iii) Conference proceedings
iv) Textbooks

Those considering a book which might be suitable for the series are strongly advised to contact the publisher or the series editors at an early stage.

§2. Categories i) and ii). These categories will be emphasized by Lecture Notes in Computational Science and Engineering. **Submissions by interdisciplinary teams of authors are encouraged.** The goal is to report new developments – quickly, informally, and in a way that will make them accessible to non-specialists. In the evaluation of submissions timeliness of the work is an important criterion. Texts should be well-rounded, well-written and reasonably self-contained. In most cases the work will contain results of others as well as those of the author(s). In each case the author(s) should provide sufficient motivation, examples, and applications. In this respect, Ph.D. theses will usually be deemed unsuitable for the Lecture Notes series. Proposals for volumes in these categories should be submitted either to one of the series editors or to Springer-Verlag, Heidelberg, and will be refereed. A provisional judgment on the acceptability of a project can be based on partial information about the work: a detailed outline describing the contents of each chapter, the estimated length, a bibliography, and one or two sample chapters – or a first draft. A final decision whether to accept will rest on an evaluation of the completed work which should include

- at least 100 pages of text;
- a table of contents;
- an informative introduction perhaps with some historical remarks which should be accessible to readers unfamiliar with the topic treated;
- a subject index.

§3. Category iii). Conference proceedings will be considered for publication provided that they are both of exceptional interest and devoted to a single topic. One (or more) expert participants will act as the scientific editor(s) of the volume. They select the papers which are suitable for inclusion and have them individually refereed as for a journal. Papers not closely related to the central topic are to be excluded. Organizers should contact Lecture Notes in Computational Science and Engineering at the planning stage.

In exceptional cases some other multi-author-volumes may be considered in this category.

§4. Category iv) Textbooks on topics in the field of computational science and engineering will be considered. They should be written for courses in CSE education. Both graduate and undergraduate level are appropriate. Multidisciplinary topics are especially welcome.

§5. Format. Only works in English are considered. They should be submitted in camera-ready form according to Springer-Verlag's specifications. Electronic material can be included if appropriate. Please contact the publisher. Technical instructions and/or TEX macros are available via http://www.springer.de/author/tex/help-tex.html; the name of the macro package is "LNCSE – LaTEX2e class for Lecture Notes in Computational Science and Engineering". The macros can also be sent on request.

General Remarks

Lecture Notes are printed by photo-offset from the master-copy delivered in camera-ready form by the authors. For this purpose Springer-Verlag provides technical instructions for the preparation of manuscripts. See also *Editorial Policy*.

Careful preparation of manuscripts will help keep production time short and ensure a satisfactory appearance of the finished book. The actual production of a Lecture Notes volume normally takes approximately 12 weeks.

The following terms and conditions hold:

Categories i), ii), and iii):
Authors receive 50 free copies of their book. No royalty is paid. Commitment to publish is made by letter of intent rather than by signing a formal contract. Springer-Verlag secures the copyright for each volume.

For conference proceedings, editors receive a total of 50 free copies of their volume for distribution to the contributing authors.

Category iv):
Regarding free copies and royalties, the standard terms for Springer mathematics monographs and textbooks hold. Please write to Peters@springer.de for details. The standard contracts are used for publishing agreements.

All categories:
Authors are entitled to purchase further copies of their book and other Springer mathematics books for their personal use, at a discount of 33,3 % directly from Springer-Verlag.

Addresses:

Professor M. Griebel
Institut für Angewandte Mathematik
der Universität Bonn
Wegelerstr. 6
D-53115 Bonn, Germany
e-mail: griebel@iam.uni-bonn.de

Professor D. E. Keyes
Computer Science Department
Old Dominion University
Norfolk, VA 23529-0162, USA
e-mail: keyes@cs.odu.edu

Professor R. M. Nieminen
Laboratory of Physics
Helsinki University of Technology
02150 Espoo, Finland
e-mail: rni@fyslab.hut.fi

Professor D. Roose
Department of Computer Science
Katholieke Universiteit Leuven
Celestijnenlaan 200A
3001 Leuven-Heverlee, Belgium
e-mail: dirk.roose@cs.kuleuven.ac.be

Professor T. Schlick
Department of Chemistry and
Courant Institute of Mathematical
Sciences
New York University
and Howard Hughes Medical Institute
251 Mercer Street, Rm 509
New York, NY 10012-1548, USA
e-mail: schlick@nyu.edu

Springer-Verlag, Mathematics Editorial
Tiergartenstrasse 17
D-69121 Heidelberg, Germany
Tel.: *49 (6221) 487-185
e-mail: peters@springer.de
http://www.springer.de/math/
peters.html

Lecture Notes in Computational Science and Engineering

Vol. 1 D. Funaro, *Spectral Elements for Transport-Dominated Equations.* 1997. X, 211 pp. Softcover. ISBN 3-540-62649-2

Vol. 2 H. P. Langtangen, *Computational Partial Differential Equations.* Numerical Methods and Diffpack Programming. 1999. XXIII, 682 pp. Hardcover. ISBN 3-540-65274-4

Vol. 3 W. Hackbusch, G. Wittum (eds.), *Multigrid Methods V.* Proceedings of the Fifth European Multigrid Conference held in Stuttgart, Germany, October 1-4, 1996. 1998. VIII, 334 pp. Softcover. ISBN 3-540-63133-X

Vol. 4 P. Deuflhard, J. Hermans, B. Leimkuhler, A. E. Mark, S. Reich, R. D. Skeel (eds.), *Computational Molecular Dynamics: Challenges, Methods, Ideas.* Proceedings of the 2nd International Symposium on Algorithms for Macromolecular Modelling, Berlin, May 21-24, 1997. 1998. XI, 489 pp. Softcover. ISBN 3-540-63242-5

Vol. 5 D. Kröner, M. Ohlberger, C. Rohde (eds.), *An Introduction to Recent Developments in Theory and Numerics for Conservation Laws.* Proceedings of the International School on Theory and Numerics for Conservation Laws, Freiburg / Littenweiler, October 20-24, 1997. 1998. VII, 285 pp. Softcover. ISBN 3-540-65081-4

Vol. 6 S. Turek, *Efficient Solvers for Incompressible Flow Problems.* An Algorithmic and Computational Approach. 1999. XVII, 352 pp, with CD-ROM. Hardcover. ISBN 3-540-65433-X

Vol. 7 R. von Schwerin, *Multi Body System SIMulation.* Numerical Methods, Algorithms, and Software. 1999. XX, 338 pp. Softcover. ISBN 3-540-65662-6

Vol. 8 H.-J. Bungartz, F. Durst, C. Zenger (eds.), *High Performance Scientific and Engineering Computing.* Proceedings of the International FORTWIHR Conference on HPSEC, Munich, March 16-18, 1998. 1999. X, 471 pp. Softcover. 3-540-65730-4

Vol. 9 T. J. Barth, H. Deconinck (eds.), *High-Order Methods for Computational Physics.* 1999. VII, 582 pp. Hardcover. 3-540-65893-9

Vol. 10 H. P. Langtangen, A. M. Bruaset, E. Quak (eds.), *Advances in Software Tools for Scientific Computing.* 2000. X, 357 pp. Softcover. 3-540-66557-9

Vol. 11 B. Cockburn, G. E. Karniadakis, C.-W. Shu (eds.), *Discontinuous Galerkin Methods.* Theory, Computation and Applications. 2000. XI, 470 pp. Hardcover. 3-540-66787-3

Vol. 12 U. van Rienen, *Numerical Methods in Computational Electrodynamics.* Linear Systems in Practical Applications. 2000. XIII, 375 pp. Softcover. 3-540-67629-5

Vol. 13 B. Engquist, L. Johnsson, M. Hammill, F. Short (eds.), *Simulation and Visualization on the Grid.* Parallelldatorcentrum Seventh Annual Conference, Stockholm, December 1999, Proceedings. 2000. XIII, 301 pp. Softcover. 3-540-67264-8

Vol. 14 E. Dick, K. Riemslagh, J. Vierendeels (eds.), *Multigrid Methods VI.* Proceedings of the Sixth European Multigrid Conference Held in Gent, Belgium, September 27-30, 1999. 2000. IX, 293 pp. Softcover. 3-540-67157-9

Vol. 15 A. Frommer, T. Lippert, B. Medeke, K. Schilling (eds.), *Numerical Challenges in Lattice Quantum Chromodynamics.* Joint Interdisciplinary Workshop of John von Neumann Institute for Computing, Jülich and Institute of Applied Computer Science, Wuppertal University, August 1999. 2000. VIII, 184 pp. Softcover. 3-540-67732-1

Vol. 16 J. Lang, *Adaptive Multilevel Solution of Nonlinear Parabolic PDE Systems.* Theory, Algorithm, and Applications. 2001. XII, 157 pp. Softcover. 3-540-67900-6

Vol. 17 B. I. Wohlmuth, *Discretization Methods and Iterative Solvers Based on Domain Decomposition.* 2001. X, 197 pp. Softcover. 3-540-41083-X

Vol. 18 U. van Rienen, M. Günther, D. Hecht (eds.), *Scientific Computing in Electrical Engineering.* Proceedings of the 3rd International Workshop, August 20-23, 2000, Warnemünde, Germany. 2001. XII, 428 pp. Softcover. 3-540-42173-4

Vol. 19 I. Babuška, P. G. Ciarlet, T. Miyoshi (eds.), *Mathematical Modeling and Numerical Simulation in Continuum Mechanics.* Proceedings of the International Symposium on Mathematical Modeling and Numerical Simulation in Continuum Mechanics, September 29 - October 3, 2000, Yamaguchi, Japan. 2002. VIII, 301 pp. Softcover. 3-540-42399-0

Vol. 20 T. J. Barth, T. Chan, R. Haimes (eds.), *Multiscale and Multiresolution Methods.* Theory and Applications. 2002. X, 389 pp. Softcover. 3-540-42420-2

For further information on these books please have a look at our mathematics catalogue at the following URL: `http://www.springer.de/math/index.html`